Jonathon D. Brown

Linear Models in Matrix Form

A Hands-On Approach
for the Behavioral Sciences

 Springer

Jonathon D. Brown
Department of Psychology
University of Washington
Seattle, WA, USA

ISBN 978-3-319-11733-1   ISBN 978-3-319-11734-8 (eBook)
DOI 10.1007/978-3-319-11734-8
Springer Cham Heidelberg New York Dordrecht London

Library of Congress Control Number: 2014949858

Printed on acid-free paper

Springer is part of Springer Science+Business Media (www.springer.com)

To Noah and Avi.
Two (very) special additions!

Linear Models in Matrix Form: A Hands-On Approach for the Behavioral Sciences

Technological advances inevitably shape the teaching of statistics. For many years, the matrix calculations required in order to do most statistical analyses were too laborious to be performed by hand, so simple algebra was used. The formula for computing an unstandardized regression coefficient in a two-variable, multiple regression analysis provides one example of this "first generation" approach.

$$b_1 = \frac{(s_{1y} * s_2^2) - (s_{12} * s_{2y})}{(s_1^2 * s_2^2) - (s_{12})^2}$$

The availability of handheld calculators in the 1970s made these calculations less tedious, but the pedagogical approach to teaching statistics remained unchanged until statistical packages designed to take advantage of high-speed computers ushered in a new, "click and run" approach in the mid-1980s. Now, all the calculations took place behind the scenes, and formulas were replaced by a more practical method that emphasized the interpretation of output rather than its production.

Unfortunately, the "click and run" approach has produced a new generation of students who know how to use computer software to perform statistical analyses without knowing much about the computations that produce the output they instantly and (apparently) magically receive. I think this lack of understanding is lamentable; I also think it is unnecessary. Instead of leading us to ignore the calculations that comprise statistical analyses, the ready accessibility of high-speed computers should have led us to reveal the matrix algebra that produces the output:

$$\mathbf{b} = \left(\mathbf{X'X}\right)^{-1}\mathbf{X'y} \ \text{ or } \ \mathbf{b} = \mathbf{R}^{-1}\mathbf{Q'y}$$

That way, students would gain valuable insight into the mathematics behind the results.

You Do the Math!

I wrote this book to provide that balance, creating an easy-to-understand, hands-on approach to statistical analysis using matrix algebra. Although many of the topics are advanced (e.g., maximum-likelihood estimation, matrix decompositions, nonparametric smoothers, and penalized cubic splines), all the analyses were performed on an open-source spreadsheet using a few built-in functions (e.g., transpose, inverse).[1] Moreover, every statistical analysis is illustrated using a (fictitious) data set comprised of only 12 observations. The sample size is kept purposefully small so that readers will not be dissuaded from performing the calculations themselves, and each data set tells a coherent story based on statistical significance and confidence intervals. In my experience, statistical analyses come alive when students know how the numbers were generated and how they can be used to make cogent arguments about everyday matters.

Intended Audience

This book is intended for use in an upper-level undergraduate course or an entry-level graduate course. I assume some knowledge of basic statistical principles (e.g., measures of central tendencies, hypothesis testing), but familiarity with matrix algebra is not needed. The first chapter introduces students to linear equations, and then covers matrix algebra, focusing on three essential operations: the sum of squares, the determinant, and the inverse. These operations are explained in common everyday language, and their calculations are demonstrated using concrete examples. The remaining chapters build on these operations, progressing from simple linear regression to mediational models with bootstrapped standard errors.

 This book is also appropriate for intellectually curious researchers. I hope to evoke an "aha" experience among those who learned only the "click and run" approach and a "so that's all there is to it?" reaction among those with only a limited background in mathematics.

---

[1] The spreadsheets accompanying each chapter are available as supplementary material with the book on SpringerLink, and can be downloaded directly at http://www.springer.com/statistics/social+sciences+%26+law/book/978-3-319-11733-1

Using R

Although this book emphasizes performing calculations with a spreadsheet rather than a statistical package, I do not expect readers to ordinarily use a spreadsheet to analyze their data once they have learned the calculations.[2] Toward that end, each chapter includes code for running all analyses in **R**, a free software programming language and software environment for statistical computing and graphics that can be downloaded at http://www.r-project.org. Whenever possible, I have used functions from the base suite of tools or have provided simple code for ones that are not available elsewhere. Occasionally, I have relied on functions associated with packages that must be installed and attached.

To the Reader

It has been said that there are three kinds of people—those who can do math and those who cannot. If you get the joke, you should be able to solve all the problems in this textbook. Stripped of the jargon and the arcane formulas, most statistical analyses are not all that complicated. We have a variable we wish to predict and a set of variables we use to fashion that prediction. There is really nothing more to it than that. In the end, then, there really is only one kind of person: a person who can do math provided it is explained clearly and illustrated with examples.

---

[2] The results of a great many calculations are presented in the text to illustrate various operations. Due to rounding error, the results obtained using the printed values do not always match the ones obtained using a spreadsheet or statistical package. Rather than confuse readers with two different values, I report the more accurate values one would get using computer software.

Contents

1   Matrix Properties and Operations   1
  1.1   Linear Equations and Matrix Algebra   1
      1.1.1   What Is a Matrix?   2
      1.1.2   Matrix Operations   3
      1.1.3   Matrix Rules of Multiplication   5
      1.1.4   Using a Spreadsheet to Perform Matrix Operations . . .   7
  1.2   Matrix Transpose and Sum of Squares   9
      1.2.1   Transposition   10
      1.2.2   Premultiplying a Matrix by Its Transpose   10
      1.2.3   Covariance Matrix   11
      1.2.4   Correlation Matrix   13
      1.2.5   Diagonal Matrix   14
      1.2.6   Summary   14
      1.2.7   R Code: Matrix Multiplication   15
  1.3   Matrix Determinants   15
      1.3.1   Visualizing the Determinant   16
      1.3.2   Using the Determinant to Solve Linear Equations   18
      1.3.3   Linear Dependencies and Singular Matrices   21
      1.3.4   Calculating the Determinant with Large Matrices   21
      1.3.5   R Code: Determinants   24
  1.4   Matrix Inverse   25
      1.4.1   Matrix Multiplication and Matrix Inverse   25
      1.4.2   Calculating the Matrix Inverse   27
      1.4.3   Using the Inverse to Solve Linear Equations   30
      1.4.4   R Code: Matrix Inverse   34
  1.5   Chapter Summary   35
  Appendix   36

2   Simple Linear Regression   39
    2.1   Mathematical Models   40
        2.1.1   What Is a Model?   40
        2.1.2   What Is a *Regression* Model?   41
        2.1.3   What Is a *Linear* Regression Model?   41
    2.2   Simple Linear Regression   42
        2.2.1   Preliminary Analysis Without an Intercept   43
        2.2.2   Complete Analysis: Adding an Intercept   45
        2.2.3   Understanding Linear Regression   49
        2.2.4   Standardized Regression Coefficients   51
        2.2.5   Correlation Coefficient   53
        2.2.6   R Code: Simple Linear Regression   54
    2.3   Population Estimation and Statistical Significance   55
        2.3.1   The Logic Behind Null Hypothesis Testing   55
        2.3.2   Testing the Regression Model   55
        2.3.3   Testing the Regression Coefficients   58
        2.3.4   Parameter Covariance Matrix (C)   61
        2.3.5   R Code: Hypothesis Testing   62
    2.4   Forecasting   63
        2.4.1   Average Expected Values   63
        2.4.2   Single Predicted Values   65
        2.4.3   Forecasting with Caution   66
        2.4.4   R Code: Forecasting   66
    2.5   Chapter Summary   67

3   Maximum-Likelihood Estimation   69
    3.1   Probability and Likelihood in a Normal Distribution   69
        3.1.1   Likelihood Function   71
        3.1.2   Log-Likelihood Function   72
        3.1.3   Using the Grid-Search Method to Find
                the Maximum-Likelihood Estimate   73
        3.1.4   R Code: Maximum-Likelihood Estimation
                with Normal Distribution   75
    3.2   Differential Calculus   76
        3.2.1   Differentiating a Function to Find Its Derivative   76
        3.2.2   Differentation and Maximum-Likelihood
                Estimation   84
        3.2.3   Computing the Standard Errors   87
        3.2.4   R Code: Derivatives and Standard Errors   91
    3.3   Maximum-Likelihood Estimation in Regression   92
        3.3.1   Differentiating the Function   93
        3.3.2   Standard Errors   95
    3.4   Maximum-Likelihood Estimation: Numerical Methods   96
        3.4.1   Newton–Raphson and Fisher's Method of Scoring   96
        3.4.2   Illustration Using Fisher's Method of Scoring   97

	3.4.3	R Code: Maximum-Likelihood Estimation with Fisher's Method of Scoring	99
3.5	Chapter Summary		100
Appendix			101

4 Multiple Regression 105
4.1	Multiple Regression		106
	4.1.1	Correlations	106
	4.1.2	Unstandardized Regression Coefficients	108
	4.1.3	Fitted Values and Residuals	109
	4.1.4	Testing the Regression Model	110
	4.1.5	R Code: Multiple Regression	111
4.2	Interpreting and Testing Regression Coefficients		112
	4.2.1	Comparing Regression Coefficients and Correlations	112
	4.2.2	Interpreting the Numerical Value of a Regression Coefficient	113
	4.2.3	Calculating Regression Coefficients	118
	4.2.4	Testing the Significance of Regression Coefficients	120
	4.2.5	Forecasting	122
	4.2.6	Comparing the Predictors	124
	4.2.7	R Code: Testing Regression Coefficients	126
4.3	Partitioning the Variance		127
	4.3.1	Semipartial Correlation	128
	4.3.2	Partial Correlation	132
	4.3.3	Are Regression Coefficients Semipartial Coefficients or Partial Coefficients?	135
	4.3.4	R Code: Partitioning the Variance	135
4.4	Calculating Regression Coefficients Using Cofactors		137
	4.4.1	Complete Sum of Squares	137
	4.4.2	Residual Sum of Squares and Coefficient of Determination	140
	4.4.3	Regression Coefficients	141
	4.4.4	Computing the Remaining Coefficients and Correlations	143
	4.4.5	Summary	143
	4.4.6	R Code: Regression Coefficients as Cofactors	143
4.5	Chapter Summary		144

5 Matrix Decompositions 147
5.1	Eigen Decomposition		147
	5.1.1	Matrix Multiplication with an "Ordinary" Vector	148
	5.1.2	Matrix Multiplication with an Eigenvector	149
	5.1.3	Calculating Eigenvalues	150

	5.1.4	Calculating Eigenvectors	153
	5.1.5	Eigenvalues and Variance Consolidation	156
	5.1.6	Eigen Decomposition and Matrix Recomposition	158
	5.1.7	R Code: Eigen Decomposition	160
5.2	QR Decomposition		160
	5.2.1	Computations with Householder Transformations	161
	5.2.2	Linear Regression	167
	5.2.3	QR Algorithm for Finding the Eigenpairs	168
	5.2.4	R Code: QR Decomposition	172
5.3	Singular Value Decomposition		173
	5.3.1	Preliminary Calculations	173
	5.3.2	Reconstructing X	174
	5.3.3	Regression Coefficients	175
	5.3.4	Standard Errors	175
	5.3.5	R Code: Singular Value Decomposition	176
5.4	Cholesky Decomposition		177
	5.4.1	Calculations	177
	5.4.2	Calculating the Determinant and the Inverse	179
	5.4.3	Least Squares Regression	179
	5.4.4	Using the Cholesky Decomposition to Find the Eigenvalues	180
	5.4.5	R Code: Cholesky Decomposition	180
5.5	Comparing the Decompositions		181
5.6	Chapter Summary		182
6	**Problematic Observations**		**185**
6.1	Influential Observations		186
	6.1.1	Discrepant Observations	187
	6.1.2	Illustrating Undue Influence	187
	6.1.3	Leverage and the Hat Matrix	190
	6.1.4	Residuals and Outliers	193
	6.1.5	Variance of Fitted Values and Residuals	196
	6.1.6	Quantifying Influence	197
	6.1.7	Commentary	203
	6.1.8	R Code: Regression Diagnostics	204
6.2	Departures from Normality		204
	6.2.1	Reviewing the Normality Assumption	204
	6.2.2	Assessing Normality	205
	6.2.3	Correcting Violations of Normality	209
	6.2.4	R Code: Departures from Normality	212
6.3	Collinearity		213
	6.3.1	Problems with Overly Redundant Predictors	214
	6.3.2	Matrices with a Near-Linear Dependence are Ill Conditioned	215
	6.3.3	Demonstrating Collinearity	216

6.3.4   Quantifying Collinearity with the Variance
Inflation Factor 217
6.3.5   Condition Index and Variance Proportion
Decomposition 219
6.3.6   Summary 223
6.3.7   R Code: Collinearity 225
6.4   Chapter Summary 225

7   **Errors and Residuals** 227
7.1   Errors and Their Assumed Distribution 227
7.1.1   Why It Matters 228
7.1.2   Errors and Residuals 229
7.1.3   Generalized Least Squares Estimation 231
7.2   Heteroscedasticity 232
7.2.1   Small Sample Example 232
7.2.2   Detecting Heteroscedasticity 235
7.2.3   Weighted Least Squares Estimation 237
7.2.4   Heteroscedasticity-Consistent Covariance Matrix 241
7.2.5   Summary 242
7.2.6   R Code: Heteroscedasticity 243
7.3   Autocorrelations 244
7.3.1   Mathematical Representation 244
7.3.2   Detecting Autocorrelations 247
7.3.3   Generalized Least Squares Estimation
for Managing Autocorrelation 250
7.3.4   Autocorrelation-Consistent Covariance Matrix 254
7.3.5   R Code: Autocorrelations 258
7.4   Chapter Summary 259

8   **Linearizing Transformations and Nonparametric Smoothers** 261
8.1   Understanding Linearity 262
8.1.1   Partial Derivatives and Linear Functions 263
8.1.2   Assessing Linear Relations 264
8.1.3   Options for Analyzing Nonlinear Relations 267
8.1.4   R Code: Assessing Nonlinearity 268
8.2   Transformations to Linearity 269
8.2.1   Understanding Transformations 269
8.2.2   Logarithmic Model 270
8.2.3   Exponential Model 274
8.2.4   Power Function 279
8.2.5   Box-Tidwell Transformation 282
8.2.6   Summary 285
8.2.7   R Code: Linear Transformations 285
8.3   Nonparametric Smoothers 286
8.3.1   Understanding Nonparametric Regression 287
8.3.2   Running Average 290

	8.3.3	Running Line	292
	8.3.4	Kernel Regression	294
	8.3.5	Locally Weighted Regression	298
	8.3.6	Extensions and Applications	300
	8.3.7	R Code: Nonparametric Smoothers	300
8.4		Chapter Summary	301

9   Cross-Product Terms and Interactions 303

9.1		Understanding Interactions	303
	9.1.1	Depicting an Interaction	304
	9.1.2	Modeling Interactions with Cross-Product Terms	304
	9.1.3	Testing Cross-Product Terms	308
	9.1.4	R Code: Testing a Cross-Product Term	311
9.2		Probing an Interaction	312
	9.2.1	Calculating Predicted Values	312
	9.2.2	Plotting Predicted Values	313
	9.2.3	Testing Simple Slopes	313
	9.2.4	Characterizing an Interaction	318
	9.2.5	R Code: Predicted Values and Simple Slopes	319
	9.2.6	Johnson-Neyman Technique	320
	9.2.7	R Code: Johnson-Neyman Regions of Significance	322
9.3		Higher-Order Interactions	323
	9.3.1	Testing the Regression Equation	323
	9.3.2	Probing a Three-Variable Interaction	323
	9.3.3	R Code: Three-Way Interaction	330
	9.3.4	Recentering Variables to Calculate Simple Slopes	332
	9.3.5	R Code: Three-Way Interaction Using Recentering	336
9.4		Effect Size and Statistical Power	336
	9.4.1	Effect Size	337
	9.4.2	Statistical Power	337
	9.4.3	R Code: Effect Size of Three-Way Cross-Product Term	339
9.5		Chapter Summary	340

10   Polynomial Regression 341

10.1		Simple Polynomial Regression	342
	10.1.1	Testing the Linear Component	342
	10.1.2	Adding a Quadratic Term	344
	10.1.3	Testing Other Polynomials	347
	10.1.4	R Code: Cubic Polynomial	348
10.2		Polynomial Interactions	349
	10.2.1	Regression Equations	349
	10.2.2	Testing the Regression Coefficients	351
	10.2.3	Probing a Polynomial Interaction	351
	10.2.4	R Code: Polynomial Interaction	354

10.3 Piecewise Polynomials 356
    10.3.1 Regression Splines 356
    10.3.2 Natural Cubic Splines 360
    10.3.3 R Code: Unpenalized Regression Splines 363
    10.3.4 Penalized Natural Cubic Spline 364
    10.3.5 R Code: Penalized Natural Cubic Splines 374
10.4 Chapter Summary 375

11 Categorical Predictors 377
11.1 Coding Schemes 378
    11.1.1 Analysis of Variance 378
    11.1.2 Overview of Coding Schemes 380
    11.1.3 Orthogonal Contrast Codes 381
    11.1.4 Dummy Codes 387
    11.1.5 Effect Codes 388
    11.1.6 Summary 391
    11.1.7 R Code: Coding Schemes 391
11.2 Creating Orthogonal Contrast Codes 392
    11.2.1 Helmert Contrasts 392
    11.2.2 Gram-Schmidt Orthogonalization 393
    11.2.3 Polynomial Terms in a Trend Analysis 396
    11.2.4 R Code: Creating Orthogonal Contrasts 399
11.3 Contrast Codes with Unbalanced Designs 399
    11.3.1 Analysis with Unweighted Means 400
    11.3.2 Weighted Means Analysis 403
    11.3.3 R Code: Unbalanced Designs 407
11.4 Chapter Summary 408

12 Factorial Designs 409
12.1 Basics of Factorial Designs 409
    12.1.1 Regression Analysis of a One-Way Design 409
    12.1.2 Recasting the Data as a Factorial Design 410
    12.1.3 Properties of a Factorial Design 411
    12.1.4 Sources of Variance in a Balanced Factorial
          Design 412
    12.1.5 Probing an Interaction 416
    12.1.6 R Code: Factorial Design 418
12.2 Unbalanced Factorial Designs 420
    12.2.1 Unweighted Means 421
    12.2.2 Weighted Means 422
    12.2.3 R Code: Unbalanced Factorial Design 424
12.3 Multilevel Designs 425
    12.3.1 Coding Scheme 426
    12.3.2 Regression Analysis 427
    12.3.3 ANOVA Table 428
    12.3.4 R Code: Multilevel Design 430

12.3.5   Probing the Interaction   431
12.3.6   Higher-Order Designs   439
12.3.7   R Code: Simple Effects in Multilevel Design   439
12.4   Chapter Summary   441

13   Analysis of Covariance   443
13.1   Introduction to ANCOVA   444
13.1.1   Mechanics   445
13.1.2   Preliminary Analyses   446
13.1.3   Main Analysis   450
13.1.4   R Code: ANCOVA   451
13.1.5   Adjusted Means and Simple Effects   452
13.1.6   R Code: Adjusted Means and Simple Effects   456
13.2   Extensions to More Complex Designs   457
13.2.1   Preliminary Analyses   458
13.2.2   Main Analysis   460
13.2.3   Adjusted Means   460
13.2.4   Augmented Matrix and Multiple Comparisons   462
13.2.5   R Code: ANCOVA with Multiple Covariates   463
13.3   Uses (and Misuses) of ANCOVA   464
13.3.1   A Residualized Criterion   464
13.3.2   Association with the Predictor   465
13.4   Chapter Summary   467

14   Moderation   469
14.1   Moderated Regression   469
14.1.1   Illustration   470
14.1.2   Implementation   470
14.1.3   Regression Coefficients   473
14.1.4   Plotting Predicted Values   474
14.1.5   Crossing Point   475
14.1.6   Testing the Simple Slopes   476
14.1.7   R Code: Moderation—Simple Slopes   477
14.2   Simple Effects   479
14.2.1   Augmented b Vector and C Matrix   480
14.2.2   S Matrix   481
14.2.3   Specific Tests of Interest   482
14.2.4   R Code: Moderation—Simple Effects   485
14.3   Regions of Significance   487
14.3.1   Reviewing the Johnson-Neyman Method   488
14.3.2   Extending the Johnson-Neyman Method   488
14.3.3   Illustration   489
14.3.4   R Code: Regions of Significance   490
14.4   Chapter Summary   491

15 Mediation 493
   15.1   Simple Mediation 494
        15.1.1 Analytic Strategy 495
        15.1.2 Assessing the Importance of the Mediated Effect 499
        15.1.3 Effect Sizes 503
        15.1.4 Contrasts 503
        15.1.5 Summary 504
        15.1.6 R Code: Simple Mediation 505
   15.2   Higher-Order Designs 506
        15.2.1 Mediation with Three Groups 507
        15.2.2 R Code: Mediation with Three Groups 510
        15.2.3 Multiple Mediators 511
        15.2.4 R Code: Multiple Mediators 515
        15.2.5 Mediation and Moderation 516
        15.2.6 R Code: Mediation and Moderation 522
   15.3   Mediation and Causal Inference 525
   15.4   Chapter Summary 526

References 529

Index 531

Chapter 1
Matrix Properties and Operations

In this chapter you will learn about matrix algebra. Of all the chapters in the book, this one has the most terms to memorize. Learning matrix algebra is a bit like learning a new language, and the only way to understand the language is to immerse yourself in the terminology until it becomes familiar. I promise that the time you invest will be time well spent and that the remaining chapters won't be as dense. I also strongly encourage you to perform all of the analyses yourself using a spreadsheet. The best way to understand anything is by doing it, and you will walk away with a much deeper appreciation for the material if you take the time to make the calculations come alive.

1.1 Linear Equations and Matrix Algebra

Most of the statistical analyses we will cover in this book were developed to solve a series of linear equations, with each equation representing data from one subject.[1] The solution to a series of linear equations is the sum of products formed by weighting each known variable by a single, unknown variable. To clarify this point, let's consider a problem you might have encountered in middle school:

$$5x + 7y = -11$$
$$8x + 4y = 4$$
$$5x + 5y = -5$$

Electronic Supplementary Material: The online version of this chapter (doi: 10.1007/978-3-319-11734-8_1) contains supplementary material, which is available to authorized users

[1] In this context, the term "subject" refers to anything about which we make an observation. These subjects could be living entities (people), objects (cars), or events (days).

© Springer International Publishing Switzerland 2014
J.D. Brown, *Linear Models in Matrix Form*, DOI 10.1007/978-3-319-11734-8_1

Even if you don't yet know how to solve these equations, you can probably see that $x = 2$ and $y = -3$ represent the solution:

$$5(2) + 7(-3) = -11$$
$$8(2) + 4(-3) = 4$$
$$5(2) + 5(-3) = -5$$

If you can understand the logic behind this solution, you will be able to understand most of the statistical techniques we will cover. To see why, imagine that these equations represent the answers three middle school students gave when they were asked three questions: How many friends do you have in school; how many hours of homework do you have each week; and how much do you like school in general? Table 1.1 presents their (hypothetical) responses.

Table 1.1 Hypothetical data from three middle school students

Student	How many friends do you have in school?	How much homework do you have (hours/week)?	How much do you like school?
1	5	7	−11
2	8	4	4
3	5	5	−5

If we were to predict their final answers (How much do you like school?) from their answers to the first two questions, we would conclude that the number of friends is positively related to overall liking for school (+2) and the amount of homework is negatively related to overall liking for school (−3). In other words, these kids like school more when they have lots of friends and little homework! We might also conclude that the negative value of homework (−3) is greater than the positive value of friendship (+2), but this conclusion should be treated cautiously because it is influenced by a variety of factors (e.g., can five hours of homework be directly compared with having five friends?).

The most important difference between this example and the problems you will encounter in this book is that the latter do not have an exact solution. So we will have to settle for a "best solution" without being able to achieve an exact one, and we will have to find some way of judging whether our "best solution" tells us something about a larger population (e.g., middle school students in general). Before we tackle these issues, we first need to learn a bit of matrix algebra.

1.1.1 What Is a Matrix?

Many students shudder when they are asked to learn matrix algebra, but the ideas are not difficult to grasp. A matrix is a rectangular array of numbers or symbols characterized by rows and columns (called vectors). They are not much different than a table you have probably seen in a spreadsheet or word processing program.

Consider the two matrices below: **A** is a 2 × 3 (read "2 by 3") matrix because it has 2 row vectors and 3 column vectors; **B** is a 4 × 2 matrix because it has 4 row vectors and 2 column vectors:

$$\mathbf{A} = \begin{bmatrix} 4 & 2 & 1 \\ 8 & 3 & 5 \end{bmatrix} \quad \mathbf{B} = \begin{bmatrix} 7 & 0 \\ 1 & -1 \\ 11 & -3 \\ 4 & 6 \end{bmatrix}$$

When referencing a matrix, we indicate the (I) number of rows before indicating the (J) number of columns:

$$\mathbf{A} = \begin{bmatrix} a_{11} & a_{12} & a_{13} \\ a_{21} & a_{22} & a_{23} \end{bmatrix} \quad \mathbf{B} = \begin{bmatrix} b_{11} & b_{12} \\ b_{21} & b_{22} \\ b_{31} & b_{32} \\ b_{41} & b_{42} \end{bmatrix}$$

Each entry can then be identified by its unique location. For example, $a_{ij} = a_{23}$ references the entry in the second row, third column of $\mathbf{A}(a_{23} = 5)$. Similarly, $b_{ij} = b_{31}$ references the entry in the third row, first column of $\mathbf{B}(b_{31} = 11)$. We also designate each matrix with capital letters in bold script and place brackets around their elements to indicate that their entries are to be treated as a group or unit.

You might have noticed that I said the entries in a matrix are to be treated as a group or unit. This interdependence distinguishes matrices from a mere collection of numbers. Matrices have properties that go beyond the elements that comprise them. We can begin to appreciate these emergent properties by considering the square matrices below (designated **C**, **D**, and **E**). Square matrices have an equal number of rows and columns, and the sum of the main diagonal (from top left to bottom right) is known as the trace (denoted *tr*). Even though they have no elements in common, notice that **C** and **D** have the same trace (*tr* = 7). Notice also that **C** and **E** contain identical elements but do not have the same trace (i.e., the trace of **E** = 6). In short, two matrices can have different elements but the same trace or the same elements and a different trace. This is why we say matrices have properties that go beyond their specific elements:

$$\mathbf{C} = \begin{bmatrix} 2 & 6 \\ 4 & 5 \end{bmatrix} \quad \mathbf{D} = \begin{bmatrix} 18 & 3 \\ 9 & -11 \end{bmatrix} \quad \mathbf{E} = \begin{bmatrix} 4 & 5 \\ 6 & 2 \end{bmatrix}$$

1.1.2   Matrix Operations

Some matrix operations are easy. For example, when adding or subtracting matrices, we simply add or subtract the corresponding elements in the matrices:

$$\begin{bmatrix} 2 & 6 \\ 4 & 5 \end{bmatrix} + \begin{bmatrix} 4 & 5 \\ 6 & 2 \end{bmatrix} = \begin{bmatrix} 6 & 11 \\ 10 & 7 \end{bmatrix}$$

and

$$\begin{bmatrix} 2 & 6 \\ 4 & 5 \end{bmatrix} - \begin{bmatrix} 4 & 5 \\ 6 & 2 \end{bmatrix} = \begin{bmatrix} -2 & 1 \\ -2 & 3 \end{bmatrix}$$

It is also easy to add, subtract, multiply, or divide a matrix with a single number (called a scalar) by performing the operation on each element individually:

$$\begin{bmatrix} 2 & 6 \\ 4 & 5 \end{bmatrix} * -3 = \begin{bmatrix} -6 & -18 \\ -12 & -15 \end{bmatrix}$$

1.1.2.1 Matrix Multiplication

Unfortunately, matrix multiplication is more complicated and much more important in statistical analyses. Instead of simply multiplying two numbers, we multiply the values in one row of a matrix by the values in the column of another matrix and then sum the products. Let's look at how this is done, using the following matrices, designated **A**, **B**, and **C**, with **C** being the product matrix:

$$\overset{\mathbf{A}}{\begin{bmatrix} 10 & 3 \\ 4 & 5 \end{bmatrix}} * \overset{\mathbf{B}}{\begin{bmatrix} 1 & 2 \\ 6 & 7 \end{bmatrix}} = \overset{\mathbf{C}}{\begin{bmatrix} 28 & 41 \\ 34 & 43 \end{bmatrix}}$$

In our example, the first value in **C** was derived by multiplying the values of the first row in **A** with the first column in **B** and then summing the products:

$$c_{11} = (10 * 1) + (3 * 6) = 28$$

Similar methods were used to derive the remaining values:

$$c_{12} = (10 * 2) + (3 * 7) = 41$$
$$c_{21} = (4 * 1) + (5 * 6) = 34$$
$$c_{22} = (4 * 2) + (5 * 7) = 43$$

1.1.2.2 Matrix Multiplication and Simultaneous Linear Equations

Although you will not be performing matrix multiplication by hand in this book, it is important to understand how it works and how it can be used to solve simultaneous linear equations of the type we will be considering. To help you appreciate this point, let's go back to the sample equations we discussed earlier (reproduced below):

$$5x + 7y = -11$$
$$8x + 4y = 4$$
$$5x + 5y = -5$$

If we put these equations into matrix form (again using designations of **A**, **b**, and **c**),[2]

$$\overset{\mathbf{A}}{\begin{bmatrix} 5 & 7 \\ 8 & 4 \\ 5 & 5 \end{bmatrix}} \overset{\mathbf{b}}{\begin{bmatrix} x \\ y \end{bmatrix}} = \overset{\mathbf{c}}{\begin{bmatrix} -11 \\ 4 \\ -5 \end{bmatrix}}$$

we can reconstruct our original equations through matrix multiplication:

$$5x + 7y = -11$$
$$8x + 4y = 4$$
$$5x + 5y = -5$$

If we then substitute the values we obtained earlier ($x = 2$ and $y = -3$),

$$\begin{bmatrix} 5 & 7 \\ 8 & 4 \\ 5 & 5 \end{bmatrix} \begin{bmatrix} 2 \\ -3 \end{bmatrix} = \begin{bmatrix} -11 \\ 4 \\ -5 \end{bmatrix}$$

we verify that matrix multiplication solves the problem (i.e., $\mathbf{Ab} = \mathbf{c}$):[3]

$$5(2) + 7(-3) = -11$$
$$8(2) + 4(-3) = 4$$
$$5(2) + 5(-3) = -5$$

1.1.3 Matrix Rules of Multiplication

1.1.3.1 Matrix Multiplication Is Not Commutative

Ordinary multiplication is commutative [e.g., $(7 * 6) = (6 * 7)$]. Although there are exceptions, matrix multiplication usually is not. Even when two matrices have the

---

[2] Matrices with a single row or column are referred to as row or column vectors, respectively, and are denoted with lower case letters with bold script.

[3] When multiplying matrices, it is customary to omit the product sign and simply place the two matrices side-by-side (e. g., $\mathbf{AB} = \mathbf{A} * \mathbf{B}$). When matrices are added or subtracted, the plus or minus sign is included to avoid confusion.

same dimensions (e.g., both are 3×3), **AB** will usually not $=$ **BA**. Consequently, it is important to pay attention to the order in which matrices are multiplied. We say "**B** was premultiplied by **A**" or "**A** was postmultiplied by **B**," to refer to **AB**.

1.1.3.2   Matrix Conformability

Matrices can be multiplied only if they are conformable. In this context, conformable means that the number of columns of the premultiplication matrix equals the number of rows of the postmultiplication matrix. In the preceding example, we could multiply **Ab** because the number of columns of **A**(2) equals the number of rows of **b**(2); we could not have multiplied **bA**, because the number of columns of **b**(1) does not equal the number of rows of **A**(3). When two matrices are conformable, the size of the product matrix will equal the rows of the premultiplication matrix and the columns of the postmultiplication matrix (see below for example).

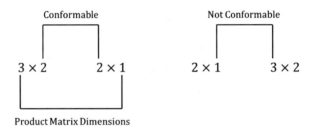

1.1.3.3   Matrix Multiplication Is Associative

Although matrix multiplication is ordinarily not commutative, it is associative (i.e., $\{AB\}C = A\{BC\}$). We can understand why by considering the conformability of three matrices.

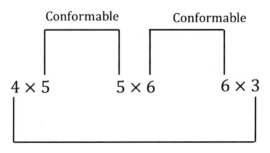

Because they are conformable, we can multiply **AB** and **BC**. When we multiply all three matrices, the product matrix will have as its dimensions the rows of the first matrix and the columns of the third matrix.

1.1.4   Using a Spreadsheet to Perform Matrix Operations

Earlier you learned that you will not be multiplying matrices by hand in this book; nor will you be asked to perform any other matrix operations without the aid of computer software. The calculations are not difficult when matrix sizes are small, but they are very difficult when matrix sizes are large (e.g., multiplying a 600×500 matrix with a $500 \times 1,000$ matrix takes a lot of time and leaves plenty of room for error!).

Fortunately, all spreadsheets include commands for performing matrix operations (see Table 1.5 at the end of this chapter for a summary). As all of the calculations in this book are available in the spreadsheet that accompanies the text, you will want to familiarize yourself with the manner in which spreadsheets handle matrix operations so that you can easily follow along and perform the analyses yourself. I will most commonly describe the operations with reference to Microsoft's EXCEL®, but you don't have to buy EXCEL in order to perform the analyses in the book. All of the calculations can be done using open-source spreadsheets that are available for free (e.g., Apache Calc; Google Sheets; Libre Office).

1.1.4.1   Working with Arrays

Spreadsheets treat matrices as arrays. Instead of pressing Enter when you finish your calculations (as you normally would do when performing simple arithmetic), you will need to press Ctrl-Shift-Enter when performing operations on an array if you are using Windows or Command-Return if you are using a Mac.[4] When you do, you will notice that your spreadsheet adds brackets around your formula in the formula bar, indicating that you have performed an array function.

1.1.4.2   Matrix Multiplication

To illustrate this process, let's multiply the two matrices we've been working with in a spreadsheet, setting up the problem as follows. Be sure to highlight the product matrix (F2:F4) as I have done below.

---

[4] Since I use a Windows computer, I will specify only the functions for that operating system hereafter.

◢	A	B	C	D	E	F	G
1	A			B		C	
2	5	7		2			
3	8	4		-3			
4	5	5					
5							
6				The dimensions of the			
7				product matrix = the rows			
8				of A (= 3) and the			
9				columns of B (= 1)			
10							

Next, we want to enter a formula. You can do this by using the formula bar, but I think it's easier to select the equal sign on your keyboard and type the following formula into the window that opens: MMULT(a2:b4,d2:d3).[5] The MMULT command performs matrix multiplication rather than simple multiplication, and the cell specifications indicate where the two matrices are located (separated by a comma in Excel and a semicolon in most open source spreadsheets). Finally, you need to press Ctrl-Shift-Enter all at once. When you do, the product matrix should appear in F2:F4.

		F2	▼	●		*fx* {=MMULT(A2:B4,D2:D3)}	
◢	A	B	C	D	E	F	
1	A			B		C	
2	5	7		2		-11	
3	8	4		-3		4	
4	5	5				-5	
5							
6				=mmult(A2:B4,D2:D3)			
7							

1.1.4.3   Naming Matrices

Spreadsheets allow you to name a range of continuous cells so that you can reference them by name rather than by cell number. In our example, I named the values in A2:B4 "A" and the values in D2:D3 "B." Thereafter, I need only to

[5] You can also drag your mouse to highlight the area of each matrix. This is often the most effective strategy as it allows you to see exactly which cells are involved in the multiplication.

type $=$ MMULT(A,B) to multiply them, which is simpler and gives a greater understanding of the operations involved.

Naming the matrices is especially useful when performing a series of calculations. Although we can only multiply two matrices at a time, we can multiply many matrices by nesting the matrices within a single formula. To illustrate, imagine we want to multiply the three matrices below (which I have named **D**, **E**, and **F**). We first need to determine whether they are conformable. Because the columns in **D** $=$ the rows in **E**, and the columns in **E** $=$ the rows in **F**, we conclude the matrices are conformable. We then highlight a 3×3 collection of cells (corresponding to D's rows and F's columns) and type $=$ MMULT(D,MMULT(E,F)).

15									
16		D				E			F
17	10	3		1	2	4		7	9	3
18	4	5		6	7	0.5		5	-2	6
19	8	9						11	-4	17
20									
21									
22				D[EF]					
23									
24									
25									

When you hit Ctrl-Shift-Enter, you should see the following values in the product matrix:

$$\begin{bmatrix} 857.5 & 4 & 1035.5 \\ 656.5 & 146 & 674.5 \\ 1230.5 & 254 & 1280.5 \end{bmatrix}$$

Take a few minutes to examine how the nested formula is constructed because we will use similar syntax throughout the text.

1.2   Matrix Transpose and Sum of Squares

Many statistical analyses involve finding a sum of squared values. By using another matrix function called the transpose, matrix multiplication offers a convenient way of computing sum of squares. The data in Table 1.2 will help us appreciate this point.

Table 1.2 Data for demonstrating a sum of squares matrix

	1	2	6
	4	8	−4
	5	6	5
	4	2	3
Mean	**3.5**	**4.5**	**2.5**
Variance	**3.00**	**9.00**	**20.33**
Standard deviation	**1.732**	**3.00**	**4.509**

To begin, we will turn the data (but not summary statistics) into matrix form to produce a matrix we will designate **A**:

$$\mathbf{A} = \begin{bmatrix} 1 & 2 & 6 \\ 4 & 8 & -4 \\ 5 & 6 & 5 \\ 4 & 2 & 3 \end{bmatrix}$$

1.2.1 Transposition

When we transpose a matrix, we exchange the columns and rows (i.e., we turn the columns into rows and the rows into columns). The new matrix is designated with a prime (**A′**) to indicate it is a transposed matrix and is referred to as the "transpose of **A**" or, more commonly, as "**A** transpose." In the example below, notice that the rows of **A** are the columns of **A′** and that the columns of **A** are the rows of **A′**:

$$\mathbf{A} = \begin{bmatrix} 1 & 2 & 6 \\ 4 & 8 & -4 \\ 5 & 6 & 5 \\ 4 & 2 & 3 \end{bmatrix} \quad \text{and} \quad \mathbf{A}' = \begin{bmatrix} 1 & 4 & 5 & 4 \\ 2 & 8 & 6 & 2 \\ 6 & -4 & 5 & 3 \end{bmatrix}$$

1.2.2 Premultiplying a Matrix by Its Transpose

Premultiplying a matrix by its transpose produces a square matrix with three very useful features:

$$\begin{array}{ccc} \mathbf{A}' & \mathbf{A} & = \quad \mathbf{B} \end{array}$$

$$\begin{bmatrix} 1 & 4 & 5 & 4 \\ 2 & 8 & 6 & 2 \\ 6 & -4 & 5 & 3 \end{bmatrix} \begin{bmatrix} 1 & 2 & 6 \\ 4 & 8 & -4 \\ 5 & 6 & 5 \\ 4 & 2 & 3 \end{bmatrix} = \begin{bmatrix} 58 & 72 & 27 \\ 72 & 108 & 16 \\ 27 & 16 & 86 \end{bmatrix}$$

1. First, the entries on the main diagonal of the product matrix (upper left to lower right) represent the sum of the squared values of each of the three columns of **A**:

$$1^2 + 4^2 + 5^2 + 4^2 = 58$$
$$2^2 + 8^2 + 6^2 + 2^2 = 108$$
$$6^2 + -4^2 + 5^2 + 3^2 = 86$$

2. Second, the off-diagonal entries of the product matrix represent cross-product sums, formed by multiplying each entry in one column of **A** by its corresponding entry in another column of **A**, and then summing the products:

$$\mathbf{b}_{12} = \mathbf{a}_1{}^*\mathbf{a}_2 = [(1^*2) + (4^*8) + (5^*6) + (4^*2)] = 72$$
$$\mathbf{b}_{13} = \mathbf{a}_1{}^*\mathbf{a}_3 = [(1^*6) + (4^* - 4) + (5^*5) + (4^*3)] = 27$$
$$\mathbf{b}_{23} = \mathbf{a}_2{}^*\mathbf{a}_3 = [(2^*6) + (8^* - 4) + (6^*5) + (2^*3)] = 16$$

3. Third, the product matrix **B** is symmetric (i.e., the lower-left half is the mirror image of the upper-right half). Formally, we define a symmetric matrix as a square matrix that is equal to its transpose:

$$\mathbf{B} = \begin{bmatrix} 58 & 72 & 27 \\ 72 & 108 & 16 \\ 27 & 16 & 86 \end{bmatrix} = \mathbf{B}' = \begin{bmatrix} 58 & 72 & 27 \\ 72 & 108 & 16 \\ 27 & 16 & 86 \end{bmatrix}$$

These properties hold true whenever we premultiply a matrix by its transpose (i.e., $\mathbf{A}'\mathbf{A}$ always produces a symmetric, square matrix with the sum of squared values on the diagonal and the sum of cross-product values on the off-diagonals). In the next section, you will learn why matrices of this form are especially important in statistical analyses.

1.2.3 Covariance Matrix

Behavioral scientists study variability. For example, we might examine the variability of IQ scores among a sample of children or the variability of docility in a sample of foxes. When we index variability, we first calculate a deviate score by subtracting the group mean from each entry. We then square the difference and sum the squared values. The average squared value is known as the variance, and the square root of the average squared value is called the standard deviation.

We also frequently try to relate the variability in one variable to the variability in other variables. For example, we might examine the associations among IQ, nutrition, socioeconomic status, and family dynamics. Formally, we are studying the extent to which two or more variables go together or covary. We assign a numerical value to the association between two variables by calculating the cross-product sum of their respective deviate scores. The average value is known as the covariance.

The transposition method we just learned provides an easy way to generate a covariance matrix. We begin by calculating a matrix of deviate scores, found by subtracting the column mean from each corresponding column entry in the original matrix. We will call this matrix $\mathbf{D}$ because the entries represent deviations from the column's mean. To illustrate, the mean of the first column in $\mathbf{A}$ is 3.5, and the entries in the first column of $\mathbf{D}$ were found by subtracting this value from each value in the first column of $\mathbf{A}$:

$$
\mathbf{A} = \begin{bmatrix} 1 & 2 & 6 \\ 4 & 8 & -4 \\ 5 & 6 & 5 \\ 4 & 2 & 3 \end{bmatrix} \quad \rightarrow \quad \mathbf{D} = \begin{bmatrix} -2.5 & -2.5 & 3.5 \\ .5 & 3.5 & -6.5 \\ 1.5 & 1.5 & 2.5 \\ .5 & -2.5 & .5 \end{bmatrix}
$$

If we premultiply $\mathbf{D}$ by its transpose and divide each term by its associated degrees of freedom $(N - 1)$, we produce a symmetric covariance matrix $\mathbf{S}$ with each variable's variance on the diagonals and the covariances (average deviation cross-product sum) on the off-diagonals:[6]

$$
\mathbf{S} = \mathbf{D}'\mathbf{D}\frac{1}{N-1} \tag{1.1}
$$

Inserting our values produces the covariance matrix:

$$
\mathbf{S} = \begin{bmatrix} 3.00 & 3.00 & -2.6667 \\ 3.00 & 9.00 & -9.6667 \\ -2.6667 & -9.6667 & 20.3333 \end{bmatrix}
$$

It is worth taking the time to confirm that the obtained values represent the average squared deviate scores (in the case of the variances on the main diagonal) and the average cross-product sum (in the case of the covariances on the off-diagonals). Below I show the relevant calculations for the variance of the middle variable (9.00) and the covariance of variables 2 and 3 (−9.6667). You can perform the calculations for the other entries if you desire:

---

[6] You might be wondering why we use $N-1$ for the denominator rather than N. The reason is that we lose 1 degree of freedom when we use the mean to create the deviate score, leaving $N-1$ degrees of freedom for computing the variability.

$$var_2 = \frac{-2.5^2 + 3.5^2 + 1.5^2 + -2.5^2}{(4-1)} = 9.00$$

$$cov_{23} = \frac{(-2.5 * 3.5) + (3.5 * -6.5) + (1.5 * 2.5) + (-2.5 * .5)}{(4-1)} = -9.6667$$

1.2.4   Correlation Matrix

A correlation matrix is a covariance matrix of standard scores. Standard scores (denoted z) are found by first subtracting the column mean from each individual score and then dividing the difference by the standard deviation of the column entries. Doing so creates a new variable with mean $= 0$ and variance $= 1$. Placing these standard scores into a matrix produces a new matrix we'll designate $\mathbf{Z}$:

$$\mathbf{A} = \begin{bmatrix} 1 & 2 & 6 \\ 4 & 8 & -4 \\ 5 & 6 & 5 \\ 4 & 2 & 3 \end{bmatrix} \rightarrow \mathbf{Z} = \begin{bmatrix} -1.4434 & -.8333 & .7762 \\ .2887 & 1.1667 & -1.4415 \\ .8660 & .50 & .5544 \\ .2887 & -.8333 & .1109 \end{bmatrix}$$

If we premultiply $\mathbf{Z}$ by its transpose and divide each term by its associated degrees of freedom $(N-1)$, we produce a correlation matrix $\mathbf{R}$:

$$\mathbf{R} = \mathbf{Z'Z}\frac{1}{N-1} \tag{1.2}$$

The standardized variances of each column are on the diagonals (and they are always 1), and the off-diagonals show the covariances among the standardized variables, which are their correlations. Plugging in our values produces the following correlation matrix:

$$\mathbf{R} = \begin{bmatrix} 1 & .5774 & -.3414 \\ .5774 & 1 & -.7146 \\ -.3414 & -.7146 & 1 \end{bmatrix}$$

As before, it's useful to confirm that a correlation is simply the average cross-product sum of two standard scores by performing the calculations using $\mathbf{Z}$. I will illustrate using columns 2 and 3:[7]

---

[7] We can also compute a correlation directly from the covariance matrix.

$$r = \frac{cov_{12}}{\sqrt{\sigma_1^2 * \sigma_2^2}}$$

$$r_{23} = \frac{(-.8333 * .7762) + (1.1667 * -1.4415) + (.50 * .5544) + (-.8333 * .1109)}{(4-1)} = -.7146$$

1.2.5   Diagonal Matrix

A matrix with 0's on the off-diagonals is called a diagonal matrix. Below is an example of a 3×3 diagonal matrix $\mathbf{A}$, but they can be of any size:

$$\mathbf{A} = \begin{bmatrix} 3 & 0 & 0 \\ 0 & 6 & 0 \\ 0 & 0 & 21 \end{bmatrix}$$

A diagonal sum of squares matrix is of especial interest. Because the off-diagonal entries represent cross-product sums, a diagonal matrix formed by premultiplying a matrix by its transpose indicates that the variables are independent (i.e., covariance/correlation $=0$). On the other hand, if the off-diagonal entries of a sum of squares matrix do not $=0$, the variables are associated, such that one predicts the other.

1.2.6   Summary

In this section you learned that premultiplying a matrix by its transpose produces a symmetric, square matrix with sums of squares on the main diagonals and cross-product sums on the off-diagonals. When we begin with deviate scores and divide by $N-1$, we get a covariance matrix; when we begin with standard scores and divide by $N-1$, we get a correlation matrix. Both matrices play an important role in statistical analyses, so it's good to take the time now to have a solid understanding of them. They are nothing but averaged squared sums on the diagonal and averaged cross-product sums on the off-diagonals.

1.2.7 R Code: Matrix Multiplication

```
#Matrix Multiplication
A <- matrix(c(10,3,4,5), nrow=2, ncol=2, byrow=TRUE)
A
B <- matrix(c(1,2,6,7), nrow=2, ncol=2, byrow=TRUE)
B
AB <-A%*%B                   # %*% indicates matrix multiplication
AB


#Matrix transpose and sum of squares
A <- matrix(c(1,2,6,4,8,-4,5,6,5,4,2,3), nrow=4, ncol=3, byrow=TRUE)
A
A.sq <-t(A)%*%A                   # t(A) indicates a transposed matrix
A.sq


#Create deviate scores and covariance matrix
N <-nrow(A)
D <-scale(A, center = TRUE, scale = FALSE)
covar <-(t(D)%*%D)/(N-1)
covar


#Create standardized scores and correlation matrix
Z <-scale(A, center = TRUE, scale = TRUE)
corr <-(t(Z)%*%Z)/(N-1)
corr
```

1.3 Matrix Determinants

When introducing matrices, we noted that they have emergent properties that go beyond their specific elements. We used the trace of a matrix to illustrate this point. The trace is not the only characteristic of a matrix that illustrates their emergent properties, however. The determinant of a square matrix provides another example.[8] The determinant quantifies the variability of a matrix. If we think of a matrix as a box, the determinant is a measure of how much stuff it holds. To illustrate, consider the following matrices with the same trace ($tr = 9$):

$$\mathbf{A} = \begin{bmatrix} 5 & 3 \\ 2 & 4 \end{bmatrix} \quad \mathbf{B} = \begin{bmatrix} 5 & 6 \\ 6 & 4 \end{bmatrix}$$

---

[8] Only square matrices have determinants, but the matrices do not have to be symmetric.

The determinant of a square matrix is a single value (called a scalar), denoted with straight line brackets. We can easily calculate the determinant of a 2×2 matrix by subtracting the product of the off-diagonal entries from the product of the main diagonal entries:

$$|\mathbf{A}| = (5 * 4) - (3 * 2) = 14$$

and

$$|\mathbf{B}| = (5 * 4) - (6 * 6) = -16$$

The absolute value of the determinant provides an index of the variability in a matrix. In our example, we would conclude that $\mathbf{B}$ has more variability than $\mathbf{A}$, even though the variability in the numbers themselves is greater in $\mathbf{A}$ than in $\mathbf{B}$. But we are not interested in the variability in the entries themselves; we are interested in the variability of the matrix as a whole.

1.3.1 Visualizing the Determinant

Thinking about the determinant in geometric terms might help you better appreciate the preceding point. Consider Fig. 1.1, which plots our $\mathbf{A}$ matrix.

The determinant equals the area inside the parallelogram formed by extending the vectors (solid lines) to the farthest point in the rectangle formed from the sum of the column vectors (dashed lines). To find this area, we first find the area of the entire rectangle by multiplying the summed values of column 1 with the summed values of column 2. In our case, both columns sum to 7, so the area of the rectangle equals 49. We then calculate the determinant by subtracting from the total area the area of the rectangle that does not lie within the parallelogram:

$$\text{Determinant} = (\text{Total Area}) - A - B - C - D - E - F \qquad (1.3)$$

$$A = x_2 y_1 \quad B = \frac{x_2 y_2}{2} \quad C = \frac{x_1 y_1}{2}$$

$$D = \frac{x_2 y_2}{2} \quad E = x_2 y_1 \quad F = \frac{x_1 y_1}{2}$$

Combining terms, we can state the formula more succinctly:

$$\text{Determinant} = [(x_1 + x_2) * (y_1 + y_2)] - 2(x_2 y_1) - x_1 y_1 - x_2 y_2 \qquad (1.4)$$

Plugging in the values from our example yields our determinant:

$$|\mathbf{A}| = [(5 + 2) * (3 + 4)] - 2(2 * 3) - (5 * 3) - (2 * 4) = 14$$

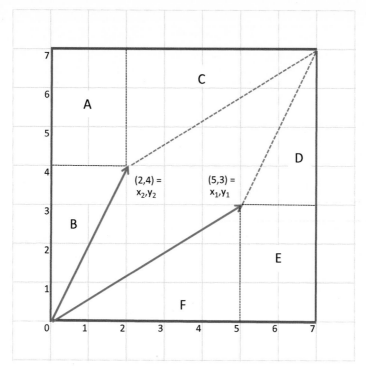

Fig. 1.1 Visual representation of a matrix determinant

In sum, the determinant is a single value of a square matrix that quantifies the matrix's variability. With a 2×2 matrix, the absolute value of the determinant represents the area within a parallelogram enclosed by the column vectors; with larger matrices, the absolute value of the determinant represents the volume of a parallelepiped formed by the column vectors.

Thinking of the determinant in geometric terms might help you remember an important property of the determinant. When we multiply two square matrices, the area or volume increases multiplicatively. As a result, the determinant of the product matrix is the product of their individual determinants:

$$|\mathbf{AB}| = |\mathbf{A}||\mathbf{B}| \tag{1.5}$$

We will confirm this property by multiplying $\mathbf{AB}$ and showing that the determinant of the product matrix equals the product of the determinants of $\mathbf{A}$ and $\mathbf{B}$:

$$|\mathbf{A}| = \begin{vmatrix} 5 & 3 \\ 2 & 4 \end{vmatrix} = 14 \quad |\mathbf{B}| = \begin{vmatrix} 5 & 6 \\ 3 & 4 \end{vmatrix} = 2 \quad |\mathbf{AB}| = \begin{vmatrix} 34 & 42 \\ 22 & 28 \end{vmatrix} = 28$$

1.3.2 *Using the Determinant to Solve Linear Equations*

Using an approach known as Cramer's rule (named after the eighteenth-century mathematician, Gabriel Cramer), determinants can be used to solve simultaneous linear equations. The rule applies only when the number of equations equals the number of unknowns, so we will illustrate its use with the following two equations with two unknowns:

$$5x + 7y = -11$$
$$8x + 4y = 4$$

Expressing the equations in matrix form yields the following:

$$\begin{bmatrix} 5 & 7 \\ 8 & 4 \end{bmatrix} \begin{bmatrix} x \\ y \end{bmatrix} = \begin{bmatrix} -11 \\ 4 \end{bmatrix}$$

1.3.2.1 Calculations

To solve for the missing values, we first calculate the determinant of a matrix formed using only the predictor variables (i.e., those to the left of the unknown quantities), which we'll designate **P**. As noted earlier, the calculations with a 2×2 matrix involve subtracting the product of the off-diagonal entries $(7 * 8)$ from the product of the diagonal entries $(5 * 4)$:

$$|\mathbf{P}| = \begin{vmatrix} 5 & 7 \\ 8 & 4 \end{vmatrix} = -36$$

Now we need to find the determinants of two additional matrices, each formed by replacing one column of **P** with the values that appear to the right of the equal sign:

- To find the value for x, we replace the first column in **P** with the values from our final column vector, creating a new matrix $\mathbf{P}_x$:

$$|\mathbf{P}_x| = \begin{vmatrix} -11 & 7 \\ 4 & 4 \end{vmatrix} = -72$$

- To find the value for y, we replace the second column in **P** with the values from our final column vector, creating a new matrix $\mathbf{P}_y$:

$$|\mathbf{P}_y| = \begin{vmatrix} 5 & -11 \\ 8 & 4 \end{vmatrix} = 108$$

Finally, we solve for x and y by forming a ratio, with one determinant from the replaced matrix in the numerator and the determinant of **P** in the denominator:

$$x = \frac{|\mathbf{P_x}|}{|\mathbf{P}|} = \frac{-72}{-36} = 2$$

$$y = \frac{|\mathbf{P_y}|}{|\mathbf{P}|} = \frac{108}{-36} = -3$$

You can verify that these are the correct values for x and y by plugging them into our original equation and doing the multiplication.

1.3.2.2   Proof

Matrix operations can seem mysterious, but they solve problems efficiently, not magically. To better understand this point, we will spend a moment proving Cramer's rule. We begin by restating our equation more abstractly:

$$5x + 7y = -11$$
$$8x + 4y = 4$$

becomes

$$ax + by = c$$
$$dx + ey = f$$

which can be expressed in matrix form:

$$\begin{bmatrix} a & b \\ d & e \end{bmatrix} \begin{bmatrix} x \\ y \end{bmatrix} = \begin{bmatrix} c \\ f \end{bmatrix}$$

Now let's return to the equations and solve for y.

- First equation:

$$ax + by = c$$
$$by = c - ax$$
$$y = \frac{c - ax}{b}$$

- Second equation:

$$dx + ey = f$$
$$ey = f - dx$$

$$y = \frac{f - dx}{e}$$

Because both equations solve for y, we can set them to be equal:

$$y = \frac{c - ax}{b} = \frac{f - dx}{e}$$

Having solved for *y*, we can solve for *x* by cross-multiplying:

$$ce - axe = bf - bdx$$

rearranging terms:

$$ce - bf = axe - bdx$$

and reducing:

$$x = \frac{ce - bf}{ae - bd} \qquad (1.6)$$

Notice that the denominator in Eq. (1.6) equals the determinant of the matrix of predictors:

$$\begin{vmatrix} a & b \\ d & e \end{vmatrix} = ae - bd$$

This is why we divide by the determinant to solve for *x* using Cramer's rule. Now recall that using Cramer's rule we find *x* by calculating the determinant of a new matrix $\mathbf{P}_x$, formed by replacing the first column of predictors with the criterion column vector:

$$\begin{vmatrix} c & b \\ f & e \end{vmatrix} = ce - bf$$

As you can see, the determinant of this matrix is the same as the numerator of our algebraic solution. So when we divide the determinant of $\mathbf{P}_x$ by the determinant of our original matrix of predictors $\mathbf{P}$, we are using the same terms we use to solve the equation algebraically:

$$x = \frac{|\mathbf{P}_x|}{|\mathbf{P}|} = \frac{\begin{vmatrix} c & b \\ f & e \end{vmatrix}}{\begin{vmatrix} a & b \\ d & e \end{vmatrix}} = \frac{ce - bf}{ae - bd} \qquad (1.7)$$

In short, Cramer's rule is simply an efficient way of solving simultaneous linear equations using the determinants of two matrices. Unsurprisingly, the approach also yields y:

$$y = \frac{|\mathbf{P}_y|}{|\mathbf{P}|} = \frac{\begin{vmatrix} a & c \\ d & f \end{vmatrix}}{\begin{vmatrix} a & b \\ d & e \end{vmatrix}} = \frac{af - cd}{ae - bd} \qquad (1.8)$$

1.3.3   Linear Dependencies and Singular Matrices

Dividing by 0 is undefined, so Cramer's rule cannot be used when the determinant of a predictor matrix is 0. To illustrate, consider the following example:

$$\begin{vmatrix} 5 & 10 \\ 2 & 4 \end{vmatrix} = 0$$

A determinant of zero reveals a linear dependence in the matrix. This term means that one (or more) of the vectors is a linear combination of at least one other vector. In our example, we can see that the second column vector is simply $2 *$ the first column vector. We could also say that the first column vector is simply $5 *$ the second or that the first row vector is $2.5 *$ the second row vector or that the second row vector is $4 *$ the first row vector. All of these descriptions are true. Matrices with a determinant of zero are called singular matrices, and we will find that we need to avoid them in order to perform most of the statistical calculations that comprise this book.

1.3.4   Calculating the Determinant with Large Matrices

With large matrices, calculating the determinant is tedious, and the operations are better left to computer programs. In a spreadsheet, we use $= \text{MDETERM}$. Still, it's not magic and it's useful to take a small (3×3) matrix and see the steps involved in calculating the determinant. We will use the matrix $\mathbf{A}$ to illustrate these steps:

$$\mathbf{A} = \begin{bmatrix} 3 & 9 & 1 \\ 10 & 8 & 4 \\ 5 & 7 & 6 \end{bmatrix}$$

and

$$|\mathbf{A}| = -270$$

1.3.4.1   Minors

The technique we will use is known as the Laplace (or cofactor) expansion. To use the technique we need to find the determinant of a smaller matrix that results when row and column elements are deleted. The determinant from this reduced matrix is known as a minor, designated m_{ij}, with the first subscript denoting the row we eliminated and the second subscript denoting the column we eliminated. I'll illustrate by eliminating the first column of our matrix, then eliminating one row at a time starting with the first:

$$m_{11} = \begin{vmatrix} 3 & 9 & 1 \\ 10 & 8 & 4 \\ 5 & 7 & 6 \end{vmatrix} = \begin{vmatrix} 8 & 4 \\ 7 & 6 \end{vmatrix} = (8 * 6) - (4 * 7) = 20$$

$$m_{21} = \begin{vmatrix} 3 & 9 & 1 \\ 10 & 8 & 4 \\ 5 & 7 & 6 \end{vmatrix} = \begin{vmatrix} 9 & 1 \\ 7 & 6 \end{vmatrix} = (9 * 6) - (1 * 7) = 47$$

$$m_{31} = \begin{vmatrix} 3 & 9 & 1 \\ 10 & 8 & 4 \\ 5 & 7 & 6 \end{vmatrix} = \begin{vmatrix} 9 & 1 \\ 8 & 4 \end{vmatrix} = (9 * 4) - (1 * 8) = 28$$

1.3.4.2   Cofactors

Our next step is to convert the minors into cofactors f_{ij} using

$$f_{ij} = \begin{cases} m_{ij} & \cdots \quad (i+j) \text{ is even} \\ -m_{ij} & \cdots \quad (i+j) \text{ is odd} \end{cases} \tag{1.9}$$

or, equivalently,

$$f_{ij} = (-1)^{(i+j)} m_{ij} \tag{1.10}$$

Following these rules, the cofactors become

$$f_{11} = 20$$
$$f_{21} = -47$$
$$f_{31} = 28$$

Finally, we calculate the determinant by multiplying each cofactor by its corresponding column entry in our original matrix and then summing the products:

$$|\mathbf{A}| = \Sigma \mathbf{A}_{ij} \mathbf{f}_{ij} \qquad (1.11)$$

Plugging in our values yields the determinant:

$$|\mathbf{A}| = (3 * 20) + (10 \quad * \quad -47) + (5 * 28) = -270$$

1.3.4.3 Creating a Cofactor Matrix

With a 3×3 matrix, only three cofactors are needed to calculate the determinant. However, later in this chapter, we will need to calculate all of the cofactors and place them into a matrix known as the cofactor matrix $\mathbf{F}$.[9] Since we are discussing the procedure now, let's go ahead and calculate the remaining cofactors, beginning with their minors:

$$m_{12} = \begin{vmatrix} 10 & 4 \\ 5 & 6 \end{vmatrix} (10 * 6) - (4 * 5) = 40$$

$$m_{22} = \begin{vmatrix} 3 & 1 \\ 5 & 6 \end{vmatrix} (3 * 6) - (1 * 5) = 13$$

$$m_{32} = \begin{vmatrix} 3 & 1 \\ 10 & 4 \end{vmatrix} (3 * 4) - (1 * 10) = 2$$

and

$$m_{13} = \begin{vmatrix} 10 & 8 \\ 5 & 7 \end{vmatrix} (10 * 7) - (8 * 5) = 30$$

$$m_{23} = \begin{vmatrix} 3 & 9 \\ 5 & 7 \end{vmatrix} (3 * 7) - (9 * 5) = -24$$

$$m_{33} = \begin{vmatrix} 3 & 9 \\ 10 & 8 \end{vmatrix} (3 * 8) - (9 * 10) = -66$$

---

[9] The cofactor matrix is often designated $\mathbf{C}$ rather than $\mathbf{F}$. I have chosen $\mathbf{F}$ because I want to reserve $\mathbf{C}$ for a covariance matrix of parameters to be described in Chap. 2.

We then form a complete cofactor matrix using Eq. (1.10):

$$\mathbf{F} = \begin{bmatrix} 20 & -40 & 30 \\ -47 & 13 & 24 \\ 28 & -2 & -66 \end{bmatrix}$$

Finally, we can find the determinant of our original matrix from any diagonal element of $\mathbf{AF'}$:

$$\begin{bmatrix} 3 & 9 & 1 \\ 10 & 8 & 4 \\ 5 & 7 & 6 \end{bmatrix} \begin{bmatrix} 20 & -47 & 28 \\ -40 & 13 & -2 \\ 30 & 24 & -66 \end{bmatrix} = \begin{bmatrix} -270 & 0 & 0 \\ 0 & -270 & 0 \\ 0 & 0 & -270 \end{bmatrix}$$

1.3.5 R Code: Determinants

```
A <- matrix(c(5,3,2,4), nrow=2, ncol=2, byrow=TRUE)
det(A)  #R function for calculating the determinant

#Using Cramer's Rule to Solve Simultaneous Linear Equations
P <- matrix(c(5,7,8,4),    nrow=2, ncol=2, byrow=TRUE)
det.P <-det(P)
Px <- matrix(c(-11,7,4,4), nrow=2, ncol=2, byrow=TRUE)
det.Px <-det(Px)
Py <- matrix(c(5,-11,8,4), nrow=2, ncol=2, byrow=TRUE)
det.Py <-det(Py)
x=det.Px/det.P
y=det.Py/det.P
xy=c(x,y)
xy

#Minor and Cofactors Function
cofactor <- function(A) {
  n <- nrow(A)
  F <- matrix(NA, n, n)
  if(n>2){
    cofactors <- function(A, i, j)
      (-1)^(i+j) * det( A[-i,-j] )
        for( i in 1:n )
          for( j in 1:n )
            F[i,j] <- cofactors(A, i, j)
F
}
  else{F<-matrix(c(A[4],-A[2],-A[3],A[1]),2,byrow=TRUE)}
F
}
```

(continued)

1.3.5   R Code: Determinants (continued)

```
A <- matrix(c(3,9,1,10,8,4,5,7,6), nrow=3, ncol=3, byrow=TRUE)
cofactor(A)

#Calculate Determinant from Cofactor Matrix
determ <-diag(A%*%t(cofactor(A)))
determ[1]
```

1.4   Matrix Inverse

The last topic we will consider in this chapter is the matrix inverse. In many respects, it is the most important topic as it plays a role in almost all of the analyses we will be performing. An inverse is a property of a square matrix, denoted with a superscript of -1, so that $\mathbf{A}^{-1}$ represents "the inverse of $\mathbf{A}$" or, equivalently, "$\mathbf{A}$ inverse." In the following sections, you will learn about the properties of an inverse, how it is calculated, and how it can be used to solve linear equations.

1.4.1   Matrix Multiplication and Matrix Inverse

We begin by noting that although we can multiply two (or more) matrices, we cannot divide them. Fortunately, we have another solution that will already be familiar to you. Consider the following equality:

$$\frac{x}{y} = \frac{1}{y} * x \qquad (1.12)$$

Equation (1.12) calls attention to the fact that dividing x by y is equivalent to multiplying x by the reciprocal of y. We might also remind ourselves that any number multiplied by its reciprocal equals $1\left(\text{e.g., } \frac{1}{y} * y = 1\right)$ and that multiplying any number by 1 leaves the number unchanged (e.g., $x * 1 = x$). Let's look at how these mathematical verities apply to matrix algebra.

1.4.1.1   Identity Matrix

An identity matrix (denoted $\mathbf{I}$) is a square, symmetric, diagonal matrix with 1's on the diagonals and 0's elsewhere. The following 3×3 matrix is an example, but an identity matrix can be any size:

$$I = \begin{bmatrix} 1 & 0 & 0 \\ 0 & 1 & 0 \\ 0 & 0 & 1 \end{bmatrix}$$

In matrix algebra, an identity matrix is the equivalent of the number 1. Just as any number multiplied by 1 returns the original number, any matrix multiplied by a conformable identity matrix returns the original matrix:

$$\mathbf{AI} = \mathbf{IA} = \mathbf{A} \tag{1.13}$$

For example,

$$\begin{bmatrix} 3 & 9 & 1 \\ 10 & 8 & 4 \\ 5 & 7 & 6 \end{bmatrix} \begin{bmatrix} 1 & 0 & 0 \\ 0 & 1 & 0 \\ 0 & 0 & 1 \end{bmatrix} = \begin{bmatrix} 3 & 9 & 1 \\ 10 & 8 & 4 \\ 5 & 7 & 6 \end{bmatrix}$$

1.4.1.2   Inverse and the Multiplicative Reciprocal

The inverse of a matrix functions as a multiplicative reciprocal. Consequently, any matrix multiplied by its inverse returns an identity matrix:

$$\mathbf{AA}^{-1} = \mathbf{A}^{-1}\mathbf{A} = \mathbf{I} \tag{1.14}$$

In a moment, we will learn how to calculate the inverse for any square matrix. For now, we will consider the inverse of a diagonal matrix, which is easy to find because its elements are simply the inverse (i.e., reciprocal) of the original elements:

$$\mathbf{A} = \begin{bmatrix} 3 & 0 & 0 \\ 0 & 8 & 0 \\ 0 & 0 & 6 \end{bmatrix} \quad \text{and} \quad \mathbf{A}^{-1} = \begin{bmatrix} .3333 & 0 & 0 \\ 0 & .1250 & 0 \\ 0 & 0 & .1667 \end{bmatrix}$$

Now let's go ahead and look at how matrix multiplication with an inverse functions like a reciprocal:

$$\text{if } \mathbf{A} = \begin{bmatrix} 3 & 0 & 0 \\ 0 & 8 & 0 \\ 0 & 0 & 6 \end{bmatrix} \quad \text{and} \quad \mathbf{b} = \begin{bmatrix} 12 \\ 24 \\ 36 \end{bmatrix}, \text{ then}$$

$$\mathbf{A}^{-1}\mathbf{b} = \begin{bmatrix} .3333 & 0 & 0 \\ 0 & .1250 & 0 \\ 0 & 0 & .1667 \end{bmatrix} \begin{bmatrix} 12 \\ 24 \\ 36 \end{bmatrix} = \begin{bmatrix} 4 \\ 3 \\ 6 \end{bmatrix}$$

Notice that the entries of $\mathbf{A}^{-1}\mathbf{b}$ represent $\mathbf{b}$ divided by the diagonal elements of $\mathbf{A}$. So when we premultiply a matrix by the inverse of a diagonal matrix, we are, in effect, dividing the post multiplication matrix by the diagonal elements of the premultiplication matrix. In a moment, we will see how these operations allow us to solve linear equations.

1.4.2   Calculating the Matrix Inverse

Before we do, we will learn how to calculate the inverse of a non-diagonal matrix. As is true with all of the topics we have covered thus far, the calculations are easy with small matrices but arduous and error-prone with large ones. Fortunately, we won't ever have to calculate the matrix of an inverse by hand in this book; we'll just use a built-in spreadsheet function: =MINVERSE. But in keeping with our attempt to understand where the numbers come from, we will describe the operations that are involved.

1.4.2.1   Calculating the Adjugate

Earlier we saw that the determinant is calculated using the cofactor matrix $\mathbf{F}$. The transpose of the cofactor matrix, called the adjugate and denoted Adj($\mathbf{A}$), is used to calculate the inverse:

$$\text{Adj}(\mathbf{A}) = \mathbf{F}' \tag{1.15}$$

Returning to an earlier example, our original matrix was

$$\mathbf{A} = \begin{bmatrix} 3 & 9 & 1 \\ 10 & 8 & 4 \\ 5 & 7 & 6 \end{bmatrix}$$

and the cofactor matrix was

$$\mathbf{F} = \begin{bmatrix} 20 & -40 & 30 \\ -47 & 13 & 24 \\ 28 & -2 & -66 \end{bmatrix}$$

To find the adjugate, we transpose the cofactor matrix:

$$\text{Adj}(\mathbf{A}) = \mathbf{F}' = \begin{bmatrix} 20 & -47 & 28 \\ -40 & 13 & -2 \\ 30 & 24 & -66 \end{bmatrix}$$

1.4.2.2   Computing the Inverse

The inverse is found by dividing the adjugate by the determinant:[10]

$$\mathbf{A}^{-1} = \frac{\mathbf{Adj}(\mathbf{A})}{|\mathbf{A}|} \tag{1.16}$$

In our example, $|\mathbf{A}| = -270$, so the inverse is found using the following operations:

$$\mathbf{A}^{-1} = \left\{ \begin{bmatrix} 20 & -47 & 28 \\ -40 & 13 & -2 \\ 30 & 24 & -66 \end{bmatrix} * \frac{1}{-270} \right\} = \begin{bmatrix} -.0741 & .1741 & -.1037 \\ .1481 & -.0481 & .0074 \\ -.1111 & -.0889 & .2444 \end{bmatrix}$$

We can verify that this is, indeed, the inverse of the original matrix by showing that $\mathbf{AA}^{-1} = \mathbf{I}$:

$$\begin{bmatrix} 3 & 9 & 1 \\ 10 & 8 & 4 \\ 5 & 7 & 6 \end{bmatrix} \begin{bmatrix} -.0741 & .1741 & -.1037 \\ .1481 & -.0481 & .0074 \\ -.1111 & -.0889 & .2444 \end{bmatrix} = \begin{bmatrix} 1 & 0 & 0 \\ 0 & 1 & 0 \\ 0 & 0 & 1 \end{bmatrix}$$

1.4.2.3   Inverse Entries are Fractions

Because an inverse is found by dividing the adjugate by the determinant, each entry in an inverse matrix is a fraction; and because a determinant represents the volume of a parallelepiped, the fraction's denominator represents the total variability in a matrix and the fraction's numerator represents the variability in a subset of the matrix (a cofactor). So the fraction is a relative measure of variability. You will want to keep this point in mind when we start using inverse matrices to solve statistical problems, because we often will be comparing the variability among one set of variables with the variability among another set of variables.

1.4.2.4   Contrasting Matrix Multiplication and the Matrix Inverse

Earlier we noted that matrix multiplication is simply an efficient way of multiplying and summing numbers. There is nothing mysterious about it (though it can take a while to get the hang of how it is done). But an inverse is different. Because the inverse is found by dividing the adjugate by the determinant, the entire variability

---

[10] The inverse can also be found from the cofactor matrix,

$$\mathbf{A}^{-1} = \frac{\mathbf{F}'}{\mathbf{AF}'[\mathrm{ii}]}$$

where the denominator indicates any diagonal element of $\mathbf{AF}'$ (which equals the determinant).

Table 1.3 Understanding the matrix inverse

Example 1a								
Pre matrix			Post matrix			Product matrix		
3	9	1	3	4	7	28	60	102
10	8	4	2	5	8	50	92	170
5	7	6	1	3	9	35	73	145
Example 1b								
Pre matrix			Post matrix			Product matrix		
3	9	1	3	4	7	28	60	102
10	8	4	2	5	8	50	92	170
5	7	2	1	3	9	31	61	109
Example 2a: Inverse of matrix 1a								
Pre matrix			Post matrix			Product matrix		
−.0741	.1741	−.1037	3	4	7	.0222	.2630	−.0593
.1481	−.0481	.0074	2	5	8	.3556	.3741	.7185
−.1111	−.0889	.2444	1	3	9	−.2667	−.1556	.7111
Example 2b: Inverse of matrix 1b								
Pre matrix			Post matrix			Product matrix		
2.00	1.8333	−4.6667	3	4	7	5.00	3.1667	−13.3333
.00	−.1667	.3333	2	5	8	.00	.1667	1.6667
−5.00	−4.0000	11.00	1	3	9	−12.00	−7.00	32.00

of the matrix contributes to the value of each entry in an inverse matrix. Put more broadly, when finding an inverse, we treat the matrix as a whole rather than considering each entry in isolation from the others.

To better appreciate this point, consider the information presented in Table 1.3. Focusing first on the two pre-matrices in Example 1, notice that the only difference between Example 1a and Example 1b is that I have changed the last value from 6 in Example 1a to 2 in Example 1b. Now notice that, except for the last row, the product matrices are also the same. So the only consequence of changing a single value in the premultiplication matrix is to change one row of the product matrix.

Now let's look at Example 2. The first matrix in Example 2a is the inverse of matrix 1a, and the second is the inverse of matrix 1b. Clearly these matrices are very different, even though the original matrices differ by only one entry. And when we then use each inverse to multiply a common matrix, their product matrices are also very different. This is because the inverse represents the entire variability of a matrix, and changing one value changes the matrix as a whole.

1.4.2.5 Singular Matrices Are Not Invertible

Because the inverse is calculated by dividing the adjugate by the determinant, we cannot invert a matrix with a determinant of zero (because dividing by zero is undefined). Earlier we noted that a matrix with a determinant of zero is known as a singular matrix because there exists at least one linear dependence among the columns or rows. The reason we need to be on the lookout for singular matrices is precisely because they cannot be inverted, and matrix inversion is a key component of many statistical analyses.

1.4.3   Using the Inverse to Solve Linear Equations

To better appreciate the importance of the preceding point, let's look at how an inverse matrix can be used to solve a system of linear equations. The calculations differ depending on whether the matrices are immediately conformable, so we will examine the calculations in two steps, starting with the conformable case.

1.4.3.1   Solving Linear Equations When the Matrices Are Conformable

Consider the two equations below:

$$5x + 7y = -11$$
$$8x + 4y = 4$$

Earlier, we learned that we could solve these equations using Cramer's rule. But this approach will only work when the number of equations equals the number of unknowns. Ultimately, we want a more general approach that can be used in all cases. The matrix inverse offers such an approach.

We begin by putting the problem in matrix form, designating the first matrix $\mathbf{X}$, the second vector $\mathbf{b}$, and the third vector $\mathbf{y}$:[11]

$$
\overset{\mathbf{X}}{\begin{bmatrix} 5 & 7 \\ 8 & 4 \end{bmatrix}}
\overset{\mathbf{b}}{\begin{bmatrix} x \\ y \end{bmatrix}}
=
\overset{\mathbf{y}}{\begin{bmatrix} -11 \\ 4 \end{bmatrix}}
$$

We then form our equation:

$$\mathbf{Xb} = \mathbf{y}$$

If these terms weren't matrices, simple algebra tells us we could solve for b by dividing y by x (or multiplying y by the reciprocal of x). Matrix division isn't defined, so we can't divide $\mathbf{y}$ by $\mathbf{X}$, but we can premultiply $\mathbf{y}$ by the inverse of $\mathbf{X}$:

---

[11] I realize these designations are a bit confusing because the unknown quantities, x and y, are elements of $\mathbf{b}$, but these designations match ones we will use throughout the book, so we might as well get used to them now.

$$\mathbf{b} = \mathbf{X}^{-1}\mathbf{y} \tag{1.17}$$

To do that, we need to calculate $\mathbf{X}^{-1}$. Recall that the inverse of a matrix is found by dividing the adjugate by the determinant. Earlier, we found the determinant when using Cramer's rule ($|\mathbf{X}| = -36$), and the following formula can be used to quickly find the adjugate of a 2×2 matrix:

$$\text{if } \mathbf{A} = \begin{bmatrix} a & b \\ c & d \end{bmatrix} \text{ then } \mathbf{Adj}(\mathbf{A}) = \begin{bmatrix} d & -b \\ -c & a \end{bmatrix} \tag{1.18}$$

Plugging in our values produces the inverse:

$$\mathbf{X}^{-1} = \begin{bmatrix} 4 & -7 \\ -8 & 5 \end{bmatrix} * \frac{1}{-36} = \begin{bmatrix} -.1111 & .1944 \\ .2222 & -.1389 \end{bmatrix}$$

If we then carry out the multiplication $\left(\mathbf{X}^{-1}\mathbf{y} = \mathbf{b}\right)$, we find our unknown values:

$$\begin{bmatrix} -.1111 & .1944 \\ .2222 & -.1389 \end{bmatrix} \begin{bmatrix} -11 \\ 4 \end{bmatrix} = \begin{bmatrix} 2 \\ -3 \end{bmatrix}$$

When we substitute the values we obtained for $\mathbf{b}$ into the original equation, we verify that they are correct:

$$\begin{bmatrix} 5 & 7 \\ 8 & 4 \end{bmatrix} \begin{bmatrix} 2 \\ -3 \end{bmatrix} = \begin{bmatrix} -11 \\ 4 \end{bmatrix}$$

1.4.3.2   Solving Linear Equations When the Matrices Are Not Conformable

In the preceding example, we didn't need to modify $\mathbf{X}$ in any fashion because $\mathbf{X}$ and $\mathbf{y}$ were conformable. This will not always be the case, however, so we want to have a more general formula we can use when it is not. Consider the equations below, which were introduced at the start of this chapter:

$$5x + 7y = -11$$
$$8x + 4y = 4$$
$$5x + 5y = -5$$

We can't use Cramer's rule to solve these equations, because the number of equations (3) doesn't equal the number of unknowns (2). And if we put the problem into matrix form:

$$\overset{\mathbf{X}}{\begin{bmatrix} 5 & 7 \\ 8 & 4 \\ 5 & 5 \end{bmatrix}} \overset{\mathbf{b}}{\begin{bmatrix} x \\ y \end{bmatrix}} = \overset{\mathbf{y}}{\begin{bmatrix} -11 \\ 4 \\ -5 \end{bmatrix}}$$

we cannot use the inverse method we just discussed, because $\mathbf{X}$ and $\mathbf{y}$ are not conformable [i.e., the columns of $\mathbf{X}$ (2) do not match the rows of $\mathbf{y}$ (3)]. Fortunately, we can solve our equation using the following formula, which is the only one you need because it will work with conformable matrices and unconformable ones:[12]

$$(\mathbf{X'X})^{-1}\mathbf{X'y} = \mathbf{b} \qquad (1.19)$$

Our first term, $\mathbf{X'X}$, involves premultiplying a matrix by its transpose. Earlier, we learned that doing so always produces a symmetric matrix with squared column sums on the diagonals and cross-product sums on the off-diagonals. With two columns, the general form of the product matrix is as follows:

$$\mathbf{X'X} = \begin{bmatrix} \Sigma X_1^2 & \Sigma X_1 X_2 \\ \Sigma X_2 X_1 & \Sigma X_2^2 \end{bmatrix}$$

In our case, the matrix becomes

$$\mathbf{X'X} = \begin{bmatrix} 114 & 92 \\ 92 & 90 \end{bmatrix}$$

We can verify these values by performing the calculations below:

$$5^2 + 8^2 + 5^2 = 114$$
$$7^2 + 4^2 + 5^2 = 90$$
$$(5 * 7) + (8 * 4) + (5 * 5) = 92$$

To calculate the inverse, we first find the determinant, which provides a measure of the variability of a matrix:

$$|\mathbf{X'X}| = (114 * 90) - (92 * 92) = 1796$$

We then find the inverse by dividing the adjugate by the determinant:

---

[12] Equation (1.19) represents the solution to a set of equations known as the normal equations. In matrix form, the normal equations are expressed as $\mathbf{X'Xb} = \mathbf{X'y}$. In Chap. 3, we will discuss their derivation and solution in more detail.

$$(\mathbf{X'X})^{-1} = \begin{bmatrix} 90 & -92 \\ -92 & 114 \end{bmatrix} * \frac{1}{1796} = \begin{bmatrix} .0501 & -.0512 \\ -.0512 & .0635 \end{bmatrix}$$

We can verify that this matrix is the inverse of $\mathbf{X'X}$ by multiplying the two matrices to produce an identity matrix $\mathbf{I}$:

$$\begin{bmatrix} 114 & 92 \\ 92 & 90 \end{bmatrix} \begin{bmatrix} .0501 & -.0512 \\ -.0512 & .0635 \end{bmatrix} = \begin{bmatrix} 1 & 0 \\ 0 & 1 \end{bmatrix}$$

Now let's find $\mathbf{X'y}$. This vector represents a cross-product term between $\mathbf{X}$ and $\mathbf{y}$:

$$\mathbf{X'y} = \begin{bmatrix} \Sigma X_1 Y \\ \Sigma X_2 Y \end{bmatrix}$$

With our example, the values become

$$\mathbf{X'y} = \begin{bmatrix} 5 & 8 & 5 \\ 7 & 4 & 5 \end{bmatrix} \begin{bmatrix} -11 \\ 4 \\ -5 \end{bmatrix} = \begin{bmatrix} -48 \\ -86 \end{bmatrix}$$

Finally, we multiply the two product matrices to solve for $\mathbf{b}$:

$$\mathbf{b} = \left(\mathbf{X'X}\right)^{-1}\mathbf{X'y} = \begin{bmatrix} .0501 & -.0512 \\ -.0512 & .0635 \end{bmatrix} \begin{bmatrix} -48 \\ -86 \end{bmatrix} = \begin{bmatrix} 2 \\ -3 \end{bmatrix}$$

As always, we can substitute the obtained values into the original equation to verify that we have found the correct solution:

$$\begin{bmatrix} 5 & 7 \\ 8 & 4 \\ 5 & 5 \end{bmatrix} \begin{bmatrix} 2 \\ -3 \end{bmatrix} = \begin{bmatrix} -11 \\ 4 \\ -5 \end{bmatrix}$$

1.4.3.3 Summary

In this section you learned how to use the inverse of a matrix to solve a series of linear equations. In the next chapter, we will learn how these operations are used in statistical analyses. Before we do, let's pause and think about the operations themselves. Remembering that an inverse functions as a reciprocal, we are essentially dividing a cross-product term by the sum of the squared predictors:

$$b \sim \frac{X'y}{X'X}$$

The analysis is not performed in exactly this way, but thinking about the operations in this way will help you understand many of the statistical analyses you will learn in this book. Much of the time, we will be comparing the cross-product variability of $X'y$ to the variability of $X'X$.

1.4.4   R Code: Matrix Inverse

```
A <- matrix(c(10,3,4,5), nrow=2, ncol=2, byrow=TRUE);A

#Create function to calculate adjugate
adjugate <- function(A) {
  n <- nrow(A)
  F <- matrix(NA, n, n)
  if(n>2){
    cofactors <- function(A, i, j)
       (-1)^(i+j) * det( A[-i,-j] )
        for( i in 1:n )
          for( j in 1:n )
            F[i,j] <- cofactors(A, j, i)
              F
}
   else{ F <-matrix(c(A[4],-A[3],-A[2],A[1]),2,byrow=TRUE)}
F
}
adjugate(A)
A.inv <-adjugate(A)/det(A);A.inv
RA.inv <-solve(A);RA.inv    #R's solve command produces the inverse


#Use inverse to solve simultaneous equations with conformable matrices
X <- matrix(c(5,7,8,4), nrow=2, ncol=2, byrow=TRUE);X
Y <-matrix(c(-11,4),nrow=2,ncol=1,byrow=TRUE);Y
B <-solve(X)%*%Y;B


#Use inverse to solve simultaneous equations with nonconformable
matrices
X <- matrix(c(5,7,8,4,5,5), nrow=3, ncol=2, byrow=TRUE);X
Y <-matrix(c(-11,4,-5),nrow=3,ncol=1,byrow=TRUE);Y
B <-solve(t(X)%*%X)%*%t(X)%*%Y;B
```

1.5   Chapter Summary

1. Many statistical analyses involve solving a series of simultaneous linear equations, with each equation representing one subject's data. Matrix algebra can be used to find this solution.

2. A matrix is a rectangular array of numbers or symbols, with rows and columns called vectors. The size of a matrix is given by its number of rows and columns (in that order).

3. Matrices possess numerical properties that go beyond the specific values that comprise them. For example, the trace of a square matrix is found by summing the diagonal elements. Two matrices can have the same trace, even though all of their elements are different.

4. Matrices can easily be added or subtracted, but matrix multiplication is possible only when the number of columns of Matrix 1 equals the number of rows of Matrix 2. When this occurs, we say the matrices are conformable.

5. We transpose a matrix by exchanging its rows and columns. A matrix premultiplied by its transpose produces a symmetric, square matrix with each column's squared sums on the main diagonal (upper left to lower right) and cross-product sums on the off-diagonals.

6. A covariance matrix is found by first subtracting the column mean from each column entry and then premultiplying the resultant deviate matrix by its transpose. When we then divide the entries of the product matrix by $N-1$, we get variances on the diagonals and covariances on the off-diagonals.

7. A correlation matrix is found by first standardizing each matrix entry (subtract its column mean and divide by its column standard deviation) and then premultiplying the resultant matrix by its transpose. When we then divide the entries of the product matrix by $N-1$, we get a symmetric matrix with 1's on the diagonals and correlation coefficients on the off-diagonals.

8. The determinant of a square matrix provides an index of its variability. Geometrically, the absolute value of the determinant is the area or volume of a parallelepiped enclosed by the column vectors.

9. The determinant of a 2×2 matrix is found by subtracting the product of the off-diagonal entries from the product of the main diagonal entries. The determinant of larger matrices can be found by constructing a cofactor matrix of minors.

10. Using Cramer's rule, the determinant of a matrix can be used to solve simultaneous linear equations.

11. The inverse of a matrix is found by dividing the transpose of the cofactor matrix (called the adjugate) by the determinant. Singular matrices (i.e., determinant $= 0$) are not invertible.

12. The inverse of a matrix functions as a reciprocal. A matrix multiplied by its inverse produces an identity matrix, with 1's on the main diagonal and 0's elsewhere.

13. The inverse of a nonsingular matrix can be used to solve simultaneous linear equations of the form $\mathbf{Xb} = \mathbf{y}$. When $\mathbf{X}$ and $\mathbf{y}$ are immediately conformable, $\mathbf{X}^{-1}\mathbf{y} = \mathbf{b}$; when $\mathbf{X}$ and $\mathbf{y}$ are not immediately conformable, $(\mathbf{X'X})^{-1}\mathbf{X'y} = \mathbf{b}$. The latter equation, which represents the solution to a series of equations known as the normal equations, plays a role in a great many statistical analyses.

Appendix

Table 1.4 Glossary of terms and operations

Term	Description	Use
Adjugate matrix	Transpose of the cofactor matrix	Returns the inverse when divided by the determinant
Cofactor matrix	A matrix of cofactors, found by multiplying minors using Eq. (1.10)	Yields the determinant when a row or column is multiplied by a corresponding row or column in the original matrix and the products are summed
Conformable	Two matrices are said to be conformable when the columns of the first matrix equal the rows of the second	Matrix multiplication can occur only when two matrices are conformable
Correlation matrix	A covariance matrix of standard scores, with correlations on the off-diagonals	
Determinant	A measure of the variance of a matrix, consisting of the area (or volume) of a parallelepiped enclosed by matrix vectors	Used to identify linear dependencies, solve linear equations, and compute the inverse
Deviate matrix	A matrix found by subtracting the column mean from each column element	Yields a covariance matrix when premultiplied by its transpose and divided by $(N-1)$
Diagonal matrix	A (usually) square matrix with 0's on the off-diagonals	
Identity matrix	A diagonal matrix with 1's on the diagonals and 0's on the off-diagonals	
Inverse	The matrix equivalent of a reciprocal, found by dividing the adjugate by the determinant	Used to solve for unknowns in linear equations
Minors	The determinant from a submatrix formed from a larger matrix	Used to find the determinant of a large matrix when arranged in a cofactor matrix
Singular matrix	Matrices with a linear dependence (i.e., determinant $=0$)	
Square matrix	A matrix with an equal number of rows and columns	
Symmetric matrix	A square, diagonal matrix where $\mathbf{A} = \mathbf{A}'$	
Trace	Sum of the diagonal elements of a square matrix	
Transpose	The exchange of a matrix's rows and columns	When postmultiplied by the original matrix yields a square, symmetric matrix

Table 1.5 Useful spreadsheet operations

Term	Description
AVERAGE	Calculates the average or mean value
DEVSQ	Calculates the deviation sum of squares: $\sum_{i=1}^{n} \left(y_i - \overline{Y}\right)^2$
SUMSQ	Calculates the sum of squared values: $\sum y^2$
STDEV	$\sqrt{DEVSQ/(N-1)}$
VARIANCE	$DEVSQ/(N-1)$
Array function for matrix operations	
MMULT	Multiplies two (conformable) matrices
MDETERM	Calculates the determinant of a matrix
MINVERSE	Calculates the inverse of a matrix
TRANSPOSE	Transposes a matrix

Chapter 2
Simple Linear Regression

Chapter 1 introduced you to a variety of matrix properties and operations. Among other things, you learned how to use the determinant and inverse of a matrix to solve equations with unknown quantities. In the remainder of this book, you will use these operations to perform a wide variety of statistical analyses.

The first statistical procedure we will discuss is called "simple linear regression." Here, the term "simple" refers to the fact that we are concerned with a single predictor and a single criterion.[1] This situation arises often in science. For example, in the physical sciences, we might predict the rate of a chemical reaction from temperature; in medicine, we might predict heart disease from cholesterol levels; and in the social sciences, we might predict income from education. If we assume that these variables are related in a linear fashion, we can use simple linear regression to quantify the nature of their association.

The procedures we follow when conducting a regression analysis are similar to the ones we used to solve simultaneous equations, with one important difference. So far we have been dealing with equations that have an exact solution. In contrast, regression analysis is used when an exact solution is not available.[2] Consequently, we will not be able to find a single value that completely accounts for the relation between a predictor and a criterion. Instead, we will attempt to predict as much as we can without predicting everything.

To achieve this goal, we analyze patterns of variability, asking whether the variability in the criterion can be predicted from the variability in the predictor.

Electronic Supplementary Material: The online version of this chapter (doi: 10.1007/978-3-319-11734-8_2) contains supplementary material, which is available to authorized users

[1] Some textbooks refer to a criterion as a response variable or dependent variable and a predictor as an explanatory variable, independent variable, or covariate. I prefer the terms "predictor" and "criterion" because they imply only that the former predicts the latter.

[2] An exact solution is guaranteed when the number of equations is equal to or less than the number of unknowns. When the number of equations exceeds the number of unknowns, an exact solution is possible but not guaranteed.

© Springer International Publishing Switzerland 2014
J.D. Brown, *Linear Models in Matrix Form*, DOI 10.1007/978-3-319-11734-8_2

If it can, we say that the two variables are associated or related. Sometimes we say that "X explains Y," but statements of this sort are valid under only a limited range of conditions. In contrast, prediction is assured. By definition, if two variables are associated, we learn something about one of them by knowing something about the other.

2.1   Mathematical Models

Before delving into the specifics of a linear regression model, it is useful to consider mathematical models more broadly.

2.1.1   What Is a Model?

Mathematical models describe physical, biological, or social processes in numerical terms. They are simplified representations of reality that are analogical, not literal. As George Box famously remarked, "All models are wrong but some are useful" (1979, p. 202).

Broadly speaking, mathematical models describe relations among sets of variables. These relations can take one of two forms:

- With a functional relation, the variability in Y is entirely predictable from the variability in X:

$$Y = f(X) \qquad (2.1)$$

   There is no allowance for errors of prediction with a functional relation. If we know X, we can predict Y. The problems we solved in Chap. 1 described functional relations.

- A statistical relation is one that is influenced by a stochastic process (i.e., chance) and, therefore, is not entirely predictable. To model a statistical relation, we include a term that represents errors in prediction.[3] This term is typically denoted e, although v and u are sometimes used instead:

$$Y = f(X) + e \qquad (2.2)$$

   Building on these differences, we can distinguish two mathematical models. A deterministic model is comprised entirely of functional relations. Thus, accurate prediction is assured in a deterministic model. A statistical model includes at least one statistical relation. Thus, accurate prediction is not assured in a statistical model. Linear regression models are statistical models.

---

[3] The error term is sometimes called the disturbance.

2.1.2 *What Is a* **Regression** *Model?*

Regression is a technique for modeling the statistical relation between pairs of variables. One member of the pair is designated as the predictor (X) and the other is designated as the criterion (Y). These designations are not arbitrary or interchangeable. The values of the predictors are presumed to be fixed (i.e., set by the investigator) or measured with only negligible error, and each one is of particular interest (i.e., if we repeated a study, we would use similar if not identical values of x to predict y). In contrast, the criterion is treated as a random variable drawn from a probability distribution that depends on the level of its associated predictor [i. e., $p(y|x)$]. The means of these conditional probability distributions are presumed to vary across levels of X.

Given this distinction between predictor and criterion, it is tempting to conclude that X is being treated as a cause of Y. This conclusion is warranted only when the investigator is able to randomly assign subjects to various levels of the predictor. Absent this form of experimental control, no statistical technique can establish a causal relation. Thus, although we always designate a predictor and a criterion when conducting a regression analysis, this distinction does not necessarily carry any causal assumptions.

2.1.3 *What Is a* **Linear** *Regression Model?*

A linear regression model specifies the form of the mathematical function that relates Y to X. With linear regression, the criterion is modeled as an additive function of weighted predictors (bX) and error (e):

$$Y = b_1X_1 + b_2X_2 + \cdots + b_kX_k + e \tag{2.3}$$

Moreover, the weights associated with each predictor must be of the first order (i.e., no exponent other than 1), and they cannot be multiplied or divided by any other weight. Expressed in matrix form, the linear regression model becomes:

$$\mathbf{y} = \mathbf{Xb} + \mathbf{e} \tag{2.4}$$

where

$\mathbf{y}$ = a vector of known values for a given set of xy observations
$\mathbf{X}$ = a matrix of predictors for a given set of xy observations
$\mathbf{b}$ = a vector of regression weights used to predict $\mathbf{y}$ from $\mathbf{X}$
$\mathbf{e}$ = a vector of errors expressing the portion of $\mathbf{y}$ that cannot be predicted from $\mathbf{Xb}$.

Notice that the linear model in Eq. (2.4) has two components, a functional component (**Xb**) and a stochastic one (**e**). The combination makes the model a statistical model. Relatedly, a linear model is not exact. We assume only that, within the range of the variables being studied, the relation between the predictors and the criterion can be well approximated by a linear function. We do not, however, necessarily assume that the function is strictly linear.

Finally, Eq. (2.4) is expressed in terms of sample values, as it describes the properties of a specific set of observations. In many cases, we are interested in extrapolating beyond the specific sample we have observed to a broader population from which the sample is drawn. Ordinarily, we do not know the true values of the regression weights in the population, so we use our sample data to estimate them. In this case, we replace Roman letters with Greek ones to indicate that we are inferring unknown population parameters rather than describing observed sample quantities:

$$\mathbf{y} = \mathbf{X}\boldsymbol{\beta} + \boldsymbol{\varepsilon} \tag{2.5}$$

2.2   Simple Linear Regression

With this discussion as background, we are ready to tackle an example. Throughout this text, we will analyze "small sample examples" to demonstrate various statistical techniques. The sample sizes are purposely small ($N = 12$) to keep the computations from being too laborious, and the data are hypothetical (which is a fancy way of saying I made them up), so we will not take them seriously. The point is to learn the analyses, and small sample examples suffice.

With that in mind, imagine I ask 12 students who are enrolled in a matrix algebra class to indicate how competent they believe they are in math using a 9-point scale ($1 =$ not at all competent, $9 =$ very competent). These beliefs, which constitute our predictor, are commonly called "self-efficacy beliefs." At the end of the academic term, I measure each student's actual performance in terms of their class rank (rounded to the nearest decile). Table 2.1 shows the (contrived) data, along with several other variables of interest.

Table 2.1 Small sample example for simple linear regression

Student	Self-efficacy x	Class rank y	Deviate x	Deviate y	Standard x	Standard y
1	1	3	−4.00	−1.8333	−1.3683	−.7392
2	9	8	4.00	3.1667	1.3683	1.2768
3	1	2	−4.00	−2.8333	−1.3683	−1.1424
4	5	8	0.00	3.1667	0.00	1.2768
5	6	5	1.00	.1667	.3421	.0672
6	8	9	3.00	4.1667	1.0263	1.6800
7	2	4	−3.00	−.8333	−1.0263	−.3360
8	4	5	−1.00	.1667	−.3421	.0672
9	2	2	−3.00	−2.8333	−1.0263	−1.1424
10	8	4	3.00	−.8333	1.0263	−.3360
11	7	2	2.00	−2.8333	.6842	−1.1424
12	7	6	2.00	1.1667	.6842	.4704
Mean	5.00	4.8333	0	0	0	0
Sum of squares	394	348	94	67.6667		
Standard deviation	2.9233	2.4802	2.9233	2.4802	1.00	1.00
Variance	8.5455	6.1515	8.5455	6.1515	1.00	1.00
Covariance			4.4545		.6144	

2.2.1 Preliminary Analysis Without an Intercept

Earlier we noted that a linear regression model assumes that $\mathbf{Y} = \mathbf{Xb} + \mathbf{e}$. To solve for the unknown quantity $\mathbf{b}$, we place our data in matrix form, just as we did in Chap. 1 when we solved simultaneous linear equations:

$$
\begin{bmatrix} 1 \\ 9 \\ 1 \\ 5 \\ 6 \\ 8 \\ 2 \\ 4 \\ 2 \\ 8 \\ 7 \\ 7 \end{bmatrix} [\mathbf{b}] = \begin{bmatrix} 3 \\ 8 \\ 2 \\ 8 \\ 5 \\ 9 \\ 4 \\ 5 \\ 2 \\ 4 \\ 2 \\ 6 \end{bmatrix}
$$

Because the columns of $\mathbf{X}$ do not equal the rows of $\mathbf{y}$, the matrices are not conformable. Consequently, we use Eq. (1.19) to solve for $\mathbf{b}$:

$$\mathbf{b} = \left(\mathbf{X}'\mathbf{X}\right)^{-1}\mathbf{X}'\mathbf{y} = [394]^{-1}[339] = .8604$$

2.2.1.1   Fitted Values

Our next step is to compute fitted values of y (denoted $\hat{y}$ and pronounced "y hat"):

$$\hat{\mathbf{y}} = \mathbf{Xb} \tag{2.6}$$

$$
\begin{bmatrix} 1 \\ 9 \\ 1 \\ 5 \\ 6 \\ 8 \\ 2 \\ 4 \\ 2 \\ 8 \\ 7 \\ 7 \end{bmatrix} [.8604] \;=\;
\begin{bmatrix} .8604 \\ 7.7437 \\ .8604 \\ 4.3020 \\ 5.1624 \\ 6.8832 \\ 1.7208 \\ 3.4416 \\ 1.7208 \\ 6.8832 \\ 6.0228 \\ 6.0228 \end{bmatrix}
$$

When we compare the fitted values with the actual values for y, we see that they are not the same (i.e., $\hat{y} \neq y$). This is because the relation is a statistical one, not a functional one.

2.2.1.2   Residuals

We can gauge the discrepancy between y and $\hat{y}$ by computing a variable called the residual, denoted e:

$$\mathbf{e} = \mathbf{y} - \hat{\mathbf{y}} \tag{2.7}$$

$$
\begin{bmatrix} 3 \\ 8 \\ 2 \\ 8 \\ 5 \\ 9 \\ 4 \\ 5 \\ 2 \\ 4 \\ 2 \\ 6 \end{bmatrix} -
\begin{bmatrix} .8604 \\ 7.7437 \\ .8604 \\ 4.3020 \\ 5.1624 \\ 6.8832 \\ 1.7208 \\ 3.4416 \\ 1.7208 \\ 6.8832 \\ 6.0228 \\ 6.0228 \end{bmatrix} =
\begin{bmatrix} 2.1396 \\ .2563 \\ 1.1396 \\ 3.6980 \\ -.1624 \\ 2.1168 \\ 2.2792 \\ 1.5584 \\ .2792 \\ -2.8832 \\ -4.0228 \\ -.0228 \end{bmatrix}
$$

The residuals represent errors of prediction.[4] If we sum their squared values, we derive an overall estimate of error known as the residual sum of squares, denoted SS_{res}:

$$SS_{res} = \Sigma e^2 = \Sigma(y - \hat{y})^2 \tag{2.8}$$

Plugging in our values yields the residual sum of squares for our data set:

$$SS_{res} = \Sigma \begin{bmatrix} 2.1396^2 \\ .2563^2 \\ 1.1396^2 \\ 3.6980^2 \\ -.1624^2 \\ 2.1168^2 \\ 2.2792^2 \\ 1.5584^2 \\ .2792^2 \\ -2.8832^2 \\ -4.0228^2 \\ -.0228^2 \end{bmatrix} = 56.3223$$

As we will see later, large values of this term indicate a poor fit to the data, so we would like this term to be as small as possible.

2.2.2   Complete Analysis: Adding an Intercept

There is a problem with the preceding analysis. In a regression model, we assume that the errors cancel out, yielding a mean (and sum) of 0. Yet if we average the unsquared residuals, we find that this is not true in our example ($\bar{e} = .5313$). We can fix this problem by adding a column of 1's to $\mathbf{X}$, as shown below. Because everyone in the sample gets the same score on this vector, it yields a constant known as the intercept, b_0. The intercept represents the average expected value of y when $x = 0$. Whether this term has any substantive meaning will depend on whether our predictor can meaningfully assume this score. But even when it doesn't, including the intercept ensures that the mean of the residuals equals 0:

---

[4] Strictly speaking, residuals are not the same as errors. Residuals are discrepancies between fitted values and observed, sample values, whereas errors are discrepancies between observed, sample values and their true population values. However, because the term residual is also used to refer to "errors of prediction," we will use the two terms more or less interchangeably throughout this text unless indicated otherwise.

$$\begin{bmatrix} 1 & 1 \\ 1 & 9 \\ 1 & 1 \\ 1 & 5 \\ 1 & 6 \\ 1 & 8 \\ 1 & 2 \\ 1 & 4 \\ 1 & 2 \\ 1 & 8 \\ 1 & 7 \\ 1 & 7 \end{bmatrix} \begin{bmatrix} b_0 \\ b_1 \end{bmatrix} \begin{bmatrix} 3 \\ 8 \\ 2 \\ 8 \\ 5 \\ 9 \\ 4 \\ 5 \\ 2 \\ 4 \\ 2 \\ 6 \end{bmatrix}$$

Now when we perform our calculations, we derive two values for **b** instead of one:[5]

$$\mathbf{b} = \begin{bmatrix} 12 & 60 \\ 60 & 394 \end{bmatrix}^{-1} \begin{bmatrix} 58 \\ 339 \end{bmatrix} = \begin{bmatrix} 2.2270 \\ .5213 \end{bmatrix}$$

The first, $b_0 = 2.2270$, is the intercept. As just noted, it describes the average expected value of y when $x = 0$. The second value is called the unstandardized regression coefficient ($b_1 = .5213$). It is the weight we give to x when using it to calculate a fitted value of y.

2.2.2.1   Calculating Fitted Values and Residuals

As before, we use Eq. (2.6) to compute fitted values:

$$\begin{bmatrix} 1 & 1 \\ 1 & 9 \\ 1 & 1 \\ 1 & 5 \\ 1 & 6 \\ 1 & 8 \\ 1 & 2 \\ 1 & 4 \\ 1 & 2 \\ 1 & 8 \\ 1 & 7 \\ 1 & 7 \end{bmatrix} \begin{bmatrix} 2.2270 \\ .5213 \end{bmatrix} = \begin{bmatrix} 2.7482 \\ 6.9184 \\ 2.7482 \\ 4.8333 \\ 5.3546 \\ 6.3972 \\ 3.2695 \\ 4.3121 \\ 3.2695 \\ 6.3972 \\ 5.8759 \\ 5.8759 \end{bmatrix}$$

---

[5] We will use matrix algebra to calculate the intercept, but with one predictor it can easily be found as $b_0 = \bar{y} - b\bar{x}$.

and Eq. (2.7) to compute the residuals:

$$\begin{bmatrix} 3 \\ 8 \\ 2 \\ 8 \\ 5 \\ 9 \\ 4 \\ 5 \\ 2 \\ 4 \\ 2 \\ 6 \end{bmatrix} - \begin{bmatrix} 2.7482 \\ 6.9184 \\ 2.7482 \\ 4.8333 \\ 5.3546 \\ 6.3972 \\ 3.2695 \\ 4.3121 \\ 3.2695 \\ 6.3972 \\ 5.8759 \\ 5.8759 \end{bmatrix} = \begin{bmatrix} .2518 \\ 1.0816 \\ -.7482 \\ 3.1667 \\ -.3546 \\ 2.6028 \\ .7305 \\ .6879 \\ -1.2695 \\ -2.3972 \\ -3.8759 \\ .1241 \end{bmatrix}$$

Now the mean (and sum) of the residuals does equal 0, and when we use Eq. (2.8) to calculate the residual sum of squares, we find that its value is smaller than the one we obtained before we modeled the intercept:[6]

$$SS_{res} = 42.1241$$

So not only does the intercept center our residuals; it also reduces the discrepancy between the fitted values and the observed ones. For these reasons, we will add a leading column of 1's to $\mathbf{X}$ throughout this text unless indicated otherwise.[7]

2.2.2.2   The Line of Best Fit

Figure 2.1 plots the observed and fitted values as a function of x. You will notice that there is also a line running through the fitted values. This line is known as the "line of best fit," and its slope matches the value of our regression coefficient ($b_1 = .5213$). For every one unit increase in x, we expect a .5213 unit increase in the average value of y. Notice also that the observed values of y are scattered around the line of best fit. Because of this scatter, the plot is known as a scatterplot.

---

[6] If you were to actually square each residual and sum the squares, the obtained value ($\sum e^2 = 42.1244$) would differ slightly from the one reported in the text. The discrepancy is due to rounding error and, as first indicated in the preface, I present the more accurate (computer-generated) value whenever discrepancies like these arise. As a consequence, calculations by hand will sometimes produce only approximate results.

[7] An intercept produces these benefits by centering x and y around their respective means. You can verify this is true by separately regressing each variable on a vector of 1's. The residuals will be deviate scores. If you then regress $deviate_y$ on $deviate_x$, you will find that the mean of the errors $= 0$ and $SS_{res} = 42.1241$.

Fig. 2.1 Scatterplot with line of best fit

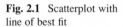

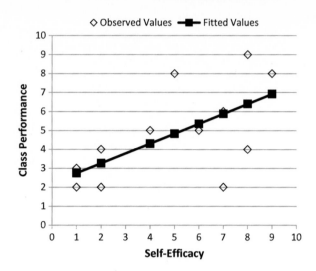

The line of best fit has an important property. More than any other line we could have drawn through the data, this line minimizes the squared discrepancy between our observed values and our fitted ones (i.e., it minimizes SS_{res}). Consequently, it represents the best solution to our set of simultaneous equations. This principle, known as least squares estimation or ordinary least squares (OLS), underlies linear regression analysis.[8]

2.2.2.3 Summary of Linear Regression with an Intercept

Table 2.2 summarizes some other properties of a simple linear regression model with an intercept. These properties can easily be verified using our example, and I encourage you to work through the math to confirm them.

Table 2.2 Properties of linear regression with an intercept

The mean of the fitted values is equal to the mean of the criterion	$\hat{\bar{Y}} = \bar{Y}$
The sum and mean of the residuals equal zero	$\Sigma_{ei} = \bar{e} = 0$
The residuals are uncorrelated with the predictors and the fitted values	$r(e, X) = r(e, \hat{Y}) = 0$
The line of best fit passes through the means of both variables	when $x = \bar{x}$, $\hat{y} = \bar{y}$

---

[8] Formally, OLS estimates are *unbiased* (i.e., the estimated regression coefficient is an accurate approximation of the population parameter) and *efficient* (i.e., the variance from the estimate is smaller than all other unbiased estimators). Because of these properties, the regression coefficients are termed BLUE: Best Linear Unbiased Estimators.

2.2.3   *Understanding Linear Regression*

It is a relatively simple matter to perform a regression analysis using matrix algebra, but understanding and interpreting the output requires careful thought. In this section, we will closely examine the meaning of the fitted values and residuals.

2.2.3.1   Fitted Values as Conditional Averages

Earlier we noted that with a linear regression model, we assume that y is a random variable with a probability distribution associated with each value of x. The fitted value represents the expected average of the probability distribution at a given level of x, and the scatter represents random variations around the mean. Figure 2.2 presents a schematic depiction of this assumption, using only three values of x for clarity. The circles represent fitted values and the distributions surrounding them represent observed values of y. As you can see, at each level of x, the fitted value lies at the center (mean) of the distribution, with the observed values scattered around it.[9] Moreover, the distributions are normal and identical.

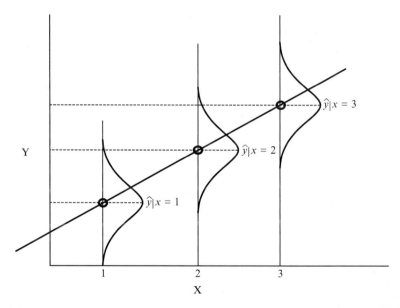

Fig. 2.2   Schematic representation of observed and fitted values in a linear regression model

---

[9] Fitted values are sometimes called point estimators of the mean response of the conditional distribution.

We can formally summarize these properties with respect to the mean and variance of these probability distributions.

- **y** is a random variable drawn from a probability distribution whose expected value (but not shape) depends on x:

$$E\{\mathbf{y}|\mathbf{X}\} = \hat{y} = \mathbf{XB} \tag{2.9}$$

- The conditional probability distributions from which **y** is drawn are normal and identically distributed, with variance σ^2:

$$\text{Var}\{\mathbf{y}|\mathbf{X}\} = \sigma^2 \tag{2.10}$$

2.2.3.2 Assumptions Regarding the Errors

Because they are calculated from the fitted and observed values, the errors also possess the distributional properties of y.[10] For this reason, it is customary to state the assumptions of a linear regression model with respect to the errors, substituting ε (the representation of a population parameter) for e (an observed sample value).

- ε is a normally distributed random variable with mean 0:

$$E\{\boldsymbol{\varepsilon}|\mathbf{X}\} = 0 \tag{2.11}$$

and variance σ^2:

$$\text{Var}\{\boldsymbol{\varepsilon}|\mathbf{X}\} = \sigma^2 \tag{2.12}$$

- The distribution of errors is identical across levels of x:

$$\sigma_i^2 = \sigma_j^2 = \cdots = \sigma_k^2 \tag{2.13}$$

and independent (i.e., their covariances $= 0$):

$$\text{Cov}\{(\varepsilon_i, \varepsilon_j)|\mathbf{X}\} = 0 \tag{2.14}$$

These assumptions are frequently combined to yield the following statement:
- In a linear regression model, the errors are assumed to be independent and normally and identically distributed random variables, with mean 0 and variance σ^2:[11]

---

[10] You might wish to verify that at each level of x, the variance of y equals the variance of e.

[11] In subsequent chapters, we will see that it is possible to relax some of these assumptions and still retain the central features of a linear regression model.

$$\varepsilon \sim NID\left(0, \sigma^2\right) \tag{2.15}$$

2.2.4 Standardized Regression Coefficients

The unstandardized regression coefficient indicates how much weight x is given in the prediction of y. The weight is expressed in raw scores (i.e., for every one unit increase in x, we predict a b unit increase in the average value of y). Sometimes it is desirable to compute a standardized regression coefficient using standardized scores rather than raw ones.

2.2.4.1 Calculating a Standardized Regression Coefficient

Although there are several ways to compute standardized regression coefficients, we will continue to use our inverse matrix method, substituting standard scores for raw scores. In Chap. 1 we learned that we standardize variables by subtracting the column mean from each score and dividing the difference by the standard deviation. Such scores, known as z scores, are shown in Table 2.1. Using our understanding of the role of an inverse, we can predict z_y from z_x using the following formula:[12]

$$\boldsymbol{\beta} = \left(\mathbf{Z}_x'\mathbf{Z}_x\right)^{-1}\mathbf{Z}_x'\mathbf{Z}_y \tag{2.16}$$

Because the mean of both distributions is 0, we dispense with adding a vector of 1's to our $\mathbf{Z}_x$ matrix and solve the problem in now familiar fashion:

$$\boldsymbol{\beta} = [11]^{-1}[6.7583] = .6144$$

The obtained value indicates that a one standard deviation increase in x predicts a .6144 standard deviation increase in y.

2.2.4.2 Verifying the Meaning of a Standardized Regression Coefficient

It is worth taking the time to verify the meaning of a standardized regression coefficient. Table 2.1 shows that the standard deviation of $y = 2.4802$. Multiplying the standard deviation by the standardized regression coefficient reveals the expected change in y associated with a one standard deviation change in x:

$$2.4802 * .6144 = 1.5238$$

---

[12] Even though we use $\boldsymbol{\beta}$ rather than $\mathbf{b}$ to denote the standardized coefficient matrix, we are still concerned with a sample value, not a population estimate.

In words, the average expected value of y is predicted to increase by 1.5238 units with every one standard deviation increase in x. To see whether this is true, let's compute fitted values for y when x increases by one standard deviation. Table 2.1 shows that the standard deviation of $x = 2.9233$. For our predicted values, we will choose $x = 3$ and $x = [3 + s_x] = 5.9233$:

$$\hat{y} = 2.2270 + 3(.5213) = 3.7908$$

$$\hat{y} = 2.2270 + 5.9233(.5213) = 5.3146$$

The difference between the two predicted scores is found using subtraction:

$$5.3146 - 3.7908 = 1.5238$$

As you can see, the obtained difference is identical to the one we calculated earlier, confirming that a one standard deviation increase in x predicts a .6144 standard deviation increase in y.

2.2.4.3   Comparing the Unstandardized and Standardized Coefficient

It is also informative to compare unstandardized and standardized regression coefficients. Three pieces of information in Table 2.1 will help us do so: the covariance between x and y ($s_{xy} = 4.4545$) and the variance of each variable ($s_x^2 = 8.5455$; $s_y^2 = 6.1515$).

With one only predictor, the following formulae can be used to calculate the unstandardized regression coefficient:

$$b = \frac{s_{xy}}{s_x^2} \tag{2.17}$$

and the standardized regression coefficient:

$$\beta = \frac{s_{xy}}{\sqrt{s_x^2} * \sqrt{s_y^2}} \tag{2.18}$$

When we examine both equations, we see that they differ only with respect to their denominator. Whereas the unstandardized coefficient is calculated using the variance of x in the denominator, the denominator for the standardized coefficient uses the standard deviations of x and y. If we go ahead and plug in the relevant numbers, we confirm that the formulae produce the correct values:

$$b = \frac{4.4545}{8.5455} = .5213$$

and

$$\beta = \frac{4.4545}{\sqrt{8.5455} * \sqrt{6.1515}} = .6144$$

2.2.4.4   Contrasting the Unstandardized and Standardized Coefficient

Because of their close connection, it is a relatively simple matter to compute one regression coefficient from the other:

$$b = \beta \frac{s_y}{s_x} \quad \text{and} \quad \beta = b \frac{s_x}{s_y} \tag{2.19}$$

The connection between the two coefficients should not blind you to an important difference. Ordinarily, we are concerned with the variance of a cross-product term relative to the variance of a predictor. By including the standard deviations of both variables in the denominator, standardized regression coefficients do not provide this information. For this reason, we will devote little attention to the standardized coefficient and focus, instead, on the unstandardized one.

2.2.5   Correlation Coefficient

Relatedly, we will spend comparatively little time discussing the correlation between x and y. Unlike a regression model that treats x as a fixed factor and y as a random variable, a correlational model makes no distinction between the predictor and the criterion. Instead, a correlation is simply the average cross product between pairs of standardized scores. Applying Eq. (1.2) to the standardized scores reported in Table 2.1 yields the correlation matrix, with standardized variances (1's) on the diagonals and the correlation (standardized covariance) on the off-diagonals:

$$\mathbf{R} = \begin{bmatrix} 1 & .6144 \\ .6144 & 1 \end{bmatrix}$$

As you might have noticed, the correlation coefficient assumes the same value as the standardized regression weight. With one predictor, this will always be true, but it will ordinarily not be true when two or more variables are used to predict y. We can understand why the two terms are equivalent by considering an alternative formula for computing a correlation coefficient:

$$r = \frac{s_{xy}}{\sqrt{s_x^2} * \sqrt{s_y^2}} \tag{2.20}$$

Comparing Eq. (2.20) with Eq. (2.18) shows that the formula for computing the correlation is the same as the formula for computing the standardized regression coefficient. This convergence occurs only when there is one predictor. Whereas Eq. (2.20) can always be used to compute the correlation between two variables, Eq. (2.18) can only be used to calculate the standardized regression coefficient with simple linear regression.

2.2.6   R Code: Simple Linear Regression

```
x <-c(1,9,1,5,6,8,2,4,2,8,7,7)
y <-c(3,8,2,8,5,9,4,5,2,4,2,6)

#Preliminary Model Without Intercept
model.1 <-lm(y~x-1)
summary(model.1)
fitted(model.1)
resid(model.1)
SSres.1 <-sum(resid(model.1)^2);SSres.1

#Complete Model With Intercept
model.2 <-lm(y~x)
summary(model.2)
fitted(model.2)
resid(model.2)
SSres.2 <-sum(resid(model.2)^2);SSres.2

#Scatterplot with line of best fit
plot(x, y, main="Scatterplot: Simple Linear Regression",
     xlab="x ", ylab="y", pch=19)
abline(lm(y~x), col="red")

#Standardized coefficients (Beta)
zmodel <-lm(scale(y)~scale(x)-1)
summary(zmodel)

#Unstandardized and standardized coefficients using simple algebra
b.coef <- cov(x,y)/var(x);b.coef
beta.coef <-cov(x,y)/(sd(x)*sd(y));beta.coef

#Correlation coefficient
correl <-cor(x,y);correl
```

2.3   Population Estimation and Statistical Significance

Earlier we noted that one function of linear regression is population estimation. We observe and describe a sample, but we often wish to determine whether our sample tells us something about the population from which it is drawn. In our case, it is unlikely that I care only whether self-efficacy beliefs predict performance among 12 students. Instead, I wonder whether I can generalize my findings to students in general.

2.3.1   The Logic Behind Null Hypothesis Testing

As you probably know from an introductory statistical course, one way to make inferences about a population is to engage in null hypothesis testing. With this procedure, we start by assuming that chance is the only factor influencing our data. We then ask "How likely are our sample data given that chance is the only operating factor?" By convention, we conclude that "not very likely" means we would observe data at least as extreme as ours less than 5 % of the time ($p < .05$) when chance variations are the only factor of importance. When that occurs, we "tentatively" conclude that factors other than chance are operating. Note that we never accept or reject an alternative hypothesis because we are not testing it; we are testing a null hypothesis and we only reject it or fail to reject it.

2.3.2   Testing the Regression Model

We can test the statistical significance of a simple linear regression model in two ways. First, we can test the overall significance of the regression model; second, we can test the significance of each regression coefficient. With a single predictor, the two approaches are identical. But in subsequent chapters we will discuss designs with multiple predictors. Consequently, we will learn both approaches now, beginning with a test of our overall regression model.

2.3.2.1   Partitioning the Sum of Squares

The first step in testing our model's significance is to partition the variability in our criterion, known as the total sum of squares and designated SS_{tot}, into two parts: a portion that can be predicted from our regression equation and a portion that cannot be predicted from our regression equation. We refer to the former as the regression sum of squares (SS_{reg}) and the latter as the residual sum of squares (SS_{res}):

$$SS_{tot} = SS_{reg} + SS_{res} \tag{2.21}$$

In matrix notation, we calculate the terms as follows:

$$SS_{tot} = (\mathbf{y} - \bar{\mathbf{y}})'(\mathbf{y} - \bar{\mathbf{y}}) \tag{2.22}$$

$$SS_{reg} = (\mathbf{Xb} - \bar{\mathbf{y}})'(\mathbf{Xb} - \bar{\mathbf{y}}) \tag{2.23}$$

$$SS_{res} = (\mathbf{y} - \mathbf{Xb})'(\mathbf{y} - \mathbf{Xb}) \tag{2.24}$$

Considering that we have already calculated and discussed the residual sum of squares,

$$SS_{res} = 42.1241$$

we will direct our attention to the other two terms. The total sum of squares is the deviate sum of squares of our criterion variable. To find it, we subtract the mean of y (Mean = 4.833) from every value and sum the squares [see Eq. (2.22)]. Table 2.1 shows that, in our example, $SS_{tot} = 67.6667$.

The regression sum of squares provides an index of how well our fitted values reproduce the mean of y. To find the term, we subtract the mean of y from every fitted value and sum the squares [see Eq. (2.23)]. Alternatively, we can use simple arithmetic; once we know two of the three values, we can easily calculate the third by subtraction:

$$SS_{reg} = 67.6667 - 42.1241 = 25.5426$$

2.3.2.2   Coefficient of Determination

One way to use these terms is to calculate a squared correlation coefficient, referred to as the coefficient of determination:

$$R^2 = \frac{SS_{reg}}{SS_{tot}} \tag{2.25}$$

Remembering that the total variability of $\mathbf{y}$ has been partitioned into two parts, we can see that the coefficient of determination represents the proportion of the total variability in y that can be explained by our regression equation:

$$R^2 = \frac{25.5426}{67.6667} = .3775$$

In our example, we say that approximately 38 % of the variability in y can be predicted from x.[13]

We can test the statistical significance of this value (i.e., determine whether it differs from 0), by using the following formula, where k = number of predictors (in our case, 1) and N refers to the sample size (in our case, 12):

$$F = \frac{R^2/k}{(1 - R^2)/(N - k - 1)} \tag{2.26}$$

Substituting our values

$$F = \frac{.3775/1}{(1 - .3775)/(12 - 1 - 1)} = 6.0643$$

Using a spreadsheet formula we can find the probability of obtaining an F value of this size or more given that only chance factors are operating [(=FDIST, 6.0643, 1, 10)=.0335]. Since this value is less than the conventional .05 level, we reject the hypothesis that the only factors operating in the situation were chance and conclude that self-efficacy beliefs (probably) predict test performance in the population.

2.3.2.3   F Test of Significance

An equivalent approach to testing the model's overall significance is to directly calculate an F value from our partitioned sum of squares without first calculating R^2:

$$F = \frac{SS_{reg}/k}{SS_{res}/(N - k - 1)} \tag{2.27}$$

Plugging our values into the equation yields an F value that matches (within rounding error) the one we obtained earlier using R^2:

$$F = \frac{25.5426/1}{42.1241/(12 - 1 - 1)} = \frac{25.5426}{4.2124} = 6.0637$$

---

[13] Because we have only one predictor, the square root of this value equals the correlation between x and y ($r = .6144$), as well as the standardized regression coefficient. Interestingly, it also equals the correlation between y and $\hat{y}$.

The denominator in Eq. (2.27) is known as the Mean Square Residual (MS_{res}):[14]

$$MS_{res} = \frac{SS_{res}}{(N - k - 1)} \qquad (2.28)$$

2.3.3   Testing the Regression Coefficients

As noted earlier, with only one predictor, testing the significance of the regression coefficient is equivalent to testing the significance of the regression model. This won't be true when we have more than one predictor, however, so let's examine the formula for testing the coefficient itself.

2.3.3.1   Standard Error

Our first step is to find the standard error of the regression coefficient (SE_b). To understand this term, let's imagine we repeat our study 10,000 times (we won't, of course, but let's imagine anyway). We expect that our regression coefficient (b_1=.5213) will equal the average coefficient across the 10,000 studies, but we certainly don't expect that every time we conduct the study we will find exactly the same estimate. The standard error is the standard deviation of the sampling distribution of regression coefficients. The smaller the value, the more we expect the 10,000 coefficients to cluster tightly around the average; the larger the value, the more we expect the 10,000 coefficients to vary around the mean.

   In a moment we will discuss ways of finding the standard error using a matrix inverse. But with only one variable it's useful to look at a simple formula:

$$se_b = \sqrt{\frac{MS_{res}}{SS_x}} \qquad (2.29)$$

Here we see that the standard error depends on two things: a term in the numerator that indexes error (i.e., the average squared difference between each fitted value and its corresponding observed value) and the variability in the predictor itself (as indexed by SS_x). From our earlier calculations, we know that $MS_{res}=4.2124$,

---

[14] The Mean Square Residual is sometimes called the Mean Square Error. It is also known as the variance of estimate, and its square root is called the standard error of estimate. I figure you already have enough terms to learn so I won't be using these terms in the book. If we ever need to take the square root of the Mean Square Residual, we'll just take the square root of the Mean Square Residual without giving it another name!

so the only term we need to calculate is the denominator, SS_x. This value is shown in Table 2.1 ($SS_x = 94$). Plugging in our values yields the standard error:

$$se_b = \sqrt{\frac{4.2124}{94}} = .2117$$

2.3.3.2 Statistical Significance

We can use the standard error to test the statistical significance of the regression coefficient by creating a fraction, with the regression coefficient in the numerator and the standard error in the denominator. We then refer the fraction to a t-distribution with $N - k - 1$ degrees of freedom:

$$t = \frac{b}{se_b} \tag{2.30}$$

Plugging in our values yields the test statistic:

$$t = \frac{.5213}{.2117} = 2.4626$$

Looking up the value in a spreadsheet for a two-tailed test (=TDIST,2.4626,10,2) shows that the probability of getting a value of this size or more when only chance factors are operating in the situation is .0335. Consequently, we reject the null hypothesis. Notice also that the probability level here is identical to the one we found when testing our overall regression model. As discussed earlier, with only one predictor, this will always be true. If you square our t value, you will see that it matches, within rounding error, the F value we found earlier:

$$2.4626^2 = 6.0642$$

2.3.3.3 Confidence Interval

Having calculated the standard error, we can also compute a confidence interval around our regression coefficient. A confidence interval represents the range of values on either size of a point estimate given a specified probability level. It is formed by finding the critical, two-tailed value of the t-distribution for a given level of significance (termed alpha and designated α) and then using the following equation:

$$b \pm \{t(\alpha/2, df) * se_b\} \tag{2.31}$$

To illustrate, we will calculate the 95 % confidence interval for the regression coefficient. Using a spreadsheet function, we find the critical, two-tailed t-value at the .05 level of significance with 10 degrees of freedom [(=TINV,.05, 10) = 2.2281]. Plugging the value into our equation yields the confidence interval:

$$.5213 \pm \{2.2281 * .2117\} = .0496 \text{ and } .9930$$

If we did repeat our study 10,000 times, we would expect to find a b value within this range 95 % of the time; the other 5 % of the time, we would expect to find values of b that fall outside of this range. Notice that 0 is not included in this range, constituting another way of rejecting the null hypothesis that the true population parameter equals 0.

2.3.3.4 Testing the Significance of the Intercept

For completeness, we will also test whether the intercept is significantly different from 0. Before we do, we should note that this test is often of little interest. The intercept represents the average expected value of y when $x=0$. If our x variable does not have a true 0 point (e.g., if age is a predictor, a person cannot be 0 years old) or our rating scale does not include 0, as was true in our example, it is of no interest to know the predicted value of y when $x=0$.

That being said, there may be times when this information is useful, so we will spend a moment familiarizing ourselves with the formula. Moreover, the formula we use to calculate the standard error of the intercept is a special case of a more general formula we use to calculate the standard error of any average expected value, so learning the procedures will be of general use.

As with our regression coefficient, we calculate a t-statistic using the coefficient in the numerator and a standard error in the denominator. With only one predictor, the formula for the standard error of an average expected value is

$$se_{\hat{y}} = \sqrt{MS_{res} * \left\{\frac{1}{n} + \frac{(x - \bar{x})^2}{SS_x}\right\}} \tag{2.32}$$

where x refers to the value of x for which prediction is being sought. For the intercept, we are interested in the average expected value of y when $x=0$, in which case Eq. (2.32) reduces to

$$se_{b_0} = \sqrt{MS_{res} * \left(\frac{1}{n} + \frac{\bar{x}^2}{SS_x}\right)} \tag{2.33}$$

Plugging in the relevant values from our data yields the standard error:

$$se_{b_0} = \sqrt{4.2124 * \left(\frac{1}{12} + \frac{5.00^2}{94}\right)} = 1.2130$$

and the t-statistic:

$$t = \frac{2.2270}{1.2130} = 1.8359$$

With $N - k - 1$ degrees of freedom, this value is not significant. Consequently, we are unable to reject the null hypothesis that the true intercept $= 0$. Calculating a confidence interval around our attained value reveals a similar story, since it includes 0:

$$2.2270 \pm \{2.2281 * 1.2130\} = -.4757 \text{ and } 4.9297$$

2.3.4   Parameter Covariance Matrix (C)

With a single predictor, the algebraic calculations for finding our standard errors are manageable. But when we have multiple predictors, as we will in subsequent chapters, we will find matrix algebra to be more expedient than ordinary algebra. In this section, we will learn how to calculate a covariance matrix of parameter estimates, $\mathbf{C}$:

$$\mathbf{C} = \left(\mathbf{X}'\mathbf{X}\right)^{-1} * MS_{res} \tag{2.34}$$

Several things about this formula merit attention. First, bear in mind that this is a covariance matrix of parameter estimates, not a covariance matrix of deviate scores derived from raw scores. In other words, don't confuse $\mathbf{C}$ and $\mathbf{S}$. Second notice that one term represents the variability of our predictors ($\mathbf{X}'\mathbf{X}$), and the other provides an index of how well our regression model reproduces our original values (MS_{res}). Because we are multiplying the latter term by the inverse of the former, we are essentially dividing MS_{res} by SS_x. This calculation matches our algebraic formula for finding the standard error with one predictor [see Eq. (2.29)]. Finally, the standard errors are found by taking the square root of the diagonal elements of the covariance matrix (denoted c_{ii}):

$$se_b = \sqrt{c_{ii}} \tag{2.35}$$

To demonstrate these points, let's plug in our numbers.

- First, we calculate $\mathbf{C}$:

$$\mathbf{C} = \begin{bmatrix} 12 & 60 \\ 60 & 394 \end{bmatrix}^{-1} * 4.2124$$

$$= \begin{bmatrix} .3493 & -.0532 \\ -.0532 & .0106 \end{bmatrix} * 4.2124 = \begin{bmatrix} 1.4714 & -.2241 \\ -.2241 & .0448 \end{bmatrix}$$

- Then, we take the square root of the diagonal elements of **C** to find our standard errors:

$$se_{b0} = \sqrt{1.4714} = 1.2130$$

$$se_{b1} = \sqrt{.0448} = .2117$$

These values match the ones we found earlier using our algebraic formulae. So this matrix gives us a very convenient way of finding our standard errors, and we will use it extensively throughout this text.

2.3.5   R Code: Hypothesis Testing

```
x <-c(1,9,1,5,6,8,2,4,2,8,7,7)
y <-c(3,8,2,8,5,9,4,5,2,4,2,6)
model <-lm(y~x)
summary(model)

#Sum of Squares
ssreg <-sum(((model$fitted-mean(y))^2);ssreg
ssres <-sum(((y-model$fitted)^2);ssres
sstot <-ssreg+ssres;sstot

#Coefficient of determination
r.squared <-ssreg/sstot
r.squared
F.test <-r.squared/1/((1-r.squared)/(12-1-1))
F.test
1 - pf(F.test,1,10)

#F test using Sum of Squares
msreg =ssreg/1
msres = ssres/10
ftest = msreg/msres
ftest

#Covariance Matrix, Standard Errors and Test Coefficients
X <-cbind(1,x)
covariance <-(solve(t(X)%*%X)*msres)
covariance
std.err <-sqrt(diag(covariance))
std.err
t.intercept <-model$coef[1]/std.err[1]
t.intercept
t.slope <-model$coef[2]/std.err[2]
t.slope
```

(continued)

2.3.5   R Code: Hypothesis Testing (continued)

```
#Confidence Intervals for Regression Coefficient
t.crit <-abs(qt(.025,length(x)-2))
slopeCI.low <-model$coef[2]-((t.crit)*std.err[2])
slopeCI.high <-model$coef[2]+((t.crit)*std.err[2])
slope.CI <-c(slopeCI.low,slopeCI.high)
slope.CI

#R Functions for Covariance Matrix and Confidence Intervals
cov <-vcov(model);cov
confint(model)
```

2.4   Forecasting

Linear regression is often used for purposes of forecasting. For example, insurance companies predict how long you are likely to live based on your health habits, and admission officers use your SAT scores to predict your likely success in college. In our example, we might want to predict a student's performance in a matrix algebra class based on knowledge of the student's self-efficacy beliefs.

The name we use to describe these estimates and the standard error we use to test them differ depending on whether we are forecasting an average value or an individual one. We will use the term "expected value" when we are forecasting an average expected value of y and the term "predicted value" when we are forecasting an individual value of y. In this section, we will discuss both approaches.

2.4.1   Average Expected Values

As it turns out, we have already covered the procedures involved in testing an average expected value when we tested the statistical significance of the intercept. The intercept, you will recall, represents the average expected value of y when $x=0$. The procedures we used to test its significance extend to any value of x, not just $x=0$.

To illustrate, suppose we wish to predict the average class rank of students whose self-efficacy beliefs $= 3$. Using matrix algebra, we begin by creating a column vector we will designate $\mathbf{p}$. The first value in the column vector will always be 1 (to model the intercept), and the other value will reflect the score we have chosen for x. With our example, $\mathbf{p}$ assumes the following form:

$$\mathbf{p} = \begin{bmatrix} 1 \\ 3 \end{bmatrix}$$

We then use matrix multiplication to compute our predicted value:

$$\hat{y} = \mathbf{p}'\mathbf{b} \qquad (2.36)$$

and its standard error:

$$se_{\bar{\hat{y}}} = \sqrt{\mathbf{p}'\mathbf{C}\mathbf{p}} \qquad (2.37)$$

Notice that I have placed a "hat" and a "bar" over y in the standard error formula to indicate that we are looking for the standard error of an average expected value. Performing the operations, we find our expected value:

$$[1 \quad 3] \begin{bmatrix} 2.2270 \\ .5213 \end{bmatrix} = 3.7908$$

and its standard error:

$$\sqrt{[1 \quad 3] \begin{bmatrix} 1.4714 & -.2241 \\ -.2241 & .0448 \end{bmatrix} \begin{bmatrix} 1 \\ 3 \end{bmatrix}} = .7282$$

We can then form a 95 %, two-tailed confidence interval around the expected value in the usual manner:

$$CI_{\bar{\hat{y}}} = 3.7908 \pm \{2.2281 * .7282\} = 2.1682 \, \text{and} \, 5.4133$$

These values tell us what to expect if we conducted our study a great many times. For example, if we conducted our 12-subject study 10,000 times, we would expect that the average class rank of students with a score of 3 on the self-efficacy scale would fall between 2.1682 and 5.4133 in 9,500 of them; the other 500 samples would produce average values outside of this range.

Finally, we can test the expected average value for its statistical significance:

$$t = \frac{3.7908}{.7282} = 5.2057$$

With $N - k - 1$ degrees of freedom, the probability of getting a value at least this large given that only chance is operating is .0004, indicating a statistically significant result. Whether this test is meaningful depends on whether 0 is a meaningful score on the criterion. In our example, a person can't be in the 0th percentile, so knowing that the fitted value is significantly different from 0 is of limited value.

2.4.2 Single Predicted Values

The preceding calculations tell us about an average predicted score, but suppose we would also like to know the likely score of a single student with our particular predictor profile (i. e., $x = 3$). In this case, we use the term "predicted value," rather than "expected value". Because the best estimate of an individual value is always the average value, the calculations we use to find the predicted value are the same as the ones we used to find the average expected value [see Eq. (2.36)]. But the standard errors are not calculated using the same formula. Instead of using Eq. (2.37), we use the equation shown below:

$$se_{\hat{y}} = \sqrt{\left[1 + \left\{\mathbf{p}'\left(\mathbf{X}'\mathbf{X}\right)^{-1}\mathbf{p}\right\}\right] * MS_{res}} \tag{2.38}$$

As you can see, rather than using the covariance matrix, we use the inverse of the sum of squares matrix from which it is derived, adding 1 to our product before multiplying by MS_{res}. This modification ensures that the standard error of an individual predicted value will always be greater than the standard error of an average expected value.

The formula looks imposing, but it's not all that complicated, so let's work through the math using values reported earlier in this chapter:

$$se_{\hat{y}} = \sqrt{\left[1 + \left\{\begin{bmatrix} 1 & 3 \end{bmatrix}\begin{bmatrix} .3493 & -.0532 \\ -.0532 & .0106 \end{bmatrix}\begin{bmatrix} 1 \\ 3 \end{bmatrix}\right\}\right] * 4.2124} = 2.1778$$

If we compare the two standard errors, we can see that the standard error of the individual case is much larger than the standard error of the average case. This makes sense because our ability to predict a single value will always be more subject to error than our ability to predict an average value across many instances. We can see this most clearly by using this standard error to construct a prediction interval around our predicted value[15]:

$$PI_{\hat{y}} = 3.7908 \pm \{2.2281 * 2.1778\} = -1.0616 \text{ and } 8.6432$$

This range is much greater than the confidence interval we calculated for our average expected value. If desired, we can also assess the statistical significance of an individual predicted value:

$$t = \frac{3.7908}{2.1778} = 1.7407, \ p = .1124$$

---

[15] It is customary to use the term "prediction interval" when considering the individual case and "confidence interval" when considering the average.

2.4.3   Forecasting with Caution

Before concluding this section, let's consider two more issues. Because the size of a standard error increases with distance from the mean of the predictor, forecasting is more certain the closer x is to $\bar{x}$. To illustrate, we will calculate the standard error of an average expected value when x is equal to the mean of all x values (i.e., $x = \bar{x} = 5$):

$$\sqrt{\begin{bmatrix} 1 & 5 \end{bmatrix}\begin{bmatrix} 1.4714 & -.2241 \\ -.2241 & .0448 \end{bmatrix}\begin{bmatrix} 1 \\ 5 \end{bmatrix}} = .5925$$

Notice that this standard error is smaller than the one we calculated using $x = 3$. In fact, no value of x will produce a smaller standard error for a forecasted value than the mean. This point is best appreciated by studying the algebraic formula for finding the standard error of an average value given in Eq. (2.32). Looking the equation over, we can see that the final term will be smallest when $x = \bar{x}$. This same is true for the standard error of a single predicted value [see Eq. (2.38)].

Second, both types of forecasts are appropriate only for values of x that fall within the range of data we have collected. In our example it is appropriate to forecast a score for $x = 3$, because even though this particular value was not observed in our sample, it lies within the range of values we did observe. It would not be appropriate, however, to predict values of y for values of x that lie outside the range of data we observed. This is because a regression equation is descriptive—it describes the association among the variables we observed in our sample. Within that range, we can use it to make inferences about the population and to forecast expected and predicted scores, but we should not make forecasts for values of x that lie outside our observed range.

2.4.4   R Code: Forecasting

```
x <-c(1,9,1,5,6,8,2,4,2,8,7,7)
y <-c(3,8,2,8,5,9,4,5,2,4,2,6)
model <-lm(y~x)
p <-c(1,3)
yhat <-t(p)%*%coef(model)
df <-length(x)-2

#Forecasting Average Values
std.error.ave <-sqrt(t(p)%*%vcov(model)%*%p);std.error.ave
t.crit <-abs(qt(.025,df))
CI.lo.ave <-yhat-(t.crit*std.error.ave )
CI.hi.ave <-yhat+(t.crit*std.error.ave )
CI.ave <-cbind(CI.lo.ave,CI.hi.ave );CI.ave
```

(continued)

2.4.4 R Code: Forecasting (continued)

```
#Forecasting Individual Values
msres <-(sum((y-model$fitted)^2)/df)
X <-cbind(1,x)
std.error.ind <- sqrt(msres*(1+t(p)%*%solve(t(X)%*%X)%*%p));
std.error.ind
CI.lo.ind <-yhat-(t.crit*std.error.ind)
CI.hi.ind <-yhat+(t.crit*std.error.ind)
CI.ind <-cbind(CI.lo.ind,CI.hi.ind);CI.ind
```

2.5 Chapter Summary

1. In a linear regression model, a criterion is modeled as an additive function of a set of predictors, plus a term that represents errors of prediction. The criterion and errors are treated as random variables, drawn from a conditional probability distribution at each level of the predictor. The errors surrounding each value of x are assumed to be independent and normally and identically distributed with mean 0 and variance σ^2.

2. Linear regression is performed by solving a series of simultaneous linear equations using $\mathbf{b} = (\mathbf{X'X})^{-1}\mathbf{X'y}$. A column of leading 1's is added to $\mathbf{X}$ to model the intercept.

3. Fitted values of y represent the average expected value of y at a particular level of x. For this reason, they are considered average conditional values.

4. Residuals are found by subtracting the fitted values from the observed values. When we square each residual, and sum the squares, we derive an index known as the residual sum of squares.

5. A line of best fit confirms that $\mathbf{b} = (\mathbf{X'X})^{-1}\mathbf{X'y}$ minimizes the residual sum of squares. Consequently, the procedure is known as ordinary least squares estimation.

6. The total variability of $\mathbf{y}$ can be partitioned into two parts: a portion that can be explained by the weighted predictors and a portion that cannot be explained by the weighted predictors.

7. A coefficient of determination (R^2) represents the percentage of the variability in y that can be explained by the weighted predictors. It can be tested for its statistical significance.

8. The statistical significance of a regression coefficient can be tested by dividing the coefficient by its standard error. The standard errors can be found by taking the square root of the diagonal elements of a covariance matrix of parameter estimates.

9. Predicted values can be calculated for the average case and for the individual case. Their point estimates are identical, but the standard errors used to construct confidence intervals around them differ, with the standard error of the individual case being larger.

Chapter 3
Maximum-Likelihood Estimation

In Chap. 2 you learned that ordinary least squares (OLS) estimation minimizes the squared discrepancy between observed values and fitted ones. This procedure is primarily a descriptive tool, as it identifies the weights we use in our sample to best predict y from x. Sample description is not the only function that regression analyses serve, however; they can also be used to identify population parameters. As it happens, the least squares solution coincides with a more general method for estimating population parameters known as maximum-likelihood estimation (MLE). With MLE, we ask, "What population parameters are most likely to have produced our sample data?" In this chapter, you will learn about MLE and its application to linear regression.

3.1 Probability and Likelihood in a Normal Distribution

Our first step in understanding MLE is to understand probability more generally. We will begin by finding the probability of a single outcome in a normal distribution using a formula developed by the German mathematician, Carl Friedrich Gauss:

$$f(x) = \frac{1}{\sqrt{2\pi\sigma^2}} * e^{-\frac{(x-\mu)^2}{2\sigma^2}} \tag{3.1}$$

Electronic Supplementary Material: The online version of this chapter (doi: 10.1007/978-3-319-11734-8_3) contains supplementary material, which is available to authorized users

Equation (3.1) describes the probability density function of a normally distributed random variable.[1] To illustrate its use, suppose we wish to calculate the probability of selecting a score of 75 from a normal distribution with $\mu = 100$ and $\sigma = 15$. We can solve the problem using a spreadsheet function [=NORMDIST(75,100,15)] or by using Eq. (3.1) (where $e =$ natural log ~ 2.71828):

$$f(x) = \frac{1}{\sqrt{2\pi(15^2)}} * e^{-\frac{(75 - 100)^2}{2(15^2)}} = .0066$$

If we wanted, we could select more scores, calculate their probabilities, and create a table or figure of the various probabilities. If we continued to select and plot the probabilities of a great many values, we would eventually have a smooth, normal curve (see Fig. 3.1).

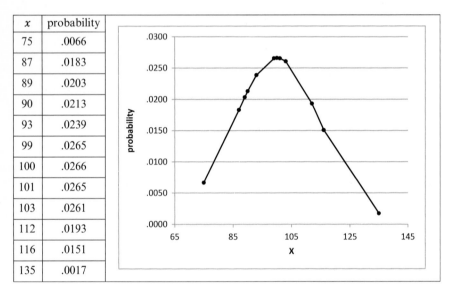

x	probability
75	.0066
87	.0183
89	.0203
90	.0213
93	.0239
99	.0265
100	.0266
101	.0265
103	.0261
112	.0193
116	.0151
135	.0017

Fig. 3.1 Density curve heights for 12 values chosen from a random distribution with known population parameters: $\mu = 100$; $\sigma = 15$

---

[1] The first term in Eq. (3.1) is sometimes written as

$$\frac{1}{\sigma\sqrt{2\pi}}$$

and the final term can be written in three, mathematically equivalent ways:

$$\frac{(x - \mu)^2}{2\sigma^2} = \frac{.5(x - \mu)^2}{\sigma^2} = \frac{1}{2\sigma^2}(x - \mu)^2$$

3.1.1 Likelihood Function

So far you have learned how to calculate the probability of an observed outcome given a known set of population parameters. A likelihood function calculates the inverse. Instead of asking "what is the probability of a sample value given a set of known population parameters?" we ask "what population parameters are most likely to have produced the sample data we observed?" The latter approach involves maximum-likelihood estimation, as now we are using sample data to estimate the most probable population mean and standard deviation.

To answer our likelihood question, we select hypothesized values for the mean ($\hat{\mu}$) and standard deviation ($\hat{\sigma}$) and compute a likelihood function:

$$L = \prod_{i=1}^{N} \left\{ \frac{1}{\sqrt{2\pi\hat{\sigma}^2}} * e^{-\frac{(x-\hat{\mu})^2}{2\hat{\sigma}^2}} \right\} \tag{3.2}$$

The portion of the equation in brackets repeats the formula used to find the probability of an observed score given a known mean and standard deviation, except here we are using hypothesized values rather than known ones. The operator preceding the formula indicates that we are to calculate the product of the likelihood scores (i.e., multiply them). An equivalent formula appears below.

$$L = \left(\frac{1}{\sqrt{2\pi\hat{\sigma}^2}} \right)^N * e^{-\left(\frac{\sum_{i=1}^{n}(x-\hat{\mu})^2}{2\hat{\sigma}^2} \right)} \tag{3.3}$$

Table 3.1 displays the likelihoods for the 12 values shown in Fig. 3.1 across five hypothesized means.[2] The first likelihood value was found for a score of 75 in a distribution with $\hat{u} = 98$:

$$L = \frac{1}{\sqrt{2\pi(15^2)}} * e^{-\frac{(75-98)^2}{2*15^2}} = .0082$$

The other column values were found in a similar fashion. Notice that the values when the hypothesized mean equals 100 are the same as the ones shown in Fig. 3.1. Finally, the values in the final row can be derived by computing the product of the 12 values in each column or by applying Eq. (3.3). To illustrate, the final row value in column 1 was found by first calculating the sum of all squared deviations around the hypothesized mean of 98 ($= 2,748$) and then using Eq. (3.3):

---

[2] For purposes of illustration, I have kept the standard deviation at 15 for all five possibilities.

Table 3.1 Likelihood function for five hypothesized means with a standard deviation of 15

x	$\hat{\mu} = 98$	$\hat{\mu} = 99$	$\hat{\mu} = 100$	$\hat{\mu} = 101$	$\hat{\mu} = 102$
75	.0082	.0074	.0066	.0059	.0053
87	.0203	.0193	.0183	.0172	.0161
89	.0222	.0213	.0203	.0193	.0183
90	.0231	.0222	.0213	.0203	.0193
93	.0252	.0246	.0239	.0231	.0222
99	.0265	.0266	.0265	.0264	.0261
100	.0264	.0265	.0266	.0265	.0264
101	.0261	.0264	.0265	.0266	.0265
103	.0252	.0257	.0261	.0264	.0265
112	.0172	.0183	.0193	.0203	.0213
116	.0129	.0140	.0151	.0161	.0172
135	.0013	.0015	.0017	.0020	.0024
Π	2.7908E-22	3.0233E-22	3.1050E-22	3.0233E-22	2.7908E-22

$$L = \left[\frac{1}{\sqrt{2\pi(15^2)}} \right]^{12} * e^{-\left[\frac{2748}{2(15^2)} \right]} = 2.7908E - 22$$

The likelihood is expressed in scientific notion, indicating that we are to move the decimal to the left 22 places. In our case, the likelihood of the sample given a population mean of 98 and a standard deviation of $15 = .000000000000000000000027908!$

3.1.2   Log-Likelihood Function

We could continue using this likelihood function, but you have probably realized that likelihood estimates this small are difficult to work with and are prone to error. For this, and other reasons, it is customary to use a (natural) log transformation of each value and then sum (not multiply) the obtained values. Alternatively, we can use the following formula, known as the log-likelihood function:[3]

$$\ln L = -\frac{N}{2}\ln(2\pi) - \frac{N}{2}\ln(\hat{\sigma}^2) - \sum_{i=1}^{n} \frac{(x - \hat{\mu})^2}{2\hat{\sigma}^2} \tag{3.4}$$

---

[3] The log-likelihood function represents a monotonic transformation of the original likelihood function. Because a monotonic transformation preserves the original order, the maximum values of the two functions coincide.

Table 3.2 Likelihood
and log-likelihood estimates
for a hypothesized $\mu = 98$ and
$\sigma = 15$

x	Likelihood	ln likelihood
75	.0082	−4.8025
87	.0203	−3.8959
89	.0222	−3.8070
90	.0231	−3.7692
93	.0252	−3.6825
99	.0265	−3.6292
100	.0264	−3.6359
101	.0261	−3.6470
103	.0252	−3.6825
112	.0172	−4.0625
116	.0129	−4.3470
135	.0013	−6.6692
	$\Pi = 2.7908E - 22$	$\Sigma = - 49.6305$

Table 3.2 compares the two likelihoods for $\hat{\mu} = 98$ and $\sigma = 15$, and the following calculations show that Eq. (3.4) produces the same value we get from summing the log likelihoods:

$$\text{ln}L = -\frac{12}{2}\text{ln}(2\pi) - \frac{12}{2}\text{ln}(15^2) - \frac{2748}{2(15^2)} = -49.6305$$

3.1.3   Using the Grid-Search Method to Find the Maximum-Likelihood Estimate

At this point we know how to calculate the likelihood function, but we still don't know how to find its maximum value (which, after all, is the whole point of maximum-likelihood estimation). We have three possibilities: (1) the grid search method, (2) differential calculus, and (3) a hybrid approach that combines the first two methods.

The grid search method involves trying out different values of $\hat{\mu}$ and $\hat{\sigma}$ until we identify the combination that produces the largest maximum-likelihood estimate. Table 3.3 shows the log-likelihood estimates for our five hypothesized means (keeping $\hat{\sigma} = 15$), and Fig. 3.2 plots them. Looking at the figure, it appears that the curve is at its maximum when our hypothesized value of μ equals our sample mean.

There is nothing wrong with using this "guess and guess again" approach to find the maximum-likelihood estimate, but it isn't very efficient. After all, we have only considered five possible values for the mean, but there are many other values we could test. We have also kept our estimate of the variance constant, but allowing this value to change along with the mean would create even more combinations to test. In short, although it is a useful way of showing what it is we are looking for, the grid search method is an inefficient way of finding the maximum-likelihood estimate.

Table 3.3 Log-likelihood estimates for five hypothesized means with standard deviation $= 15$

x	$\hat{\mu} = 98$	$\hat{\mu} = 99$	$\hat{\mu} = 100$	$\hat{\mu} = 101$	$\hat{\mu} = 102$
75	−4.8025	−4.9070	−5.0159	−5.1292	−5.2470
87	−3.8959	−3.9470	−4.0025	−4.0625	−4.1270
89	−3.8070	−3.8492	−3.8959	−3.9470	−4.0025
90	−3.7692	−3.8070	−3.8492	−3.8959	−3.9470
93	−3.6825	−3.7070	−3.7359	−3.7692	−3.8070
99	−3.6292	−3.6270	−3.6292	−3.6359	−3.6470
100	−3.6359	−3.6292	−3.6270	−3.6292	−3.6359
101	−3.6470	−3.6359	−3.6292	−3.6270	−3.6292
103	−3.6825	−3.6625	−3.6470	−3.6359	−3.6292
112	−4.0625	−4.0025	−3.9470	−3.8959	−3.8492
116	−4.3470	−4.2692	−4.1959	−4.1270	−4.0625
135	−6.6692	−6.5070	−6.3492	−6.1959	−6.0470
ln*L*	**−49.6305**	**−49.5505**	**−49.5239**	**−49.5505**	**−49.6305**

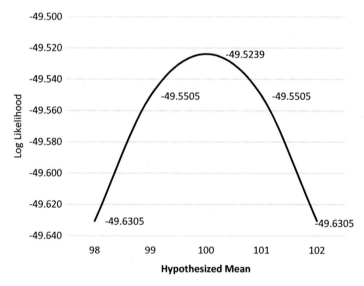

Fig. 3.2 Log-likelihood estimates for five hypothesized means with standard deviation $= 15$

3.1.4   R Code: Maximum-Likelihood Estimation with Normal Distribution

```
#Normal probabilities
mu=98
sigma=15
x=75
dnorm(x, mu, sigma)

#Log probabilities
dnorm(x,mu,sigma,log=T)

#Log likelihood functions given specific values for mu and sigma
x=c(75,87,89,90,93,99,100,101,103,112,116,135)
mu=98
sigma=15
like <-prod(dnorm(x, mu, sigma))
like
loglike <-sum(dnorm(x, mu, sigma,log=T))
loglike

#GRID SEARCH METHOD USING LOG LIKELIHOOD
normprob = function (x,mu,sigma) {
  sum(-0.5*(log(2*pi))-0.5*log(sigma^2)-((x-mu)^2)/(2*sigma^2))
}
X=c(75,87,89,90,93,99,100,101,103,112,116,135)
means = seq(95,105,by=.1)
vars = seq(10,20,by=.5)
mean.times=length(means)
mean.likes=sapply(means,function(y)prod(normprob(x=X,mu=y,
sigma=10)))
var.likes=sapply(vars,function(y)prod(normprob(x=X,mu=95,
sigma=y)))
plot(means,mean.likes,type="b")
windows()
plot(vars,var.likes,type="b")
means=rep(means,each=length(vars))
vars=rep(vars,mean.times)
mv.mat=cbind(means,vars)
mv.likes = apply(mv.mat,1,function(y)prod(normprob(x=X,mu=y[1],
sigma=y[2])))
mv.mat[mv.likes==max(mv.likes)]
best.combo<-(1:length(mv.likes))[mv.likes==max(mv.likes)]
best.combo
max.log <-mv.likes[mv.likes==max(mv.likes)]
max.log
```

(continued)

3.1.4 R Code: Maximum-Likelihood Estimation with Normal Distribution (continued)

```
#MLE in R
norm.fit<-function(mu,sigma){
-sum(dnorm(x,mu,sigma,log=T))
}
library(bbmle)                    #attach bbmle package
x=c(75,87,89,90,93,99,100,101,103,112,116,135)
mle.results<-mle2(norm.fit,start=list(mu=95,sigma=10),data=list(x))
mle.results
```

3.2 Differential Calculus

Fortunately, there is another method for finding a maximum-likelihood estimate that is faster and more precise than the grid search method. Instead of trying out lots of different values, we can use calculus to differentiate the maximum-likelihood function. Don't let the language scare you; it's not as difficult as it sounds. Just take a deep breath and remember that the whole point of this book is to show you that statistical analyses are not nearly as complicated as they are often made to seem.

3.2.1 Differentiating a Function to Find Its Derivative

With that in mind, let's start by defining some terms.

1. The derivative of a function represents its instantaneous rate of change. For example, if you throw a ball into the air, the derivative represents its velocity at a given instant.
2. The process of calculating a derivative is called differentiation (i.e., to differentiate a function is to find its instantaneous rate of change).
3. We can consider the derivative in two ways:

    3a. First, we can treat the derivative as the instantaneous rate of change at specific input values (e.g., how fast is the ball falling 2 s after it reaches its maximum height?). The following notation represents the instantaneous rate of change at "a" (with "a" being a specific value):

    $$\frac{dy}{dx}(a) \tag{3.5}$$

    3b. Second, we can treat the derivative as a function that describes the instantaneous rate of change across a range of input values (i.e., what is the formula for calculating the ball's speed at any given instant?). We use the following notation when we are calculating the derivative function across a range of values:[4]

$$\frac{df}{dx} \qquad (3.6)$$

3.2.1.1 Finding the Instantaneous Rate of Change at Specific Input Values

To help you better understand the processes involved in finding both types of derivatives, we'll examine a squaring function of the following form:

$$y = x^2$$

Figure 3.3 shows the solution across a range of values for x. As you can see, when $x = -5$, $y = 25$, when $x = -4$, $y = 16$, and so on. The figure also shows a line tangent to the curve that we will discuss momentarily.

Now suppose we want to calculate the instantaneous rate of change (i.e., the derivative) at a particular value of x. Let's try $x = 3$. There are several ways to do

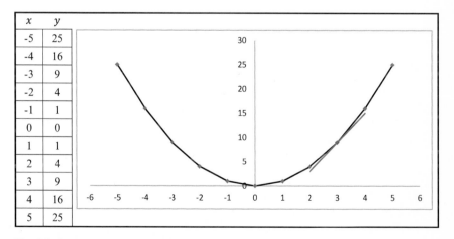

x	y
-5	25
-4	16
-3	9
-2	4
-1	1
0	0
1	1
2	4
3	9
4	16
5	25

Fig. 3.3 Squaring function for various values of x

---

[4] To confuse matters even more, the derivative can be written in different ways. For example, the formula $f'(a)$ is also used to express the derivative at a specific value.

this, but we will use a formula for finding the slope of a line, formally termed the delta method and colloquially referred to as "rise over run":

$$\text{Slope} = \frac{y_2 - y_1}{x_2 - x_1} \tag{3.7}$$

Why can we use this formula to find a derivative? The answer is that geometrically, the derivative represents the slope of a line tangent to a point on a curve. A tangent line touches the curve at a single point without crossing it, so it, too, represents the (near) instantaneous rate of change at a single point. Mathematically, we can never calculate the exact instantaneous rate of change at a single point, so we approximate it as the slope found when the change in x approaches 0:

$$\frac{dy}{dx} = \lim_{\Delta x \to 0} \frac{\Delta y}{\Delta x} \tag{3.8}$$

Figure 3.3 displays the tangent line when $x = 3$. Now we will use Eq. (3.7) to approximate its slope. The only problem is that in order to use the equation, we need to designate two values of x, not just one. The solution, known as the limit method of differentiation, is to choose two values that lie very close to x, one on either side, allowing us to approximate the slope. We will use $x_1 = 2.9999$ and $x_2 = 3.0001$. First, we compute the predicted values from our squaring function:

$$x_1 = 2.9999, \quad y_1 = 8.99994$$
$$x_2 = 3.0001, \quad y_2 = 9.00006$$

Then we plug the values into our slope formula:

$$\text{Slope} = \frac{9.0006 - 8.9994}{3.0001 - 2.9999} = 6.0000$$

So the (near) instantaneous rate of change when $x = 3$ is 6. We could also say that the derivative of y with respect to $x = 3$ is 6 or that the slope of the tangent line when $x = 3$ is 6. All of these terms and phrases refer to the same thing. Finally, we can express the derivative symbolically:

$$\frac{dy}{dx}(a = 3) = 6$$

Now let's use the limit method to find the derivative when $x = 5$:

$$x_1 = 4.9999, \quad y_1 = 24.9990$$
$$x_2 = 5.0001, \quad y_2 = 25.0010$$

and

$$\text{Slope} = \frac{25.0010 - 24.9990}{5.0001 - 4.9999} = 10$$

Here we see that the (near) instantaneous rate of change when $x = 5$ is 10, or that the derivative of y with respect to $x = 5$ is 10, or that the slope of the tangent line when $x = 5$ is 10:

$$\frac{dy}{dx}(a = 5) = 10$$

3.2.1.2   Derivative Function: Instantaneous Rate of Change Across a Range of Values

Having calculated the instantaneous rate of change at two specific values, we are ready to find a more general rule that describes the derivative function. Perhaps you can already see what it is. The derivative function of a squaring function is a doubling function. When $x = 3$, the derivative is 6 ($2x$); when $x = 5$, the derivative is 10 ($2x$). Formally, we write

$$\frac{df}{dx} = 2x$$

"the derivative function with respect to $x = 2x$."

Applying the derivative function, we can calculate the instantaneous rate of change without using the slope formula. For example, the instantaneous rate of change when $x = 4$ is 8, and the instantaneous rate of change when $x = 8$ is 16, and so on for as many values as we care to compute.

3.2.1.3   Analytic Solutions to Differentiating the Derivative Function

If you are familiar with calculus, you are probably aware that we could have saved ourselves some time by using the power rule to differentiate the function:

$$\frac{df}{dx}x^n = nx^{n-1} \ (for \ n \neq 0) \tag{3.9}$$

In words, "for every exponent other than zero, the derivative of a power function can be found by placing the exponent in front of the differentiated term and reducing the exponent by 1." So if

$$y = x^2,$$

$$\frac{df}{dx} = (2)x^{(2-1)},$$

which equals

$$\frac{df}{dx} = 2x.$$

Several more rules for differentiating functions, such as the quotient rule, log rule, product rule, and chain rule, can be found in all elementary calculus textbooks. These rules, which provide an analytic solution to finding the derivative function, are more precise and are quicker to use than is the method of limits. But you don't need to learn these rules to work through the material in this book, because I will provide them when they are needed.[5] Instead, all you need to know is that the derivative function quantifies the instantaneous rate of change in a function across a range of input values.

3.2.1.4   Derivative Orders

To this point we have considered only first-order derivatives (aka first derivatives), but a derivative function has its own derivative, which is called the second-order derivative (or the second derivative). Whereas the first derivative function quantifies the instantaneous rate of change, the second derivative function quantifies the change in the first derivative function. In terms of our earlier example, the velocity of a falling object is its first derivative and its acceleration is its second derivative. In statistics, the first derivative is used to estimate a population parameter (called a point estimate) and the second derivative is used to find its standard error. Symbolically, we express the second derivative by adding a superscript:

$$\frac{d^2 f}{d^2 x} \tag{3.10}$$

3.2.1.5   Mixed Derivatives and Partial Derivatives

When a function contains more than one parameter, we say we have a mixed derivative and calculate a partial derivative. A partial derivative approximates the instantaneous rate of change in a function as one input value changes and the rest remain constant. The first-order partial derivative function is denoted:

$$\frac{\partial f}{\partial x} \tag{3.11}$$

and the second-order partial derivative function is designated.

$$\frac{\partial^2 f}{\partial^2 x} \tag{3.12}$$

As we will see, the partial derivative is most relevant to linear regression analysis.

---

[5] As of this writing, an online derivative calculator can also be found at http://www.derivative-calculator.net/

3.2.1.6 Complete Example

To help solidify your understanding of derivatives, Table 3.4 works through an example by differentiating the following function:

$$z = 3x^3 - 4x^2 + 5y^2 - 6y$$

Only three rules are needed to calculate two first partial derivatives and two second partial derivatives:[6]

1. Ignore all terms that do not involve the differentiated term (because the derivative of a constant $= 0$).
2. When a differentiated term appears in an equation without an exponent, change the term to "1" (because the derivative of a derivative $= 1$).
3. Apply the power rule (discussed earlier) when a differentiated term appears with an exponent.

Table 3.4 Steps for computing partial derivatives of a function

First partial derivative of the function with respect to x	
Disregard all terms in which x does not appear	$3x^3 - 4x^2$
Applying power rule, move exponents to front and subtract 1 from each exponent	$(3)3x^{(3-1)} - (2)4x^{(2-1)}$
Reduce to find first partial derivative of the function with respect to x	$\dfrac{\partial f}{\partial x} = 9x^2 - 8x$
Second partial derivative of the function with respect to x	
Starting with the first partial derivative with respect to x, apply power rule to first term, and replace x with 1 in the second term (because the derivative of a variable with respect to itself equals 1)	$(2)9x^{(2-1)} - 8(1)$
Reduce to find second partial derivative of the function with respect to x	$\dfrac{\partial^2 f}{\partial^2 x} = 18x - 8$
First partial derivative of the function with respect to y	
Disregard all terms in which y does not appear	$5y^2 - 6y$
Apply power rule to first term, and substitute 1 for y in the second term	$(2)5y^{(2-1)} - 6(1)$
Reduce	$\dfrac{\partial f}{\partial y} = 10y - 6$
Second partial derivative of the function with respect to y	
Starting with the first partial derivative with respect to y, replace y with 1 and disregard constant	$\dfrac{\partial^2 f}{\partial^2 y} = 10$

---

[6] These rules are a subset of calculus differentiation rules and are expressed in language that I believe will best convey their implementation. Please consult a calculus textbook for more information or when differentiating functions other than the ones used in this book.

3.2.1.7   Limit Method Revisited

These rules are handy and we will use them throughout the remainder of this text, but it's also good to remember that we could have used the limit method to get (almost) the same results. To prove it, let's calculate the partial derivative of the function with respect to $x = 8$. Using the equations above, we find our first partial derivative:

$$\frac{\partial f}{\partial x} = 9x^2 - 8x = 512$$

and our second partial derivative:

$$\frac{\partial^2 f}{\partial^2 x} = 18x - 8 = 136$$

Now let's use the limit method to find the first and second partial derivatives when $x = 8$. Table 3.5 uses the limit method to calculate the derivatives with three input values: 7, 8, and 9. All I am doing is using values that lie very close to those whole numbers as input values for our function. If you then look at the middle values, you will see that the first partial derivative when $x = 8$ is found by calculating:

$$\frac{1,280.0512 - 1,279.9488}{8.0001 - 7.9999} = 512.00000003070$$

Obviously, this value is very close to the value we found using our formula for the first partial derivative. Now let's do the same thing with the second partial derivative:

$$\frac{657.00000002988 - 385.00000003045}{9.0001 - 7.0001} = 135.99999999972$$

This value, too, is a very close approximation to the one we found using our formula for the second partial derivative. The analytic solutions are quicker and more precise, but both methods yield the partial derivatives: a measure of how much a function changes with a change in one of its input values.

Table 3.5 Illustration of limit method for calculating partial derivatives with respect to x

x	y	$3x^3$	$-4x^2$	$5y^2$	$-6y$	Sum	First derivative	Second derivative
6.9999	0	1,028.9559	−195.9944	0	0	832.9615		
7.0001	0	1,029.0441	−196.0056	0	0	833.0385	385.00000003045	
7.9999	0	1,535.9424	−255.9936	0	0	1,279.9488		
8.0001	0	1,536.0576	−256.0064	0	0	1,280.0512	512.00000003070	135.99999999972
8.9999	0	2,186.9271	−323.9928	0	0	1,862.9343		
9.0001	0	2,187.0729	−324.0072	0	0	1,863.0657	657.00000002988	

3.2.1.8   Summary

- To differentiate a function is to find its derivative.
- A derivative represents the instantaneous rate of change of a function as its input values change.
- We can find the derivative at a particular input value by using the limit method or by using a derivative function.
- A derivative function provides a general formula for calculating a derivative at any input value and is found using various rules of differentiation.
- A derivative function can itself be differentiated, yielding a second-order derivative.
- When a function has multiple input values, we can calculate a partial derivative function by holding some input values constant and differentiating the function with respect to the input values that are free to vary.

3.2.2   Differentiation and Maximum-Likelihood Estimation

Having taken the time to learn about differentiation, we are ready to see how it can be applied to maximum-likelihood estimation. To begin, the log-likelihood estimates first shown in Fig. 3.2 are shown again in Fig. 3.4. Here I have modified our earlier figure by adding three tangent lines, each one representing a derivative of the log-likelihood function. Notice that the curve is at its maximum when the tangent line is horizontal, indicating that the slope (and derivative) equals 0 (see solid black line in Fig. 3.4). So maximizing the likelihood function entails finding the point on the curve where the tangent line is horizontal or, equivalently, finding the input values of the log-likelihood function for which the derivative equals 0. This mathematical principle holds the key to using calculus to find maximum-likelihood

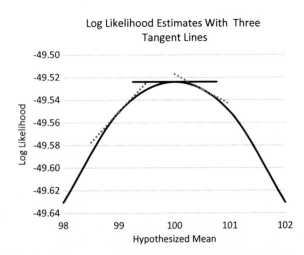

Fig. 3.4 Representation of Fig. 3.2 with three added tangent lines

estimates. We create a likelihood function and then identify our most likely population values by setting the first derivative equal to zero.

3.2.2.1   Maximum-Likelihood Estimate of μ

To find the maximum-likelihood estimate for the mean, we begin by finding the first partial derivative of the log-likelihood function with respect to the mean. Table 3.9 in the Appendix shows the steps we use to differentiate the log-likelihood equation, and Eq. (3.13) shows the first partial derivative:

$$\frac{\partial \ln L}{\partial \mu} = \frac{1}{\sigma^2} \sum_{i=1}^{n} (x_i - \mu) \tag{3.13}$$

Now we need to find the value of μ that makes the equation equal 0. We do so using the following steps:

- Multiply terms in parentheses by Σ:

$$\frac{\partial \ln L}{\partial \mu} = \frac{1}{\sigma^2} (\Sigma x - \Sigma \mu)$$

- Substitute $N\mu$ for $\Sigma\mu$ (because μ is a constant, its sum $= N\mu$):

$$\frac{\partial \ln L}{\partial \mu} = \frac{1}{\sigma^2} (\Sigma x - N\mu)$$

- Disregard the denominator (because the fraction equals 0 when the numerator equals 0), and set the numerator to equal 0:

$$\Sigma x - N\mu = 0$$

- Rearrange terms:

$$-N\mu = -\Sigma x$$

- Solve for μ:

$$\mu_{mle} = \frac{\Sigma x}{N} \tag{3.14}$$

Notice that this equation is identical to the one used to compute a sample mean. Formally, we have used calculus to prove that the sample mean provides the best estimate of the population mean.

3.2.2.2   Maximum-Likelihood Estimate of σ^2

We can repeat the process to find the maximum-likelihood estimate for the variance. Table 3.9 in the Appendix describes the steps needed to differentiate the
equation, and Eq. (3.15) shows the first partial derivative:

$$\frac{\partial \ln L}{\partial \sigma^2} = -\frac{N}{2\sigma^2} + \frac{1}{2\sigma^4}\sum_{i=1}^{n}(x_i - \mu)^2 \tag{3.15}$$

To find the maximum-likelihood estimate of the variance, we

- Set the equation to 0:

$$-\frac{N}{2\sigma^2} + \frac{1}{2\sigma^4}\sum_{i=1}^{n}(x_i - \mu)^2 = 0$$

- Rearrange terms:

$$-\frac{N}{2\sigma^2} = -\frac{1}{2\sigma^4}\sum_{i=1}^{n}(x_i - \mu)^2$$

- Cross-multiply:

$$N2\sigma^4 = \frac{1}{2\sigma^2}\sum_{i=1}^{n}(x_i - \mu)^2$$

- Divide both sides by $N2\sigma^2$:

$$\sigma^2_{mle} = \frac{\Sigma(x - \mu)^2}{N} \tag{3.16}$$

Notice that the maximum-likelihood estimate for the variance closely resembles
but is not identical to the formula for finding the sample variance. Whereas the
sample variance uses $N-1$ in the denominator, our estimate of the population
variance is made using N in the denominator. Formally, we say that the two estimates
are "asymptotically equivalent;" informally, this means that the estimates converge
as the sample size approaches infinity. With small sample sizes, the sample variance
provides a larger (more conservative) estimate of the population variance.

3.2.3   Computing the Standard Errors

Earlier I noted that first derivatives are used to find point estimates for population parameters and that second derivatives are used to compute their standard errors. Before discussing the procedures we use to calculate the standard error, let's review its meaning. As first discussed in Chap. 2, every estimated population parameter has a sampling distribution, with an average value and a standard deviation. The standard error of the mean is the standard deviation of the sampling distribution of the mean. If we repeated our study many times, the standard deviation would quantify the variability of our estimated means.

Perhaps you were introduced to the standard error of the mean in an introductory statistics course. If so, you were probably taught to find its value using the following formula:

$$se_\mu = \sqrt{\frac{\sigma^2}{N}} \qquad\qquad (3.17)$$

In this section, you will learn how this formula is derived from the second partial derivative of the likelihood function with respect to the mean. There are quite a few steps involved, but the process is worth learning because all standard errors are calculated using second partial derivatives.

3.2.3.1   Second Derivatives and the Slope of a Curve

Recall that the second derivative provides an index of how fast the first derivative function changes. The easiest way to understand this point is to look at a curve. Figure 3.5 plots two log-likelihood functions. The function labeled "low variance" comes from the data first shown in Table 3.1, with a mean of 100 and a standard deviation of 15; the function labeled "high variance" comes from a data set with the same mean (100) but a standard deviation of 30.[7]

Both curves reach their maximum at $\hat{\mu} = 100$, but they have different shapes. When the variance is low, the curve is steep and changes quickly near its maximum height; when the variance is high, the curve is flat and changes slowly near its maximum height. Because the second partial derivative quantifies the curvature of the graph of the log-likelihood function, the second partial derivative of each curve assumes a different value.

The top portion of Table 3.10 in the Appendix shows how we differentiate the log-likelihood function to find the second partial derivative with respect to the mean, and Eq. (3.18) shows the second partial derivative that differentiation produces:

---

[7] Because we are trying to identify the population values that are most likely to have produced our sample data, the standard deviations for these samples are calculated using N in the denominator rather than $N - 1$.

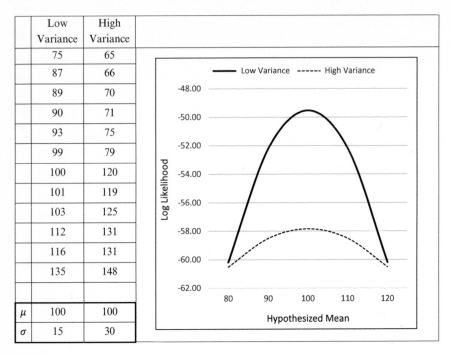

	Low Variance	High Variance
	75	65
	87	66
	89	70
	90	71
	93	75
	99	79
	100	120
	101	119
	103	125
	112	131
	116	131
	135	148
μ	100	100
σ	15	30

Fig. 3.5 Two log-likelihood functions with identical means but different variances

$$\frac{\partial^2 \ln L}{\partial^2 \mu} = -\frac{N}{\sigma^2} \tag{3.18}$$

Plugging in numbers from the two distributions shown in Fig. 3.5 yields the following values:

$$\text{Low Variance} = -\frac{12}{15^2} = -.0533$$

$$\text{High Variance} = -\frac{12}{30^2} = -.0133$$

3.2.3.2   Information

The second partial derivative indicates how much information our sample data provides about our population parameters. Referring back to Fig. 3.5, notice that the values surrounding the maximum are more distinct when the variance is low than when the variance is high. Consequently, the low variance curve provides more information about the population parameter than does the high variance curve, and we are more confident we have found the best estimate in the former case than in the latter.

The size of the second partial derivative reflects these informational differences. When the variance is low and the curve is steep, the second partial derivative is relatively large and the information it provides about the population parameter is substantial. Conversely, when the variance is high and the curve is flat, the second partial derivative is relatively small and the information it provides about the population parameter is comparatively slight.

3.2.3.3   Hessian Matrix

The next step in calculating the standard errors involves creating a matrix of the second partial derivatives known as a Hessian matrix (or sometimes, "the Hessian") (**H**). The left-hand side of Table 3.6 shows the matrix in its general form. Notice it is a symmetric matrix: the diagonal entries represent the second partial derivatives of the mean and variance (respectively), and the off-diagonal entries represent the cross-derivatives. Table 3.10 in the Appendix describes the steps we use to calculate the derivatives, and the differentiated terms appear in the right-hand side of Table 3.6.

Table 3.6   General form and derivatives for Hessian matrix (**H**)

$$
\mathbf{H} =
\begin{bmatrix}
\dfrac{\partial^2 \ln L}{\partial^2 \mu} & \dfrac{\partial^2 \ln L}{\partial \mu \partial \sigma^2} \\[2mm]
\dfrac{\partial^2 \ln L}{\partial \sigma^2 \partial \mu} & \dfrac{\partial^2 \ln L}{\partial^2 \sigma^2}
\end{bmatrix}
=
\begin{bmatrix}
-\dfrac{N}{\sigma^2} & -\dfrac{1}{\sigma^4}\sum_{i=1}^{n}(x_i - \mu) \\[3mm]
-\dfrac{1}{\sigma^4}\sum_{i=1}^{n}(x_i - \mu) & \dfrac{N}{2\sigma^4} - \dfrac{1}{\sigma^6}\sum_{i=1}^{n}(x_i - \mu)^2
\end{bmatrix}
$$

3.2.3.4   Fisher Information Matrix

We use the Hessian matrix to create another matrix called the Fisher information matrix (named after the statistician, Sir Ronald Fisher). The Fisher information matrix (**I**) is the negative of the expected value of the Hessian matrix:

$$
\mathbf{I} = -E(\mathbf{H}) \tag{3.19}
$$

The expected values, shown in Table 3.7, are calculated using our maximum-likelihood estimates.

Table 3.7   Fisher information matrix

$$
\mathbf{I} = -
\begin{bmatrix}
-\dfrac{N}{\sigma^2} & 0 \\[3mm]
0 & -\dfrac{N}{2\sigma^4}
\end{bmatrix}
=
\begin{bmatrix}
\dfrac{N}{\sigma^2} & 0 \\[3mm]
0 & \dfrac{N}{2\sigma^4}
\end{bmatrix}
$$

The information matrix is easier to work with than the Hessian. Because the expected value of $(x - \mu) = 0$, the cross-derivatives in the information matrix $= 0$, and because the expected value of $\sum_{i=1}^{n}(x_i - \mu)^2 = N\sigma^2$, the final term can be greatly simplified:

$$E(\sigma^2) = \left\{ \frac{N}{2\sigma^4} - \frac{N\sigma^2}{\sigma^6} \right\} = \left\{ \frac{N}{2\sigma^4} - \frac{2N}{2\sigma^4} \right\} = -\frac{N}{2\sigma^4}$$

Inserting numbers from our two sample distributions yields the information matrix for the low variance distribution and high variance distribution:

$$\mathbf{I}_{\text{low variance}} = \begin{bmatrix} .05333333 & 0 \\ 0 & .00011852 \end{bmatrix} \quad \mathbf{I}_{\text{high variance}} = \begin{bmatrix} .01333333 & 0 \\ 0 & .00000741 \end{bmatrix}$$

3.2.3.5   Standard Errors

The standard errors of the maximum-likelihood estimators are found from the Fisher information matrix. First, we take the inverse of the Fisher information matrix to find the variance/covariance of our parameter estimates. The standard errors are then found by taking the square root of the diagonal entries of the inverse matrix:

$$se_{ii} = \sqrt{\mathbf{I}_{ii}^{-1}} \tag{3.20}$$

Performing the calculations, we find our inverse matrices:

$$\mathbf{I}_{\text{low variance}} = \begin{bmatrix} 18.75 & 0 \\ 0 & 8437.50 \end{bmatrix} \quad \mathbf{I}_{\text{high variance}} = \begin{bmatrix} 75 & 0 \\ 0 & 135000 \end{bmatrix}$$

and our standard errors:

$$se_{low} = \sqrt{18.75} = 4.3301 \quad \text{and} \quad se_{high} = \sqrt{75} = 8.6603$$

It is informative to compare these values with ones we found using the usual formula for computing the standard error of the mean [see Eq. (3.17)]. Plugging in our values, we see that the two formulae produce identical results:

$$se_{low} = \sqrt{\frac{225}{12}} = 4.3301 \quad \text{and} \quad se_{high} = \sqrt{\frac{900}{12}} = 8.6603$$

That's because the standard error *is* the square root of the inverse of the Fisher information matrix.

3.2.3.6 Using the Standard Error to Calculate Confidence Limits

We can further appreciate the informational value of the second partial derivative by remembering that standard errors are used to compute confidence intervals around our mean. A confidence interval is another measure of information. The smaller the standard error, the narrower is our confidence interval and the more certain we are that our estimated value represents the true population parameter. Using a two-tailed, .05 level of significance, we compute the confidence interval after finding the critical t value for $N = 12$ ($t = 2.1788$):

$$\text{Small variance}: CL = 100 \pm (2.1788^*4.3301) = 90.5656 - 109.4344$$
$$\text{Large variance}: CL = 100 \pm (2.1788^*8.6603) = 81.1309 - 118.8691$$

Notice that the confidence interval is relatively narrow when the variance is small but comparatively broad when the variance is large. In short,

- Low variance = steep maximum-likelihood curve = large second derivative = high information = small standard error = narrow confidence interval.
- High variance = flat maximum-likelihood curve = small second derivative = low information = large standard error = broad confidence interval.

3.2.4 R Code: Derivatives and Standard Errors

```
#Define function
funct = expression(3*x^3-4*x^2+5*y^2-6*y, 'x,y')
derivx1 <-D(funct,"x")
derivx1
derivx2 <-D(derivx1,"x")
derivx2
derivy1 <-D(funct,"y")
derivy1
derivy2 <-D(derivy1,"y")
derivy2

#Standard Errors for High and Low Variance Distributions
#Log likelihood function for normal distribution
loglike <- function(theta) {
sum ( 0.5*(x - theta[1])^2/theta[2] + 0.5* log(theta[2]) )
}

#Standard errors for Low Variance distribution
x =c(75,87,89,90,93,99,100,101,103,112,116,135)
lik.est <-nlm(loglike, theta <- c(100,15),hessian=TRUE)
std.err.low <-sqrt(solve(lik.est$hessian[1]))
```

(continued)

3.2.4   R Code: Derivatives and Standard Errors (continued)

```
std.err.low
t.crit <-abs(qt(.025,12))
CI.low <-theta[1]-(t.crit*std.err.low[1])
CI.high <-theta[1]+(t.crit*std.err.low[1])
both.low <-cbind(CI.low,CI.high)
both.low

#Standard errors for High Variance distribution
x =c(65,66,70,71,75,79,120,119,125,131,131,148)
lik.est <-nlm(loglike, theta <- c(100,15),hessian=TRUE)
std.err.high <-sqrt(solve(lik.est$hessian[1]))
std.err.high
CI.low <-theta[1]-(t.crit*std.err.high[1])
CI.high <-theta[1]+(t.crit*std.err.high[1])
both.high <-cbind(CI.low,CI.high)
both.high
```

3.3   Maximum-Likelihood Estimation in Regression

Applying the procedures we have just learned to linear regression is a relatively simple matter. Equation (3.21) presents the log-likelihood formula we use. As you can see, it closely resembles the log-likelihood function for a normal distribution. The only difference is that the numerator in the final term represents the squared deviation of our observed y values from our estimated ones ($\mathbf{Xb}$), rather than the squared deviation of all x values from a hypothesized population mean ($\hat{\mu}$):

$$\ln L = -\frac{N}{2}\ln(2\pi) - \frac{N}{2}\ln(\sigma^2) - \left[\frac{(\mathbf{y} - \mathbf{Xb})^2}{2\sigma^2}\right] \tag{3.21}$$

To illustrate the processes involved, we will revisit the data first shown in Table 2.1, now displayed in Table 3.8. Here we are predicting test performance (y) from self-efficacy beliefs (x).

OLS estimation produces the regression coefficients:

$$\mathbf{b} = \left(\mathbf{X}'\mathbf{X}\right)^{-1}\mathbf{X}'\mathbf{y} = \begin{bmatrix} 12 & 60 \\ 60 & 394 \end{bmatrix}^{-1} \begin{bmatrix} 58 \\ 339 \end{bmatrix} = \begin{bmatrix} 2.2270 \\ .5213 \end{bmatrix}$$

Table 3.8 Small sample example predicting test performance from self-efficacy beliefs (reproduced from Table 2.1)

Student	Self-efficacy (x)	Performance (y)
1	1	3
2	9	8
3	1	2
4	5	8
5	6	5
6	8	9
7	2	4
8	4	5
9	2	2
10	8	4
11	7	2
12	7	6
Mean	**5.00**	**4.83**
Standard Deviation	**2.9233**	**2.4802**

After calculating the residuals:

$$SS_{res} = \Sigma(\mathbf{y} - \mathbf{Xb})^2 = 42.1241$$

and estimating the (population) variance:

$$\sigma^2 = \frac{42.1241}{12} = 3.5103$$

we compute our log-likelihood estimate:

$$-\frac{12}{2}\ln(2\pi) - \frac{12}{2}\ln(3.5103) - \left[\frac{42.1241}{2(3.5103)}\right] = -24.5615$$

3.3.1   Differentiating the Function

Now we want to use differential calculus to prove that OLS estimation maximizes the likelihood function. Table 3.11 in the Appendix describes the steps involved in finding the first partial derivatives of the log-likelihood function. If you compare the operations with the ones shown in Table 3.9, you will see that they are nearly identical. Our first partial derivative with respect to β is shown below:

$$\frac{\partial \ln L}{\partial \beta} = \frac{1}{\sigma^2}\left[\mathbf{X'y} - \mathbf{X'Xb}\right] \tag{3.22}$$

To find our maximum-likelihood estimate, we need to set the equation to zero and solve. Only the numerator of the term matters, so we proceed as follows:

- Set the equation to 0:

$$\mathbf{X'y} - \mathbf{X'Xb} = 0 \tag{3.23}$$

- Rearrange the terms to produce the matrix form of the normal equations (see Chap. 1):

$$\mathbf{X'y} = \mathbf{X'Xb} \tag{3.24}$$

- Solve the normal equations for **b**, producing the OLS solution:

$$\left(\mathbf{X'X}\right)^{-1}\mathbf{X'y} = \mathbf{b} \tag{3.25}$$

As you can see, the matrix inverse method that yields the smallest residual sum of squares also yields the most likely estimate of the population parameters that produced our sample data.

We use a similar process to find the maximum-likelihood estimate of the variance. The bottom portion of Table 3.11 in the Appendix shows the steps we use to differentiate the function, and the resultant partial derivative is shown in Eq. (3.26):

$$\frac{\partial \ln L}{\partial \sigma^2} = -\frac{N}{2\sigma^2} + \frac{1}{2\sigma^4}\left[(\mathbf{y} - \mathbf{Xb})'(\mathbf{y} - \mathbf{Xb})\right] \tag{3.26}$$

To find the maximum-likelihood estimate, we

- Set the derivative to 0:

$$-\frac{N}{2\sigma^2} + \frac{1}{2\sigma^4}\left[(\mathbf{y} - \mathbf{Xb})'(\mathbf{y} - \mathbf{Xb})\right] = 0$$

- Rearrange terms:

$$-\frac{N}{2\sigma^2} = -\frac{1}{2\sigma^4}\left[(\mathbf{y} - \mathbf{Xb})'(\mathbf{y} - \mathbf{Xb})\right]$$

- Cross-multiply and divide both sides by $N2\sigma^2$:

$$\sigma^2 = \frac{1}{N}\left[(\mathbf{y} - \mathbf{Xb})'(\mathbf{y} - \mathbf{Xb})\right] = \frac{SS_{res}}{N}$$

Once again, the variance estimate is asymptotically equivalent to the one found using OLS estimation.

3.3.2   Standard Errors

The procedures used to find the standard errors of the regression coefficients mirror the ones we used to find the standard error of the mean in a normal distribution. Table 3.12 in the Appendix shows the steps needed to compute the second partial derivatives, and the Hessian matrix is shown below:

$$
\mathbf{H} =
\begin{bmatrix}
\dfrac{\partial^2 \ln L}{\partial^2 \beta} & \dfrac{\partial^2 \ln L}{\partial \beta \partial \sigma^2} \\[2ex]
\dfrac{\partial^2 \ln L}{\partial \sigma^2 \partial \beta} & \dfrac{\partial^2 \ln L}{\partial^2 \sigma^2}
\end{bmatrix}
=
\begin{bmatrix}
\dfrac{-(\mathbf{X'X})}{\sigma^2} & -\dfrac{1}{\sigma^4}\left[\mathbf{X'y} - \mathbf{X'Xb}\right] \\[2ex]
-\dfrac{1}{\sigma^4}\left[\mathbf{X'y} - \mathbf{X'Xb}\right] & -\dfrac{N}{2\sigma^4}
\end{bmatrix}
$$

$Scalar$ (3.27)

- Plugging in our values produces the Hessian:

$$
\mathbf{H} =
\begin{bmatrix}
-3.4185 & -17.0923 & 0 \\
-17.0923 & -112.2398 & 0 \\
0 & 0 & -.4869
\end{bmatrix}
$$

- Taking the negative expected value of each term yields the information matrix:

$$
\mathbf{I} =
\begin{bmatrix}
3.4185 & 17.0923 & 0 \\
17.0923 & 112.2398 & 0 \\
0 & 0 & .4869
\end{bmatrix}
$$

- Finding the inverse of this matrix creates a variance/covariance matrix:

$$
\mathbf{I}^{-1} =
\begin{bmatrix}
1.2261 & -.1867 & 0 \\
-.1867 & .0373 & 0 \\
0 & 0 & 2.0538
\end{bmatrix}
$$

- The square root of the first two diagonal entries yields the standard errors:

$$
se_{b0} = \sqrt{1.2261} = 1.1073
$$
$$
se_{b1} = \sqrt{.0373} = .1932
$$

If you look back to Chap. 2, you will see that these standard errors are close to, but not identical with, the ones found using OLS estimation. As previously noted, the discrepancy arises because OLS estimation computes the sample variance as the deviation sum of squares divided by $N - k - 1$, whereas MLE computes the population variance as the deviation sum of squares divided by N. With large samples, the two estimates converge.

3.4   Maximum-Likelihood Estimation: Numerical Methods

So far, we have discussed two methods of finding a maximum-likelihood estimate. The first approach, known as the *grid search method*, involves trying out lots of parameter values and then finding the one that produces the maximum-likelihood estimate. This method is time-intensive and is impractical if there are many parameters to estimate. The second technique, known as the *analytic method*, uses differential calculus to set the first partial derivatives equal to zero and then finds the standard errors using the second partial derivatives. This procedure works only when there is an analytical (aka closed-form) solution, which is not the case in all instances in which maximum-likelihood estimation is applicable.

There is a third approach, known as the *numerical method*, that we have yet to learn. Here, we start with an approximate value and find the most likely estimate through an iterative (repetitive) process involving differentiation. So this approach is sort of a hybrid of the other two, combining guessing and calculus. Computer programs that solve MLE problems generally use a numerical method.

3.4.1   Newton–Raphson and Fisher's Method of Scoring

There are several iterative techniques available, but we are going to learn two: the Newton–Raphson method and a closely related technique known as Fisher's method of scoring. Both approaches follow a similar logic and use terms we have just learned:

- We begin with an initial guess regarding our population parameters, placed in a vector we call **k**.
- Using the methods described earlier, we calculate the first and second partial derivatives from the initial estimates.
- We place the first partial derivatives in a vector known as the gradient (and sometimes called the score) (**g**) and the second partial derivatives in a Hessian matrix (**H**) if we are using the Newton–Raphson method, or the Fisher information matrix (**I**) if we are using Fisher's method of scoring.
- Finally, we use the following formula to derive a new approximation for our population parameters and continue doing so until the difference between the old and new estimates becomes smaller than a designated level (e. g., $< .00000001$):

$$\text{Newton} - \text{Raphson} : \mathbf{k}_{+n} = \mathbf{k} - \left\{\mathbf{H}^{-1}\mathbf{g}\right\}$$

or

$$\text{Fisher's Method of Scoring} : \quad \mathbf{k}_{+n} = \mathbf{k} + \left\{\mathbf{I}^{-1}\mathbf{g}\right\}$$

3.4.2   Illustration Using Fisher's Method of Scoring

As always, it's less complicated than it seems. To show you how it's done, I will demonstrate the process using Fisher's method of scoring for our regression data.

1. We begin with an initial estimate of three parameters (intercept, regression coefficient, and variance), placed in a vector **k**. For purposes of illustration, I will guess the values shown below. Notice that the first two values in k estimate the regression coefficients (**b**) and the third value estimates the variance (σ^2):

$$\mathbf{k} = \begin{bmatrix} b_0 = 2 \\ b_1 = 3 \\ \sigma^2 = 4 \end{bmatrix}$$

2. We then use these guesses to calculate:

   2a. The first partial derivative with respect to β using Eq. (3.22):

$$\frac{\partial \ln L}{\partial \beta} = \frac{1}{\sigma^2}\left[\mathbf{X}'\mathbf{y} - \mathbf{X}'\mathbf{X}\mathbf{b}\right] = \begin{bmatrix} -36.50 \\ -240.75 \end{bmatrix}$$

   2b. The first partial derivative with respect to σ^2 using Eq. (3.26):

$$\frac{\partial \ln L}{\partial \sigma^2} = -\frac{N}{2\sigma^2} + \frac{1}{2\sigma^4}\left[(\mathbf{y} - \mathbf{X}\mathbf{b})'(\mathbf{y} - \mathbf{X}\mathbf{b})\right] = 73.3750$$

   2c. Finally, we merge these values to form a gradient matrix of first partial derivatives:

$$\mathbf{g} = \begin{bmatrix} -36.50 \\ -240.75 \\ 73.3750 \end{bmatrix}$$

3. Next, we need to find Fisher's information matrix **I**.

   3a. We start by computing the second partial derivative of the log-likelihood function with respect to β using our guesses as input values:

$$\frac{\partial^2 \ln L}{\partial^2 \beta} = \frac{-(\mathbf{X}'\mathbf{X})}{\sigma^2} = \begin{bmatrix} -3 & -15 \\ -15 & -98.50 \end{bmatrix}$$

3b. Next, we find the second partial derivative with respect to the variance using our guesses as input values:

$$\frac{\partial^2 \ln L}{\partial^2 \sigma^2} = -\frac{N}{2\sigma^4} = -.3750$$

3c. Finally, we combine these values to form the information matrix:

$$I = \begin{bmatrix} 3 & 15 & 0 \\ 15 & 98.50 & 0 \\ 0 & 0 & .3750 \end{bmatrix}$$

4. If we then compute $k + \{I^{-1}g\}$, we get a new k matrix we will call k_1:

$$k_1 = k + \{I^{-1}g\}$$

$$\begin{bmatrix} b_0 = 2 \\ b_1 = 3 \\ \sigma^2 = 4 \end{bmatrix} + \left\{ \begin{bmatrix} 1.3972 & -.2128 & 0 \\ -.2128 & .0426 & 0 \\ 0 & 0 & 2.6667 \end{bmatrix} \begin{bmatrix} -36.50 \\ -240.75 \\ 73.3750 \end{bmatrix} \right\} = \begin{bmatrix} b_0 = 2.2270 \\ b_1 = .5213 \\ \sigma^2 = 199.6667 \end{bmatrix}$$

Notice that the procedure has already identified the maximum-likelihood estimates for the regression coefficients, although the variance estimate is still incorrect.

5. We then repeat this process using k_1 as a starting point for calculating g and I:

$$\overset{k_1}{\begin{bmatrix} b_0 = 2.2270 \\ b_1 = .5213 \\ \sigma^2 = 199.6667 \end{bmatrix}} + \left\{ \overset{\{I^{-1}g\}}{\begin{bmatrix} 69.7417 & -10.6206 & 0 \\ -10.6206 & 2.1241 & 0 \\ 0 & 0 & 6,644.4630 \end{bmatrix} \begin{bmatrix} 0 \\ 0 \\ -.0295 \end{bmatrix}} \right\}$$

$$= \overset{k_1}{\begin{bmatrix} b_0 = 2.2270 \\ b_1 = .5213 \\ \sigma^2 = 3.5103 \end{bmatrix}}$$

After three iterations, the estimates converge on the OLS regression coefficients and variance.[8]

---

[8] As noted throughout this chapter, although the estimate of the population variance (3.5103) differs from the sample variance (3.2124), the two values are asymptotically equivalent.

3.4.3   R Code: Maximum-Likelihood Estimation with Fisher's Method of Scoring

```
x=c(1,9,1,5,6,8,2,4,2,8,7,7)
y=c(3,8,2,8,5,9,4,5,2,4,2,6)
N=length(y)
X=cbind(1,x)

#Initialize
theta <-c(2,3,4)
res0<-1
tol<-1e-8
norm<-10
iter<-0
change<-1

#Iterate
while (change>tol) {
grad.1=(t(X)%*%y-t(X)%*%X%*%theta[1:2])/theta[3]
grad.2=-N/(2*theta[3])+1/(2*theta[3]^2)*(t(y-X%*%theta[1:2])%*%(y-
X%*%theta[1:2]))
g=rbind(grad.1,grad.2)
hess.1 = -(t(X)%*%X)/theta[3]
hess.2 = -N/(2*theta[3]^2)
#Create Information Matrix
p=ncol(X)
n=p+1
I = matrix(0,n,n)
I[1:p,1:p]=hess.1
I[n*n]=hess.2
I=-I
theta =theta+solve(I)%*%g
norm.1=norm(theta, type = "F")  #F=Frobenius; can also use type ="2"
change <-abs(norm-norm.1)
norm=norm.1
iter<-iter+1
}
theta; iter; change;I

#Same Result Using R's Optimization Function
loglike<-function(theta){-sum(dnorm(y,mean=theta[1]+theta[2]*x,
sd=sqrt(theta[3]),log=T))
}
maxmod <-optim( theta <- c(2,3,4), loglike, hessian=T, method = "BFGS")
maxmod
```

3.5   Chapter Summary

1. Maximum-likelihood estimation identifies the population parameters that are most likely to have produced our sample data. In most cases, we work with a log-likelihood function.
2. Three strategies can be used to find the maximum value of the likelihood function:

   2a. The grid-search method involves auditioning numerous parameters until we find ones that maximize the likelihood function. This approach is not very efficient.
   2b. Differential calculus can also be used to find the maximum-likelihood estimates. We differentiate the log-likelihood function to find its first partial derivatives and then set the derivatives to zero to find the maximum-likelihood estimates.
   2c. Numerical methods can also be used to identify population parameters that maximize the log-likelihood function. These methods combine the grid-search method with differential calculus to produce an iterative technique. With each successive iteration, the parameter values are estimated more accurately until further attempts produce only trivial improvements.

3. The second partial derivatives of the likelihood function are used to find the standard errors of the population parameters. The following steps are involved:

   3a. Calculate the second partial derivatives.
   3b. Combine the second partial derivatives into a matrix called the Hessian matrix.
   3c. Take the negative expected value of the Hessian matrix to form an information matrix.
   3d. Invert the information matrix and take the square root of the diagonal entries to find the standard errors.

4. The procedures used to find the maximum likelihood of the mean and variance in a univariate distribution are easily extended to find the regression coefficients and variance of a bivariate distribution.

Appendix

First Partial Derivatives of the Log-Likelihood Function for a Normal Distribution

Table 3.9 Steps for calculating the first partial derivatives of the log-likelihood function for a normal distribution

Log-likelihood function	$\ln L = -\dfrac{N}{2}\ln(2\pi) - \dfrac{N}{2}\ln(\sigma^2) - \dfrac{1}{2\sigma^2}\sum_{i=1}^{n}(x_i - \mu)^2$
	First partial derivative with respect to μ
Disregard first two terms because they do not involve μ	$-\dfrac{1}{2\sigma^2}\sum_{i=1}^{n}(x_i - \mu)^2$
Using the chain rule, apply the power rule to outer layer and differentiate inner layer	$-\dfrac{1}{2\sigma^2} * \left[2\sum_{i=1}^{n}(x_i - \mu)^{(2-1)}\right] * \left[\dfrac{\partial \ln L}{\partial \mu}\sum_{i=1}^{n}(x_i - \mu)\right]$
Subtract exponents in outer layer and differentiate inner layer by replacing x with 0 (because the derivative of a constant $=0$) and μ with 1 (because the derivative of a variable with respect to itself $=1$)	$-\dfrac{1}{2\sigma^2} * \left[2\sum_{i=1}^{n}(x_i - \mu)\right] * [0 - 1]$
Multiply and reduce	$\dfrac{\partial \ln L}{\partial \mu} = \dfrac{1}{\sigma^2}\sum_{i=1}^{n}(x_i - \mu)$
	First partial derivative with respect to σ^2
Disregard first term because it does not involve σ^2, use log rule to move σ^2 to denominator of second term, and move exponent in the denominator of the third term to the numerator	$-\dfrac{N}{2\sigma^2} - \dfrac{1(\sigma^2)^{-1}}{2}\sum_{i=1}^{n}(x_i - \mu)^2$
Use power rule to rewrite second term	$-\dfrac{N}{2\sigma^2} - \dfrac{(-1)1(\sigma^2)^{-2}}{2}\sum_{i=1}^{n}(x_i - \mu)^2$
Return exponent to the denominator and multiply	$\dfrac{\partial \ln L}{\partial \sigma^2} = -\dfrac{N}{2\sigma^2} + \dfrac{1}{2\sigma^4}\sum_{i=1}^{n}(x_i - \mu)^2$

Second Partial Derivatives of the Log-Likelihood Function for a Normal Distribution

Table 3.10 Steps for calculating the second partial derivatives of the log-likelihood function for a normal distribution

Second partial derivative with respect to μ	
Rewrite first partial derivative with respect to μ by multiplying by Σ and substituting $N\mu$ for $\Sigma\mu$	$\dfrac{\partial \ln L}{\partial \mu} = \dfrac{1}{\sigma^2}(\Sigma x - N\mu)$
Set $\Sigma x = 0$ (since it doesn't involve μ) and set $\mu = 1$ (since a derivative of a derivative $= 1$)	$\dfrac{1}{\sigma^2}(-N1 + 0)$
Solve	$\dfrac{\partial^2 \ln L}{\partial \mu \partial \mu} = -\dfrac{N}{\sigma^2}$
Second partial derivative with respect to σ^2	
Starting with first partial derivative with respect to σ^2, move exponents to the numerator	$-\dfrac{N(\sigma^2)^{-1}}{2} + \dfrac{(\sigma^2)^{-2}}{2}\sum_{i=1}^{n}(x_i - \mu)^2$
Apply power rule to both equations	$-\dfrac{(-1)N(\sigma^2)^{-2}}{2} + \dfrac{(-2)(\sigma^2)^{-3}}{2}\sum_{i=1}^{n}(x_i - \mu)^2$
Return exponents to denominator	$\dfrac{\partial^2 \ln L}{\partial \sigma^2 \partial \sigma^2} = \dfrac{N}{2\sigma^4} - \dfrac{1}{\sigma^6}\sum_{i=1}^{n}(x_i - \mu)^2$
Cross-derivatives for Hessian matrix	
Starting with first partial derivative with respect to μ, move exponent to the numerator	$(\sigma^2)^{-1}\sum_{i=1}^{n}(x_i - \mu)$
Apply power rule	$(-1)(\sigma^2)^{(-1-1)}\sum_{i=1}^{n}(x_i - \mu)$
Return exponent to denominator	$\dfrac{\partial^2 \ln L}{\partial \mu \partial \sigma^2} = -\dfrac{1}{\sigma^4}\sum_{i=1}^{n}(x_i - \mu)$

First Partial Derivatives of the Log-Likelihood Function for Linear Regression

Table 3.11 Steps for calculating the first partial derivatives of the log-likelihood function for a linear regression model

Log-likelihood function	$\ln L = -\dfrac{N}{2}\ln(2\pi) - \dfrac{N}{2}\ln(\sigma^2) - \left[\dfrac{(y - Xb)^2}{2\sigma^2}\right]$

First partial derivative with respect to β	
Disregard first two terms because they do not involve **B**	$-\dfrac{1}{2\sigma^2}\left[(y - Xb)'(y - Xb)\right]$
Multiply matrix terms	$-\dfrac{1}{2\sigma^2}\left[y'y - 2X'b'y + b'bX'X\right]$
Disregard terms that do not involve **B**, change **B** = 1, when it is by itself, and apply power rule to **B'B**	$-\dfrac{1}{2\sigma^2}\left[-2X'y + X'X2b\right]$
Factor -2 and cancel terms	$\dfrac{\partial \ln L}{\partial \beta} = \dfrac{1}{\sigma^2}\left[X'y - X'Xb\right]$

First partial derivative with respect to σ^2	
Disregard first term because it does not involve σ^2, use log rule to move σ^2 to denominator of second term, and move exponent to the numerator of the third term	$-\dfrac{N}{2\sigma^2} - \dfrac{1(\sigma^2)^{-1}}{2}\left[(y - Xb)'(y - Xb)\right]$
Use power rule to rewrite last term	$-\dfrac{N}{2\sigma^2} - \dfrac{(-1)1(\sigma^2)^{-2}}{2}\left[(y - Xb)'(y - Xb)\right]$
Return exponent to the denominator	$\dfrac{\partial \ln L}{\partial \sigma^2} = -\dfrac{N}{2\sigma^2} + \dfrac{1}{2\sigma^4}\left[(y - Xb)'(y - Xb)\right]$

Second Partial Derivatives of the Log-Likelihood Function for Linear Regression

Table 3.12 Steps for calculating the second partial derivatives of the log-likelihood function for a linear regression model

Second partial derivative with respect to β	
Starting with first partial derivative with respect to β, disregard terms that do not involve **B** and change **B** to 1 when it is by itself	$\dfrac{\partial^2 \ln L}{\partial^2 \beta} = \dfrac{-(\mathbf{X}'\mathbf{X})}{\sigma^2}$

Second partial derivative with respect to σ^2	
Starting with first partial derivative with respect to σ^2, move exponents to the numerator	$-\dfrac{N(\sigma^2)^{-1}}{2} + \dfrac{(\sigma^2)^{-2}}{2}\left[(\mathbf{y}-\mathbf{Xb})'(\mathbf{y}-\mathbf{Xb})\right]$
Apply power rule to both equations	$-\dfrac{(-1)N(\sigma^2)^{-2}}{2} + \dfrac{(-2)(\sigma^2)^{-3}}{2}(\mathbf{y}-\mathbf{Xb})'(\mathbf{y}-\mathbf{Xb})$
Return exponents to denominator	$\dfrac{\partial^2 \ln L}{\partial \sigma^2 \partial \sigma^2} = \dfrac{N}{2\sigma^4} - \dfrac{1}{\sigma^6}\left[(\mathbf{y}-\mathbf{Xb})'(\mathbf{y}-\mathbf{Xb})\right]$
To simplify, substitute $[(\mathbf{y}-\mathbf{Xb})'(\mathbf{y}-\mathbf{Xb})] = n\sigma^2$ and reduce	$\dfrac{\partial^2 \ln L}{\partial \sigma^2 \partial \sigma^2} = -\dfrac{N}{2\sigma^4}$

Cross-derivatives for Hessian matrix	
Starting with first partial derivative with respect to β, move exponent to the numerator	$\sigma^{2-1}[\mathbf{X}'\mathbf{y} - \mathbf{X}'\mathbf{Xb}]$
Apply power rule	$(-1)\sigma^{2-2}[\mathbf{X}'\mathbf{y} - \mathbf{X}'\mathbf{Xb}]$
Return exponent to denominator	$\dfrac{\partial^2 \ln L}{\partial \beta \partial \sigma^2} = -\dfrac{1}{\sigma^4}\left[\mathbf{X}'\mathbf{y} - \mathbf{X}'\mathbf{Xb}\right]$

Chapter 4
Multiple Regression

In previous chapters, you have learned how to calculate regression coefficients and related terms using a single predictor. Yet most phenomena of interest to scientists involve many variables, not just one. Heart disease, for example, is associated with diet, stress, genetics, and exercise, and school performance is associated with aptitude, motivation, family environment, and a host of sociocultural factors. To better capture the complexity of the phenomena they study, scientists use multiple regression to examine the association among several predictors and a criterion. In this chapter, you will learn how to perform multiple regression analysis using tools you mastered in previous chapters.

Before presenting the technique, it's informative to consider why we just don't perform several simple linear regression analyses. There are two reasons. First, predictor variables are often correlated. For example, people who frequently exercise are also likely to maintain a healthy diet. By taking this overlap into account, multiple regression allows us to assess the unique contribution each variable makes to the prediction of a criterion. Second, multiple regression allows us to determine whether the effect of one variable changes at different levels of another variable. For example, diet might be a more important predictor of heart disease among people who don't exercise often than among people who do. The present chapter considers the first of these issues, and Chap. 9 discusses the second.

As we did in Chap. 2, we can learn about multiple regression by first solving a series of simultaneous linear equations. Imagine we are asked to find b_1 in the following equations.

$$-6b_1 = -3$$
$$4b_1 = 8$$
$$2b_1 = 2$$

Electronic Supplementary Material: The online version of this chapter (doi: 10.1007/978-3-319-11734-8_4) contains supplementary material, which is available to authorized users

© Springer International Publishing Switzerland 2014
J.D. Brown, *Linear Models in Matrix Form*, DOI 10.1007/978-3-319-11734-8_4

These equations do not have an exact solution, but if you were to solve them using the matrix inverse method we learned in Chap. 2, you would find that $b_1 = .9643$.

Now imagine we insert a second set of unknown quantities into our original equation.

$$-6b_1 - 8b_2 = -3$$
$$4b_1 + 3b_2 = 8$$
$$2b_1 + 5b_2 = 2$$

If you are thinking that adding these terms will change the weight we give to b_1, you have identified the essence of multiple regression. Unless the two sets of equations are uncorrelated (in statistical jargon, we say "orthogonal"), adding a variable to a series of linear equations will always alter the weight of other variables. In our case, b_1 now equals 2.0952 and $b_2 = -.9048$.

4.1   Multiple Regression

With this discussion as background, we will return to the fictitious example we used in Chap. 2. There, we used students' self-efficacy beliefs to predict their class rank in a matrix algebra course. We are going to keep using these data, but now we are going to add a second predictor: how well students scored on a nine-item practice test given at the start of the academic term. We will refer to this predictor as "math aptitude" and designate it x_2 to distinguish it from self-efficacy beliefs (which we will now designate x_1). Table 4.1 presents the (invented) findings. If you compare the values with the ones reported in Table 2.1, you will see that self-efficacy beliefs and class performance have not changed.

4.1.1   Correlations

Let's begin by examining the correlations among the three variables. From previous chapters, we know that a correlation is the average cross product of standardized variables, computed using the following formula.

$$\mathbf{R} = \mathbf{Z}'\mathbf{Z}\frac{1}{N-1}$$

Table 4.1 Small sample example of multiple regression. In this hypothetical example, 12 students indicate how proficient they think they are at math (x_1) and take a test that measures their actual math aptitude (x_2). We then use these measures to predict performance in a matrix algebra class (y)

Student	Self-efficacy x_1	Aptitude x_2	Performance y
1	1	2	3
2	9	7	8
3	1	1	2
4	5	8	8
5	6	5	5
6	8	7	9
7	2	4	4
8	4	7	5
9	2	1	2
10	8	4	4
11	7	3	2
12	7	4	6
Mean	**5.00**	**4.42**	**4.83**
Standard deviation	**2.9233**	**2.4293**	**2.4802**
Deviation sum of squares	**94.00**	**64.9167**	**67.6667**

After standardizing our predictors, we perform the calculations and obtain the correlation matrix.

$$\mathbf{R} = \begin{bmatrix} r_{11} & r_{12} & r_{1y} \\ r_{21} & r_{22} & r_{2y} \\ r_{y1} & r_{y2} & r_{yy} \end{bmatrix} = \begin{bmatrix} 1 & .5889 & .6144 \\ .5889 & 1 & .8877 \\ .6144 & .8877 & 1 \end{bmatrix}$$

Equation (4.1) can be used to test the statistical significance of a correlation,

$$F = \frac{r^2(N - k - 1)}{1 - r^2} \tag{4.1}$$

where $k = 1$ because we are examining the association between one predictor and one criterion.

$$r_{12} = .5889, \ F(1, 10) = 5.3084, \ p = .0440$$
$$r_{y1} = .6144, \ F(1, 10) = 6.0636, \ p = .0335$$
$$r_{y2} = .8877, \ F(1, 10) = 37.1657, \ p = .0001$$

Here, we see that all three correlations are statistically significant, indicating that self-efficacy and aptitude independently predict math performance and that the predictors themselves are significantly associated. This is precisely the situation multiple regression was designed to handle. As noted earlier, if the two predictors were uncorrelated (i.e., orthogonal), multiple regression wouldn't be needed because simple regression would suffice.

4.1.2 Unstandardized Regression Coefficients

Having calculated the correlations among the variables, we will now calculate the
unstandardized regression coefficients using the procedures we learned in Chap. 2.
Because the procedures are extensions of ones we learned earlier, we will move
rapidly through the calculations.

- We first calculate $\mathbf{X'X}$, adding a column of leading 1's to model the intercept.

$$
\begin{bmatrix}
1 & 1 & 1 & 1 & 1 & 1 & 1 & 1 & 1 & 1 & 1 & 1 \\
1 & 9 & 1 & 5 & 6 & 8 & 2 & 4 & 2 & 8 & 7 & 7 \\
2 & 7 & 1 & 8 & 5 & 7 & 4 & 7 & 1 & 4 & 3 & 4
\end{bmatrix}
\begin{bmatrix}
1 & 1 & 2 \\
1 & 9 & 7 \\
1 & 1 & 1 \\
1 & 5 & 8 \\
1 & 6 & 5 \\
1 & 8 & 7 \\
1 & 2 & 4 \\
1 & 4 & 7 \\
1 & 2 & 1 \\
1 & 8 & 4 \\
1 & 7 & 3 \\
1 & 7 & 4
\end{bmatrix}
$$

$$
=
\begin{bmatrix}
12 & 60 & 53 \\
60 & 394 & 311 \\
53 & 311 & 299
\end{bmatrix}
$$

- Next, we calculate $\mathbf{X'y}$.

$$
\begin{bmatrix}
1 & 1 & 1 & 1 & 1 & 1 & 1 & 1 & 1 & 1 & 1 & 1 \\
1 & 9 & 1 & 5 & 6 & 8 & 2 & 4 & 2 & 8 & 7 & 7 \\
2 & 7 & 1 & 8 & 5 & 7 & 4 & 7 & 1 & 4 & 3 & 4
\end{bmatrix}
\begin{bmatrix}
3 \\
8 \\
2 \\
8 \\
5 \\
9 \\
4 \\
5 \\
2 \\
4 \\
2 \\
6
\end{bmatrix}
=
\begin{bmatrix}
58 \\
339 \\
315
\end{bmatrix}
$$

- Finally, we solve for **b**,

$$\mathbf{b} = \left(\mathbf{X}'\mathbf{X}\right)^{-1}\mathbf{X}'\mathbf{y} = \begin{bmatrix} 12 & 60 & 53 \\ 60 & 394 & 311 \\ 53 & 311 & 299 \end{bmatrix}^{-1} \begin{bmatrix} 58 \\ 339 \\ 315 \end{bmatrix} = \begin{bmatrix} .6079 \\ .1191 \\ .8219 \end{bmatrix}$$

and form our sample regression equation.

$$\hat{y} = .6079 + .1191x_1 + .8219x_2$$

4.1.3 Fitted Values and Residuals

Before considering each regression coefficient separately, we will consider the regression equation as a whole. As with simple linear regression, these coefficients minimize the squared discrepancy between the observed values and the fitted values. To better appreciate this fact, we will use them to calculate fitted values,

$$\hat{y} = \mathbf{Xb} = \begin{bmatrix} 1 & 1 & 2 \\ 1 & 9 & 7 \\ 1 & 1 & 1 \\ 1 & 5 & 8 \\ 1 & 6 & 5 \\ 1 & 8 & 7 \\ 1 & 2 & 4 \\ 1 & 4 & 7 \\ 1 & 2 & 1 \\ 1 & 8 & 4 \\ 1 & 7 & 3 \\ 1 & 7 & 4 \end{bmatrix} \begin{bmatrix} .6079 \\ .1191 \\ .8219 \end{bmatrix} = \begin{bmatrix} 2.3708 \\ 7.4329 \\ 1.5489 \\ 7.7786 \\ 5.4318 \\ 7.3138 \\ 4.1337 \\ 6.8376 \\ 1.6679 \\ 4.8480 \\ 3.9071 \\ 4.7290 \end{bmatrix}$$

and residuals.

$$\mathbf{e} = \mathbf{y} - \hat{y} = \begin{bmatrix} 3 \\ 8 \\ 2 \\ 8 \\ 5 \\ 9 \\ 4 \\ 5 \\ 2 \\ 4 \\ 2 \\ 6 \end{bmatrix} - \begin{bmatrix} 2.3708 \\ 7.4329 \\ 1.5489 \\ 7.7786 \\ 5.4318 \\ 7.3138 \\ 4.1337 \\ 6.8376 \\ 1.6679 \\ 4.8480 \\ 3.9071 \\ 4.7290 \end{bmatrix} = \begin{bmatrix} .6292 \\ .5671 \\ .4511 \\ .2214 \\ -.4318 \\ 1.6862 \\ -.1337 \\ -1.8376 \\ .3321 \\ -.8480 \\ -1.9071 \\ 1.2710 \end{bmatrix}$$

If we square each residual, we can sum the squares to derive our residual sum of squares. This value quantifies the discrepancy between our fitted values and our observed ones; it also represents the part of y that our two predictors fail to predict.

$$SS_{res} = \Sigma e^2 = (\mathbf{y} - \hat{\mathbf{y}})^{'}(\mathbf{y} - \hat{\mathbf{y}}) = 13.4762$$

Because we have used ordinary least squares estimation to find our coefficients, we know that no other set of coefficients would produce a smaller residual sum of squares.

4.1.4   Testing the Regression Model

Our next step is to test the statistical significance of our regression model. The last entry in Table 4.1 shows that the deviation sum of squares for $y = 67.6667$. With this value in hand, we can

- Use the subtraction method [see Eq. (2.21)] to calculate the regression sum of squares

$$SS_{reg} = 67.6667 - 13.4762 = 54.1905$$

- Find the coefficient of determination using Eq. (2.25)

$$R^2 = \frac{54.1905}{67.6667} = .8008$$

and test the significance of the regression model using Eq. (2.26)

$$F = \frac{.8008/2}{(1 - .8008)/(12 - 2 - 1)} = 18.0954$$

Referring this value to an F distribution with 2 and 9 degrees of freedom shows a significant result ($p=.0007$). Consequently, we reject the null hypothesis that the only factors operating in the situation are chance.

4.1.5   R Code: Multiple Regression

```
x1 <-c(1,9,1,5,6,8,2,4,2,8,7,7)
x2 <-c(2,7,1,8,5,7,4,7,1,4,3,4)
y <-c(3,8,2,8,5,9,4,5,2,4,2,6)
X <-cbind(1,x1,x2)

#Create standardized variables
z1 <-scale(x1, center = T, scale = T)
z2 <-scale(x2, center = T, scale = T)
zy <-scale(y,center=T,scale=T)
Z <-cbind(z1,z2,zy)

#Correlation matrix
corr <-(t(Z)%*%Z)/(length(y)-1)
corr

#Function for finding Correlation Probabilities
cor.prob <- function(X, dfr = nrow(X) - 2) {
    R <- cor(X)
    above <- row(R) < col(R)
    r2 <- R[above]^2
    Fstat <- r2 * dfr / (1 - r2)
    R[above] <- 1 - pf(Fstat, 1, dfr)
    R
}
cor.prob(cbind(x1,x2,y))

#Multiple Regression
mod <-lm(y~x1+x2)
summary(mod)
fitted(mod)
resid(mod)

#Sum of Squares
SS.res <-sum(mod$resid^2)
SS.tot <-(var(y)*(length(y)-1))
SS.reg <-SS.tot-SS.res
df=length(y)-ncol(X)
MS.res <-SS.res/df
F.test <-(SS.reg/2)/MS.res; F.test
1-pf(F.test,2,df)
```

4.2   Interpreting and Testing Regression Coefficients

Knowing that an overall regression model is significant does not tell us anything about the statistical significance of each predictor. Instead, we know only that, taken together, the variables allow us to predict the variance in a criterion beyond what would be expected by chance alone. In this section, you will learn how to test the significance of each regression coefficient using procedures first introduced in Chap. 2. Before we do, however, we are going to spend some time learning how to interpret regression coefficients from a multiple regression analysis.

4.2.1   Comparing Regression Coefficients and Correlations

We can begin to understand the meaning of a regression coefficient by contrasting it with a correlation coefficient. In Chap. 2, we learned that, with a single predictor, the correlation coefficient and standardized regression coefficient are identical. So there is a close correspondence between a correlation and a regression coefficient with simple linear regression. This is not so with multiple regression. With multiple regression, correlations and regression coefficients (standardized or unstandardized) ordinarily have different values and a different interpretation. The reason is that correlations treat each x variable as an independent predictor of y, but regression coefficients treat the x variables as a system of predictors. Consequently, the weight of each regression coefficient depends on the weight of the other predictors.

The following demonstration will help clarify the preceding point. Imagine that we have gathered data from three students.

Student	Self-efficacy (x_1)	Aptitude (x_2)	Performance (y)
1	5	2	6
2	1	3	5
3	3	1	3

If we then calculate the regression coefficients and correlations, we get the following values.

$$b_1 = .6667 \quad \text{and} \quad r_{y1} = .3273$$
$$b_2 = 1.6667 \quad\quad\quad r_{y2} = .6547$$

Now imagine we change only one number, switching the first student's aptitude score from "2" to "6."

Student	Self-efficacy (x_1)	Aptitude (x_2)	Performance (y)
1	5	6	6
2	1	3	5
3	3	1	3

If we then recalculate the regression coefficients and correlations, we discover something interesting.

$$b_1 = -.2857 \quad \text{and} \quad r_{y1} = .3273$$
$$b_2 = .7143 \quad\quad\quad\quad r_{y2} = .9538$$

Notice that both of the regression coefficients have changed, but the correlation between x_1 and y remains unchanged (i.e., $r_{y1}=.3273$ in both cases). The reason is that a correlation is simply the average standardized cross product between two variables, with the variables treated in isolation from all others. Since I didn't change x_1 or y, the correlation between x_1 and y didn't change either.

In contrast, a regression coefficient is the weight that each variable is given to predict our criterion, and the weight of one variable is affected by the weight of other variables. Formally, we can say that whereas correlations are characterized by independence, regression coefficients are characterized by interdependence. This interdependence is especially apparent in this demonstration, as the b_1 coefficient changes from a positive value in Example 1 ($b_1=.6667$) to a negative value in Example 2 ($b_1=-.2857$), even though I didn't change the value of x_1 or y!

4.2.2   Interpreting the Numerical Value of a Regression Coefficient

Now that we know that a regression coefficient must be considered in the context of other predictors, we are ready to interpret its numerical value. There are at least three ways to think about this issue, and we will discuss them all, beginning with the one that I think is most informative.

4.2.2.1   The Unique Weight of Each Predictor

One way to define a regression coefficient is as the unique weight given to a predictor in the prediction of a criterion. This interpretation is best understood by creating residualized predictors. Recall that a residualized variable represents the part of a criterion that a predictor cannot predict (i.e., $e = y - \hat{y}$). When we create residualized predictors, we create predictors that are uncorrelated. As a result, we are able to identify the unique contribution each predictor makes to the prediction of a criterion.

To illustrate, we will regress x_2 on x_1 (i.e., we will designate x_2 as the criterion and x_1 as the predictor) and then use the residual to predict y.

- We first calculate $\mathbf{X'X}$, adding a column of leading 1's to model the intercept and including only the values for x_1.

$$
\begin{bmatrix} 1 & 1 & 1 & 1 & 1 & 1 & 1 & 1 & 1 & 1 & 1 & 1 \\ 1 & 9 & 1 & 5 & 6 & 8 & 2 & 4 & 2 & 8 & 7 & 7 \end{bmatrix}
\begin{bmatrix} 1 & 1 \\ 1 & 9 \\ 1 & 1 \\ 1 & 5 \\ 1 & 6 \\ 1 & 8 \\ 1 & 2 \\ 1 & 4 \\ 1 & 2 \\ 1 & 8 \\ 1 & 7 \\ 1 & 7 \end{bmatrix}
=
\begin{bmatrix} 12 & 60 \\ 60 & 394 \end{bmatrix}
$$

- Next we calculate $\mathbf{X'y}$, using x_1 as the predictor and x_2 as the criterion.

$$
\begin{bmatrix} 1 & 1 & 1 & 1 & 1 & 1 & 1 & 1 & 1 & 1 & 1 & 1 \\ 1 & 9 & 1 & 5 & 6 & 8 & 2 & 4 & 2 & 8 & 7 & 7 \end{bmatrix}
\begin{bmatrix} 2 \\ 7 \\ 1 \\ 8 \\ 5 \\ 7 \\ 4 \\ 7 \\ 1 \\ 4 \\ 3 \\ 4 \end{bmatrix}
=
\begin{bmatrix} 53 \\ 311 \end{bmatrix}
$$

- We then solve for $\mathbf{b}$,

$$
\mathbf{b} = \begin{bmatrix} 12 & 60 \\ 60 & 394 \end{bmatrix}^{-1} \begin{bmatrix} 53 \\ 311 \end{bmatrix} = \begin{bmatrix} 1.9699 \\ .4894 \end{bmatrix}
$$

compute fitted values,

$$
\mathbf{\hat{y}} = \mathbf{Xb} =
\begin{bmatrix} 1 & 1 \\ 1 & 9 \\ 1 & 1 \\ 1 & 5 \\ 1 & 6 \\ 1 & 8 \\ 1 & 2 \\ 1 & 4 \\ 1 & 2 \\ 1 & 8 \\ 1 & 7 \\ 1 & 7 \end{bmatrix}
\begin{bmatrix} 1.9699 \\ .4894 \end{bmatrix}
=
\begin{bmatrix} 2.4592 \\ 6.3741 \\ 2.4592 \\ 4.4167 \\ 4.9060 \\ 5.8848 \\ 2.9486 \\ 3.9273 \\ 2.9486 \\ 5.8848 \\ 5.3954 \\ 5.3954 \end{bmatrix}
$$

and calculate the residuals.

$$\mathbf{e} = \mathbf{y} - \hat{\mathbf{y}} = \begin{bmatrix} 2 \\ 7 \\ 1 \\ 8 \\ 5 \\ 7 \\ 4 \\ 7 \\ 1 \\ 4 \\ 3 \\ 4 \end{bmatrix} - \begin{bmatrix} 2.4592 \\ 6.3741 \\ 2.4592 \\ 4.4167 \\ 4.9060 \\ 5.8848 \\ 2.9486 \\ 3.9273 \\ 2.9486 \\ 5.8848 \\ 5.3954 \\ 5.3954 \end{bmatrix} = \begin{bmatrix} -.4592 \\ .6259 \\ -1.4592 \\ 3.5833 \\ .0940 \\ 1.1152 \\ 1.0514 \\ 3.0727 \\ -1.9486 \\ -1.8848 \\ -2.3954 \\ -1.3954 \end{bmatrix}$$

These residuals are uncorrelated with x_1. So when we use them to predict y (as we are about to do), we are identifying the unique weight given to x_2 in the prediction of y. We find the weight using the usual calculations:

- Calculate $\mathbf{X}'\mathbf{X}$,

$$\begin{bmatrix} 1 & 1 & 1 & 1 & 1 & 1 & 1 & 1 & 1 & 1 & 1 & 1 \\ -.4592 & .6259 & -1.4592 & 3.5833 & .0940 & 1.1152 & 1.0514 & 3.0727 & -1.9486 & -1.8848 & -2.3954 & -1.3954 \end{bmatrix}$$

$$\times \begin{bmatrix} 1 & -.4592 \\ 1 & .6259 \\ 1 & -1.4592 \\ 1 & 3.5833 \\ 1 & .0940 \\ 1 & 1.1152 \\ 1 & 1.0514 \\ 1 & 3.0727 \\ 1 & -1.9486 \\ 1 & -1.8848 \\ 1 & -2.3954 \\ 1 & -1.3954 \end{bmatrix} = \begin{bmatrix} 12 & 0 \\ 0 & 42.4060 \end{bmatrix}$$

calculate $\mathbf{X}'\mathbf{y}$,

$$\begin{bmatrix} 1 & 1 & 1 & 1 & 1 & 1 & 1 & 1 & 1 & 1 & 1 & 1 \\ -.4592 & .6259 & -1.4592 & 3.5833 & .0940 & 1.1152 & 1.0514 & 3.0727 & -1.9486 & -1.8848 & -2.3954 & -1.3954 \end{bmatrix}$$

$$\times \begin{bmatrix} 3 \\ 8 \\ 2 \\ 8 \\ 5 \\ 9 \\ 4 \\ 5 \\ 2 \\ 4 \\ 2 \\ 6 \end{bmatrix} = \begin{bmatrix} 58 \\ 34.8546 \end{bmatrix}$$

and solve for **b**.

$$\mathbf{b} = \begin{bmatrix} 12 & 0 \\ 0 & 42.4060 \end{bmatrix}^{-1} \begin{bmatrix} 58 \\ 34.8546 \end{bmatrix} = \begin{bmatrix} 4.8333 \\ .8219 \end{bmatrix}$$

Our second **b** value represents the regression coefficient associated with the residualized predictor. Notice that the value of the coefficient matches the value of b_2 in our main analysis ($b=.8219$). This is no coincidence. Each regression coefficient in the original analysis represents the unique weight of a variable in the prediction of the criterion. You can verify this is so by performing the comparable analyses for x_1.

4.2.2.2   A Predictor's Weight After Controlling for All Other Variables

A regression coefficient also represents the weight given to a predictor after all other predictors have been taken into account. In essence, each predictor is treated as if it were considered last, and we ask "what does this predictor add that other predictors fail to predict?"

To clarify, recall that, in Chap. 2, we originally used self-efficacy beliefs to predict performance without considering aptitude. Our regression coefficient was $b=.5213$. But if we enter self-efficacy beliefs into a predictive equation after first entering aptitude, its associated regression coefficient $=.1191$. This latter value matches the one reported earlier in this chapter when the two variables were entered together. Because each coefficient in a multiple regression analysis is adjusted for all of the other predictors, "entered together" is the same as "entered after," which is the same as "entered last."

Finally, let's consider why this regression coefficient changed from .5213 to .1191. Because self-efficacy beliefs and aptitude are correlated, much of the variance we initially attributed to self-efficacy beliefs is attributable to the overlap between the two predictors. In multiple regression, the regression coefficients represent only the unique contribution a variable makes to the prediction of a criterion, not its total contribution.

4.2.2.3   The Effect of One Variable Holding Other Variables Constant

A final way to define a regression coefficient is as follows: With all other predictors held constant, a regression coefficient specifies the expected one unit increase in y with every one unit increase in the predictor with which it is associated. Using our data, we would say "with aptitude held constant, a one unit increase in self-efficacy beliefs predicts a .1191 unit increase in expected class performance," and "with self-efficacy beliefs held constant, a one unit increase in aptitude predicts a .8219 unit increase in expected class performance."

Table 4.2 Fitted values of y holding one predictor constant

x_1	x_2	Fitted value	Change (within rounding error)
Holding x_2 constant and changing x_1			
1	2	2.3708	
2	2	2.4898	.1191
3	2	2.6089	.1191
1	3	3.1927	
2	3	3.3118	.1191
3	3	3.4308	.1191
Holding x_1 constant and changing x_2			
x_1	x_2	Fitted value	Change
2	1	1.6679	
2	2	2.4898	.8219
2	3	3.3118	.8219
3	1	1.7870	
3	2	2.6089	.8219
3	3	3.4308	.8219

Plugging some numbers into our regression equation and examining our fitted values will illustrate this interpretation. The top half of Table 4.2 shows what happens when we hold x_2 constant (first at 2, then at 3) and vary our values for x_1 (from 1 to 3).

The final column in the table shows the change from one fitted value to the next. Notice that, independent of the values we select for x_2, a one unit increase in x_1 produces a .1191 increase in the fitted value of y. Notice also that $b_1 = .1191$. This is no coincidence. A regression coefficient represents the expected change in y associated with a one unit change in a predictor holding all other predictors constant. So no matter what value we choose for x_2, the expected difference in y associated with a one unit change in x_1 will always equal b_1. The bottom half of Table 4.2 shows that a comparable pattern occurs when we hold x_1 constant and vary x_2.

Viewing a regression coefficient from this perspective leads to the following conclusion: if everyone had the same aptitude, a one unit increase in self-efficacy beliefs would predict a .1191 unit increase in performance, or, if everyone had the same self-efficacy beliefs, a one unit increase in aptitude would predict a .8219 increase in performance. This is what it means to say we are holding a variable constant.

Finally, notice the parallels between this description of a regression coefficient and a first-order partial derivative. In Chap. 3, we computed first-order partial derivatives by calculating the instantaneous rate of change in a variable as one input value changes and the remaining values stay constant. We do the same thing when calculating regression coefficients, so there is a close connection between the two terms.

4.2.3   Calculating Regression Coefficients

A final way to learn about a regression coefficient is to consider how it is calculated.

4.2.3.1   Algebraic Formula for Computing Multiple Regression Coefficients

In Chap. 2, we learned that with one predictor, we find a regression coefficient by creating a fraction with a covariance term in the numerator and the variance of the predictor in the denominator [see Eq. (2.17), repeated below].

$$b = \frac{s_{xy}}{s_x^2} \tag{4.2}$$

We use a similar approach with multiple regression, except now we take the variance of all predictors into account. Equation (4.3) shows the formula for finding b_1 with two predictors.

$$b_1 = \frac{(s_{1y} * s_2^2) - (s_{12} * s_{2y})}{(s_1^2 * s_2^2) - (s_{12})^2} \tag{4.3}$$

Applying Eq. (1.1), we can quickly compute the variances and covariances of our variables by creating a covariance matrix from their deviate scores.

$$\mathbf{S} = \begin{bmatrix} 8.5455 & 4.1818 & 4.4545 \\ 4.1818 & 5.9015 & 5.3485 \\ 4.4545 & 5.3485 & 6.1515 \end{bmatrix} = \begin{bmatrix} s_1^2 & s_{12} & s_{1y} \\ s_{21} & s_2^2 & s_{2y} \\ s_{y1} & s_{y2} & s_y^2 \end{bmatrix}$$

Inserting the relevant values into Eq. (4.3) produces the same solution we found using the matrix inverse method.

$$b_1 = \frac{(4.4545 * 5.9015) - (4.1818 * 5.3485)}{(8.5455 * 5.9015) - (4.1818)^2} = \frac{3.9220}{32.9438} = .1191$$

4.2.3.2   Regression Coefficients and Determinants

It might not be obvious, but Eq. (4.3) represents a fraction of two determinants from $\mathbf{S}$.[1] The denominator is the determinant of the predictors.

---

[1] As discussed in Chap. 1, these determinants are known as "minors."

$$\begin{vmatrix} s_1^2 & s_{12} \\ s_{21} & s_2^2 \end{vmatrix} = \begin{vmatrix} 8.5455 & 4.1818 \\ 4.1818 & 5.9015 \end{vmatrix} = 32.9438$$

The numerator in Eq. (4.3) is also a determinant if we follow Cramer's Rule. Recall from Chap. 1 that Cramer's Rule uses determinants to solve simultaneous linear equations by replacing a predictor column with a criterion column. In our case, we find the numerator of b_1 by replacing the first two rows of column 1 with the first two rows of column 3.

$$\begin{vmatrix} s_{1y} & s_{12} \\ s_{2y} & s_2^2 \end{vmatrix} = \begin{vmatrix} 4.4545 & 4.1818 \\ 5.3485 & 5.9015 \end{vmatrix} = 3.9220$$

Then we find our regression coefficient by dividing one determinant by another, just as we did using Cramer's Rule.

$$b_1 = \frac{\begin{vmatrix} s_{1y} & s_{12} \\ s_{2y} & s_2^2 \end{vmatrix}}{\begin{vmatrix} s_1^2 & s_{12} \\ s_{21} & s_2^2 \end{vmatrix}} = \frac{3.9220}{32.9438} = .1191$$

The formula for finding b_2 is similar.

$$b_2 = \frac{\left(s_{2y} * s_1^2\right) - \left(s_{12} * s_{1y}\right)}{\left(s_1^2 * s_2^2\right) - \left(s_{12}\right)^2} \tag{4.4}$$

Here, we find our numerator by replacing the first two rows of column 2 with the first two rows of column 3.

$$b_2 = \frac{\begin{vmatrix} s_1^2 & s_{1y} \\ s_{21} & s_{2y} \end{vmatrix}}{\begin{vmatrix} s_1^2 & s_{12} \\ s_{21} & s_2^2 \end{vmatrix}} = \frac{\begin{vmatrix} 8.5455 & 4.4545 \\ 4.1818 & 5.3485 \end{vmatrix}}{\begin{vmatrix} 8.5455 & 4.1818 \\ 4.1818 & 5.9015 \end{vmatrix}} = \frac{27.0778}{32.9438} = .8219$$

In sum, if we remember that a determinant is a measure of variability (i.e., the volume of a parallelepiped), we can see that, just as we did with a single predictor, we find a regression coefficient by creating a fraction with a cross-product term in the numerator and an index of the variability of the predictors in the denominator.

4.2.4   Testing the Significance of Regression Coefficients

Now that we understand how to interpret and compute regression coefficients we
will learn how to test their statistical significance.

4.2.4.1   Computing the Standard Errors

Testing the significance of a regression coefficient involves creating a fraction with
the coefficient in the numerator and its standard error in the denominator. With only
one predictor, we used the following formula to compute the standard error [see
Eq. (2.29), repeated below].

$$se_b = \sqrt{\frac{MS_{res}}{ss_x}} \tag{4.5}$$

With multiple predictors, we modify the formula by including a term in the
denominator that represents the multiple correlation among all of the predictors,
with the particular coefficient being tested designated as the criterion.

$$se_b = \sqrt{\frac{MS_{res}}{ss_{x_i} * \left(1 - R^2_{x_i}\right)}} \tag{4.6}$$

Looking over the formula, we see that the size of a standard error is inversely
related to the variability of the predictor, but positively related to the overlap among
the predictors. Consequently, a standard error will be relatively small when the
variability of its associated predictor is high and its covariance with the other
predictors is low. In fact, if our predictors are entirely uncorrelated, then the
denominator in Eq. (4.6) reduces to ss_x, which is the same term we use in
Eq. (4.5) with a single predictor.

   Returning now to our data, earlier we found that the correlation between the two
predictors was $r = .5889$. Each variable's deviation sum of squares is given in
Table 4.1 ($ss_{x1} = 94.00$ and $ss_{x2} = 64.9167$), so we can easily calculate our standard
errors,

$$se_{b1} = \sqrt{\frac{1.4974}{94.00 * \left(1 - .5889^2\right)}} = .1562$$

and

$$se_{b2} = \sqrt{\frac{1.4974}{64.9167 * \left(1 - .5889^2\right)}} = .1879$$

and t values.

$$t_{b_1} = \frac{.1191}{.1562} = .7624$$

and

$$t_{b_2} = \frac{.8219}{.1879} = 4.3741$$

After referring these values to a t-distribution with $(N - k - 1) = 9$ degrees of freedom, we find that the latter is significant but the former is not. Notice what has happened here. Independently, each predictor was significantly associated with the criterion. But because the two predictors are correlated, the regression weights are not the same as the correlations. In this case, math aptitude is a unique predictor of class performance, but self-efficacy beliefs are not.

4.2.4.2 Parameter Covariance Matrix (C)

As first discussed in Chap. 2, rather than computing standard errors one by one using ordinary algebra, it is faster to find them all at once using matrix algebra. The formula we use to find the parameter covariance matrix $\mathbf{C}$ is repeated below.

$$\mathbf{C} = \left(\mathbf{X'X}\right)^{-1} * MS_{res} \tag{4.7}$$

The diagonal entries of this matrix represent the variances of the bs, and the off-diagonals represent covariances among the bs. The standard errors are found by taking the square root of the diagonal elements (denoted c_{ii}).

$$se_b = \sqrt{c_{ii}} \tag{4.8}$$

We already calculated $\mathbf{X'X}$ when we first found our regression coefficients, so now we need to find its inverse,

$$\left(\mathbf{X'X}\right)^{-1} = \begin{bmatrix} .4408 & -.0305 & -.0465 \\ -.0304 & .0163 & -.1015 \\ -.0465 & -.01115 & .0236 \end{bmatrix}$$

and perform the relevant calculations.

$$\mathbf{C} = \begin{bmatrix} .4408 & -.0305 & -.0465 \\ -.0304 & .0163 & -.1015 \\ -.0465 & -.01115 & .0236 \end{bmatrix} * 1.4974 = \begin{bmatrix} .6600 & -.0456 & -.0696 \\ -.0456 & .0244 & -.0173 \\ -.0696 & -.0173 & .0353 \end{bmatrix}$$

We can then find our standard errors, including the standard error of the intercept, by taking the square root of the diagonal entries.

$$se_{b_0} = \sqrt{.6600} = .8124$$
$$se_{b_i} = \sqrt{.0244} = .1562$$
$$se_{b_2} = \sqrt{.0353} = .1879$$

The values of the last two terms match the ones we found with our earlier calculations, and the first term gives us the denominator we need to test the statistical significance of our intercept.

$$t_{b_0} = \frac{.6079}{.8124} = .7482$$

With 9 *df*, this term is not significant, so we cannot reject the null hypothesis that the true intercept $=0$ in the population.

4.2.4.3 Confidence Intervals

We now have all the information we need to construct confidence intervals around all three coefficients using procedures first described in Chap. 2 [see Eq. (2.31)]. Using a spreadsheet function (=TINV), we find that the critical *t*-value for 9 *df* at .05, two-tailed level of significance $=2.2622$. Plugging the value into our equation yields the confidence intervals for the intercept and our two regression coefficients.

$$CI_{b_0} = .6079 \pm \{2.2622 * .8124\} = -1.2299 \text{ and } 2.4457$$
$$CI_{b_1} = .1191 \pm \{2.2622 * .1562\} = -.2344 \text{ and } .4725$$
$$CI_{b_2} = .8219 \pm \{2.2622 * .1879\} = .3968 \text{ and } 1.2470$$

Only the final value doesn't include 0, providing another indication that we cannot reject the null hypothesis for the intercept and first regression coefficient, but we can reject the null hypothesis for the second regression coefficient.

4.2.5 *Forecasting*

As with simple linear regression, multiple regression is often used for prediction. In our example, we can predict performance in a matrix algebra class based on a student's self-efficacy beliefs and aptitude. The steps we use to make this prediction are similar to the ones described in Chap. 2, so we will move quickly through the procedures.

4.2.5.1 Expected Average Scores

Suppose we wish to calculate the average expected score of students with a self-efficacy value $=3$ and an aptitude value $=4$. We first create a column vector we will designate $\mathbf{p}$. The first value in the column vector will always equal 1 (to model the intercept), and the other two values will reflect the scores we have chosen for x_1 and x_2, respectively.

$$\mathbf{p} = \begin{bmatrix} 1 \\ 3 \\ 4 \end{bmatrix}$$

We then compute our fitted value using $\mathbf{p'b}$,

$$\hat{y} = \begin{bmatrix} 1 & 3 & 4 \end{bmatrix} \begin{bmatrix} .6079 \\ .1191 \\ .8219 \end{bmatrix} = 4.2527$$

and its standard error by taking the square root of $\mathbf{p'Cp}$.

$$se_{\hat{y}} = \sqrt{ \begin{bmatrix} 1 & 3 & 4 \end{bmatrix} \begin{bmatrix} .6600 & -.0456 & -.0696 \\ -.0456 & .0244 & -.0173 \\ -.0696 & -.0173 & .0353 \end{bmatrix} \begin{bmatrix} 1 \\ 3 \\ 4 \end{bmatrix} } = .4468$$

We can then form a 95 %, two-tailed confidence interval around the predicted value in the usual manner.

$$CI_{\hat{y}} = 4.2527 \pm \{2.2622 * .4468\} = 3.2420 \text{ and } 5.2635$$

These values tell us what to expect if we reran our study many times. To illustrate, if we conducted our 12-subject study 10,000 times, we expect that the average class rank of students with a score of 3 on the self-efficacy scale and a score of 4 on the aptitude scale would fall between 3.2420 and 5.2635 in 9,500 of them; the other 500 samples would (probably) produce average values outside this range.

4.2.5.2 Predicted Single Scores

If we want to predict a single score, rather than an average one, we change our formula for finding its standard error, using $(\mathbf{X'X})^{-1}$ rather than the covariance matrix [see Eq. (2.38), repeated below].

$$se_{\hat{y}} = \sqrt{ \left[1 + \left\{ \mathbf{p'} (\mathbf{X'X})^{-1} \mathbf{p} \right\} \right] * MS_{res} } \tag{4.9}$$

Plugging in our values produces the standard error for an individual score.

$$se_{\hat{y}} = \sqrt{\left[1 + \left\{[1 \quad 3 \quad 4] \begin{bmatrix} .4408 & -.0305 & -.0465 \\ -.0304 & .0163 & -.1015 \\ -.0465 & -.01115 & .0236 \end{bmatrix} \begin{bmatrix} 1 \\ 3 \\ 4 \end{bmatrix} \right\}\right]} * 1.4974 = 1.3027$$

If we compare the two standard errors, we can see that the standard error of the individual case is much larger than the standard error of the average case. As noted in Chap. 2, this makes sense because our ability to predict a single value is always more subject to error than is our ability to predict an average value across many instances. We can see this most graphically by using the standard error to construct a prediction interval around our predicted value.

$$PI_{\hat{y}} = 4.2527 \pm (2.2622 * 1.3027) = 1.3059 \text{ and } 7.1996$$

Here, we see that the width of the prediction interval far exceeds the width of the confidence interval.

4.2.6   Comparing the Predictors

Earlier we found that x_2 is a unique predictor of math performance but x_1 is not. Given these findings, it is tempting to conclude that x_2 is important but x_1 isn't, but this conclusion is not warranted. The fact that one coefficient is significant and the other is not provides scant evidence that one is more important than the other. After all, if one coefficient was significant at the .0499 level and the other was not significant at the .0501 level, it is unlikely that the difference between them would be due to anything other than chance variation.

Even when two unstandardized coefficients are of very different magnitudes, assessing their importance is fraught with interpretive difficulty. Because they are interdependent, we can't interpret one regression coefficient in isolation from the others. Self-efficacy might turn out to be a weak predictor of performance when paired with math aptitude, but a very strong predictor of performance when paired with some other variable or when considered alone.

The interdependence that characterizes regression coefficients has an important consequence. By choosing particular predictors, we can alter the magnitude of the other regression coefficients in our predictive equation. In our example, I added a variable that was correlated with self-efficacy beliefs and performance. As a result, self-efficacy beliefs became less important. Had I added a different variable, I could have magnified the importance of self-efficacy beliefs. So researchers have a lot of power in determining the magnitude of their coefficients, and we should only characterize the importance of a variable with reference to the particular set of variables with which it is embedded.

4.2.6.1 Standardized Coefficients

Keeping these cautions in mind, we will discuss two additional ways to compare the relative importance of our predictors. The first is to use standardized coefficients. In Chap. 2, we learned that we can calculate standardized coefficients using Eq. (2.16) (reproduced below).

$$\boldsymbol{\beta} = \left(\mathbf{ZX'ZX} \right)^{-1} \mathbf{ZX'Zy}$$

Applying the formula to our data yields our standardized coefficients.

$$\boldsymbol{\beta} = \begin{bmatrix} .1403 \\ .8051 \end{bmatrix}$$

Because all variables have been standardized, the comparative size of these coefficients is informative. With aptitude held constant, a standard deviation increase in self-efficacy beliefs predicts a .1403 standard deviation increase in performance; with self-efficacy beliefs held constant, a standard deviation increase in aptitude predicts a .8051 standard deviation increase in performance. From this perspective, aptitude is a better predictor of performance than is self-efficacy.

4.2.6.2 Directly Comparing Regression Coefficients

We can use the following formulae to directly test the difference between two unstandardized regression coefficients.

$$t = \frac{b_1 - b_2}{se_{(b_1 - b_2)}} \tag{4.10}$$

and

$$se_{(b_1 - b_2)} = \sqrt{c_{11} + c_{22} - 2c_{12}} \tag{4.11}$$

Remembering that the first column and row in $\mathbf{C}$ represent the intercept, we insert the values from the second and third diagonal elements and their cross product to derive our standard error.

$$se_{(b_1 - b_2)} = \sqrt{.0244 + .0353 - (2 * -.0173)} = .3070$$

and

$$t = \frac{.1191 - .8219}{.3070} = -2.2894, \; p < .05$$

With $N - k - 1$ degrees of freedom, the obtained value exceeds the significance threshold, indicating that the difference between the two coefficients is unlikely to be due to chance alone.

4.2.7   R Code: Testing Regression Coefficients

```
x1 <-c(1,9,1,5,6,8,2,4,2,8,7,7)
x2 <-c(2,7,1,8,5,7,4,7,1,4,3,4)
y <-c(3,8,2,8,5,9,4,5,2,4,2,6)
X <-cbind(1,x1,x2)
mod <-lm(y~x1+x2)

#Regress x1 on x2 and save residual
reg.1 <-lm(x2~x1); res2=(resid(reg.1))
unique2 <-lm(y~res2);summary(unique2)

#Regress x2 on x1 and save residual
reg.2 <-lm(x1~x2); res1=(resid(reg.2))
unique1 <-lm(y~res1);summary(unique1)

#Covariance Matrix and Standard Errors
SS_res <-sum(mod$resid^2)
df <-length(y)-ncol(X)
MS_res <-SS_res/df
C <-solve(t(X)%*%X)*MS_res
std.err <-cbind(sqrt(C[1,1]), sqrt(C[2,2]), sqrt(C[3,3]))
std.err

#Confidence Intervals
confint(mod)

#Forecasting with x1=3 and x2=4
p <-c(1,3,4)
yhat <-t(p)%*% coef(mod)
```

(continued)

4.2.7   R Code: Testing Regression Coefficients (continued)

```
#Forecasting Average Values
std.error.ave <-sqrt(t(p)%*%vcov(mod)%*%p);std.error.ave
t.crit <-abs(qt(.025,df))
CI.lo.ave <-yhat-(t.crit*std.error.ave )
CI.hi.ave <-yhat+(t.crit*std.error.ave )
CI.ave <-cbind(CI.lo.ave,CI.hi.ave );CI.ave

#Forecasting Individual Values
msres <-(sum((y-mod$fitted)^2)/df)
std.error.ind <- sqrt((1+t(p)%*%solve(t(X)%*%X)%*%p)*msres);
std.error.ind
CI.lo.ind <-yhat-(t.crit*std.error.ind)
CI.hi.ind <-yhat+(t.crit*std.error.ind)
CI.ind <-cbind(CI.lo.ind,CI.hi.ind);CI.ind

#Standardized Regression Coefficients
z1 <-scale(x1, center = T, scale = T)
z2 <-scale(x2, center = T, scale = T)
zy <-scale(y,center=T,scale=T)
Z <-cbind(z1,z2,zy)
summary(zmod <-lm(zy~z1+z2-1))

#Test difference between two coefficients
b.dif <-(mod$coef[2]-mod$coef[3])/(sqrt(vcov(mod)[2,2]+vcov(mod)
[3,3]-2*(vcov(mod)[2,3])))
b.dif
pt(b.dif,df)*2
```

4.3   Partitioning the Variance

Directly comparing regression coefficients provides one way of gauging their relative importance; another way is to partition the variance in y among them. Earlier we saw that $R^2_{Y.12}=.8008$, indicating that, in combination, self-efficacy and math aptitude predict $\sim 80\%$ of the variance in math performance. In this section, we will learn how to partition (divide) the amount of explained variance into three parts: (1) the variance uniquely predicted by x_1, (2) the variance uniquely predicted by x_2, and (3) the variance attributable to the combination of the two predictors.

Figure 4.1, known as a Ballentine Venn diagram, presents a way to visualize our task. The figure shows overlapping circles representing the variability of three variables: x_1 (self-efficacy), x_2 (math aptitude), and y (class performance). For illustrative purposes, we will assume that these are standardized variables, so the area of each circle $= 1$, and the overlapping areas represent their squared correlation.

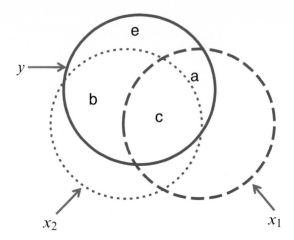

Fig. 4.1 Ballentine Venn diagram for partitioning the variance in y

The lowercase letters identify four sections of interest. The section labeled "e" represents the portion of y that is not predicted by either variable. In our example, this area is quite small ($1 - .8008 = .1992$). The section labeled "a" shows the portion of y that uniquely overlaps with x_1; the section labeled "b" shows the portion of y that uniquely overlaps with x_2; and the section labeled "c" shows the portion of y that overlaps with both predictors. If we add up sections a, b, and c, we get our value for $R^2_{Y.12}$, and if we then add section e, we get the total variance of 1.

In the following section, you will learn to quantify the area of these partitions by calculating two terms—a semipartial correlation and a partial correlation. Each term can also be squared, so you will want to be pay attention to when we are squaring each value.

4.3.1   Semipartial Correlation

A *squared* semipartial correlation provides an index of the unique contribution a variable makes to the prediction of y. In terms of our Ballentine diagram, areas "a" and "b" represent the squared semipartial correlations of x_1 and x_2, respectively.

4.3.1.1   Correlation Between a Residualized Predictor and a Raw Criterion

Although there are easier ways to calculate the term, I believe it is best to think of the squared semipartial correlation as a squared correlation between a residualized predictor and a raw criterion. This definition parallels one of the ways we defined a regression coefficient. Just as a regression coefficient represents the unique weight a variable is given in the prediction of a criterion, a squared semipartial correlation represents the unique variance in a criterion that is attributable to a predictor.

To better understand these parallels, consider the information shown in the left-hand side of Table 4.3. Here, we have the residualized variable we calculated earlier when we regressed x_2 on x_1. The values in the column labeled $x_{2.1}$ represent the portion of x_2 that x_1 cannot predict. Now let's calculate the correlation between this residual variable and our criterion by adapting Eq. (2.20).[2]

$$r_{y(i.k)} = \frac{s_{x_{i.k}\,y}}{s_{x_{i.k}} * s_y} \tag{4.12}$$

Plugging in our values yields the (unsquared) semipartial correlation.

$$r_{y(2.1)} = \frac{3.1686}{\sqrt{(3.8551)} * \sqrt{(6.1515)}} = .6507$$

If we square the semipartial correlation, we find the proportion of variance in y that can be uniquely attributed to x_2 ($sr^2 = .4234$). Section "b" in the Ballentine diagram represents this value. In our example, ~42 % of the variance in performance can be uniquely predicted by aptitude.

4.3.1.2 ΔR^2 and the Squared Semipartial Correlation

An equivalent way to view a squared semipartial correlation is to think of it as the change in R^2 that results from adding or eliminating one variable from the overall regression equation (denoted ΔR^2). Since we earlier found the semipartial correlation for x_2, we will illustrate this approach by finding the squared semipartial correlation for x_1 (see right-hand side of Table 4.3). The formula appears below, and the notation, $r^2_{y(1.2)}$, indicates that y is the criterion and we are partialing x_2 from x_1.

$$r^2_{y(1.2)} = R^2_{y.12} - R^2_{y.2} \tag{4.13}$$

Looking over the formula, we see that we first calculate the multiple R^2 that results when y is regressed on both predictors $R^2(R^2_{Y.12} = .8008)$. We then eliminate x_1 and calculate R^2 when we regress y on x_2 only. Because we know the zero-order correlation between y and $x_2 = .8877$, we can square the value to calculate the term: $R^2_{Y.2} = .7880$. Now we subtract the second term from the first one to derive the squared semipartial correlation for x_1.

---

[2] The notation $r_{y(i.k)}$ indicates that we are calculating the correlation between y and x_i with the variance of x_k removed from x_i.

Table 4.3 Residuals, regression coefficients, and semipartial and partial correlations

	Removing the variance attributable to x_1			Removing the variance attributable to x_2		
	$x_{2.1}$	y	$y_{.1}$	$x_{1.2}$	y	$y_{.2}$
	-.4592	3	.2518	-2.2875	3	.3569
	.6259	8	1.0816	2.1694	8	.8254
	-1.4592	2	-.7482	-1.5789	2	.2632
	3.5833	8	3.1667	-2.5392	8	-.0809
	.0940	5	-.3546	.5866	5	-.3620
	1.1152	9	2.6028	1.1694	9	1.8254
	1.0514	4	.7305	-2.7047	4	-.4557
	3.0727	5	.6879	-2.8306	5	-2.1746
	-1.9486	2	-1.2695	-.5789	2	.2632
	-1.8848	4	-2.3972	3.2953	4	-.4557
	-2.3954	2	-3.8759	3.0039	2	-1.5494
	-1.3954	6	.1241	2.2953	6	1.5443
Variance	**3.8551**	**6.1515**	**3.8295**	**5.5822**	**6.1515**	**1.3042**
Covariance		**3.1686**	**3.1686**		**.6646**	**.6646**
		Semipartial	Partial		Semipartial	Partial
	r	.6507	.8247	r	.1134	.2463
	$r^2 = \Delta R^2$	.4234	.6801	$r^2 = \Delta R^2$	.0129	.0607
	b	.8219	.8219	b	.1191	.1191

$$r^2_{y(1.2)} = .8008 - .7880 = .0129$$

The size of the section labeled "a" in Fig. 4.1 represents this value, indicating that less than 2 % of the variance in performance can be uniquely predicted by self-efficacy beliefs. If we take the square root of this value, we find the (unsquared) semipartial correlation.

$$r_{y(1.2)} = \sqrt{.0129} = .1134$$

We can then repeat these steps to calculate $r^2_{y(2.1)}$, except this time we eliminate x_2 from the regression equation. Since the zero-order correlation between x_1 and $y = .6144$, we can square the value to determine that $R^2_{Y.1} = .3775$ and find our squared semipartial correlation by subtraction.

$$r^2_{y(2.1)} = .8008 - .3775 = .4234$$

This value matches the value we found earlier using the residuals. That's because a squared semipartial correlation can be conceptualized in two, equivalent ways: as the squared correlation between a residualized predictor and a criterion, and as the change in R^2 that results from adding or subtracting a predictor from an overall regression equation.

Finally, because areas a, b, and c in Fig. 4.1 sum to $R^2_{Y.12}$, we can compute the size of "c" by subtraction.

$$c = .8008 - .0129 - .4234 = .3646$$

This value indicates that $\sim 36\%$ of the variance in performance is predictable from the overlap between our two predictors.

4.3.1.3   Tests of Statistical Significance

The easiest way to test the statistical significance of a squared semipartial correlation is to test the significance of the change in R^2 that arises when a variable is added or subtracted from the overall regression equation. Since we used the exclusion method earlier, we will use it to test the significance of our squared semipartial correlations.

$$F_{y(i.k)} = \frac{\left(R^2_{y.ik} - R^2_{y.k}\right)/\left(k_{y.ik} - k_{y.k}\right)}{\left(1 - R^2_{y.ik}\right)/\left(N - k_{y.ik} - 1\right)} \tag{4.14}$$

Substituting our values, we obtain the following values for our two semipartial correlations.

$$F_{y(1.2)} = \frac{(.8008 - .7880)/(2 - 1)}{(1 - .8008)/(12 - 2 - 1)} = .5813$$

and

$$F_{y(2.1)} = \frac{(.8008 - .3775)/(2 - 1)}{(1 - .8008)/(12 - 2 - 1)} = 19.1323$$

With 1 and 9 df, the former value is not significant, but the latter value is. If you take the square root of these F values, you will find that the result matches the t values we found when we tested the b coefficients using their standard errors. This will always be true: the test of a semipartial correlation (squared or not) is identical to the test of the b coefficient with which it is associated; both tests assess the unique contribution of a variable to the overall prediction of y.

4.3.1.4 Using Zero-Order Correlations to Calculate Semipartial Correlations

There are other ways to think about and compute a semipartial correlation and its square. Later we are going to learn how to calculate these terms and several others using matrix algebra; but for now, we'll note that we can also compute the semipartial correlation using zero-order correlations.

$$r_{y(i.k)} = \frac{r_{yi} - r_{ik} * r_{yk}}{\sqrt{1 - r_{ik}^2}} \tag{4.15}$$

These calculations are straightforward, but you might want to perform them to verify that they also produce the semipartial correlation.

4.3.2 Partial Correlation

So far, you have learned that a squared semipartial correlation quantifies the unique association between a predictor and a criterion. In terms of our example, the value we obtain for $r_{y(1.2)}^2$ answers the question: apart from the variance it shares with self-efficacy beliefs, does aptitude predict math performance? A slightly different method of partitioning the variance is to calculate a squared partial correlation. Instead of finding the variance in y that is uniquely predicted by x_i, a squared partial correlation finds the *unique* variance in y that is uniquely predicted by x_i. In other words, the question becomes "Does x_i predict y once x_k is statistically removed from both variables?" To denote the squared partial correlation, we remove the parentheses in the subscript, indicating that the control variable is partialed from the criterion and the predictor (e.g., $r_{y1.2}^2$ represents the squared partial correlation between x_1 and y, controlling for x_2).

4.3.2.1 Correlation Between a Residualized Predictor and a Residualized Criterion

As before, there are several ways to calculate a partial correlation and its square. Although not the easiest, the most clarifying approach is to revisit the role of residuals. Whereas a semipartial correlation represents the correlation between a residualized predictor and a raw criterion, the partial correlation represents the correlation between a residualized predictor and a residualized criterion. To illustrate, if we wanted to calculate the partial correlation between x_2 and y after controlling for x_1, we would first regress x_2 on x_1 (and save the residuals) and then regress y on x_1 (and save the residuals). Finally, we correlate the two residuals to calculate our partial correlation. If we square this value, we get the squared partial correlation.

The left-hand side of Table 4.3 provides the relevant values. Using our variance/covariance method, we find the partial correlation for x_2.

$$r_{y2.1} = \frac{s_{x_{2.1}\ y.1}}{s_{x_{2.1}} * s_{y.1}} = \frac{3.1686}{\sqrt{(3.8551)} * \sqrt{(3.8295)}} = .8247$$

If we then square this value, we find the squared partial correlation.

$$r_{y2.1}^2 = .8247^2 = .6801$$

4.3.2.2 ΔR^2 and the Squared Partial Correlation

We can also find a squared partial correlation by using changes in R^2.

$$r_{yi.k}^2 = \frac{R_{y.ik}^2 - R_{y.k}^2}{1 - R_{y.k}^2} \tag{4.16}$$

Here, we see that the numerator is the same term we used when calculating the squared semipartial correlation, but we have now added a denominator. The denominator is a residual, representing the portion of y that cannot be explained by the predictor we are controlling. As long as our control variable is at least somewhat correlated with our criterion, the denominator will be less than 1, and the value of a partial correlation will be larger than the value of a semipartial correlation. Inserting our values yields the squared partial correlations.

$$r^2_{y1.2} = \frac{.8008 - .7880}{1 - .7880} = .0607$$

and

$$r^2_{y2.1} = \frac{.8008 - .3775}{1 - .3775} = .6801$$

4.3.2.3   Using Zero-Order Correlations to Calculate Partial Correlations

Finally, we can calculate the partial correlations by taking the square root of the values we just calculated or by using the zero-order correlations.

$$r_{yi.k} = \frac{r_{yi} - r_{ik} * r_{yk}}{\sqrt{1 - r^2_{ik}} \sqrt{1 - r^2_{yk}}} \qquad (4.17)$$

4.3.2.4   Tests of Significance

Now that you have learned so many ways to calculate a partial correlation, you are probably wondering how you test its significance. Curiously, you don't need to. The test of a partial correlation yields the same value as the test of a semipartial correlation, and this value is the same as a test of an unstandardized regression coefficient, a standardized regression coefficient, and the change in R^2. Yes, you read correctly. All of the coefficients we have calculated—unstandardized and standardized regression coefficients, semipartial and partial correlations and their squares, and changes in R^2—have exactly the same statistical significance.

Given these equivalences, you might be wondering why we need all of these terms (or at least why you are being asked to learn them all!). The reason is that they illuminate different issues.

- A regression coefficient represents the unique weight of each predictor in the prediction of a criterion. With unstandardized regression coefficients, the weight is described in raw units, and with standardized regression coefficients, the weight is described in standardized units.
- The change in R^2 and squared semipartial correlation shows the proportion of variance in a raw criterion that is uniquely predicted by each variable.
- The squared partial correlation describes the proportion of variance in a residualized criterion that is uniquely predicted by each variable.

In short, the different terms provide overlapping, but not entirely redundant, information regarding the association between the predictors and a criterion.

4.3.3 Are Regression Coefficients Semipartial Coefficients or Partial Coefficients?

Finally, let's return to a consideration of a regression coefficient. In one interpretation, we noted that a regression coefficient represents the weight given to a residualized predictor in the prediction of a raw criterion. This sounds like a *semipartial* coefficient, as here we have a residualized predictor and a raw criterion. But we also noted that a regression coefficient represents the weight given to a predictor after all other predictors have been taken into account. This sounds like a *partial* coefficient, as here we have a residualized predictor and a residualized criterion. So which is the right way to think about a regression coefficient—as a semipartial coefficient or as a partial coefficient?

Interestingly, both descriptions are accurate because they yield the same value. To see why, return to the information presented at the bottom of Table 4.3. Here, we see that the covariance between $x_{2.1}$ and y equals the covariance between $x_{2.1}$ and $y_{.1}$ (both values = 3.1686). Now remember that a regression coefficient is calculated using the covariance in the numerator and the variance of the predictor in the denominator. Since a semipartial and partial regression coefficient use the same (residualized) predictor, they yield the same value.

$$b = \frac{s_{xy}}{s_x^2} = \frac{3.1686}{3.8551} = .8219$$

For this reason, we can refer to a regression coefficient as a semipartial coefficient or as a partial coefficient. My preference is to think of it as a semipartial coefficient. My reasoning is this: If you get the same value using a raw criterion as a residualized one, then you don't need to residualize the criterion to understand the effect. So I refer to these coefficients as semipartial coefficients rather than partial ones.

4.3.4 R Code: Partitioning the Variance

```
x1 <-c(1,9,1,5,6,8,2,4,2,8,7,7)
x2 <-c(2,7,1,8,5,7,4,7,1,4,3,4)
y <-c(3,8,2,8,5,9,4,5,2,4,2,6)
X <-cbind(1,x1,x2)
mod <-lm(y~x1+x2)
rsqr.mod <-summary(mod)$r.squared

#Regress x1 on x2 and save residual
reg.1 <-lm(x2~x1)
summary(x1only <-lm(y~ resid(reg.1)))
```

(continued)

4.3.4 R Code: Partitioning the Variance (continued)

```
#Regress x2 on x1 and save residual
reg.2 <-lm(x1~x2)
summary(x2only <-lm(y~ resid(reg.2)))

#Semi-partial correlations
semi.1 <-cor (resid(reg.1),y)
semi.2 <-cor (resid(reg.2),y)

#Squared semi-partial correlations
semi.1.sqr <-semi.1^2
semi.2.sqr <-semi.2^2

#Squared semi-partials as changes in R^2
just1 <-lm(y~x1)
just2 <-lm(y~x2)
anova(mod,just1,test="F")
anova(mod,just2,test="F")
rsqr.just1 <-summary(just1)$r.squared
rsqr.just2 <-summary(just2)$r.squared
x1.adds <-rsqr.mod-rsqr.just1
x2.adds <-rsqr.mod-rsqr.just2
allsemi <-cbind(semi.1, semi.1.sqr, x1.adds,semi.2,semi.2.sqr,x2.adds)
allsemi

#Partial Correlations
#regress y on x1 and save residual
reg.y1 <-lm(y~x1)
#regress y on x2 and save residual
reg.y2 <-lm(y~x2)

partial.1 <-cor (resid(reg.1),resid(reg.y1))
partial.2 <-cor (resid(reg.2), resid(reg.y2))
partial.1.sqr <-partial.1^2
partial.2.sqr <-partial.2^2
allpartial <-cbind(partial.1, partial.1.sqr, partial.2,partial.2.sqr)
allpartial
```

4.4   Calculating Regression Coefficients Using Cofactors

In this chapter, we have discussed numerous ways of calculating and testing the distribution of variance among our variables. Many of the calculations help us understand the logic of the operations we are performing, but remembering all of the formulae is difficult. Fortunately, all of these calculations can be made using a few simple matrix operations. Moreover, the matrix method can handle any number of predictors, so it provides a more efficient way to calculate the various terms.

4.4.1   Complete Sum of Squares

We begin by creating a matrix we will designate $\mathbf{V}$ because it is comprised of all of our raw variables (including the leading 1's). For clarity, Table 4.4 reproduces the data from Table 4.1, with the leading 1's included.

Table 4.4 Raw variables including leading 1's

	Self-efficacy x_1	Aptitude x_2	Performance y
1	1	2	3
1	9	7	8
1	1	1	2
1	5	8	8
1	6	5	5
1	8	7	9
1	2	4	4
1	4	7	5
1	2	1	2
1	8	4	4
1	7	3	2
1	7	4	6

4.4.1.1   Compute $\mathbf{V'V}$

When we compute $\mathbf{V'V}$, we get a symmetric matrix, with squared sums on the diagonals and cross-product sums on the off-diagonals. Because we have used a column of leading 1's, the cross-product sums for the first row and column represent the sum of each variable.

$$\mathbf{V'V} = \begin{bmatrix} 12 & 60 & 53 & 58 \\ 60 & 394 & 311 & 339 \\ 53 & 311 & 299 & 315 \\ 58 & 339 & 315 & 348 \end{bmatrix} = \begin{bmatrix} N & \Sigma X_1 & \Sigma X_2 & \Sigma Y \\ \Sigma X_1 & \Sigma X_1^2 & \Sigma X_1 X_2 & \Sigma X_1 Y \\ \Sigma X_2 & \Sigma X_2 X_1 & \Sigma X_2^2 & \Sigma X_2 Y \\ \Sigma Y & \Sigma Y X_1 & \Sigma Y X_2 & \Sigma Y^2 \end{bmatrix}$$

We will use the following formulae to extract each variable's deviation sum of squares.[3]

$$ss = \Sigma X^2 - \frac{(\Sigma X)^2}{N} \tag{4.18}$$

Plugging in the data from our example yields the following values.

$$ss_{x1} = 394 - \frac{60^2}{12} = 94.00$$

$$ss_{x2} = 299 - \frac{53^2}{12} = 64.9168$$

and

$$ss_y = 348 - \frac{58^2}{12} = 67.6667$$

4.4.1.2   Calculate the Determinant of V'V

Our next step is to calculate the determinant of $\mathbf{V'V}$, which we can easily do using a spreadsheet function (=MDETERM).

$$|\mathbf{V'V}| = \begin{vmatrix} 12 & 60 & 53 & 58 \\ 60 & 394 & 311 & 339 \\ 53 & 311 & 299 & 315 \\ 58 & 339 & 315 & 348 \end{vmatrix} = 644621$$

4.4.1.3   Calculate Cofactors

Finally, we need to find nine cofactors from two adjugate matrices. Six of the cofactors come from the adjugate of $\mathbf{V'V}$, and the other three come from the adjugate of $\mathbf{X'X}$. Most spreadsheets don't come with a function for calculating the adjugate, but since an inverse is found by dividing the adjugate by the determinant, we can find the adjugate by multiplying the inverse by the determinant

---

[3] The spreadsheet function (=DEVSQ) also produces the deviation sum of squares.

Table 4.5 Matrices needed for a complete regression analysis

V'V				X'X		
12	60	53	58	12	60	53
60	394	311	339	60	394	311
53	311	299	315	53	311	299
58	339	315	348	Determinant = **47834**		
ss	94	64.9167	67.6667			
		Determinant = **644621**				
v = Adjugate (**V'V**)				x = Adjugate (**X'X**)		
301821	−16173	−6045	−29077	21085	−1457	−2222
−16173	11176	−2758	−5695	−1457	779	−552
−6045	−2758	47516	−39316	−2222	−552	1128
−29077	−5695	−39316	47834			
v_{11}	v_{12}	v_{13}	v_{14}	x_{11}	x_{12}	x_{13}
v_{21}	v_{22}	v_{23}	v_{24}	x_{21}	x_{22}	x_{23}
v_{31}	v_{32}	v_{33}	v_{34}	x_{31}	x_{32}	x_{33}
v_{41}	v_{42}	v_{43}	v_{44}			

(adjugate = MINVERSE*MDETERM). Table 4.5 presents the adjugate matrices we need, and the nine cofactors of interest are shaded to help you identify them.

Before we learn how to use these cofactors to find our regression coefficients, let's review how they are calculated. In Chap. 1, we learned that a cofactor is found by computing the determinant of a submatrix formed from a larger matrix (called a minor) and then calculating the sign of the cofactor using Eq. (1.9) (reproduced below).

$$f_{ij} = \begin{cases} m_{ij} & \dots & (i+j) \text{ is even} \\ -m_{ij} & \dots & (i+j) \text{ is odd} \end{cases} \tag{4.19}$$

To illustrate, the minor, v_{14}, represents the determinant of a matrix formed by eliminating the first row and fourth column of **V'V**.

$$v_{14} = \begin{vmatrix} 12 & 60 & 53 & 58 \\ 60 & 394 & 311 & 339 \\ 53 & 311 & 299 & 315 \\ 58 & 339 & 315 & 348 \end{vmatrix} = \begin{vmatrix} 60 & 394 & 311 \\ 53 & 311 & 299 \\ 58 & 339 & 315 \end{vmatrix} = 29077$$

When we apply the rule above, we see that $(1+4)$ is an odd number, so we change the sign.

$$v_{14} = -29077$$

We use a similar logic to calculate the other values in the final column of our adjugate matrix.

$$m_{24} = \begin{vmatrix} 12 & 60 & 53 & 58 \\ 60 & 394 & 311 & 339 \\ 53 & 311 & 299 & 315 \\ 58 & 339 & 315 & 348 \end{vmatrix} = \begin{vmatrix} 12 & 60 & 53 \\ 53 & 311 & 299 \\ 58 & 339 & 315 \end{vmatrix} = -5695$$

We do not change the sign here, because $(2+4)$ is an even number.

$$v_{24} = -5695$$

Continuing on, we will compute the other two cofactors.

$$m_{34} = \begin{vmatrix} 12 & 60 & 53 & 58 \\ 60 & 394 & 311 & 339 \\ 53 & 311 & 299 & 315 \\ 58 & 339 & 315 & 348 \end{vmatrix} = \begin{vmatrix} 12 & 60 & 53 \\ 60 & 394 & 311 \\ 58 & 339 & 315 \end{vmatrix} = 39316$$

$$v_{34} = -39316$$

$$m_{44} = \begin{vmatrix} 12 & 60 & 53 & 58 \\ 60 & 394 & 311 & 339 \\ 53 & 311 & 299 & 315 \\ 58 & 339 & 315 & 348 \end{vmatrix} = \begin{vmatrix} 12 & 60 & 53 \\ 60 & 394 & 311 \\ 53 & 311 & 299 \end{vmatrix} = 47834$$

$$v_{44} = 47834$$

Notice that this last cofactor represents the determinant of the predictors. Consequently, we can designate this value as v_{44} or $|\mathbf{X}'\mathbf{X}|$. As you will see, this value plays a role in all of the calculations you are going to learn.

4.4.2   Residual Sum of Squares and Coefficient of Determination

We will begin by calculating the residual sum of squares—the portion of the variance in y that our predictors cannot explain.

$$SS_{res} = \frac{|\mathbf{V}'\mathbf{V}|}{|\mathbf{X}'\mathbf{X}|} \tag{4.20}$$

Plugging in our values, we find.

$$SS_{res} = \frac{644621}{47834} = 13.4762$$

If you look back, you will see that this is the same value we found earlier by summing the squared residuals.

It's instructive to think about the operations here. Determinants are measures of variability (i.e., the volume of a parallelepiped). The $\mathbf{V'V}$ determinant will always be greater than the $\mathbf{X'X}$ determinant, but if the two are similar then the residual sum of squares will be small and the regression sum of squares will be correspondingly large. So we can gauge how well our predictors perform by forming a fraction with the overall determinant in the numerator and the determinant of the predictors in the denominator.

We can also find our coefficient of determination by using a slightly modified version of Eq. (4.20).

$$R^2 = 1 - \frac{|\mathbf{V'V}|}{|\mathbf{X'X}| * SS_y} \tag{4.21}$$

Inserting our values, we find our multiple squared correlation.

$$R^2 = 1 - \frac{644621}{47834 * 67.6667} = .8008$$

We can then easily compute our F value.

$$F = \frac{.8008/2}{(1 - .8008)/(12 - 2 - 1)} = 18.0954$$

In short, we extract a great deal of information by dividing the determinant of all variables by the determinant of the predictors.

4.4.3 Regression Coefficients

Table 4.6 shows that we can also use the cofactors and sum of squares to compute all of the regression coefficients and correlations we have covered in this chapter.

Table 4.6 Using cofactors and sum of squares to find regression coefficients and correlations

Formula	x_1	x_2
$b_i = \frac{-v_{ij}}{v_{jj}}$	$b_1 = \frac{5695}{47834} = .1191$	$b_2 = \frac{39316}{47834} = .8219$
$\beta_i = \frac{-v_{ij} * \sqrt{SS_{x_i}}}{v_{jj} * \sqrt{SS_y}}$	$\beta_1 = \frac{5695 * \sqrt{94}}{47834 * \sqrt{67.6667}} = .1403$	$\beta_2 = \frac{39316 * \sqrt{64.9167}}{47834 * \sqrt{67.6667}} = .8051$
$sr_i = \frac{-v_{ij}}{\sqrt{v_{jj} * x_{ii} * SS_y}}$	$sr_1 = \frac{5695}{\sqrt{47834 * 779 * 67.6667}} = .1134$	$sr_2 = \frac{39316}{\sqrt{47834 * 1128 * 67.6667}} = .6507$
$pr_i = \frac{-v_{ij}}{\sqrt{v_{jj} * v_{ii}}}$	$pr_1 = \frac{5695}{\sqrt{47834 * 11176}} = .2463$	$pr_2 = \frac{39316}{\sqrt{47834 * 47516}} = .8247$
$se_{b_i} = \sqrt{MS_e * \frac{x_{ii}}{v_{jj}}}$	$se_{b_1} = \sqrt{1.4974 * \frac{779}{47834}} = .1562$	$se_{b_2} = \sqrt{1.4974 * \frac{1128}{47834}} = .1879$

4.4.3.1 Unstandardized Regression Coefficients

The steps used to calculate the various terms in Table 4.6 are similar, so I will use the unstandardized regression coefficient to illustrate them,

$$b = \frac{-v_{ij}}{v_{jj}} \tag{4.22}$$

with v_{jj} denoting the final column entry. As you can see, our formula involves taking one entry from the last column of the adjugate matrix, changing its sign, and dividing it by the final entry in our adjugate matrix (which represents the determinant of the predictors). Performing the calculations produces our regression coefficients.[4]

$$b_1 = \frac{-v_{24}}{v_{44}} = \frac{-(-5695)}{47834} = .1191$$

and

$$b_2 = \frac{-v_{34}}{v_{44}} = \frac{-(-39316)}{47834} = .8219$$

If you look back to the beginning of the chapter, you will see that these values match the ones we found earlier.

4.4.3.2 Cramer's Rule and the Cofactor Matrix

Unsurprisingly, the preceding calculations represent another illustration of Cramer's Rule. Recall that Cramer's Rule uses the determinant of the predictors as a denominator. This term (v_{44}) appears in the denominator of Eq. (4.22). We find the numerator using Cramer's Rule by replacing one of the columns in the original matrix with the values in the criterion column. This is what we have done for each of our regression coefficients, changing the sign when needed according to Eq. (4.19).

$$b_1 = \begin{vmatrix} 12 & 58 & 53 \\ 60 & 339 & 311 \\ 53 & 315 & 299 \end{vmatrix} = 5695$$

and

---

[4] Because the first row in the adjugate matrix represents the intercept, we use the second row for the b_1 coefficients and the third row for the b_2 coefficients.

$$b_2 = \begin{vmatrix} 12 & 60 & 58 \\ 60 & 394 & 339 \\ 53 & 311 & 315 \end{vmatrix} = 39316$$

All we need to do to find our regression coefficients, then, is change the sign of each cofactor before dividing it by the determinant of the predictors. So it's not magical. The cofactor matrix simply provides an efficient way of using Cramer's Rule to perform the necessary calculations.

4.4.4 Computing the Remaining Coefficients and Correlations

Computing the standardized regression coefficients and partial and semipartial correlations is a straightforward extension of the procedures we used to find our unstandardized regression coefficients. Table 4.6 provides the necessary formulae and calculations, and you should take the time to verify that they produce the same values we found earlier in this chapter.

4.4.5 Summary

Using a few cofactors and the deviation sum of squares, we were able to test our overall regression model and calculate all of the coefficient terms we have covered in this chapter. I don't expect you to memorize all of the formulae, but you should take a close look at how they are constructed. They all involve creating a fraction of determinants, with a cross-product term in the numerator and the determinant of the predictors in the denominator. Since the determinant is an index of variability, we are always comparing the variability of a cross-product term with the variability of our predictors.

4.4.6 R Code: Regression Coefficients as Cofactors

```
x1 <-c(1,9,1,5,6,8,2,4,2,8,7,7)
x2 <-c(2,7,1,8,5,7,4,7,1,4,3,4)
y <-c(3,8,2,8,5,9,4,5,2,4,2,6)
V <-cbind(1,x1,x2,y)
VV <-t(V)%*%V
X =cbind(1,x1,x2)
XX =t(X)%*%X
```

(continued)

4.4.6   R Code: Regression Coefficients as Cofactors (continued)

```
Vadj <-solve(VV)*det(VV); Vadj
Xadj <-solve(XX)*det(XX); Xadj
df = length(y)-1

#Compute Sum of Squares and Coefficient of Determination
SS.res <-det(VV)/det(XX); SS.res
SS.tot <-var(y)*df
SS.reg <-SS.tot-SS.res
R2 <-1-(det(VV)/(det(XX)*SS.tot)); R2

#Coefficients for x1
b.1 <- -Vadj[2,4]/det(XX)
beta.1 <- -Vadj[2,4]*sqrt(var(x1)*df)/(det(XX)*sqrt(SS.tot))
semi.1 <- -Vadj[2,4]/sqrt(det(XX)*(SS.tot)*Xadj[2,2])
partial.1 <-  -Vadj[2,4]/sqrt(det(XX)*Vadj[2,2])
stderr.1 <- sqrt(SS.res/9*(Xadj[2,2]/Vadj[4,4]))
all.1 <-cbind(b.1,beta.1,semi.1,partial.1,stderr.1)
all.1

#Coefficients for x2
b.2 <- -Vadj[3,4]/det(XX)
beta.2 <- -Vadj[3,4]*sqrt(var(x2)*df)/(det(XX)*sqrt(SS.tot))
semi.2 <- -Vadj[3,4]/sqrt(det(XX)*(SS.tot)*Xadj[3,3])
partial.2 <-  -Vadj[3,4]/sqrt(det(XX)*Vadj[3,3])
stderr.2 <- sqrt(SS.res/9*(Xadj[3,3]/Vadj[4,4]))
all.2 <-cbind(b.2,beta.2,semi.2,partial.2,stderr.2)
all.2
```

4.5   Chapter Summary

1. Multiple regression involves solving a series of linear equations with multiple unknown quantities. As with simple linear regression, solving the normal equations using ordinary least squares estimation yields the best estimate of our population parameters.
2. Regression coefficients are characterized by interdependence. Consequently, a change in any of the variables used to predict a criterion will ordinarily produce changes in the value of all of the coefficients in the equation.
3. There are three ways to conceptualize a regression coefficient:

   3a. It represents the unique contribution of each predictor to the prediction of a criterion.

3b. It represents the contribution of a predictor to a criterion after all other predictors have been taken into account.

3c. Holding all other predictors constant, it represents the one unit change in a criterion with a one unit change in a predictor.

4. The size of a standard error is inversely related to the variance of the predictor and positively related to the overlap among the predictors. It can be found using ordinary algebra or by taking the square root of the diagonal entries of the parameter covariance matrix.

5. The regression equation can be used to predict future values. The standard error we use to construct confidence intervals around those values depends on whether we are predicting an average fitted value or an individual one.

6. With two predictors, the variance in a criterion can be partitioned into three parts: the part uniquely due to variable 1, the part uniquely due to variable 2, and the part that is due to the overlap between the two predictors.

7. A semipartial correlation represents the correlation between a residualized predictor and a raw criterion. A squared semipartial correlation represents the proportion of the variance in a raw criterion that is uniquely due to a residualized predictor. A squared semipartial correlation also represents the change in R^2 that is produced when a variable is added or eliminated from a regression analysis.

8. A partial correlation represents the correlation between a residualized predictor and a residualized criterion. A squared partial correlation represents the proportion of the variance in a residualized criterion that is uniquely due to a residualized predictor.

9. All of the regression coefficients and correlations can be found from two adjugate matrices and each variable's deviation sum of squares. There are many formulae to learn, but they are all fractions, with the determinant of a cross-product term in the numerator and the determinant of the predictors in the denominator. In short, all terms assess the variability between a predictor and a criterion, relative to the variability among the predictor themselves.

Chapter 5
Matrix Decompositions

Multiple regression finds a fitted value for a criterion from a linear combination of the predictors. But suppose we have a collection of variables without a criterion. This state of affairs characterizes our design matrix, $\mathbf{X}$, as none of the predictors is a criterion. Is there a way to create a linear combination of these variables? There is, but it's not as simple as predicting to a criterion. To understand how it's done, we take up the study of matrix decompositions. There are many varieties, but all decompose a matrix into two or more smaller matrices. Their value is twofold: they highlight variables that share common variance, and they offer computationally efficient ways of solving linear equations and performing least squares estimation.

5.1 Eigen Decomposition

The first decomposition method we will consider is the eigen decomposition.[1] Here, a square matrix $\mathbf{A}$ is decomposed into a set of eigenvalues (λ) and eigenvectors ($\mathbf{v}$). When the eigenvalues are stored in a diagonal matrix $\mathbf{\Lambda}$ and the eigenvectors are consolidated into a matrix $\mathbf{V}$, the eigenpairs (as they are collectively known) reproduce the original matrix.

$$\mathbf{A} = \mathbf{V}\mathbf{\Lambda}\mathbf{V}^{-1} \tag{5.1}$$

Eigen decomposition is used extensively in many areas of science, technology, engineering, and mathematics. In this text, we focus on their use in statistical

---

Electronic Supplementary Material: The online version of this chapter (doi: 10.1007/978-3-319-11734-8_5) contains supplementary material, which is available to authorized users

[1] Eigen decomposition is sometimes referred to as spectral decomposition.

© Springer International Publishing Switzerland 2014
J.D. Brown, *Linear Models in Matrix Form*, DOI 10.1007/978-3-319-11734-8_5

analyses, omitting many important topics and glossing over some important details. Readers interested in a more thorough treatment should consult additional sources (e.g., Stewart 2001; Strang 2009).

5.1.1   Matrix Multiplication with an "Ordinary" Vector

We begin our discussion of eigen decomposition by reviewing some basic properties of matrix multiplication. Consider the following equation.

$$7 * \begin{bmatrix} 2 \\ 6 \end{bmatrix} = \begin{bmatrix} 14 \\ 42 \end{bmatrix}$$

In this equation, we multiply a vector by a scalar. Notice that the product vector represents a linear transformation of the original vector (i.e., each value in the product vector is 7* the value in the first vector). We can see this linearity clearly in Panel A of Fig. 5.1. Notice that the product vector simply extends the original vector. Formally, we say that scalar multiplication produces a linear transformation of our original vector.

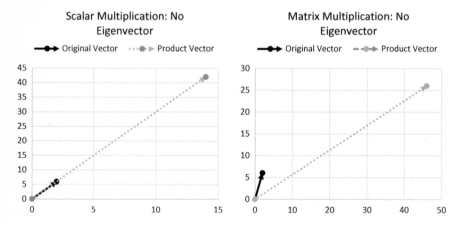

Fig. 5.1 Scalar and matrix multiplication using an "ordinary" vector

Now let's consider matrix multiplication. Below, we premultiply the same column vector by a matrix.

$$\begin{bmatrix} 5 & 6 \\ 1 & 4 \end{bmatrix} \begin{bmatrix} 2 \\ 6 \end{bmatrix} = \begin{bmatrix} 46 \\ 26 \end{bmatrix}$$

In this case, the product vector does not constitute a linear transformation of the original column vector (see Panel B of Fig. 5.1). Instead, the product vector changes length and direction.

5.1.2 Matrix Multiplication with an Eigenvector

So far, we have seen that only a scalar multiplication produces a linear transformation of a vector. There is an exception, however. When a vector is an *eigenvector* of a matrix, matrix multiplication also produces a linear transformation of the vector.[2] To demonstrate, let's redo our calculations with a different vector.

$$\begin{bmatrix} 6 \\ 2 \end{bmatrix}$$

First, we will perform our scalar multiplication and confirm that our new product vector still produces a linear transformation of the original vector (see Panel A in Fig. 5.2).

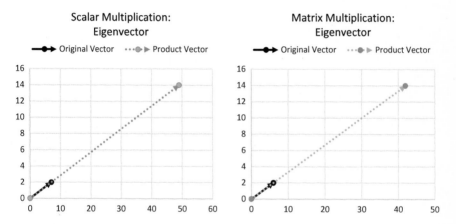

Fig. 5.2 Scalar and matrix multiplication using an eigenvector

$$7 * \begin{bmatrix} 6 \\ 2 \end{bmatrix} = \begin{bmatrix} 42 \\ 14 \end{bmatrix}$$

This is not surprising because scalar multiplication always produces a linear transformation. Now let's perform our matrix multiplication with this new vector.

$$\begin{bmatrix} 5 & 6 \\ 1 & 4 \end{bmatrix} \begin{bmatrix} 6 \\ 2 \end{bmatrix} = \begin{bmatrix} 42 \\ 14 \end{bmatrix}$$

---

[2] The German word eigen is loosely translated to mean "of one's own."

Here, we do find something surprising, as the product vector is also a linear transformation of the original vector (see Panel B in Fig. 5.2). Moreover, the product vector is again 7* the original vector.

To summarize:

- Scalar multiplication of a vector always produces a linear transformation, but matrix multiplication of a vector does so only when a specific vector, called an eigenvector, is used.
- The values of the eigenvector vary with the composition of the premultiplication matrix. Consequently, $\begin{bmatrix} 6 \\ 2 \end{bmatrix}$ is not always an eigenvector; it's an eigenvector only when used with matrices that have properties similar to our premultiplication matrix $\begin{bmatrix} 5 & 6 \\ 1 & 4 \end{bmatrix}$.
- Note, however, that any scalar multiple of an eigenvector is also an eigenvector. So we could divide each element of our eigenvector by 3 $\begin{bmatrix} 2 \\ .6667 \end{bmatrix}$ or multiply each entry by 2,000 $\begin{bmatrix} 12,000 \\ 4,000 \end{bmatrix}$ and still observe a linear transformation.
- Finally, in our example, the linear transformation we observe will always be 7* the eigenvector. This is because 7 is an eigen*value* for this premultiplication matrix and eigenvector. We denote the eigenvalue with the Greek letter lambda (λ) and define an eigenvector as any vector that satisfies the following equality, where $\mathbf{A}$ designates the premultiplication matrix, $\mathbf{v}$ designates the eigenvector, and λ designates the eigenvalue.[3]

$$\mathbf{A}\mathbf{v} = \lambda\mathbf{v} \tag{5.2}$$

5.1.3   Calculating Eigenvalues

To calculate an eigenvector, we first need to find its corresponding eigenvalue. Eigenvalues are single values (i.e., scalars) that transform a nonsingular matrix into a singular one by introducing a linear dependence. As first discussed in Chap. 1, a singular matrix has a determinant of 0 and, consequently, cannot be inverted. Below is an example of a singular matrix. As you can see, the second column is simply 2* the first.

---

[3] Even though it will satisfy Eq. (5.1), a vector of 0's is not considered an eigenvector.

$$\begin{bmatrix} 5 & 10 \\ 1 & 2 \end{bmatrix}$$

You can easily verify that the determinant of the matrix is 0 by subtracting the product of the off-diagonal entries from the product of the main diagonal entries.

$$\begin{vmatrix} 5 & 10 \\ 1 & 2 \end{vmatrix} = (5 * 2) - (10 * 1) = 0$$

5.1.3.1 Characteristic Equation

Eigenvalues are found by solving an equation known as the *characteristic equation*,

$$|\mathbf{A} - \lambda\mathbf{I}| = 0 \tag{5.3}$$

where $\mathbf{A}$ is our original matrix, and $\mathbf{I}$ is an identity matrix with the same dimensions as $\mathbf{A}$. What we are looking for, then, is a value that produces a singular matrix when it is subtracted from the diagonal entries of a matrix. This value is known as an eigenvalue.

With a 2×2 matrix, we can solve the equation using simple algebra without too much difficulty. With larger matrices, we will need to learn an iterative technique that will be more complicated. We will start with the simple case, as it provides the most insight into the underlying logic, and learn the more difficult technique later in this chapter.

5.1.3.2 Solving the Characteristic Equation

Returning to our earlier matrix, we first note that $\mathbf{A}$ is not singular.

$$\begin{vmatrix} 5 & 6 \\ 1 & 4 \end{vmatrix} = 14$$

Now we solve the characteristic equation.

$$\left| \begin{bmatrix} 5 & 6 \\ 1 & 4 \end{bmatrix} - \lambda \begin{bmatrix} 1 & 0 \\ 0 & 1 \end{bmatrix} \right| = 0$$

After some simple multiplication, we get

$$\begin{vmatrix} 5 - \lambda & 6 \\ 1 & 4 - \lambda \end{vmatrix} = 0$$

Because we know the determinant of a 2×2 matrix can be found from the products of the two diagonals, we can set up the following equation to find our eigenvalues.

$$\{(5 - \lambda) * (4 - \lambda)\} - (6 * 1) = 0$$

Working through the multiplication gives us the following polynomial (quadratic) equation.

$$\lambda^2 - (5 + 4)\lambda + (5 * 4) - (6 * 1) = 0$$

After collecting terms,

$$\lambda^2 - 9\lambda + 14 = 0$$

we solve the equation using the quadratic function

$$\lambda = \frac{-b \pm \sqrt{b^2 - 4ac}}{2a} \tag{5.4}$$

with $a = 1$, $b = -9$, and $c = 14$.

$$\frac{-(-9) + \sqrt{(-9)^2 - (4 * 14)}}{2} = 7$$

and

$$\frac{-(-9) - \sqrt{(-9)^2 - (4 * 14)}}{2} = 2$$

We now have the two eigenvalues for this 2×2 matrix.[4] This will always be the case (i.e., there will always be as many eigenvalues as there are columns in **A**). In a moment, we will see that a unique set of eigenvectors is associated with each eigenvalue, but first let's confirm that each eigenvalue turns our nonsingular **A** matrix into a singular one.

$$\begin{vmatrix} 5 - 7 & 6 \\ 1 & 4 - 7 \end{vmatrix} = \begin{vmatrix} -2 & 6 \\ 1 & -3 \end{vmatrix} = 0$$

---

[4] By convention, the largest eigenvalue is designated first, with the size of each successive eigenvalue decreasing in magnitude.

and

$$\begin{vmatrix} 5-2 & 6 \\ 1 & 4-2 \end{vmatrix} = \begin{vmatrix} 3 & 6 \\ 1 & 2 \end{vmatrix} = 0$$

5.1.4   Calculating Eigenvectors

Having calculated the eigenvalues, we can find the eigenvectors associated with each one by solving the equality shown in Eq. (5.2). Rearranging terms yields the following.

$$(\mathbf{A} - \lambda)\mathbf{v} = 0 \tag{5.5}$$

Notice that $\mathbf{A} - \lambda$ is a subtraction matrix, so we are looking for a vector that, when premultiplied by a subtraction matrix, returns a product vector of 0's.

5.1.4.1   Calculating the Eigenvectors Associated with the First Eigenvalue

We will use Eq. (5.5) along with our first subtraction matrix to find our first set of eigenvectors.

$$\begin{bmatrix} -2 & 6 \\ 1 & -3 \end{bmatrix} \begin{bmatrix} v_{1.1} \\ v_{2.1} \end{bmatrix} = \begin{bmatrix} 0 \\ 0 \end{bmatrix}$$

After multiplying, we produce two equations with two unknowns.

$$-2v_{1.1} + 6v_{2.1} = 0$$

and

$$1v_{1.1} + -3v_{2.1} = 0$$

The easiest way to solve these equations is to set one of the unknown terms equal to 1 and then solve for the other unknown. To illustrate, we will use the top equation and determine the value of $v_{1.1}$ after setting $v_{2.1} = 1$.

$$-2v_{1.1} + 6 = 0$$
$$-2v_{1.1} = -6$$

and

$$v_{1.1} = \frac{-6}{-2} = 3$$

Since we have designated $v_{2.1} = 1$, we have found the eigenvector associated with our first eigenvalue ($\lambda_1 = 7$).

$$\begin{bmatrix} v_{1.1} \\ v_{2.1} \end{bmatrix} = \begin{bmatrix} 3 \\ 1 \end{bmatrix}$$

5.1.4.2 Converting Eigenvectors to Unit Length

Earlier, we noted that any scalar multiple of an eigenvector is also an eigenvector. Consequently, this particular eigenvector is only one of an infinite number associated with our first eigenvalue. To avoid confusion, it is customary to express eigenvectors in unit length, such that the sum of their squared values $= 1$. To effect this conversion, we divide each eigenvector by the square root of the sum of the squared values. This conversion creates a vector known as the Euclidean norm of the eigenvector.

$$\frac{v_i}{\sqrt{\Sigma v_i^2}} \tag{5.6}$$

Applying this rule produces unit length eigenvectors associated with our first eigenvalue.

$$\frac{3}{\sqrt{3^2 + 1^2}} = .9487$$

and

$$\frac{1}{\sqrt{3^2 + 1^2}} = .3162$$

$$\begin{bmatrix} v_{1.1} \\ v_{2.1} \end{bmatrix} = \begin{bmatrix} .9487 \\ .3162 \end{bmatrix}$$

5.1.4.3 Calculating the Eigenvectors Associated with the Second Eigenvalue

Now we will find the eigenvectors associated with $\lambda_2 = 2$. Using our second subtraction matrix produces

$$\begin{bmatrix} 3 & 6 \\ 1 & 2 \end{bmatrix} \begin{bmatrix} v_{1.2} \\ v_{2.2} \end{bmatrix} = \begin{bmatrix} 0 \\ 0 \end{bmatrix}$$

yielding two simultaneous equations.

$$3v_{1.2} + 6v_{2.2} = 0$$
$$1v_{1.2} + 2v_{2.2} = 0$$

This time we will set $v_{1.2} = 1$ and solve for $v_{2.2}$.

$$3 + 6v_{2.2} = 0$$
$$6v_{2.2} = -3$$

and

$$v_{2.2} = \frac{-3}{6} = -.5$$

$$\begin{bmatrix} v_{1.2} \\ v_{2.2} \end{bmatrix} = \begin{bmatrix} 1 \\ -.5 \end{bmatrix}$$

If we use the second equation, we get the same result.[5]
 Finally, we can create unit vectors following the procedures described earlier.

$$\frac{1}{\sqrt{1^2 + -.5^2}} = .8944$$

and

$$\frac{-.5}{\sqrt{1^2 + -.5^2}} = -.4472$$

producing

$$\begin{bmatrix} v_{1.2} \\ v_{2.2} \end{bmatrix} = \begin{bmatrix} .8944 \\ -.4472 \end{bmatrix}$$

Just to be thorough, let's prove that premultiplying this eigenvector by our original matrix satisfies the conditions of an eigenvector, such that the product vector is λv (which in our case $= 2v$).

---

[5] Because any scalar multiple of an eigenvector is an eigenvector, we could flip the signs of these eigenvectors by multiplying both values by -1. For this reason, which of the two values receives a negative sign is arbitrary.

$$\begin{bmatrix} 5 & 6 \\ 1 & 4 \end{bmatrix} \begin{bmatrix} .8944 \\ -.4472 \end{bmatrix} = \begin{bmatrix} 1.7888 \\ -.8944 \end{bmatrix}$$

Finally, we will combine our eigenvalues and eigenvectors to produce our eigenpairs.

$$\lambda = \quad 7 \qquad\qquad 2$$
$$\mathbf{V} = \begin{bmatrix} .9487 \\ .3162 \end{bmatrix} \begin{bmatrix} .8944 \\ -.4472 \end{bmatrix}$$

5.1.4.4 Summary

- A 2×2 matrix has two eigenvalues.
- When subtracted from the diagonal elements of the original matrix, each eigenvalue produces a matrix with a linear dependence (i.e., not invertible with a determinant $= 0$).
- Associated with each eigenvalue is an eigenvector that, when premultiplied by the original matrix, returns a vector that changes length but not direction. The change in length $= \lambda v$.[6]

5.1.5 Eigenvalues and Variance Consolidation

At the outset of this chapter, we asked whether there was a way to create a linear combination among a set of variables for which no variable is a criterion. Having taken the time to learn how to calculate eigenpairs, we will now see how they provide a solution to our question.

5.1.5.1 Eigenvalues and Other Properties of a Square Matrix

We will begin by examining the association between eigenvalues and three other properties of a matrix: its trace, its determinant, and its rank. To help us see these connections, let's look again at our original matrix, remembering that our eigenvalues were 7 and 2.

---

[6] The absolute value of an eigenvalue quantifies how much the eigenvector is stretched ($|\lambda| > 1$) or shrunk ($|\lambda| < 1$) when it is premultiplied by $\mathbf{A}$; the sign of an eigenvalue indicates whether the eigenvector points in the same direction ($\lambda > 0$) or the opposite direction ($\lambda < 0$) when it is premultiplied by $\mathbf{A}$. When $\lambda = 0$, the eigenvector is unchanged when premultiplied by $\mathbf{A}$.

$$A = \begin{bmatrix} 5 & 6 \\ 1 & 4 \end{bmatrix}$$

Notice that the trace of the matrix (found by summing the elements on the main diagonal) equals 9 and that this value matches the sum of the two eigenvalues ($7+2=9$). This is no coincidence, as the sum of the eigenvalues always equals the trace.

Now let's reconsider the determinant of the matrix.

$$|A| = \begin{vmatrix} 5 & 6 \\ 1 & 4 \end{vmatrix} = (5*4) - (6*1) = 14$$

This value matches the product of the eigenvalues ($7*2=14$), and this, too, will always be true (i.e., the product of the eigenvalues equals the determinant of the original matrix). Because the trace and the determinant of a matrix are measures of its variability, these associations reveal that eigenpairs redistribute (but preserve) the variability in a matrix.

Finally, we define the rank of a matrix [rank(A)] as the number of eigenvalues that are >0. An $\lambda=0$ signifies the presence of a linear dependence. There are no linear dependencies in our example, and the rank of $A=2$. When the rank of a square matrix equals its number of columns (or rows), the matrix is said to be of *full rank* and the matrix is invertible; when the rank of a square matrix is less than its number of columns (or rows), the matrix is termed *rank deficient* and cannot be inverted.

5.1.5.2   Eigenvalues of Correlation Matrices

Correlation matrices provide further evidence that eigenvalues apportion the variability in a matrix. Recall that correlations are covariances among standardized variables. The diagonals of a correlation matrix represent each variable's variance (which is always 1), and the off-diagonals show the standardized covariances.

Keeping those points in mind, let's examine the (pretend) correlation matrices in Table 5.1.

Table 5.1 Eigenvalues for six correlation matrices with varying degrees of linear dependence

	A			B			C		
	1	0		1	.25		1	.75	
	0	1		.25	1		.75	1	
λ	1	1		1.25	.75		1.75	.25	

	D			E			F		
	1	.25	0	1	.25	.25	1	.25	.25
	.25	1	0	.25	1	0	.25	1	.25
	0	0	1	.25	0	1	.25	.25	1
λ	1.25	1	.75	1.3535	1	.6465	1.5	.75	.75

- Our first matrix is an identity matrix. Because the variables are independent, there is no variance to consolidate and both eigenvalues $= 1$.
- The variables in our second matrix are correlated ($r=.25$), and the eigenvalues reflect this covariation, as the first eigenvalue is now > 1 and the second is < 1. They still sum to 2, of course, because the sum of the eigenvalues equals the trace of a matrix.
- Example C includes a stronger linear relation, and the eigenvalues change again to reflect this greater overlap.
- The examples in the bottom portion of Table 5.1 have three variables.[7] Notice that Example D is similar to Example B except that I have added a third variable that is uncorrelated with the other two. Notice that two eigenvalues remain unchanged, but now there is a third that equals 1. Thus, the variance of two variables can be consolidated, but the third remains independent.
- In Example E, Variable 3 is correlated with Variable 1 but uncorrelated with Variable 2. The size of the dominant eigenvalue increases, but one of the eigenvalues still remains 1, because Variables 2 and 3 are uncorrelated.
- Our final matrix shows the case where all three variables are correlated. The dominant eigenvalue grows to reflect this overlap, and no eigenvalue $=1$ because no variables are completely independent.

5.1.6   Eigen Decomposition and Matrix Recomposition

We can further appreciate the manner in which an eigen decomposition redistributes the variance in a matrix by using the eigenpairs to reconstruct our original matrix

$$\mathbf{A} = \mathbf{V}\mathbf{\Lambda}\mathbf{V}^{-1} \tag{5.7}$$

where $\mathbf{V}$ refers to a matrix of eigenvectors, and the Greek capital letter lambda ($\mathbf{\Lambda}$) refers to a diagonal matrix of eigenvalues with the same dimensions as $\mathbf{A}$.

$$\mathbf{\Lambda} = \begin{bmatrix} \lambda_1 & \cdots & 0 \\ \vdots & \ddots & \vdots \\ 0 & \cdots & \lambda_p \end{bmatrix}$$

Returning to our earlier example,

$$\mathbf{A} = \begin{bmatrix} .9487 & .8944 \\ .3162 & -.4472 \end{bmatrix} \begin{bmatrix} 7 & 0 \\ 0 & 2 \end{bmatrix} \begin{bmatrix} .9487 & .8944 \\ .3162 & -.4472 \end{bmatrix}^{-1} = \begin{bmatrix} 5 & 6 \\ 2 & 4 \end{bmatrix}$$

---

[7] The next section of this chapter will teach you how to calculate eigenvalues with larger matrices.

5.1.6.1 Algebraic Formula

Using matrix algebra, it is easy to lose sight of the fact that all we are doing is expressing each of the values in the original matrix as a linear combination of the eigenvalues and eigenvectors. To make this equivalence more apparent, here is the algebraic formula for computing the first term in our original matrix, where $|V| = -.70711$.

$$a_{11} = \left\{ v_{11} * \lambda_1 * \frac{v_{22}}{|V|} \right\} + \left\{ v_{12} * \lambda_2 * \frac{-v_{21}}{|V|} \right\}$$

Plugging in our values produces the expected result.

$$\left\{ .9487 * 7 * \frac{-.4472}{-.70711} \right\} + \left\{ .8944 * 2 * \frac{-.3162}{-.70711} \right\} = 5$$

The remaining values can be calculated using similar equations, underscoring that an eigen decomposition redistributes the variance in a matrix.

$$a_{21} = \left\{ v_{21} * \lambda_1 * \frac{v_{22}}{|V|} \right\} + \left\{ v_{22} * \lambda_2 * \frac{-v_{21}}{|V|} \right\}$$

$$a_{12} = \left\{ v_{11} * \lambda_1 * \frac{-v_{12}}{|V|} \right\} + \left\{ v_{12} * \lambda_2 * \frac{v_{11}}{|V|} \right\}$$

$$a_{22} = \left\{ v_{21} * \lambda_1 * \frac{-v_{12}}{|V|} \right\} + \left\{ v_{22} * \lambda_2 * \frac{v_{11}}{|V|} \right\}$$

5.1.6.2 Matrix Powers

Calculating the power series of a matrix provides another way to appreciate the variance consolidation properties of an eigen decomposition. Imagine that we wish to raise a matrix to the fifth power. We could multiply the matrix five times, but this is inefficient. Instead, we can use the following formula to find any matrix power by raising the eigenvalues to the desired power.

$$\mathbf{A^n = V\Lambda^n V^{-1}} \tag{5.8}$$

Plugging in our values, we easily raise the matrix to the desired power.

$$\mathbf{A}^5 = \begin{bmatrix} .9487 & .8944 \\ .3162 & -.4472 \end{bmatrix} \begin{bmatrix} 7^5 & 0 \\ 0 & 2^5 \end{bmatrix} \begin{bmatrix} .9487 & .8944 \\ .3162 & -.4472 \end{bmatrix}^{-1} = \begin{bmatrix} 10097 & 20130 \\ 3355 & 6742 \end{bmatrix}$$

5.1.7   R Code: Eigen Decomposition

```
#Matrix multiplication with an ordinary vector ("b")
A <- matrix(c(5,6,1,4), nrow=2, ncol=2, byrow=TRUE)
b <- matrix(c(2,6), nrow=2, ncol=1, byrow=TRUE)
scalar=7

mult.1 <-b*scalar; mult.1
mult.2 <-A%*%b; mult.2

#Matrix multiplication with an eigenvector ("c")
c <- matrix(c(6,2), nrow=2, ncol=1, byrow=TRUE)
mult.3 <-c*scalar; mult.3
mult.4 <-A%*%c; mult.4

#Algebraic function for finding eigenpairs from a 2X2 matrix
eigs <-function(A){
a <-1
b <- -(A[1,1]+A[2,2])
c <-(A[1,1]*A[2,2])-(A[1,2]*A[2,1])
lambda1 <-(-b+sqrt(b^2-4*c))/(2*a)
lambda2 <-(-b-sqrt(b^2-4*c))/(2*a)
v1 <-(-A[1,2])/(A[1,1]-lambda1)
vec1 <-matrix(cbind(v1,1)/(sqrt(sum(v1^2+1^2))),nrow=2)
v2 <-(-(A[1,1]-lambda2))/(A[1,2])
vec2 <-matrix(cbind(1,v2)/(sqrt(sum(1^2+v2^2))),nrow=2)
out <-matrix(c(lambda1,vec1,lambda2,vec2),nrow=3,ncol=2)
}

A <-matrix(c(5,6,1,4),nrow=2,ncol=2,byrow=T);A
eigpairs <-eigs(A)
dimnames(eigpairs)=list(c("lambda","v1","v2"),c("pair1","pair2"))
eigpairs

#Reconstruct A
lambda <-diag(c(eigpairs[1],eigpairs[4]))
V <-matrix(c(eigpairs[2],eigpairs[3],eigpairs[5],eigpairs[6]),nrow=2,
ncol=2)
reconstruct <-V%*%lambda%*%solve(V)
reconstruct
```

5.2   QR Decomposition

An eigen decomposition can only be performed with a square matrix, but other decompositions can be used with rectangular matrices. One widely used method is the **QR** decomposition. This method decomposes a rectangular matrix into an orthonormal matrix **Q** and an upper triangular matrix **R**.

$$\mathbf{A} = \mathbf{QR} \tag{5.9}$$

An orthonormal matrix returns an identity matrix when it is premultiplied by its transpose ($\mathbf{Q'Q} = \mathbf{I}$), and an upper triangular matrix has 0's below the main diagonal. These features make the **QR** decomposition a highly efficient and accurate method for solving a variety of statistical problems.

5.2.1 Computations with Householder Transformations

There are several ways to perform a **QR** decomposition, but we will use a method known as a Householder transformation. The calculations aren't difficult, but many steps are involved, so we will first describe them and then work through an example.

5.2.1.1 Steps of a Householder Transformation

1. Create an $N \times N$ identity matrix and designate each of the p columns with a subscript (e.g., $\mathbf{p}_1 \cdots \mathbf{p}_p$), where the subscript refers to the number of vectors being transformed. When using the **QR** decomposition for a regression analysis, it is customary to perform the transformation using only the predictors (and vector of leading 1's). So if we have two predictors, we will be transforming three column vectors ($\mathbf{p}_1, \mathbf{p}_2,$ and $\mathbf{p}_3$).
2. Use the following rule to transform each column vector of the original rectangular matrix.

$$\begin{cases} \mathbf{u} = \dfrac{\mathbf{v} + ||\mathbf{v}||\mathbf{p}_i}{||(\mathbf{v} + ||\mathbf{v}||\mathbf{p}_i)||} & \text{if first nonzero element of } \mathbf{v}_i > 0 \\[4mm] \mathbf{u} = \dfrac{\mathbf{v} - ||\mathbf{v}||\mathbf{p}_i}{||(\mathbf{v} - ||\mathbf{v}||\mathbf{p}_i)||} & \text{if first nonzero element of } \mathbf{v}_i < 0 \end{cases} \tag{5.10}$$

The value in double brackets is the Euclidean norm (i.e., it creates an N-dimensional unit vector).

$$||\mathbf{v}|| = \sqrt{\mathbf{v'v}} = 1 \tag{5.11}$$

It looks complicated, but remember it is simply the square root of the sum of squares. Notice that we calculate two such norms in Eq. 5.10, as the denominator is the Euclidean norm of the numerator.

3. Calculate a Householder matrix.

$$\mathbf{H} = \mathbf{I} - 2\mathbf{uu}'$$ (5.12)

4. Calculate R_k,

$$\mathbf{R}_k = \mathbf{HR}_{k-1}$$ (5.13)

where $\mathbf{R}_{k-1}$ initially refers to our original matrix $\mathbf{A}$.

5. Repeat for the remaining column vectors, substituting the next column of $\mathbf{R}$ for $\mathbf{A}$, and setting the value of rows already calculated to zero.

5.2.1.2   Example

Let's work through an example using the data originally reported in Table 4.1, reproduced in Table 5.2. As you may recall, in this (phony) investigation, we used two variables, self-efficacy and math aptitude, to predict performance in a matrix algebra course.

Table 5.2 Small sample example of multiple regression using the QR decomposition

	x_1	x_2	y	v_1	u_1	v_2	u_2	v_3	u_3
1	1	2	3	**1**	.802706	0	0	0	0
1	9	7	8	1	.179814	**4.89604**	.86746	0	0
1	1	1	2	1	.179814	−3.10396	−.18453	**−1.20133**	−.76957
1	5	8	8	1	.179814	.89604	.05327	3.64145	.36331
1	6	5	5	1	.179814	1.89604	.11272	.10215	.01019
1	8	7	9	1	.179814	3.89604	.23162	1.02353	.10212
1	2	4	4	1	.179814	−2.10396	−.12508	1.25937	.12565
1	4	7	5	1	.179814	−.10396	−.00618	3.18076	.31735
1	2	1	2	1	.179814	−2.10396	−.12508	−1.74063	−.17367
1	8	4	4	1	.179814	3.89604	.23162	−1.97647	−.19720
1	7	3	2	1	.179814	2.89604	.17217	−2.43716	−.24316
1	7	4	6	1	.179814	2.89604	.17217	−1.43716	−.14339

Data reproduced from Table 4.1

1. Our first step is to create an $N \times N$ identity matrix, designating the first three columns p_1, p_2, and p_3. You should be familiar with the form of a 12×12 identity matrix, so I will forgo displaying it.

2. Next we transform our vector of leading 1's to find $\mathbf{u}_1$. The first entry in $\mathbf{v}_1$ is > 0, so we use the top portion of Eq. 5.10, along with the first column of our identity matrix. To help you with the calculations, I will break them into parts.

2a. Calculate $\|v\|$

$$\sqrt{\sum \begin{bmatrix} 1^2 \\ 1^2 \\ 1^2 \\ 1^2 \\ 1^2 \\ 1^2 \\ 1^2 \\ 1^2 \\ 1^2 \\ 1^2 \\ 1^2 \\ 1^2 \end{bmatrix}} = 3.4641$$

2b. Calculate $v + \|v\|p_i$

$$\begin{bmatrix} 1 \\ 1 \\ 1 \\ 1 \\ 1 \\ 1 \\ 1 \\ 1 \\ 1 \\ 1 \\ 1 \\ 1 \end{bmatrix} + \left\{ 3.4641 * \begin{bmatrix} 1 \\ 0 \\ 0 \\ 0 \\ 0 \\ 0 \\ 0 \\ 0 \\ 0 \\ 0 \\ 0 \\ 0 \end{bmatrix} \right\} = \begin{bmatrix} 4.464102 \\ 1 \\ 1 \\ 1 \\ 1 \\ 1 \\ 1 \\ 1 \\ 1 \\ 1 \\ 1 \\ 1 \end{bmatrix}$$

2c. Calculate $\|(v + \|v\|p_i)\|$

$$\sqrt{\sum \begin{bmatrix} 4.464102^2 \\ 1^2 \\ 1^2 \\ 1^2 \\ 1^2 \\ 1^2 \\ 1^2 \\ 1^2 \\ 1^2 \\ 1^2 \\ 1^2 \\ 1^2 \end{bmatrix}} = 5.56131$$

2d. Divide $\{v + \|v\|p_i\}$ by $\{\|(v + \|v\|p_i)\|\}$ to produce the values for u_1 shown in Table 5.2.

$$
u_1 = \begin{bmatrix} 4.464102 \\ 1 \\ 1 \\ 1 \\ 1 \\ 1 \\ 1 \\ 1 \\ 1 \\ 1 \\ 1 \\ 1 \\ 1 \end{bmatrix} \div 5.56131 = \begin{bmatrix} .802706 \\ .179814 \\ .179814 \\ .179814 \\ .179814 \\ .179814 \\ .179814 \\ .179814 \\ .179814 \\ .179814 \\ .179814 \\ .179814 \\ .179814 \end{bmatrix}
$$

3. Next, we calculate $H_1 = I - 2u_1u_1'$.

$$
H_1 = \begin{bmatrix}
-.28868 & -.28868 & -.28868 & -.28868 & -.28868 & -.28868 & -.28868 & -.28868 & -.28868 & -.28868 & -.28868 & -.28868 \\
-.28868 & .93533 & -.06467 & -.06467 & -.06467 & -.06467 & -.06467 & -.06467 & -.06467 & -.06467 & -.06467 & -.06467 \\
-.28868 & -.06467 & .93533 & -.06467 & -.06467 & -.06467 & -.06467 & -.06467 & -.06467 & -.06467 & -.06467 & -.06467 \\
-.28868 & -.06467 & -.06467 & .93533 & -.06467 & -.06467 & -.06467 & -.06467 & -.06467 & -.06467 & -.06467 & -.06467 \\
-.28868 & -.06467 & -.06467 & -.06467 & .93533 & -.06467 & -.06467 & -.06467 & -.06467 & -.06467 & -.06467 & -.06467 \\
-.28868 & -.06467 & -.06467 & -.06467 & -.06467 & .93533 & -.06467 & -.06467 & -.06467 & -.06467 & -.06467 & -.06467 \\
-.28868 & -.06467 & -.06467 & -.06467 & -.06467 & -.06467 & .93533 & -.06467 & -.06467 & -.06467 & -.06467 & -.06467 \\
-.28868 & -.06467 & -.06467 & -.06467 & -.06467 & -.06467 & -.06467 & .93533 & -.06467 & -.06467 & -.06467 & -.06467 \\
-.28868 & -.06467 & -.06467 & -.06467 & -.06467 & -.06467 & -.06467 & -.06467 & .93533 & -.06467 & -.06467 & -.06467 \\
-.28868 & -.06467 & -.06467 & -.06467 & -.06467 & -.06467 & -.06467 & -.06467 & -.06467 & .93533 & -.06467 & -.06467 \\
-.28868 & -.06467 & -.06467 & -.06467 & -.06467 & -.06467 & -.06467 & -.06467 & -.06467 & -.06467 & .93533 & -.06467 \\
-.28868 & -.06467 & -.06467 & -.06467 & -.06467 & -.06467 & -.06467 & -.06467 & -.06467 & -.06467 & -.06467 & .93533
\end{bmatrix}
$$

4. Finally, we calculate $R_1 = H_1A$.

$$
R_1 = \begin{bmatrix}
-3.46410 & -17.32051 & -15.29978 \\
0 & 4.89604 & 3.12469 \\
0 & -3.10396 & -2.87531 \\
0 & .89604 & 4.12469 \\
0 & 1.89604 & 1.12469 \\
0 & 3.89604 & 3.12469 \\
0 & -2.10396 & .12469 \\
0 & -.10396 & 3.12469 \\
0 & -2.10396 & -2.87531 \\
0 & 3.89604 & .12469 \\
0 & 2.89604 & -.87531 \\
0 & 2.89604 & .12469
\end{bmatrix}
$$

Our interest lies only in the first row, as this will be the first row of R. Notice that all of the values below the first value in column 1 equal 0, which is the property we desire in an upper triangular matrix.

Now we repeat the steps, designating the second column vector in $\mathbf{R}_1$ as $\mathbf{v}_2$. Referring back to the column labeled $\mathbf{v}_2$ in Table 5.2, notice that we have replaced the first entry with a zero. This is because we have already found our first row of values, so we begin using the second row, not the first. Because the first nonzero value in the second column (4.8960) is > 0, we again use the top portion of (5.10) to calculate $\mathbf{u}_2$, this time using the second column of our identity matrix (i. e., $\mathbf{p}_2$). Table 5.2 shows the resultant vector. We then calculate another Householder matrix, $\mathbf{H}_2 = \mathbf{I} - 2\mathbf{u}_2\mathbf{u}_2'$,

$$
\mathbf{H}_2 =
\begin{bmatrix}
1 & 0 & 0 & 0 & 0 & 0 & 0 & 0 & 0 & 0 & 0 & 0 \\
0 & -.50499 & .32015 & -.09242 & -.19556 & -.40185 & .21701 & .01072 & .21701 & -.40185 & -.29870 & -.29870 \\
0 & .32015 & .93190 & .01966 & .04160 & .08548 & -.04616 & -.00228 & -.04616 & .08548 & .06354 & .06354 \\
0 & -.09242 & .01966 & .99432 & -.01201 & -.02468 & .01333 & .00066 & .01333 & -.02468 & -.01834 & -.01834 \\
0 & -.19556 & .04160 & -.01201 & .97459 & -.05222 & .02820 & .00139 & .02820 & -.05222 & -.03881 & -.03881 \\
0 & -.40185 & .08548 & -.02468 & -.05222 & .89270 & .05794 & .00286 & .05794 & -.10730 & -.07976 & -.07976 \\
0 & .21701 & -.04616 & .01333 & .02820 & .05794 & .96871 & -.00155 & -.03129 & .05794 & .04307 & .04307 \\
0 & .01072 & -.00228 & .00066 & .00139 & .00286 & -.00155 & .99992 & -.00155 & .00286 & .00213 & .00213 \\
0 & .21701 & -.04616 & .01333 & .02820 & .05794 & -.03129 & -.00155 & .96871 & .05794 & .04307 & .04307 \\
0 & -.40185 & .08548 & -.02468 & -.05222 & -.10730 & .05794 & .00286 & .05794 & .89270 & -.07976 & -.07976 \\
0 & -.29870 & .06354 & -.01834 & -.03881 & -.07976 & .04307 & .00213 & .04307 & -.07976 & .94071 & -.05929 \\
0 & -.29870 & .06354 & -.01834 & -.03881 & -.07976 & .04307 & .00213 & .04307 & -.07976 & -.05929 & .94071
\end{bmatrix}
$$

and use it to find an updated $\mathbf{R}_2$ matrix, substituting $\mathbf{R}_1$ for $\mathbf{A}$.

$$\mathbf{R}_2 = \mathbf{H}_2\mathbf{R}_1$$

$$
\mathbf{R}_2 =
\begin{bmatrix}
-3.46410 & -17.32051 & -15.2998 \\
0 & -9.69536 & -4.74454 \\
0 & 0 & -1.20133 \\
0 & 0 & 3.64145 \\
0 & 0 & .10215 \\
0 & 0 & 1.02353 \\
0 & 0 & 1.25937 \\
0 & 0 & 3.18076 \\
0 & 0 & -1.74063 \\
0 & 0 & -1.97647 \\
0 & 0 & -2.43716 \\
0 & 0 & -1.43716
\end{bmatrix}
$$

The first two rows of this matrix constitute the first two rows of $\mathbf{R}$.

We repeat the process one more time, starting with the third column vector in $\mathbf{R}_2$, placing $0's$ for the first two rows (see column labeled $\mathbf{v}_3$) in Table 5.2. The first nonzero entry is now < 0, so we use the bottom portion of (5.10), along with the third column vector from our identity matrix ($\mathbf{p}_3$) to find $\mathbf{u}_3$. We then find our third Householder matrix,

$$\mathbf{H_3} = \begin{bmatrix} 1 & 0 & 0 & 0 & 0 & 0 & 0 & 0 & 0 & 0 & 0 & 0 \\ 0 & 1 & 0 & 0 & 0 & 0 & 0 & 0 & 0 & 0 & 0 & 0 \\ 0 & 0 & -.18448 & .55919 & .01569 & .15718 & .19339 & .48845 & -.26730 & -.30351 & -.37426 & -.22069 \\ 0 & 0 & .55919 & .73601 & -.00741 & -.07420 & -.09130 & -.23060 & .12619 & .14329 & .17669 & .10419 \\ 0 & 0 & .01569 & -.00741 & .99979 & -.00208 & -.00256 & -.00647 & .00354 & .00402 & .00496 & .00292 \\ 0 & 0 & .15718 & -.07420 & -.00208 & .97914 & -.02566 & -.06482 & .03547 & .04028 & .04966 & .02929 \\ 0 & 0 & .19339 & -.09130 & -.00256 & -.02566 & .96842 & -.07975 & .04364 & .04956 & .06111 & .03603 \\ 0 & 0 & .48845 & -.23060 & -.00647 & -.06482 & -.07975 & .79858 & .11023 & .12516 & .15433 & .09101 \\ 0 & 0 & -.26730 & .12619 & .00354 & .03547 & .04364 & .11023 & .93968 & -.06849 & -.08446 & -.04980 \\ 0 & 0 & -.30351 & .14329 & .00402 & .04028 & .04956 & .12516 & -.06849 & .92223 & -.09590 & -.05655 \\ 0 & 0 & -.37426 & .17669 & .00496 & .04966 & .06111 & .15433 & -.08446 & -.09590 & .88175 & -.06973 \\ 0 & 0 & -.22069 & .10419 & .00292 & .02929 & .03603 & .09101 & -.04980 & -.05655 & -.06973 & .95888 \end{bmatrix}$$

and calculate $\mathbf{R_3} = \mathbf{H_3}\mathbf{R_2}$ to complete our upper triangular matrix $\mathbf{R}$.

$$\mathbf{R_3} = \mathbf{R} = \begin{bmatrix} -3.46410 & -17.32051 & -15.29978 \\ 0 & -9.69536 & -4.74454 \\ 0 & 0 & 6.51199 \\ 0 & 0 & 0 \\ 0 & 0 & 0 \\ 0 & 0 & 0 \\ 0 & 0 & 0 \\ 0 & 0 & 0 \\ 0 & 0 & 0 \\ 0 & 0 & 0 \\ 0 & 0 & 0 \\ 0 & 0 & 0 \end{bmatrix}$$

Finally, we compute $\mathbf{Q}$ by multiplying the $\mathbf{H_p}$ matrices.

$$\mathbf{Q} = \prod \mathbf{H_p} \tag{5.14}$$

With three matrices, $\mathbf{Q} = \mathbf{H_1}\mathbf{H_2}\mathbf{H_3}$.

$$\mathbf{Q} = \begin{bmatrix} Q1 & Q2 & Q3 \\ .28868 & -.41257 & -.07052 \\ .28868 & .41257 & .09611 \\ .28868 & -.41257 & -.22408 \\ .28868 & .00000 & .55027 \\ .28868 & .10314 & .01443 \\ .28868 & .30943 & .17126 \\ .28868 & -.30943 & .16146 \\ .28868 & -.10314 & .47185 \\ .28868 & -.30943 & -.29923 \\ .28868 & .30943 & -.28943 \\ .28868 & .20628 & -.36784 \\ .28868 & .20628 & -.21428 \end{bmatrix}$$

As noted earlier, $\mathbf{Q}$ is an orthonormal matrix. Each column is normalized $\left(\sqrt{\mathbf{q}\mathbf{q}'} = 1\right)$, and the columns are pairwise orthogonal (i.e., their inner products $= 0$). With these properties, $\mathbf{Q}'\mathbf{Q} = \mathbf{I}$. Because this is a reconstruction, you can verify that $\mathbf{A} = \mathbf{QR}$.

5.2.2 Linear Regression

Using the $\mathbf{QR}$ decomposition, it is a relatively simple matter to compute the regression coefficients and standard errors of a linear regression analysis.

- Regression coefficients

$$\mathbf{b} = \mathbf{R}^{-1}\mathbf{Q}'\mathbf{y} \tag{5.15}$$

- Fitted values

$$\hat{\mathbf{y}} = \mathbf{Q}\mathbf{Q}'\mathbf{y} \tag{5.16}$$

- Parameter covariance matrix (for standard errors)

$$\mathbf{C} = \mathbf{R}^{-1}\left(\mathbf{R}^{-1}\right)' * MS_{res} \tag{5.17}$$

If you perform these calculations, you will find that the output matches the values we obtained in Chap. 4 using the matrix inverse method. To understand why, let's compare Eq. (5.17) with Eq. (2.34), repeated below.

$$\mathbf{C} = \left(\mathbf{X}'\mathbf{X}\right)^{-1} * MS_{res}$$

Looking over the two equations, we see that $\mathbf{R}^{-1}(\mathbf{R}^{-1})' = (\mathbf{X}'\mathbf{X})^{-1}$. So there is nothing magical about the $\mathbf{QR}$ method of solving the normal equations. Like all decompositions, it simply redistributes the variance in a matrix. The advantage of using this method is efficiency and accuracy. Inverting an upper triangular matrix is easier and less prone to error than is inverting a sum of squares matrix. At the end of this chapter, we will discuss some additional advantages/disadvantages of using various methods to perform linear regression analyses. First, we will learn how the $\mathbf{QR}$ decomposition can be modified to find the eigenpairs of a large matrix.

5.2.3   QR Algorithm for Finding the Eigenpairs

Simple algebra can be used to calculate the eigenpairs when a square matrix is small, but eigenpairs can only be approximated when the size of a matrix is greater than 3×3. Consequently, iterative techniques are needed. As first discussed in Chap. 3, an iterative technique begins with an initial guess and then finds closer and closer approximations to a solution until the difference between approximations becomes less than some designated value (e.g., quit when change in approximations becomes less than $(1.00E - 8)$.

A modification of the **QR** decomposition, known as the **QR** algorithm, can be used to find the eigenpairs of a square matrix. There are several ways to perform the algorithm, but we are going to focus on a basic implementation before turning to a more advanced technique.[8]

5.2.3.1   Basic (Unshifted) QR Algorithm to Find Eigenvalue

- Perform a **QR** decomposition on **A** using the procedures described earlier.
- Compute $\mathbf{A}_1 = \mathbf{RQ}$. Notice the order of the multiplication, as here we are reversing the order in which we multiply our two derived matrices.
- Perform a **QR** decomposition on $\mathbf{A}_1$ and compute $\mathbf{A}_2 = \mathbf{RQ}$.
- Continue the iterations until the eigenvalues appear on the diagonals of **RQ**.

$$\Lambda = \mathbf{RQ}_{ii} \tag{5.18}$$

We will illustrate the algorithm using the $\mathbf{X'X}$ matrix from Table 5.2.

$$\mathbf{A} = \mathbf{X'X} = \begin{bmatrix} 12 & 60 & 53 \\ 60 & 394 & 311 \\ 53 & 311 & 299 \end{bmatrix}$$

The eigenvalues of the matrix, reported to 10 decimals, are as follows.

$$\lambda_1 = 670.8324605415; \quad \lambda_2 = 31.9346873364; \quad \lambda_3 = 2.2328521221$$

Table 5.3 shows how these values can be found using the **QR** algorithm. To help orient you, the table also displays the **QR** decomposition of the matrix, although our interest lies in **RQ** not **QR**. As you can see, the eigenvalues begin appearing on the diagonals of **RQ** after only a few iterations. Convergence will not always be this rapid, but it usually doesn't take long for the eigenvalues to emerge. Ten iterations produce the values displayed above.

---

[8] The procedures presented here represent a simplified version of the one statistical packages use. Detailed descriptions of the complete algorithm can be found in Golub and van Loan (2013) and Watkins (2010).

Table 5.3 Four iterations of the **QR** algorithm

A = X'X =			12	60	53
			60	394	311
			53	311	299

QR decomposition					
Q			R		
.148239	−.469168	−.870578	80.950602	504.542263	434.128951
.741193	.635499	−.216273	0	31.529435	−10.574342
.654720	−.613206	.441950	0	0	18.741331

First four iterations of QR algorithm					
RQ$_1$			RQ$_2$		
670.19609	16.44615	−12.27033	670.83215	.44544	.04556
16.44615	26.52118	11.49230	.44544	31.89955	−1.02541
−12.27033	11.49230	8.28273	.04556	−1.02541	2.26830
RQ$_3$			RQ$_4$		
670.83246	.02112	−.00015	670.83246	.00101	.00000
.02112	31.93451	.07178	.00101	31.93469	−.00502
−.00015	.07178	2.23303	.00000	−.00502	2.23285

5.2.3.2 Explicitly Shifted QR Algorithm for Finding Eigenvalues

Convergence can be speeded up by applying a shift s to the diagonal values of **A**. The following steps are involved.

- Subtract s from the diagonal elements of **A**.

$$A_s = A − sI \tag{5.19}$$

- Perform a **QR** decomposition on A_s and calculate **RQ**.

$$RQ_s = QR[A_s] \tag{5.20}$$

- Create a new matrix for the next iteration by adding the subtracted diagonal elements to RQ_s, and continue iterating until convergence is reached.

$$A_{s+1} = RQ_s + sI \tag{5.21}$$

Several rules exist for finding a suitable shift value, but the most effective strategy is to use a value that lies close to one of the eigenvalues. Because the eigenvalues begin to emerge after only a few iterations, one approach is to forgo shifting for the first few iterations and then use an approximate eigenvalue for a shift once it appears.

The benefits of shifting are most apparent when the eigenvalues of a matrix are close together. Consider the following matrix with eigenvalues ($\lambda_1 = 3.90$, $\lambda_2 = 2.85$, $\lambda_3 = 1.80$).

$$\mathbf{A} = \begin{bmatrix} 3.025 & -.35 & -.875 \\ -.35 & 2.50 & -.35 \\ -.875 & -.35 & 3.025 \end{bmatrix}$$

If you were to perform the **QR** algorithm on this matrix without shifting, you would need 50 iterations to identify the eigenvalues within the level of round-off error (10^{-16}). Shifting greatly accelerates the algorithm's convergence. Table 5.4 shows how. We deploy no shift to start, but then use the value in the lower right-hand corner of the **RQ+sI** matrix as our shift value for our second iteration $(s = 2.379618)$. This entry provides a suitable approximation of our smallest eigenvalue. We continue to use the shift from the final entry of the preceding matrix for the next five iterations and then shift to the middle value for iterations 8 and 9, at which point the eigenvalues have been identified. Thus, shifting produces a ~6-fold increase in the algorithm's efficiency.

Table 5.4 Explicitly shifted **QR** algorithm

Iteration	Shift (s)	RQ − sI			RQ + sI		
1	0	3.025	−.350	−.875	3.532480	−.002941	.698596
		−.350	2.500	−.350	−.002941	2.637901	.498440
		−.875	−.350	3.025	.698596	.498440	**2.379618**
2	**2.379618**	1.152862	−.002941	.698596	3.840989	−.027552	.318713
		−.002941	.258283	.498440	−.027552	2.601067	.450361
		.698596	.498440	.000000	.318713	.450361	2.107944
3	2.107944	1.733046	−.027552	.318713	3.895668	.043554	.063025
		−.027552	.493123	.450361	.043554	2.799620	.230980
		.318713	.450361	.000000	.063025	.230980	1.854712
4	1.854712	2.040956	.043554	.063025	3.899273	.027579	.001733
		.043554	.944909	.230980	.027579	2.850559	.013229
		.063025	.230980	.000000	.001733	.013229	1.800168
5	1.800168	2.099106	.027579	.001733	3.899818	.013805	.000000
		.027579	1.050392	.013229	.013805	2.850182	.000002
		.001733	.013229	.000000	.000000	.000002	1.800000
6	1.800000	2.099818	.013805	.000000	3.899955	.006904	.000000
		.013805	1.050182	.000002	.006904	2.850045	.000000
		.000000	.000002	.000000	.000000	.000000	1.800000
7	1.800000	2.099955	.006904	.000000	3.899989	.003452	.000000
		.006904	1.050045	.000000	.003452	**2.850011**	.000000
		.000000	.000000	.000000	.000000	.000000	1.800000
8	**2.850011**	1.049977	.003452	.000000	3.90000	−3.73084E-08	3.73012E-16
		.003452	.000000	.000000	−3.73084E-08	2.850000	3.63751E-16
		.000000	.000000	1.800000	1.19881E-35	7.29288E-33	1.800000
9	2.850000	1.050000	.000000	.000000	3.900000	.000000	.000000
		.000000	.000000	.000000	.000000	2.850000	.000000
		.000000	.000000	1.800000	.000000	.000000	1.800000

5.2.3.3 Calculating Eigenvectors

Equation (5.22) can be used to find the eigenvectors of a symmetric matrix.

$$\mathbf{V} = \prod \mathbf{Q}_p \qquad (5.22)$$

The formula works, but it is not very efficient; the more iterations it require to obtain the eigenvalues, the more matrices we need to multiply to find the eigenvectors. Moreover, Eq. (5.22) won't work with nonsymmetric matrices. Because most statistical analyses are done with a symmetric sum of squares matrix, this limitation won't matter much for the problems we will be solving. Still, there might be times you need to find the eigenvectors of a nonsymmetric matrix, so we will learn another technique known as the inverse shift method that is more efficient than Eq. (5.22) and works regardless of whether the matrix is symmetric. The following steps are involved:

- Perform the **QR** algorithm of **A** until the eigenvalues converge.
- Perform one or two additional iterations of the **QR** algorithm on the inverse of a shifted matrix, formed by subtracting one of the eigenvalues from the diagonal elements of **A**.[9] The eigenvectors associated with each eigenvalue appear as the first column in **Q**. For example, to find the eigenvectors associated with the second eigenvalue, we perform the **QR** algorithm on $[(\mathbf{A} - \lambda_2\mathbf{I})^{-1}]$ and use the first column of **Q**.

We will illustrate the process with the following nonsymmetric matrix:

$$\mathbf{A} = \begin{bmatrix} 10.70 & -10.00 & 2.575 \\ 16.40 & -15.75 & 4.275 \\ 24.80 & -25.00 & 7.30 \end{bmatrix}$$

Using an explicit shift, 11 iterations were needed to calculate the eigenvalues (see diagonal elements of Λ in Table 5.5). Each of their corresponding eigenvectors was found with one additional iteration using the inverse shift method.

Table 5.5 QR Algorithm for finding eigenpairs of a nonsymmetric matrix using inverse shift method

	Eigenvalues (diagonal elements)				Eigenvectors		
					$(\mathbf{A} - \lambda_1\mathbf{I})^{-1}$	$(\mathbf{A} - \lambda_2\mathbf{I})^{-1}$	$(\mathbf{A} - \lambda_3\mathbf{I})^{-1}$
	1.00	−.31334	−44.7153		.21821789	.59702231	.37713259
Λ	0	.75	−6.9557	**V**	.43643578	.69652603	.57220116
	0	0	.50		.87287156	.39801488	.72825603

---

[9] Sometimes it is necessary to perturb the eigenvalue slightly to avoid singularity. In this case, multiplying the eigenvalue by a number very close to 1 should solve the problem.

5.2.3.4 Summary

In this section, you have learned how to calculate eigenpairs for large matrices. I don't expect you to routinely perform these calculations by hand (or even spreadsheet), but I do hope you will appreciate that they are not all that complicated and are certainly not magical. Almost all statistical packages offer built-in commands for computing eigenpairs when you need them, but now you know how it's done in case your license expires!

5.2.4   R Code: QR Decomposition

```
x1 <-c(1,9,1,5,6,8,2,4,2,8,7,7)
x2 <-c(2,7,1,8,5,7,4,7,1,4,3,4)
X <-cbind(1,x1,x2)
y <-c(3,8,2,8,5,9,4,5,2,4,2,6)

qr.house=function(A){
  m=nrow(A)
  n=ncol(A)
  B=A
  I=diag(1,m)
  Q=diag(1,m)
  R=diag(0,n)
  for (k in 1:n){
    v=matrix(c(rep(0,m)),byrow=T)
    v[k:m]=B[k:m,k]
    i=matrix(c(rep(0,m)),byrow=T)
    i[k]=1
    u = sign(v[k])*norm(v,type="2")*i + v
    u = u/norm(u,type="2")
    Hk = I - 2*u%*%t(u)
    R=Hk%*%B
    Q=Q%*%Hk
    B=R
  }
  RQ <- list("R" = round(R[1:k,1:k],15), "Q" = Q[1:m,1:k]);RQ
}
qr.house(X)
A= qr.house(X)$Q%*%qr.house(X)$R;A

#QR Regression using R's QR function
R.QR <-qr(X)
R <-qr.R(R.QR)
Q <-qr.Q(R.QR)
QR.beta <-solve(R)%*%t(Q)%*%y;QR.beta
QR.fit <-Q%*%t(Q)%*%y;QR.fit
```

(continued)

5.2.4 R Code: QR Decomposition (continued)

```
df <-length(y)-ncol(X)
MSe <-(sum((y-QR.fit)^2)/df)
covar <-MSe*solve(R)%*%t(solve(R));covar

#QR Algorithm for Finding Eigen pairs (default method in R)
XX <-t(X)%*%X
eigs <-eigen(XX);eigs
```

5.3 Singular Value Decomposition

Eigen decomposition is a special case of a more general decomposition technique called singular value decomposition (SVD) that can be used to reconstruct any rectangular matrix $\mathbf{A}$ into three matrices,

$$\mathbf{A} = \mathbf{UDV}'\tag{5.23}$$

where $\mathbf{V}$ = eigenvectors of $\mathbf{A}'\mathbf{A}$, $\mathbf{D}$ = diagonal matrix of the square root of the eigenvalues of $\mathbf{A}'\mathbf{A}$ (called singular values), and $\mathbf{U} = \mathbf{A}\mathbf{V}\mathbf{D}^{-1}$. As with the **QR** decomposition, these matrices possess desirable properties that make them well suited for performing least squares estimation (e.g., $\mathbf{U}'\mathbf{U} = \mathbf{I}$). We will illustrate the computations using the data from Table 5.2, substituting $\mathbf{X}$ for $\mathbf{A}$.

5.3.1 Preliminary Calculations

We calculate our matrices in the usual manner.

1. Compute $\mathbf{X}'\mathbf{X}$.

$$\mathbf{X}'\mathbf{X} = \begin{bmatrix} 12 & 60 & 53 \\ 60 & 394 & 311 \\ 53 & 311 & 299 \end{bmatrix}$$

2. Use the QR algorithm to find the eigenvalues of $\mathbf{X}'\mathbf{X}$, and create a diagonal matrix, $\mathbf{D}$, with their square root.

$$\mathbf{D} = \begin{bmatrix} \sqrt{670.8325} & 0 & 0 \\ 0 & \sqrt{31.9347} & 0 \\ 0 & 0 & \sqrt{2.2329} \end{bmatrix} = \begin{bmatrix} 25.90043 & 0 & 0 \\ 0 & 5.65108 & 0 \\ 0 & 0 & 1.49427 \end{bmatrix}$$

3. Compute **V**, the eigenvectors of $\mathbf{X'X}$, using the **QR** algorithm.

$$
\mathbf{V} = \begin{bmatrix}
.12061 & .03695 & .99201 \\
.75294 & -.65466 & -.06716 \\
.64695 & .75502 & -.10678
\end{bmatrix}
$$

4. Calculate $\mathbf{U}(\mathbf{U} = \mathbf{XVD}^{-1})$.

$$
\mathbf{U} =
\begin{bmatrix}
1 & 1 & 2 \\
1 & 9 & 7 \\
1 & 1 & 1 \\
1 & 5 & 8 \\
1 & 6 & 5 \\
1 & 8 & 7 \\
1 & 2 & 4 \\
1 & 4 & 7 \\
1 & 2 & 1 \\
1 & 8 & 4 \\
1 & 7 & 3 \\
1 & 7 & 4
\end{bmatrix}
\begin{bmatrix}
.12061 & .03695 & .99201 \\
.75294 & -.65466 & -.06716 \\
.64695 & .75502 & -.10678
\end{bmatrix}
\begin{bmatrix}
25.90043 & 0 & 0 \\
0 & 5.65108 & 0 \\
0 & 0 & 1.49427
\end{bmatrix}^{-1}
=
\begin{bmatrix}
.08368 & .15791 & .47601 \\
.44114 & -.10083 & -.24086 \\
.05871 & .02430 & .54747 \\
.34983 & .49616 & -.13255 \\
.30397 & -.02051 & .03690 \\
.41207 & .01501 & -.19592 \\
.16271 & .30927 & .28814 \\
.29579 & .47840 & -.01614 \\
.08778 & -.09155 & .50252 \\
.33713 & -.38581 & .01847 \\
.28308 & -.40357 & .13488 \\
.30806 & -.26996 & .06342
\end{bmatrix}
$$

5.3.2   Reconstructing X

Before turning to the calculation of the regression coefficients and their standard errors, we will confirm that our three matrices reconstruct our original, rectangular matrix, **X**.

$$
\mathbf{X} = \mathbf{UDV'}
$$

$$
\mathbf{X} =
\begin{bmatrix}
.08368 & .15791 & .47601 \\
.44114 & -.10083 & -.24086 \\
.05871 & .02430 & .54747 \\
.34983 & .49616 & -.13255 \\
.30397 & -.02051 & .03690 \\
.41207 & .01501 & -.19592 \\
.16271 & .30927 & .28814 \\
.29579 & .47840 & -.01614 \\
.08778 & -.09155 & .50252 \\
.33713 & -.38581 & .01847 \\
.28308 & -.40357 & .13488 \\
.30806 & -.26996 & .06342
\end{bmatrix}
\begin{bmatrix}
25.90043 & 0 & 0 \\
0 & 5.65108 & 0 \\
0 & 0 & 1.49427
\end{bmatrix}
\begin{bmatrix}
.12061 & .03695 & .99201 \\
.75294 & -.65466 & -.06716 \\
.64695 & .75502 & -.10678
\end{bmatrix}'
=
\begin{bmatrix}
1 & 1 & 2 \\
1 & 9 & 7 \\
1 & 1 & 1 \\
1 & 5 & 8 \\
1 & 6 & 5 \\
1 & 8 & 7 \\
1 & 2 & 4 \\
1 & 4 & 7 \\
1 & 2 & 1 \\
1 & 8 & 4 \\
1 & 7 & 3 \\
1 & 7 & 4
\end{bmatrix}
$$

5.3.3 Regression Coefficients

We use the following formula to calculate the regression coefficients.

$$\mathbf{b} = \mathbf{V}\mathbf{D}^{-1}\mathbf{U}'\mathbf{y} \tag{5.24}$$

Notice that here we are inverting our diagonal matrix of singular values instead of the matrix of predictors. This is advantageous because, as with the upper triangular matrix $\mathbf{R}$, inverting a diagonal matrix is more economical and less prone to error than is inverting a sum of squares matrix.

5.3.4 Standard Errors

We can also compute our standard errors using the eigenpairs. First, we must calculate MS_{res}. Since we have our matrix of regression coefficients, we can calculate the residuals in the usual manner ($\mathbf{e} = \mathbf{y} - \mathbf{X}\mathbf{b}$) and then divide the sum of their squared values by our degrees of freedom ($df = 9$). Doing so yields the $MS_{res} = 1.4974$. The following formula is then used to find the standard errors.

$$se_{b_j} = \sqrt{MS_{res} * \sum_{k=1}^{p} \frac{v_{jk}^2}{\lambda_k}} \tag{5.25}$$

Looking over the formula, we see that, within each eigenvector row, we square each value, divide each squared value by its corresponding eigenvalue, and then sum the quotients. To illustrate, the standard error of b_1 is found as follows.

$$se_{b_1} = \sqrt{1.4974 * \left[\frac{.75294^2}{670.8325} + \frac{-.65466^2}{31.9347} + \frac{-.06716^2}{2.2329} \right]} = .15616$$

If you look back to Chap. 4, you will find that this value matches the one we found using Eq. (4.6) (reproduced below).

$$se_{b_j} = \sqrt{MS_{res} * (\mathbf{X}'\mathbf{X})_{ii}^{-1}} \tag{5.26}$$

The equivalence between the two methods reveals an important similarity. Comparing Eq. (5.25) with Eq. (5.26), we see that the diagonal elements of $(\mathbf{X}'\mathbf{X})^{-1}$ equal $\sum_{k=1}^{p} \frac{v_{jk}^2}{\lambda_k}$. To prove this is so, we will perform the calculations. First, we will find the inverse of $\mathbf{X}'\mathbf{X}$.

$$\mathbf{X'X} = \begin{bmatrix} 12 & 60 & 53 \\ 60 & 394 & 311 \\ 53 & 311 & 299 \end{bmatrix} \text{ and } \mathbf{X'X}^{-1} = \begin{bmatrix} .44080 & -.03046 & -.04645 \\ -.03046 & .01629 & -.01154 \\ -.04645 & -.01154 & .02358 \end{bmatrix}$$

Then we will derive the diagonal entries using the eigenpairs of $\mathbf{X'X}$.

$$\left(\mathbf{X'X}\right)_{11}^{-1} = \frac{.12061^2}{670.8325} + \frac{.03695^2}{31.9347} + \frac{.99201^2}{2.2329} = .44080$$

$$\left(\mathbf{X'X}\right)_{22}^{-1} = \frac{.75294^2}{670.8325} + \frac{-.65466^2}{31.9347} + \frac{-.06716^2}{2.2329} = .01629$$

$$\left(\mathbf{X'X}\right)_{33}^{-1} = \frac{.64695^2}{670.8325} + \frac{.75502^2}{31.9347} + \frac{-.10678^2}{2.2329} = .02358$$

Remembering that the diagonal entries of $\mathbf{X'X}^{-1}$ represent a ratio, with a cofactor in the numerator and the variability of the entire matrix in the denominator (i.e., the determinant), we see again that eigenpairs simply redistribute the variance in a matrix, such that the variance of each regression coefficient can be found by summing the ratio of squared eigenvectors to their corresponding eigenvalues.

5.3.5   R Code: Singular Value Decomposition

```
#SVD for Regression
x0 <-rep(1,12)
x1 <-c(1,9,1,5,6,8,2,4,2,8,7,7)
x2 <-c(2,7,1,8,5,7,4,7,1,4,3,4)
y <-c(3,8,2,8,5,9,4,5,2,4,2,6)
X <-cbind(x0,x1,x2)
svd.reg <-svd(X)
svd.reg
D <-diag(svd.reg$d)
V <-svd.reg$v
U <-svd.reg$u

#Reconstruct X
X_rc<-U%*%D%*%solve(V)
X_rc

#Regression Coefficients
svd.beta <-V%*%solve(D)%*%t(U)%*%y
svd.beta

#Standard Errors Using Eigen Pairs
V1 <-V[1, ]
c1 <-sum(V1^2/svd.reg$d^2)
```

(continued)

5.3.5   *R Code: Singular Value Decomposition (continued)*

```
V2 <-V[2, ]
c2 <-sum(V2^2/svd.reg$d^2)
V3 <-V[3, ]
c3 <-sum(V3^2/svd.reg$d^2)
C <-cbind(c1,c2,c3)
reg <-lm(y~x1+x2)
mse <-summary(reg)$sigma^2

std.errors <- sqrt(mse*C)
std.errors
```

5.4   Cholesky Decomposition

The final decomposition we will learn is the Cholesky decomposition. Unlike the two previous methods, which can be performed with any rectangular matrix, the Cholesky decomposition can be performed only on a symmetric matrix with all eigenvalues greater than 0 (termed symmetric, positive definite). Although this might seem limiting, a sum of squares matrix is symmetric, positive definite, so the Cholesky decomposition is applicable to linear regression.

The Cholesky decomposition proceeds by noting that every symmetric, positive definite matrix $\mathbf{A}$ can be decomposed into the product of two lower triangular matrices.

$$\mathbf{A} = \mathbf{LL}' \tag{5.27}$$

A lower triangular matrix has 0's above the main diagonal, and $\mathbf{L}$ is known as the Cholesky factor of $\mathbf{A}$. It can be thought of as a square root matrix, a term that will become clearer after we learn how the matrix is created.[10]

5.4.1   *Calculations*

Imagine we wish to perform a Cholesky decomposition on a 3×3 matrix.

---

[10] The Cholesky decomposition can be modified to yield an upper triangular matrix, $\mathbf{R}$. This matrix is identical to the $\mathbf{R}$ matrix generated by a $\mathbf{QR}$ decomposition of a symmetric, positive definite matrix, and the reconstruction becomes $\mathbf{A} = \mathbf{R}'\mathbf{R}$.

$$A = \begin{bmatrix} a_{11} & a_{12} & a_{13} \\ a_{21} & a_{22} & a_{23} \\ a_{31} & a_{32} & a_{33} \end{bmatrix} = \begin{bmatrix} l_{11} & 0 & 0 \\ l_{21} & l_{22} & 0 \\ l_{31} & l_{32} & l_{33} \end{bmatrix} \begin{bmatrix} l_{11} & 0 & 0 \\ l_{21} & l_{22} & 0 \\ l_{31} & l_{32} & l_{33} \end{bmatrix}'$$

We perform the following calculations to form the decomposition.

$$L = \begin{bmatrix} l_{11} = \sqrt{a_{11}} & 0 & 0 \\ l_{21} = a_{21}/l_{11} & l_{22} = \sqrt{a_{22} - l_{21}^2} & 0 \\ l_{31} = a_{31}/l_{11} & l_{32} = \dfrac{a_{32} - (l_{21} * l_{31})}{l_{22}} & l_{33} = \sqrt{a_{33} - l_{31}^2 - l_{32}^2} \end{bmatrix}$$

Expressed in more formal terms, the following formulae describe the computations for the diagonal elements

$$l_{kk} = \sqrt{a_{kk} - \sum_{j=1}^{k-1} l_{kj}^2} \tag{5.28}$$

and the elements below the diagonal.

$$l_{ik} = \frac{1}{l_{kk}} \left(a_{ik} - \sum_{j=1}^{k-1} l_{ij} l_{kj} \right) \tag{5.29}$$

To make things less abstract, let's perform a Cholesky decomposition on our sum of squares matrix.

$$X'X = \begin{bmatrix} 12 & 60 & 53 \\ 60 & 394 & 311 \\ 53 & 311 & 299 \end{bmatrix}$$

$$L = \begin{bmatrix} \sqrt{12} = 3.4641 & 0 & 0 \\ \dfrac{60}{3.4641} = 17.3205 & \sqrt{394 - 17.3205^2} = 9.6954 & 0 \\ \dfrac{53}{3.4641} = 15.2998 & \dfrac{311 - (17.3205 * 15.2998)}{9.6954} = 4.7445 & \sqrt{299 - 15.2998^2 - 4.7445^2} = 6.5120 \end{bmatrix}$$

$$L = \begin{bmatrix} 3.4641 & 0 & 0 \\ 17.3205 & 9.6954 & 0 \\ 15.2998 & 4.7445 & 6.5120 \end{bmatrix}$$

You can go ahead and verify that $LL' = A$.

5.4.2 Calculating the Determinant and the Inverse

The Cholesky factor can be used to efficiently compute the determinant of a matrix. With a square matrix, we would normally need to compute many cofactors in order to find the determinant, but with a lower triangular matrix we need only square the product of the diagonal values.

$$|\mathbf{A}| = \left(\prod l_{kk}\right)^2 \tag{5.30}$$

Using the formula, we can quickly find the determinant of our sum of squares matrix.

$$|\mathbf{X}'\mathbf{X}| = (3.4641 * 9.6954 * 6.5120)^2 = 47834$$

Similarly, the Cholesky factor can be used to efficiently find the inverse of a matrix.

$$\mathbf{A}^{-1} = \left(\mathbf{L}^{-1}\right)'\mathbf{L}^{-1} \tag{5.31}$$

5.4.3 Least Squares Regression

There are several ways to use the Cholesky factor for linear regression analyses, including techniques involving forward and backward substitution. In my opinion, the easiest method is to use the following formulae to find the regression coefficients, fitted values, and covariance matrix.

$$\mathbf{b} = \left(\mathbf{L}'\right)^{-1}\left[\mathbf{X}\left(\mathbf{L}'\right)^{-1}\right]'\mathbf{y} \tag{5.32}$$

$$\hat{\mathbf{y}} = \mathbf{X}(\mathbf{L}')^{-1}\left[\mathbf{X}(\mathbf{L}')^{-1}\right]'\mathbf{y} \tag{5.33}$$

$$\mathbf{C} = \left(\mathbf{L}^{-1}\right)'\mathbf{L}^{-1}*\mathrm{MS_e} \tag{5.34}$$

Using our sample data, you can readily verify that these equations produce their intended values.

5.4.4 Using the Cholesky Decomposition to Find the Eigenvalues

Using an iterative technique, the Cholesky decomposition can be used to find the eigenvalues of a symmetric matrix. We will use $\mathbf{X'X}$ to illustrate the steps.

- Form a Cholesky decomposition of $\mathbf{X'X}$.
- Calculate a sum of squares matrix from the Cholesky factor ($\mathbf{L'L}$).
- Form the Cholesky decomposition of $\mathbf{L'L}$ and continue iterating until the change in eigenvalues falls below a designated level.

Table 5.6 shows the first three iterations for our data set.

If you continue performing the iterations, you will find that the eigenvalues appear on the diagonals of the sum of squares ($\mathbf{L'_n L_n}$) matrix. For our example, sufficient precision is reached in 10 iterations.

Table 5.6 First three iterations of a Cholesky decomposition algorithm for finding eigenvalues of $\mathbf{X'X}$

Iteration	Matrix to be decomposed	Cholesky factor
1	$\mathbf{X'X} = \begin{bmatrix} 12 & 60 & 53 \\ 60 & 394 & 311 \\ 53 & 311 & 299 \end{bmatrix}$	$\mathbf{L_1} = \begin{bmatrix} 3.4641 & 0 & 0 \\ 17.3205 & 9.6954 & 0 \\ 15.2998 & 4.7445 & 6.5120 \end{bmatrix}$
2	$\mathbf{L'_1 L_1} = \begin{bmatrix} 546.0833 & 240.5189 & 99.6321 \\ 240.5189 & 116.5106 & 30.8964 \\ 99.6321 & 30.8964 & 42.4060 \end{bmatrix}$	$\mathbf{L_2} = \begin{bmatrix} 23.3684 & 0 & 0 \\ 10.2925 & 3.2520 & 0 \\ 4.2635 & -3.9932 & 2.8780 \end{bmatrix}$
3	$\mathbf{L'_2 L_2} = \begin{bmatrix} 670.1961 & 16.4462 & 12.2703 \\ 16.4462 & 26.5212 & -11.4923 \\ 12.2703 & -11.4923 & 8.2827 \end{bmatrix}$	$\mathbf{L_3} = \begin{bmatrix} 25.8882 & 0 & 0 \\ .6353 & 5.1105 & 0 \\ .4740 & -2.3077 & 1.6531 \end{bmatrix}$

$$L'_{10}L_{10} = \begin{bmatrix} 670.8325 & 0 & 0 \\ 0 & 31.9347 & 0 \\ 0 & 0 & 2.2328 \end{bmatrix}$$

5.4.5 R Code: Cholesky Decomposition

```
#Cholesky Decomposition
x0 <-rep(1,12)
x1 <-c(1,9,1,5,6,8,2,4,2,8,7,7)
x2 <-c(2,7,1,8,5,7,4,7,1,4,3,4)
y <-c(3,8,2,8,5,9,4,5,2,4,2,6)
X <-cbind(x0,x1,x2)
XX <-t(X)%*%X
```

(continued)

5.4.5 R Code: Cholesky Decomposition (continued)

```
M=XX
n = max(dim(M))
L = matrix(0, nrow=n, ncol=n)
for (i in 1:n) {
  L[i,i] = sqrt(M[i,i] - L[i,,drop=FALSE] %*% t(L[i,,drop=FALSE]))
  if (i < n) {
    for (j in (i+1):n) {
      L[j,i] = (M[j,i] - L[i,,drop=FALSE] %*% t(L[j,,drop=FALSE]))/L[i,i]
    }
  }
}
L

#Calculate Determinant
determ <-prod(diag(L)^2)
determ

#Calculate Inverse
XXinv <-t(solve(L))%*%solve(L)
XXinv

#Cholesky Regression
chol.beta <-solve(t(L))%*%t(X%*%solve(t(L)))%*%y
chol.beta
chol.fitted <-X%*%solve(t(L))%*%t(X%*%solve(t(L)))%*%y
chol.fitted
covar <- t(solve(L))%*%solve(L)*sum((y-chol.fitted)^2)/9
covar

#Cholesky Eigenvalues (10 iterations)
CC <- t(X)%*%X;
for(i in 1:10)
{
  CC <- chol(CC);
  CC <-CC%*%t(CC);
}
lambda <-diag(diag(CC))
lambda
```

5.5 Comparing the Decompositions

In this chapter, you have learned several procedures that can be used to decompose a matrix, solve linear regression analyses, find the eigenvalues, and, in some cases, find the eigenvectors of a square matrix. Table 5.7 summarizes the calculations for each method, along with the sum of squares inverse method we used in earlier chapters.

It is natural at this point to wonder which method is best for least squares estimation. Seber and Lee (2003) considered this question and concluded that the Cholesky decomposition is the most efficient, the SVD is the most accurate, and the **QR** decomposition best balances efficiency and accuracy. When we include its ability to find the eigenvalues and eigenvectors of a matrix, it is easy to see why most statistical packages use the **QR** decomposition for least squares estimation and eigen decomposition.

Table 5.7 Comparing four methods for performing linear regression

Method	$\mathbf{b}$	$\hat{\mathbf{y}}$	$\mathbf{C}$
Sum of squares	$(\mathbf{X'X})^{-1}\mathbf{X'y}$	$\mathbf{X}(\mathbf{X'X})^{-1}\mathbf{X'y}$	$MS_{res} * (\mathbf{X'X})^{-1}$
QR	$\mathbf{R}^{-1}\mathbf{Q'y}$	$\mathbf{QQ'y}$	$MS_{res} * \mathbf{R}^{-1}(\mathbf{R}^{-1})'$
SVD	$\mathbf{V}\mathbf{D}^{-1}\mathbf{U'y}$	$\mathbf{UU'y}$	$MS_{res} * \sum_{k=1}^{p} \frac{v_{jk}^2}{\lambda_k}$
Cholesky	$(\mathbf{L'})^{-1}[\mathbf{X}(\mathbf{L'})^{-1}]'\mathbf{y}$	$\mathbf{X}(\mathbf{L'})^{-1}[\mathbf{X}(\mathbf{L'})^{-1}]'\mathbf{y}$	$MS_{res} * (\mathbf{L}^{-1})'\mathbf{L}^{-1}$

Does the superiority of the **QR** decomposition mean that we wasted our time learning the matrix inverse, sum of squares method? Not at all. First, the sum of squares method provides more insight into the processes involved in computing a regression coefficient (i.e., divide a cross-product term by a sum of squares) than does the **QR** decomposition, so it's better from a pedagogical standpoint. Moreover, the other decompositions outperform the sum of squares method only when $\mathbf{X'X}$ is nearly singular. There may be times when you encounter such a matrix, but they will be rare, and the examples you will find in this book will not be among them (see Trefethen and Schreiber 1990, for a related argument). Consequently, I will continue illustrating various techniques using the sum of squares matrix and rely on more sophisticated decomposition techniques only when they are needed.

5.6   Chapter Summary

1. Matrix decompositions factor a matrix into two or more smaller matrices that can be combined to reproduce the original matrix. Their values lie in identifying underlying patterns of variability and offering computationally efficient methods of performing statistical analyses, such as linear regression.
2. An eigen decomposition redistributes the variance of a square matrix $\mathbf{A}$ into a set of eigenvalues (λ) and their associated eigenvectors ($\mathbf{v}$).

$$\mathbf{Av} = \lambda\mathbf{v}$$

3. When the eigenvalues are stored in a diagonal matrix $\mathbf{\Lambda}$ and the eigenvectors are consolidated into a matrix $\mathbf{V}$, they reproduce the original matrix.

$$\mathbf{A} = \mathbf{V}\mathbf{\Lambda}\mathbf{V}^{-1}$$

4. When subtracted from the diagonal elements of a matrix, eigenvalues create a singular matrix with a linear dependence. The eigenvalues are found by solving a polynomial equation known as the "characteristic equation."

$$|\mathbf{A} - \lambda\mathbf{I}| = 0$$

5. Associated with each eigenvalue is an eigenvector $\mathbf{v}$. When premultiplied by the original matrix, an eigenvector changes length but not direction. The magnitude of the change is determined by the eigenvalue. Eigenvectors are found from the eigenvalues.

$$(\mathbf{A} - \lambda)\mathbf{v} = 0$$

6. The sum of the eigenvalues equals the trace of the matrix, and the product of the eigenvalues equals the determinant of the matrix. A matrix in which all eigenvalues are greater than zero is invertible and of full rank.

7. The eigenvalues of a correlation matrix reflect the linear dependencies among the variables. The more highly correlated the variables are, the larger is the dominant eigenvalue and the smaller are successive eigenvalues.

8. The **QR** decomposition factors a rectangular matrix into an orthonormal matrix $\mathbf{Q}$ and an upper triangular matrix $\mathbf{R}$.

$$\mathbf{A} = \mathbf{QR}$$

9. The **QR** decomposition can be used to perform least squares estimation.

$$\mathbf{b} = \mathbf{R}^{-1}\mathbf{Q}'\mathbf{y}$$

10. A modification of the **QR** decomposition, the **QR** algorithm, uses an iterative technique to find the eigenpairs of a large, square matrix. The algorithm can be speeded up by performing an explicit shift and modified for use with nonsymmetric matrices.

11. Using the eigenpairs from a sum of squares matrix, singular value decomposition reconstructs a rectangular matrix into three matrices.

$$\mathbf{A} = \mathbf{UDV}'$$

12. Singular value decomposition can be used to perform least squares estimation.

$$\mathbf{b} = \mathbf{VD}^{-1}\mathbf{U}'\mathbf{y}$$

21. The Cholesky decomposition factors a matrix into two, lower triangular matrices.

$$\mathbf{A} = \mathbf{L}\mathbf{L}'$$

22. The Cholesky decomposition provides an economical way to find the determinant and inverse of a matrix and can perform least squares estimation. An iterative modification of the decomposition yields the eigenvalues of a matrix.

23. The **QR** decomposition offers the best balance between efficiency and precision, and most statistical packages use it to perform problems involving least squares estimation and eigen decomposition.

Chapter 6
Problematic Observations

The suitability of a linear regression model depends on several factors. We briefly covered some of these factors in Chap. 2, and we are now ready for a more detailed discussion. Table 6.1 organizes them into three categories of increasing importance. The first category refers to properties that are desirable but not required; the second refers to properties that are required when ordinary least squares (OLS) estimation is used to find a least squares solution; and the final category refers to properties that are required for all methods of linear estimation.

Checking our data against these assumptions is an important part of conducting a linear regression analysis, and the term "regression diagnostics" is often applied to this process. Many researchers approach this endeavor with trepidation, but assumption violations rarely constitute an insurmountable problem. Most can be accommodated by gathering more data, transforming our variables, or modifying our regression model. So discovering that an assumption has been violated is not a death sentence; rather, it represents an opportunity to improve your model and strengthen your confidence in your findings. Moreover, you will learn a lot about regression analysis by carefully considering the assumptions that a linear model makes, the problems that arise when these assumptions are violated, and the remedial measures we can take to minimize problems when they occur.

The next five chapters concern issues of this nature. The current chapter discusses the first category in Table 6.1—problematic observations. We begin here because problems often arise from discrepant values that are disproportionately influential. Chapter 7 considers the distribution of the errors, and Chaps. 8, 9, and 10 take up specification errors. Because our coverage of these issues is spread out over several chapters, we defer a detailed consideration of each one until we come to the chapter in which it is most prominently discussed.

Electronic Supplementary Material: The online version of this chapter (doi: 10.1007/978-3-319-11734-8_6) contains supplementary material, which is available to authorized users

Table 6.1 Assumptions underlying simple linear regression (SLR) and multiple regression (MR)

Category	Assumed property	Description	Applies to SLR	Applies to MR	Affects meaning of the regression coefficients	Affects size of the standard errors
Desiderata						
Problematic observations	Equal influence	Predictor values are not too discrepant from each other, and each criterion lies near its fitted value	✔	✔		✔
	No collinearity	Predictors are not too highly correlated		✔		✔
Requirements for using ordinary least squares estimation						
Errors and their distribution	Normally distributed	Error terms are normally distributed with mean 0	✔	✔		✔
	Constant variance	Error terms have common variance	✔	✔		✔
	Independent	Error terms are uncorrelated	✔	✔		✔
Requirements of linear estimation						
Model specification errors	Additivity	The criterion is an additive function of the weighted predictors and the disturbance term	✔	✔	✔	
	Linearity	With all other variables held constant, a one unit change in x_i always predicts a b_i change in y	✔	✔	✔	

6.1 Influential Observations

When we conduct a linear regression analysis, we assume that the regression coefficient expresses a uniform relation between a set of predictors and a criterion, with each pair of observations (i.e., x and y) contributing equally to its value. In fact, this ideal does not always hold. To illustrate, consider the data shown in

Table 6.2 [adapted from Anscombe (1973)]. Notice that all but one of the predictor values $= 8$, yet a line of best fit shows a steep slope relating x to y ($b_1 = .4999$). Clearly, this slope does not characterize the sample as whole; rather it is unduly influenced by one observation. This is the problem with a least squares analysis. Even with a large sample, discrepant values can be disproportionately influential.

Table 6.2 Illustration of an overly influential data point

Observation #	x	y
1	8	6.58
2	8	5.76
3	8	7.71
4	8	8.84
5	8	8.47
6	8	7.04
7	8	5.25
8	19	12.50
9	8	5.56
10	8	7.91
11	8	6.89

6.1.1 Discrepant Observations

Discrepant observations come in two varieties.

- First, the value of a predictor can be unusually large or small in comparison with the other predictors; when this occurs, the observation is said to have high leverage.
- Second, a criterion can be unusually discrepant from its fitted value; in this case, the observation is said to be an outlier.

An observation that is characterized by only one of these qualities is unlikely to be problematic, but an observation that possesses both of them exerts undue influence on the regression coefficients and distorts their standard errors.

6.1.2 Illustrating Undue Influence

Identifying values of undue influence proceeds through a process of elimination. From a complete set of observations, we eliminate one observation at a time, reestimating our model using $N - 1$ observations. If the recalculated model differs substantially from the full one, we conclude that the eliminated observation is disproportionately influential.

6.1.2.1 Analysis Using All Observations

To illustrate this process, consider the data shown in Table 6.3. Here we have recruited five students for a (hypothetical) study between a predictor (how many hours/week do you study matrix algebra) and a criterion (how well did you perform on a final exam).[1] Notice that the line of best fit is positive in sign but not significantly different from 0 ($b_1 = .1667$, $p = .7177$).

Table 6.3 Hypothetical data illustrating regression diagnostics

Observation	x	y		b_1	p			
1	2	3	Excluding Observation 1	−.1538	.7519			
2	4	4	Excluding Observation 2	.1667	.7610			
3	5	5	Excluding Observation 3	.1791	.7682			
4	7	6	Excluding Observation 4	−.1111	.9024			
5	2	7	Excluding Observation 5	.6154	.0077			
Mean	**4**	**5**						
Deviation Sum of Squares	**18**	**10**						

Significance test of regression model							
	SS	df	MS	R^2	F	p	
Regression	.50	1	.50	.05	.1579	.7177	
Residual	9.50	3	3.1667				
Total	10.00						

Regression coefficients								
	b	se_b	t	p	$(\mathbf{X'X})^{-1}$		$\mathbf{C}$	
b_0	4.3333	1.8569	2.3336	.1018	1.0889	−.2222	3.4481	−.7037
b_1	.1667	.4194	.3974	.7177	−.2222	.0556	−.7037	.1759

6.1.2.2 Deleting Observations

One way to gauge an observation's influence is to perform a regression analysis with the observation excluded. Figure 6.1 plots the regression lines for the entire sample and then successively with one observation removed. Looking over the scatterplots, we see a clear linear pattern when Observation #5 is excluded, such that performance improves as study time increases. As this pattern emerges only after this observation is eliminated, the observation seems unduly influential.

---

[1] Because we are interested in individual cases, the sample size is purposely small to highlight each observation's influence.

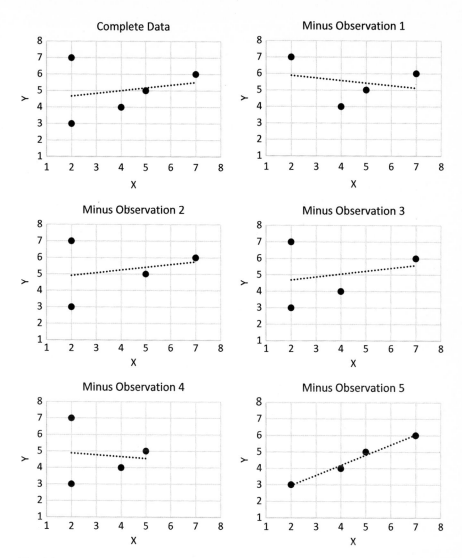

Fig. 6.1 Regression lines of best fit for a complete sample and with one observation eliminated

6.1.2.3 Impracticality of the Deletion Method

Inspecting each regression line is informative, but it suffers from some limitations. First, it is tedious with large sample sizes. Second, we often have more than one predictor, making it even more impractical to create multiple graphs. Finally, saying that the regression lines "look a lot different" is too impressionistic for science. Instead, we need some way of quantifying the discrepancies. Fortunately, we can solve all of these problems by calculating a few preliminary statistics and combining them to form some summary values. That being said, you should also be aware that there are no hard-and-fast rules for gauging undue influence. We will discuss some guidelines, but ultimately researchers are free to decide whether an observation is so extreme that it needs to be removed.

6.1.3 Leverage and the Hat Matrix

We begin by examining the property of a predictor known as its leverage. As noted earlier, predictor values that lie far from the mean of the predictor (in the case of simple linear regression) or far from the center of the vector space of all predictors (in the case of multiple regression) possess high leverage and are potentially problematic.

6.1.3.1 Calculating the Hat Matrix

Quantifying an observation's leverage involves examining the diagonal entries of a matrix known as the hat matrix $\mathbf{H}$.

$$\mathbf{H} = \mathbf{X}\left(\mathbf{X'X}\right)^{-1}\mathbf{X'} \tag{6.1}$$

The formula looks imposing, but it's really quite simple.[2] Throughout the text, we have used the following formula to calculate fitted values,

$$\hat{\mathbf{y}} = \mathbf{Xb}$$

and regression coefficients.

$$\mathbf{b} = \left(\mathbf{X'X}\right)^{-1}\mathbf{X'y}$$

Substituting terms yields an equivalent formula for finding the fitted values.

$$\hat{\mathbf{y}} = \mathbf{X}\left(\mathbf{X'X}\right)^{-1}\mathbf{X'y}$$

---

[2] The hat matrix can also be calculated using the $\mathbf{QR}$ decomposition presented in Chap. 5: $\mathbf{H} = \mathbf{QQ'}$.

Now notice that the **X** terms in this equation constitute the hat matrix, which is why it is called the hat matrix: it puts the hat (fitted values) on **y**.[3] Inserting values from Table 6.3 produces the hat matrix for our data set.

$$\mathbf{X'X} = \begin{bmatrix} 1 & 1 & 1 & 1 & 1 \\ 2 & 4 & 5 & 7 & 2 \end{bmatrix} \begin{bmatrix} 1 & 2 \\ 1 & 4 \\ 1 & 5 \\ 1 & 7 \\ 1 & 2 \end{bmatrix} = \begin{bmatrix} 5 & 20 \\ 20 & 98 \end{bmatrix}$$

and

$$\mathbf{H} = \begin{bmatrix} 1 & 2 \\ 1 & 4 \\ 1 & 5 \\ 1 & 7 \\ 1 & 2 \end{bmatrix} \begin{bmatrix} 5 & 20 \\ 20 & 98 \end{bmatrix}^{-1} \begin{bmatrix} 1 & 1 & 1 & 1 & 1 \\ 2 & 4 & 5 & 7 & 2 \end{bmatrix}$$

$$= \begin{bmatrix} .4222 & .2000 & .0889 & -.1333 & .4222 \\ .2000 & .2000 & .2000 & .2000 & .2000 \\ .0889 & .2000 & .2556 & .3667 & .0889 \\ -.1333 & .2000 & .3667 & .7000 & -.1333 \\ .4222 & .2000 & .0889 & -.1333 & .4222 \end{bmatrix}$$

6.1.3.2 Hat Matrix and Fitted Values

Having computed the hat matrix, we can use it find the fitted values.

$$\hat{\mathbf{y}} = \mathbf{Hy} \tag{6.2}$$

$$\hat{\mathbf{y}} = \begin{bmatrix} .4222 & .2000 & .0889 & -.1333 & .4222 \\ .2000 & .2000 & .2000 & .2000 & .2000 \\ .0889 & .2000 & .2556 & .3667 & .0889 \\ -.1333 & .2000 & .3667 & .7000 & -.1333 \\ .4222 & .2000 & .0889 & -.1333 & .4222 \end{bmatrix} \begin{bmatrix} 3 \\ 4 \\ 5 \\ 6 \\ 7 \end{bmatrix} = \begin{bmatrix} 4.6667 \\ 5.0000 \\ 5.1667 \\ 5.5000 \\ 4.6667 \end{bmatrix}$$

It is informative to work through the multiplication underlying each fitted value. Using our first row of data, we derive our first fitted value.

$$(.4222 * 3) + (.2000 * 4) + (.0889 * 5) + (-.1333 * 6) + (.4222 * 7) = 4.6667$$

---

[3] The hat matrix is idempotent, $\mathbf{H}^n = \mathbf{H}$. The term, idempotent, means "of the same power," and no matter how many times you multiply an idempotent matrix by itself, the product matrix is always the same as the original matrix. This peculiar result occurs because all of the eigenvalues of an idempotent matrix are 1 or 0.

As you can see, each fitted value is a linear combination of an observed value weighted by a corresponding entry in the hat matrix. Thus, each entry in the hat matrix (h_{ij}) indicates how much weight an observed value (y_j) receives in the calculation of a fitted value ($\hat{y}_i$). When h_{ij} is large, y_j has a large effect on $\hat{y}_i$; when h_{ij} is small, y_j has a small effect on $\hat{y}_i$. In terms of our example, notice that h_{34} is the largest entry in the third row of the hat matrix ($h_{34}=.3667$). Consequently, y_4 exerts the strongest influence on $\hat{y}_3$.

$$(.0889 * 3) + (.2000 * 4) + (.2556 * 5) + (.3667 * 6) + (.0889 * 7) = 5.1667$$

6.1.3.3   Calculating Hat Values

The diagonal elements in the hat matrix, called hat values, indicate how much weight each observed value is given in the calculation of its own fitted value (e. g., if h_{44} is large, the fourth observation has a large effect on the fourth fitted value). Often our interest centers on the hat values, allowing us to use a somewhat easier formula than Eq. (6.1). First, let's consider the situation when we have a single predictor, in which case the hat values can easily be found using ordinary algebra.

$$h_i = \frac{1}{N} + \frac{\left(x_i - \overline{X}\right)^2}{\sum x^2} \tag{6.3}$$

Notice that the hat value is the sum of two fractions. The first is the inverse of the sample size, and the second represents the squared deviation of a single value from the mean of all predictors $\left[\left(x_i - \overline{X}\right)^2\right]$, divided by the sum of all squared deviations from the mean ($\sum x^2$). So a hat value is really a weighted index that quantifies how much of the total variability in the predictors is attributable to the specific observation of interest. The following calculations produce the first hat value for our data set.

$$\frac{1}{5} + \frac{(2 - 4)^2}{18} = .4222$$

The remaining hat values are found in a similar fashion. Notice that the first and last observations have identical hat values ($h=.4222$). This is because they have the same value for the predictor (i.e., 2 h/week of studying). We see, then, that the hat value depends only on the predictors and does not take the criterion into account.

With multiple predictors, we use matrix algebra to find our hat values.

$$h_i = \mathbf{x}_i' \left(\mathbf{X}'\mathbf{X}\right)^{-1} \mathbf{x}_i \tag{6.4}$$

Notice that the first and last terms in this equation are vectors, not matrices. To illustrate, we will recalculate our first diagonal hat value using the matrix formula.

$$[1 \quad 2]\begin{bmatrix} 5 & 20 \\ 20 & 98 \end{bmatrix}^{-1}\begin{bmatrix} 1 \\ 2 \end{bmatrix} = .4222$$

6.1.3.4   Using the Hat Values to Quantify Leverage

The size of a hat value constitutes its leverage, with large values potentially exerting more influence than small values. Since large values are ones that make a large contribution to the overall variability of the predictors, it follows that extreme values possess high leverage. The smallest hat value occurs when a predictor equals the average predictor value (i.e., the mean of the predictors). In this case, its value will always equal $1/n$. In our sample, the predictor for the second observation lies at the mean of all predictors (4), and its hat value $= 1/5 = .20$.[4]

Although there are no strict standards as to what constitutes a large hat value, the sum of the hat values equals the number of parameters in our model (including the intercept). Consequently, the average hat value $= p/n$. Belsley, Kuh, and Welsch (1980) recommend that values twice that size should be characterized as having high leverage when sample sizes are relatively large (> 30) and that values three times the average should be characterized as large when sample sizes are relatively small (≤ 30). Our sample size is small, so the application of this rule would lead us to characterize hat values greater than 1.20 as having high leverage. Although none of our hat values exceeds this standard, bear in mind that leverage is just one component of influence.

6.1.4   Residuals and Outliers

The need to consider factors other than an observation's leverage when determining its influence is particularly apparent in our data set. Notice that the suspicious case in our data set (i.e., Observation #5) is not very discrepant from the others on either the predictor or the criterion. Instead, it is the discrepant combination of low study time and a high test score that makes it so influential.

---

[4] Some statistical packages report scaled hat values, such that the sum of all values equals 1 ($h_{scaled} = h - 1/N$). In this case, each term can be treated as a percentage of the whole. In our example, the first scaled hat value $= .2222$ and the second $= 0$.

The residuals reflect this combination. Large residuals, known as outliers, indicate that a fitted value lies far from its corresponding observation. The residuals can be easily calculated from the hat matrix:

$$\mathbf{e} = (\mathbf{I} - \mathbf{H})\mathbf{y} \tag{6.5}$$

and plugging in our values produces our residuals. Looking them over, the residual associated with Observation #5 seems unduly large.

$$
e = \left\{
\begin{bmatrix}
1 & 0 & 0 & 0 & 0 \\
0 & 1 & 0 & 0 & 0 \\
0 & 0 & 1 & 0 & 0 \\
0 & 0 & 0 & 1 & 0 \\
0 & 0 & 0 & 0 & 1
\end{bmatrix}
-
\begin{bmatrix}
.4222 & .2000 & .0889 & -.1333 & .4222 \\
.2000 & .2000 & .2000 & .2000 & .2000 \\
.0889 & .2000 & .2556 & .3667 & .0889 \\
-.1333 & .2000 & .3667 & .7000 & -.1333 \\
.4222 & .2000 & .0889 & -.1333 & .4222
\end{bmatrix}
\right\}
\begin{bmatrix}
3 \\
4 \\
5 \\
6 \\
7
\end{bmatrix}
=
\begin{bmatrix}
-1.6667 \\
-1.0000 \\
-.1667 \\
.5000 \\
2.3333
\end{bmatrix}
$$

6.1.4.1   Scaling the Residuals

Because residuals are expressed in raw units, it is often difficult to determine what constitutes an outlier simply by surveying their absolute size. To manage this problem, we customarily scale them by dividing them by a variance estimate. Three different scaling procedures are available (see Table 6.4).[5]

1. First, we can create a standardized residual by dividing each residual by the square root of MS_{res}:

$$e_{\text{standardized}} = \frac{e_i}{\sqrt{MS_{res}}} \tag{6.6}$$

Table 6.3 shows the mean square residual for our data set ($MS_{res} = 3.1667$), and the first standardized residual shown in Table 6.4 was calculated as follows:

$$e_{1.\text{standardized}} = \frac{-1.6667}{\sqrt{3.1667}} = -.9366$$

The obtained value shows the distance between the residual and the line of best fit, expressed in residual standard deviations. In our case, our first residual is nearly one standard deviation below the regression line.

---

[5] Unfortunately, terminology isn't standard and these residuals go by different names in different statistical packages.

Table 6.4 Leverage and residual values for hypothetical data

x	y	$\hat{y}$	e	h	s^2	Standardized residual	Studentized residual	Deleted studentized residual
2	3	4.6667	−1.6667	.4222	2.3462	−.9366	−1.2322	−1.4315
4	4	5.0000	−1.0000	.2000	4.1250	−.5620	−.6283	−.5505
5	5	5.1667	−.1667	.2556	4.7313	−.0937	−.1086	−.0888
7	6	5.5000	.5000	.7000	4.3333	.2810	.5130	.4385
2	7	4.6667	2.3333	.4222	.0385	1.3112	1.7250	15.6525

2. Standardized residuals assume that all residuals have the same variance. When this is not the case, we can use a different denominator with each observation to form a studentized residual.

$$e_{studentized} = \frac{e_i}{\sqrt{MS_{res} * (1 - h_i)}} \tag{6.7}$$

With our first observation, the studentized residual assumes the following value.

$$e_{1.studentized} = \frac{-1.6667}{\sqrt{3.1667 * (1 - .4222)}} = -1.2322$$

3. A third scaled residual is known as the deleted studentized residual.

$$e_{deleted} = \frac{e_i}{\sqrt{s_i^2 * (1 - h_i)}} \tag{6.8}$$

Here, we replace the estimate of the overall variance (MS_{res}) in Eq. (6.7) with one that excludes the observation of interest.

$$s_i^2 = \frac{[(N - p) * MS_e] - e_i^2/(1 - h_i)}{(N - p - 1)} \tag{6.9}$$

Using our data, we compute the deleted residual variance for our first observation,

$$s_1^2 = \frac{[(5 - 2) * 3.1667] - [-1.6667^2/(1 - .4222)]}{(5 - 2 - 1)} = 2.3462$$

and then calculate its deleted studentized residual.

$$e_{1.deleted} = \frac{-1.6667}{\sqrt{2.3462 * (1 - .4222)}} = -1.4315$$

The statistical significance of a deleted studentized residual can be found by referring it to a t-distribution with $N - p - 1$ degrees of freedom. Values that are statistically different from 0 are considered outliers. Clearly, Observation #5 exceeds this threshold (see last entry in Table 6.4).

6.1.4.2 Plotting the Residuals

After calculating the residuals, it is useful to plot them to examine their distribution. The most common approach is to create a scatterplot with standardized fitted values on the x axis and one of the three residuals on the y axis. Figure 6.2 shows such a scatterplot using deleted studentized residuals. Ideally, across all levels of the standardized fitted value, the residuals should fall in a random pattern around 0. We can't expect to find this pattern with a sample size as small as ours but deleted studentized residuals with absolute values of 3 or more deserve scrutiny. In our sample, it is evident that the value for Observation #5 is very discrepant from the others.

Fig. 6.2 Deleted studentized residuals as a function of standardized fitted values

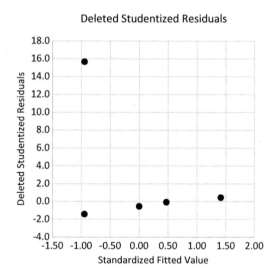

6.1.5 *Variance of Fitted Values and Residuals*

As first discussed in Chap. 2, fitted values in a linear regression model represent means from a conditional probability distribution [i. e., $E(y|x = i)$]. Thus, each fitted value (and, therefore, its corresponding residual) has its own variance. The hat matrix yields the variance/covariance matrix of the fitted values

$$Var_{(\hat{y})} = \mathbf{H} * MS_{res} \qquad (6.10)$$

and residuals.

$$Var_{(e)} = (\mathbf{I} - \mathbf{H}) * MS_{res} \qquad (6.11)$$

The diagonal elements of each matrix display the variances, and the off-diagonal elements represent the covariances.

Remembering from Chap. 3 that the variance of an estimate represents its information value, we can see that the fitted value of y from an x with high leverage is imprecise (i.e., high variance), whereas the fitted value of y from an x with low leverage is precise (i.e., low variance). This is hardly surprising, as we ordinarily expect extreme values to be less informative than moderate ones. But now consider the variances of the residuals. Here, large hat values are associated with small residual estimates, and small hat values are associated with large residual estimates. The inverse relation between the size of a hat value and its residual is the reason why the hat value is given the name "leverage." As h approaches its maximum value of 1, the discrepancy between its observed value and fitted value approaches 0. So high values of h *pull* the regression line towards them, virtually demanding that their fitted values match their observed value.

6.1.6   *Quantifying Influence*

So far we have learned how to quantify the leverage of a predictor and the extremity of a criterion. Observations with extreme values on both of these dimensions are likely to be unduly influential, and several statistics are available to quantify this influence. Table 6.5 shows these statistics, using nicknames that have become standard in the field.

6.1.6.1   DFFIT (Difference in Fitted Values with One Observation Excluded)

Our first summary statistic compares the fitted values using all observations vs. the fitted values with one observation omitted. For example, if before deletion, $\hat{y}\,|x = 3$ is 6 but after an observation is omitted, $\hat{y}\,|x = 3$ is 2, the DFFIT value would be $(6 - 2) = 4$. Thus, DFFIT values provide an indication of how influential each observation is in determining its own fitted value. Fortunately, we don't have to perform many regression analyses to find these values; instead, they can be found algebraically.

Table 6.5 Summary of regression diagnostics for hypothetical data

Case	DFFIT	DFFIT_S	$DFBETA_0$	$DFBETA_1$	se_{b0}	se_{b1}	$DFBETA\_S_0$	$DFBETA\_S_1$	Cook's D	COVRATIO	R_i^2
1	-1.2179	-1.2237	-1.8590	.3205	1.5983	.3610	-1.1631	.8878	.5547	.9501	.0615
2	-.2500	-.2752	-.2500	.0000	2.1194	.4787	-.1180	.0000	.0493	2.1211	.0571
3	-.0572	-.0520	.0050	-.0124	2.2698	.5127	.0022	-.0243	.0020	2.9987	.0537
4	1.1667	.6699	-.7778	.2778	2.1722	.4907	-.3581	.5661	.3070	6.2419	.0095
5	1.7051	13.3805	2.6026	-.4487	.2046	.0462	12.7173	-9.7073	1.0873	.0003	.9846

$$DFFIT = \frac{h_i e_i}{(1 - h_i)} \qquad (6.12)$$

Plugging in our values, we find the DFFIT value for our first observation.

$$DFFIT_1 = \frac{.4222 * -1.6667}{(1 - .4222)} = -1.2179$$

As with the residuals, DFFIT values are expressed in raw units, making it desirable to scale them. The following formula finds the scaled DFFIT values (known as DFFIT_S).

$$DFFIT\_S_i = \sqrt{\frac{h_i}{(1 - h_i)}} * \frac{e_i}{\sqrt{s_i^2 * (1 - h_i)}} \qquad (6.13)$$

Applying Eq. (6.13) produces the scaled DFFIT value for our first observation.

$$DFFIT\_S_1 = \sqrt{\frac{.4222}{(1 - .4222)}} * \frac{-1.6667}{\sqrt{2.3462 * (1 - .4222)}} = -1.2237$$

Absolute scaled DFFIT values greater than $2\sqrt{p/N}$ deserve scrutiny. Using our sample values, we are wary of DFFIT_S values that exceed $|1.2649|$. Table 6.5 shows that Observation #5 greatly surpasses this threshold.

6.1.6.2   DFBETA (Difference in Regression Coefficients with One Observation Excluded)

Out next summary value, DFBETA, quantifies how much the regression coefficients change when an observation is eliminated. If, for example, $b_1 = .75$ when all observations are included, but $b_1 = .50$, when Observation #1 is excluded, the DFBETA value for Observation #1 would be $(.75 - .50) = .25$. As before, we don't need to run separate regression models to find these values; instead, we use the following formula (remembering that $\mathbf{x_i}$ references a vector not a matrix).

$$DFBETA_i = \frac{(\mathbf{X'X})^{-1}\mathbf{x_i}e_i}{(1 - h_i)} \qquad (6.14)$$

Plugging in the values produces the DFBETA for Observation #1.

$$DFBETA_i = \frac{\begin{bmatrix} 5 & 20 \\ 20 & 98 \end{bmatrix}^{-1} \begin{bmatrix} 1 \\ 2 \end{bmatrix} - 1.6667}{(1 - .4222)} = \begin{bmatrix} -1.8590 \\ .3205 \end{bmatrix}$$

The top entry in the product vector shows the change in the intercept, and the bottom entry shows the change in the slope of the regression line. We can verify that the bottom value is correct by revisiting Table 6.3. The penultimate column shows the regression coefficients with each observation deleted.[6] For Observation #1, the value was $b_1 = -.1538$. Remembering now that $b_1 = .1667$ from the complete data set, we find that the difference between the two regression coefficients matches the value for DBETA using Eq. (6.14).

$$.1667 - (-.1538) = .3205$$

As with all unstandardized regression coefficients, DFBETA values are expressed in raw units rather than standardized ones. To scale them, we find the standard error for each coefficient using the following formula, where the final term refers to the diagonal entry of $(\mathbf{X'X})^{-1}$.

$$sb_i = \sqrt{s_i^2 (\mathbf{X'X})_{ii}^{-1}} \tag{6.15}$$

Notice that this formula matches the one we have used throughout the book to find our parameter standard errors, except instead of using MS_{res} to represent our residual variance, we substitute the variance estimate for each deleted observation. For Observation #1, our standard errors become

$$\sqrt{2.3462 * 1.0889} = 1.5983$$

and

$$\sqrt{2.3462 * .0556} = .3610$$

for the intercept and slope, respectively.

We then divide DFBETA by the standard errors to compute scaled DFBETA values.

$$DFBETA\_S = \frac{DFBETA}{sb_i} \tag{6.16}$$

Plugging in values for Observation #1 in our example produces the scaled intercept:

$$DFBETA\_S_{b0.1} = \frac{-1.8590}{1.5983} = -1.1631$$

and slope.

---

[6] Penultimate means "next to the last."

$$DFBETA\_S_{b1.1} = \frac{.3205}{.3610} = .8878$$

Scaled DFBETA values greater than $2/\sqrt{N}$ are considered extreme. With our sample size, values greater than .8944 deserve attention. Table 6.5 shows that the values for $DFBETA\_S$ Observation #5 exceed this standard.

6.1.6.3 Cook's D

Calculating scaled DFBETA values is useful when we have a small number of predictors, but suppose we have many predictors. Ideally, we would like to have a way of integrating the scaled discrepancies across all of them to create an overall discrepancy index. Cook's D provides such a measure (Cook 1977). It quantifies how much the vector of scaled regression coefficients changes with the deletion of a single observation. The formula appears below. As you can see, it uses components of leverage, outliers, and the residual sum of squares. For this reason, it provides the most comprehensive indication of how much each observation affects the least squares solution.[7]

$$D_i = \frac{e_i^2 h_i}{(p * MS_{res}) * (1 - h_i)^2} \qquad (6.17)$$

Plugging in values for Observation #1 produces its D value.

$$D_1 = \frac{-1.6667^2 * .4222}{2 * 3.1667 * (1 - .4222)^2} = .5547$$

Opinions vary regarding the critical value for Cook's D. Some statisticians recommend referring the obtained value to an F distribution with p and $N - p$ degrees of freedom and then judging all values that approach or exceed the 50th percentile value as problematic. Other analysts believe that all values greater than $4/(N - p)$ merit scrutiny. In our small data set, Observation #5 would be considered problematic by the former criterion, but not by the latter.

---

[7] The "D" in Cook's D stands, for distance, and the statistic can also be calculated with a formula that measures how far the vector of fitted values moves when a case is deleted:

$$D = \frac{\Sigma(\mathbf{Xb} - \mathbf{Xb_1})^2}{MS_{res}*p}$$

(where b_1 refers to the coefficients with the ith values excluded, and p refers to the number of predictors).

6.1.6.4   COVRATIO (Changes in Standard Errors)

Cook's D quantifies how much the regression coefficients change when an observation is eliminated, but we might also want to know how much the standard errors change when an observation is excluded. This information is provided by the diagonal elements of the parameter covariance matrix, so it's informative to compare the covariance matrix when all of the observations are used with the covariance matrix when one observation is omitted. One way to do this is to calculate a fraction, with the determinant from the complete covariance matrix in the denominator and the determinant from the covariance matrix with one observation excluded in the numerator. This statistic is called COVRATIO.

$$COVRATIO = \frac{\left|(\mathbf{X'X})_i^{-1} * s_i^2\right|}{\left|(\mathbf{X'X})^{-1} * MS_{res}\right|} \tag{6.18}$$

Unsurprisingly, we do not have to directly calculate the determinants for each term. Instead, we can use the following formula to calculate the COVRATIO for each observation in our sample.[8]

$$COVRATIO = \left(\frac{s_i^2}{MS_{res}}\right)^p * \frac{1}{(1 - h_i)} \tag{6.19}$$

Plugging in values from our first observation produces its COVRATIO.

$$COVRATIO_1 = \left(\frac{2.3462}{3.1667}\right)^2 * \frac{1}{(1 - .4222)} = .9501$$

Belsley et al. (1980) suggest using the following formula to identify COVRATIO values that deserve scrutiny.

$$|COVRATIO - 1| \geq \frac{3p}{N} \tag{6.20}$$

Applying the formula, the cutoff value with our example is 1.2.

Table 6.5 shows the COVRATIO value for all five observations. When COVRATIO is < 1, the deleted observation inflates the variance estimates (i.e., makes them less precise); when COVRATIO is > 1, the deleted observation shrinks the variance estimates (i.e., makes them more precise). Using these guidelines, Table 6.5 shows that the first and last observations inflate the standard errors, and the middle three observations reduce them.

---

[8] The COVRATIO can also be used to find each observation's leverage:
$h_i = 1 - \left\{\frac{1}{COVRATIO_i} * \left(\frac{s_i^2}{MS_e}\right)^p\right\}$

6.1.6.5 R^2 Following Omitted Values

In some situations, it might be desirable to calculate the coefficient of determination for a regression model with a particular observation excluded. We can readily compute SS_{res_i} from s_i^2:

$$SS_{resi} = s_i^2 * (N - p - 1) \tag{6.21}$$

so we need to only find SS_{y_i}.

$$SS_{y_i} = SS_y - \left\{ (y_i - \overline{Y})^2 * \frac{N}{N-1} \right\} \tag{6.22}$$

With those values in hand, it's easy to calculate R_i^2 for a subset of observations.

$$R_i^2 = \frac{SS_{y_i} - SS_{resi}}{SS_{y_i}} \tag{6.23}$$

Using Eq. (6.23) with our first observation produces its R_i^2 value.

$$R_1^2 = \frac{\left[10 - \left\{ (3-5)^2 * \frac{5}{4} \right\} \right] - [2.3462 * 2]}{\left[10 - \left\{ (3-5)^2 * \frac{5}{4} \right\} \right]} = \frac{.3076}{5} = .0615$$

The coefficient's significance can be calculated using Eq. (2.26), with k and $N - k - 1$ degrees of freedom. If desired, subtraction can be used to calculate changes in R^2 from the full model.

6.1.7   Commentary

Using the formulae presented above, we can calculate a variety of regression diagnostics with relative ease using a handheld calculator or, in a few cases, a spreadsheet with the capability of performing matrix algebra. The decision about how to handle influential observations, however, cannot be made as easily. As tempting as it is to eliminate observations that run counter to our hypotheses or keep an effect from reaching statistical significance, the decision to do so is warranted only if the source of the influence can be shown to be the result of an identifiable error (e.g., mistakes in data entry or a failure on the part of the participant to follow experimental procedures). More generally, regression diagnostics identify observations of undue influence, but they do not provide any information regarding the source of that influence. Excluding observations based on diagnostic information should be made rarely and only with full justification.

6.1.8   R Code: Regression Diagnostics

```
#Regression Diagnostics
x=c(2,4,5,7,2)
y=c(3,4,5,6,7)
model=lm(y~x)
summary(model)

#Returns: DFBETA DFFIT COV.R COOK.D HAT INFLUENCE
influence.measures(model)

#Compute Hat Matrix
X <-cbind(1,x)
hat <- X%*%(solve(t(X)%*%X))%*%t(X)
hat

#Calculate Residuals
resid <-resid(model)
standardized <-resid(model)/sqrt(sum(resid(model)^2)/3)
studentized <-rstandard(model)
deleted <-rstudent(model)
all.resids <-cbind(resid, standardized, studentized, deleted)
all.resids
```

6.2   Departures from Normality

Influential data points can produce residuals that are not normally distributed. Because OLS estimation assumes that the errors are normally distributed (see Table 6.1), this state of affairs is potentially problematic. Fortunately, this violation rarely poses a serious problem when sample sizes are large (~30 observations/predictor). Still, we don't always have access to large sample sizes, and even when we do, there will be times when it will be important to identify and rectify this violation, so we will learn methods for doing so.

6.2.1   Reviewing the Normality Assumption

Let's begin by reviewing the normality assumption, first discussed in Chap. 2. In a linear regression model, each fitted value represents the mean of a conditional probability distribution (i.e., $E(y|x=i)$), with the observed values of y normally distributed around its fitted value. Because the errors are computed from the observed and fitted values, the normality assumption also applies to the errors (i.e., we assume that the errors are normally distributed with mean 0 and variance σ^2).[9]

6.2.2 Assessing Normality

In practice, we don't have access to an entire population, so we use our residuals to estimate the distribution of the errors. To learn how this is done, we will consider a data set in which the normality assumption is met. I created the data in Table 6.6 by first generating two sets of random numbers between 0 and 1—one for the predictor and one for the error term (which I then centered around its mean). I then created the following regression equation:

$$y = 4 + .7x + e$$

and analyzed the data using simple linear regression. The fitted values and residuals appear in Table 6.6, along with the hat values and Cook's D. As one would expect considering how they were generated, all of the observations are of (approximately) equal influence.

Table 6.6 Small sample example of "perfect" data

x	y	$\hat{y}$	e	h	Cook's D	e^2	e^3	e^4
.5161	3.7293	4.3762	−.6469	.0834	.0233	.4185	−.2708	.1752
.4688	3.3527	4.2784	−.9257	.0845	.0484	.8569	−.7932	.7343
.0237	3.0408	3.3575	−.3167	.2921	.0328	.1004	−.0318	.0101
.2626	4.2383	3.8519	.3864	.1363	.0153	.1493	.0577	.0223
.5866	4.1130	4.5220	−.4091	.0893	.0101	.1673	−.0685	.0280
.2475	4.9917	3.8205	1.1712	.1431	.1498	1.3714	1.6060	1.8808
.9705	6.4193	5.3163	1.1030	.2784	.3645	1.2168	1.3423	1.4807
.4377	3.4769	4.2141	−.7372	.0874	.0320	.5434	−.4006	.2953
.7817	3.4688	4.9256	−1.4568	.1522	.2519	2.1224	−3.0920	4.5046
.7573	6.1731	4.8753	1.2978	.1406	.1797	1.6846	2.1865	2.8379
.0665	3.9843	3.4462	.5381	.2566	.0754	.2895	.1558	.0838
.9430	5.2553	5.2594	−.0041	.2560	.0000	.0000	.0000	.0000
Mean						**.7434**	**.0576**	**1.0044**

---

[9] In a linear regression model, the errors are also assumed to be independent and identically distributed (see Table 6.1). Violations of these assumptions are discussed in Chap. 7.

6.2.2.1 Normal Probability (QQ) Plot

One way to determine whether the residuals are normally distributed is to create a histogram or stem-and-leaf display and compare the distribution to a normal distribution. These graphical displays are useful when sample sizes are large, but a smooth bell curve is not going to appear when the sample size is small.

A better alternative is to create a normal probability plot (aka QQ plot). Such a plot directly compares the observed distribution of residuals to a normal one. To create the plot, we first order our residuals from smallest to largest, as shown in Table 6.7, and then compare the order statistics (aka sample quantiles) to the quantiles from a normal distribution. Doing so involves first computing the percentile rank of each sample quantile, q_i, where r_i represents each residual's rank order.

$$q_i = \frac{r_i - .5}{N} \tag{6.24}$$

To illustrate, we'll compute the percentile rank of our smallest residual (-1.4568), whose rank order equals 1:

$$q_1 = \frac{1 - .5}{12} = .0417$$

The calculated values represent areas under a standardized normal curve, so we can use a spreadsheet function to find standardized scores corresponding to each value. For example, the first value in the final column of Table 6.7 was found by entering

$$= \text{NORMINV}(.0417, 0, 1) = -1.7317,$$

indicating that in a standardized normal distribution, ~4.2 % of scores fall below a Z score of -1.7317. For the 6th variable in the ordered sequence, we find that ~46 % of all scores fall below a Z score of $-.1046$.

$$= \text{NORMINV}(.4583, 0, 1) = -.1046$$

Finally, we plot the observed residuals against their corresponding standardized scores, and compare the fit of the two lines. The more similar the residuals are to the theoretical scores, the more normally distributed are the residuals. The left-hand graph in Fig. 6.3 presents the normal probability plot for our data, and it's clear that the two lines are quite similar, suggesting that our residuals are distributed normally.

Table 6.7 Normal probability (aka QQ) plot

Rank	Ordered residuals	$\frac{r - .5}{N}$	Theoretical standardized score
1	−1.4568	.0417	−1.7317
2	−.9257	.1250	−1.1503
3	−.7372	.2083	−.8122
4	−.6469	.2917	−.5485
5	−.4091	.3750	−.3186
6	−.3167	.4583	−.1046
7	−.0041	.5417	.1046
8	.3864	.6250	.3186
9	.5381	.7083	.5485
10	1.1030	.7917	.8122
11	1.1712	.8750	1.1503
12	1.2978	.9583	1.7317

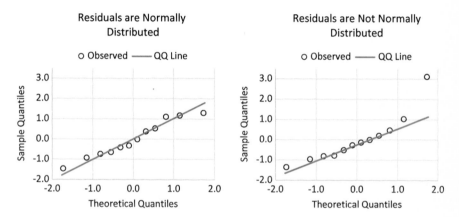

Fig. 6.3 QQ plots for normally distributed residuals and residuals that are not normally distributed

6.2.2.2 Jarque-Bera Test

Visual impressions of normality are informative, but it is also good to have a more objective test. There are several, but the one we will learn is known as the Jarque-Bera test,

$$\chi^2_{JB} = N \left[\frac{S^2}{6} + \frac{(K - 3)^2}{24} \right] \tag{6.25}$$

where S refers to the skew of the distribution (is it symmetric around its mean?),

$$S = \frac{d_3}{d_2^{1.5}} \tag{6.26}$$

K refers to the kurtosis of the distribution (is it rounded or peaked?),

$$K = \frac{d_4}{d_2^2} \tag{6.27}$$

and the various terms with d and a subscript refer to the average of deviate scores raised to the subscript's power. To illustrate, we calculate d_3 by subtracting the mean from every score, raising each deviate score to the 3rd power, and then finding the average of the cubed deviate scores. Fortunately, the mean of the residuals is always zero in an OLS regression when an intercept is included, so we can dispense with subtracting the mean from each residual. Using the averages displayed in the final three columns of Table 6.7 produces

$$S = \frac{.0576}{.7434^{1.5}} = .0899,$$

$$K = \frac{1.0044}{.7434^2} = 1.8174,$$

and the Jarque-Bera test.

$$\chi^2_{JB} = 12 * \left[\frac{.0899^2}{6} + \frac{(1.8174 - 3)^2}{24} \right] = .7154$$

The test statistic is distributed as χ^2 with 2 degrees of freedom. The critical value at the .05 level of significance is 5.9915. Our value falls far below this threshold, providing further evidence that our residuals are normally distributed.[10]

6.2.2.3 Detecting Violations of Normality in a Second Data Set

Now that we know what normally distributed residuals look like, let's consider a data set in which the normality assumption is violated. Table 6.8 displays the raw data. A regression analysis (see Table 6.9) shows a slope that falls just short of significance, and a probability plot using the calculations just described yields the figure in the right-hand column of Fig. 6.3. As you can see, the observed quantiles do not match the theoretical ones, and a Jarque-Bera test is significant at the .05 level of significance.

---

[10] Skew, kurtosis, and the Jarque-Bera test can be calculated in slightly different ways. The formulae presented here are the ones **R** uses. Other tests of normality are available in **R**, including the Shapiro-Wilk test and the Kolmogorov–Smirnov test.

$$\chi^2_{JB} = 12 * \left[\frac{1.5448^2}{6} + \frac{(5.0881 - 3)^2}{24} \right] = 6.9529$$

Thus, we conclude that the residuals are not normally distributed.

Table 6.8 Small sample example of data with residuals that violate the normality assumption

Obs.	x	y	$\hat{y}$	e	h	Cook's D	e^2	e^3	e^4
1	1.11	3.15	2.6779	.4721	.3282	.0526	.22290	.10524	.04968
2	1.25	2.72	2.7245	−.0045	.3099	.0000	.00002	.00000	.00000
3	3.25	4.42	3.3908	1.0292	.1259	.0567	1.05919	1.09009	1.12189
4	3.26	3.60	3.3942	.2058	.1254	.0023	.04237	.00872	.00180
5	4.40	2.42	3.7740	−1.3540	.0860	.0612	1.83320	−2.48207	3.36062
6	4.47	3.02	3.7973	−.7773	.0851	.0199	.60416	−.46960	.36501
7	5.52	3.34	4.1471	−.8071	.0933	.0240	.65140	−.52573	.42432
8	5.59	3.66	4.1704	−.5104	.0952	.0098	.26052	−.13297	.06787
9	6.76	3.60	4.5602	−.9602	.1545	.0647	.92199	−.88530	.85006
10	6.86	4.32	4.5935	−.2735	.1619	.0056	.07481	−.02046	.00560
11	6.95	7.74	4.6235	3.1165	.1688	.7706	9.71256	30.26916	94.33377
12	7.95	4.82	4.9567	−.1367	.2658	.0030	.01868	−.00255	.00035
Mean							**1.28348**	**2.24621**	**8.38175**

Table 6.9 Regression analysis from "imperfect" data set

		Significance test of regression model				
	SS	df	MS	R^2	F	p
Regression	6.1079	1	6.1079	.2840	3.9657	.0744
Residual	15.4018	10	1.5402			
Total	21.5097					

		Regression coefficients						
	b	se_b	t	p	$(X'X)^{-1}$		C	
b_0	2.3081	.8764	2.6336	.0250	.4987	−.0869	.7681	−.1338
b_1	.3332	.1673	1.9914	.0744	−.0869	.0182	−.1338	.0280

6.2.3   Correcting Violations of Normality

A glance back at Table 6.8 reveals the likely problem. Our 11th observation has an extremely large value for Cook's D ($D=.7706$), indicating undue influence. One way to handle this observation is to delete it. But with a sample size this small, we might prefer to lessen its influence by transforming our criterion. A transformation isn't guaranteed to produce normally distributed residuals, but it often does.[11]

6.2.3.1 Box-Cox Transformation

Numerous transformations are available. For example, we could take the square root of the criterion (y^5), calculate its reciprocal (y^{-1}), or try squaring it (y^2). A more systematic approach is to utilize a Box-Cox transformation (Box and Cox 1964). The Box-Cox procedure finds the power function of y (y^λ) that produces the smallest standardized residuals and, equivalently, the maximum likelihood estimate. Equation (6.28) presents the formula.

$$y^{(\lambda)} = \begin{cases} \dfrac{y^\lambda - 1}{\lambda \dot{Y}^{\lambda-1}} \dots (\lambda \neq 0) \\[2ex] \dot{Y}\ln(y) \dots (\lambda = 0) \end{cases} \tag{6.28}$$

The quantity $\dot{Y}$ refers to the geometric mean of the criterion, which can be found using a spreadsheet function (=GEOMEAN) or calculated directly by first multiplying all the criterion values and then raising the product to $1/n$.

$$\dot{Y} = \left\{ \prod_{i=1}^{n} Y_i \right\}^{1/n} \tag{6.29}$$

After the transformed variables have been created, we regress them on our predictors in the usual fashion. The residual sum of squares from the analysis represents the likelihood estimate for the selected value of λ. We then repeat this process, using different values of λ until we identify the value that produces the smallest residual sum of squares. Finally, notice that we substitute a natural logarithmic transformation when $\lambda = 0$.[12]

It is customary when conducting the analysis to begin by using values of λ between -2 and 2. If greater precision is required, one can then narrow the search within a smaller window. Table 6.10 illustrates the process using some selected values, with the first row of values in the middle three columns calculated as follows:

---

[11] In linear regression, the normality assumption refers to the residuals, not the criterion. It is possible, for example, that the criterion is normally distributed but the residuals are not (or vice versa). This is because the residuals are calculated from a conditional distribution (i.e., $y|x$). So although we correct the violation by transforming the criterion, we do so only as a vehicle for correcting the distribution of the residuals.

[12] Use of the Box-Cox procedure requires that all values of y are ≥ 0.

Table 6.10 Illustrative calculations for Box-Cox transformation

Observation	y	$\lambda=-2$	$\lambda=-1.5$	$\lambda=-.5$	$\lambda=0$	$\lambda=.5$	$\lambda=1.5$	$\lambda=2$
1	3.15	23.1816	14.6300	6.2695	4.2706	2.9896	1.5864	1.1986
2	2.72	22.2952	13.8452	5.6533	3.7243	2.5051	1.2046	.8596
3	4.42	24.4601	15.8995	7.5301	5.5313	4.2535	2.8656	2.4902
4	3.60	23.7905	15.2084	6.7921	4.7676	3.4625	2.0148	1.6067
5	2.42	21.3777	13.0842	5.1294	3.2893	2.1439	.9553	.6524
6	3.02	22.9531	14.4220	6.0972	4.1137	2.8468	1.4680	1.0909
7	3.34	23.4687	14.8980	6.5030	4.4885	3.1931	1.7638	1.3643
8	3.66	23.8552	15.2723	6.8544	4.8291	3.5232	2.0741	1.6652
9	3.60	23.7905	15.2084	6.7921	4.7676	3.4625	2.0148	1.6067
10	4.32	24.3983	15.8326	7.4515	5.4461	4.1612	2.7572	2.3727
11	7.74	25.3493	16.9895	9.1990	7.6166	6.8761	7.0955	7.9136
12	4.82	24.6700	16.1332	7.8197	5.8538	4.6126	3.3112	2.9867
Geometric mean	**3.7219**							
SS_{res}		**8.9161**	**8.2167**	**8.4267**	**9.5444**	**11.6894**	**21.6247**	**31.9967**

$$\lambda = -.5, \quad y_1^\lambda = \frac{\left(3.15^{-.5} - 1\right)}{-.5 * \left(3.7219^{-1.5}\right)} = 6.2695$$

$$\lambda = 0, \quad y_1^\lambda = 3.7219 * \ln(3.15) = 4.2706$$

and

$$\lambda = .5, \quad y_1^\lambda = \frac{\left(3.15^{.5} - 1\right)}{.5 * \left(3.7219^{-.5}\right)} = 2.9896$$

Looking over the values in Table 6.10, we see that the smallest residual is found when $\lambda = -1.5$. If we then narrow our search, we would find that $\lambda = -1.0707$ produces the smallest residual sum of squares and thus maximizes the likelihood function $[SS_{res} = 8.0283]$. But we don't use this value to transform our criterion because it's too difficult to interpret. Instead, we use a reciprocal transformation y^{-1} that is easier to understand. Similarly, had we found that $\lambda = 1.8$ maximizes the function, would we use a squaring transformation ($\lambda = 2$), and if we had found that $\lambda = .7$ maximizes the function, we would use a square root transformation ($\lambda = .5$). In short, we use the Box-Cox transformation to help us find the most interpretable transformation.

The results using a reciprocal transformation of y are shown in Table 6.11. As you can see, the slope of the regression line relating x to y is now significant. Moreover, the final column in Table 6.11 shows that all values for Cook's D fall within an acceptable range, and a Jarque-Bera test confirms that the distribution of errors is now normalized ($X^2_{JB}=.0367$, $p=.9819$). In this instance, then, a reciprocal transformation of our criterion normalized the residuals and improved our prediction.

Table 6.11 Regression analysis from Box-Cox transformed data

Original data		After reciprocal transformation				
x	y	y^{-1}	$\hat{y}$	e	h	Cook's D
1.11	3.15	.3175	.3515	−.0340	.3282	.1000
1.25	2.72	.3676	.3487	.0189	.3099	.0311
3.25	4.42	.2262	.3094	−.0832	.1259	.1333
3.26	3.60	.2778	.3092	−.0314	.1254	.0188
4.40	2.42	.4132	.2868	.1264	.0860	.1922
4.47	3.02	.3311	.2854	.0457	.0851	.0243
5.52	3.34	.2994	.2648	.0346	.0933	.0171
5.59	3.66	.2732	.2634	.0098	.0952	.0005
6.76	3.60	.2778	.2404	.0373	.1545	.0420
6.86	4.32	.2315	.2385	−.0070	.1619	.0021
6.95	7.74	.1292	.2367	−.1075	.1688	.3471
7.95	4.82	.2075	.2171	−.0096	.2658	.0028

Significance test of regression model						
	SS	df	MS	R^2	F	p
Regression	.0212	1	.0212	.3366	5.0745	.0480
Residual	.0419	10	.0042			
Total	.0631					

Regression coefficients								
	b	se_b	t	p	$(\mathbf{X'X})^{-1}$		$\mathbf{C}$	
b_0	.3733	.0457	8.1689	.0000	.4987	−.0869	.0021	−.0004
b_1	−.0196	.0087	−2.2527	.0480	−.0869	.0182	−.0004	.0001

6.2.4   R Code: Departures from Normality

```
x=c(1.11,1.25,3.25,3.26,4.40,4.47,5.52,5.59,6.76,6.86,6.95,7.95)
y=c(3.15,2.72,4.42,3.60,2.42,3.02,3.34,3.66,3.60,4.32,7.74,4.82)
model=lm(y~x)
summary(model)
influence.measures(model)
```

(continued)

6.2.4 R Code: Departures from Normality (continued)

```
#Studentized (called standard) and Deleted (called student) Residuals
model.student <-rstandard(model)
model.student
model.deleted <-rstudent(model)
model.deleted

#QQ Plot Using Residuals
qqnorm(model$resid);qqline(model$resid)

#Function to calculate skew, kurtosis, and Jarque-Bera test
jarque.bera <-function(model){
e <-model$resid
n <- length(e)
skew <-(sum((e - mean(e))^3)/n/(sum((e - mean(e))^2)/n)^(3/2))
kurtosis <-n* sum((e - mean(e))^4)/(sum((e - mean(e))^2)^2)
jb.test <- n*((skew^2/6)+((kurtosis-3)^2)/24)
p <- 1 - pchisq(jb.test, df = 2)
jb.table<-round(matrix(c(skew,kurtosis,jb.test,p),nrow=1,ncol=4),
digits=5)
dimnames(jb.table)=list(c(),c("skew","kurtosis","jarque-
bera","p"))
return(list(jb.table))
}
jarque.bera(model)

#Box-Cox Test
library(MASS)                   #attach MASS package
dev.new()
bc <-boxcox(y~x)
bc$x[which.max(bc$y)]

#Transform y using reciprocal and rerun regression and jarque-bera
y.1=y^-1
reg.recip <-lm(y.1~x)
summary(reg.recip)
dev.new()
qqnorm(reg.recip$resid);qqline(reg.recip$resid)
jarque.bera(reg.recip)
```

6.3 Collinearity

To this point we have considered only a simple linear regression model, but the procedures we have learned extend to multiple regression as well. In the final section of this chapter, we will take up an issue that arises only in the context of multiple regression.

As you know, multiple regression identifies the unique contribution each predictor makes to a criterion holding all other predictors constant. In the typical regression analysis, the predictors are correlated. This is entirely appropriate. In fact, if the predictors were uncorrelated (i.e., orthogonal), we wouldn't need multiple regression; we'd just conduct a series of simple linear regression analyses using one predictor at a time. So overlap among our predictors is expected in a multiple regression analysis.

6.3.1 Problems with Overly Redundant Predictors

Problems can arise, however, if the predictors are too closely associated with one another. In such cases, the predictors are said to be collinear.[13]

6.3.1.1 Singular Matrices Are Not Invertible

To appreciate the potential problems, let's start with an extreme case in which one of the predictors is perfectly correlated with another predictor. In the example below, it is obvious that x_2 is simply $3 * x_1$.

$$\begin{bmatrix} x_1 & x_2 \\ 2 & 6 \\ 4 & 12 \\ 6 & 18 \end{bmatrix}$$

When we compute $\mathbf{X'X}$ matrix, we get the following sum of squares matrix.

$$\begin{bmatrix} 56 & 168 \\ 168 & 504 \end{bmatrix}$$

To find our regression coefficients, we need to invert this matrix, and to invert this matrix, we need to divide the adjugate by the determinant. The problem is that the determinant of this matrix $= 0$, so the matrix is not invertible.

$$\begin{vmatrix} 56 & 168 \\ 168 & 504 \end{vmatrix} = (56 * 504) - (168 * 168) = 0$$

Formally, a matrix with at least one linear dependence is termed singular (or rank deficient) and is not invertible, precluding the use of multiple regression analysis.

---

[13] Some textbooks use the term "multicollinearity" to refer to the situation we are calling "collinearity." The two terms refer to the same phenomenon, so I have opted for the shorter one.

6.3.2 Matrices with a Near-Linear Dependence are Ill Conditioned

Unless we have made a mistake in coding or have inadvertently included a predictor that is a linear combination of other predictors, we are unlikely to find complete linear dependencies among our predictors. But problems can arise even when two (or more) variables are not perfectly correlated. Consider the following example:

$$\begin{bmatrix} x_1 & x_2 \\ 2 & 6 \\ 4 & 12 \\ 6 & 18.001 \end{bmatrix}$$

Here I have changed the last value slightly so that the two predictors are no longer perfectly correlated. Now when we compute our $\mathbf{X'X}$ matrix, we get

$$\begin{bmatrix} 56 & 168.006 \\ 168.006 & 504.036 \end{bmatrix}$$

and find that the determinant is no longer zero.

$$\begin{vmatrix} 56 & 168.006 \\ 168.006 & 504.036 \end{vmatrix} = (56 * 504.036) - (168.006 * 168.006) = .00002$$

Consequently, the matrix is invertible.

$$\mathbf{X'X}^{-1} = \begin{vmatrix} 25201795 & -8400298 \\ -8400298 & 2799999 \end{vmatrix}$$

Such a matrix is termed, "ill conditioned"; however, because the regression estimates it yields are prone to error (e.g., small variations, such as the number of decimal points to which a value is reported, produce large differences in the estimates). So even though a perfect linear dependency no longer exists, the regression output is unstable.

6.3.2.1 Ill-Conditioned Matrices Inflate Standard Errors

The effect of an ill-conditioned matrix is most apparent when we consider the calculation of standard errors. Recall that, with multiple predictors, Eq. (4.6) (repeated below) can be used to compute the standard errors.

$$se_b = \sqrt{\frac{MS_{res}}{ss_x * \left(1 - R_{xx}^2\right)}} \qquad (6.30)$$

Looking at the denominator, we see that when the predictors are highly correlated, the term $(1 - R_{xx}^2)$ becomes increasingly small, thereby inflating the standard errors. This is the problem with collinear predictors—they create large, imprecise standard errors. Consequently, our regression coefficients are less likely to attain statistical significance and our findings are less likely to replicate in a different sample.

6.3.3   Demonstrating Collinearity

Keeping these points in mind, we will learn some tools for quantifying the degree of collinearity and the effect it has on the standard errors. As always, we start with an example (Table 6.12).

Table 6.12 Hypothetical data set for collinearity diagnostics

	x_1	x_2	x_3	x_4	x_5	y
	.52	.86	.14	.51	.33	.08
	.19	.39	.78	.65	.48	.60
	.81	.94	.35	.47	.52	.47
	.80	.22	.53	.44	.39	.39
	.62	.05	.65	.67	.31	.58
	.46	.62	.28	.01	.32	.11
	.42	.70	.61	.47	.45	.42
	.26	.88	.76	.28	.47	.37
	.70	.41	.38	.07	.37	.11
	.64	.60	.69	.23	.48	.38
	.17	.72	.22	.79	.28	.09
	.08	.28	.71	.26	.27	.20
Deviation sum of squares	.6984	.8925	.5542	.6347	.0849	.4025
Sum of squares	3.3775	4.5999	3.6550	2.5949	1.9023	1.6058

6.3.3.1   Correlations

Before we find our regression coefficients, let's examine the bivariate correlations. With 12 observations, correlations greater than .50 are statistically significant. Table 6.13 shows that none of the predictors is significantly correlated with another but that x_3 and x_5 are significantly correlated with the criterion.

Table 6.13 Correlations for hypothetical data set for collinearity diagnostics

	x_1	x_2	x_3	x_4	x_5	y
x_1	1	−.0482	−.2674	−.1933	.3098	.1643
x_2	−.0482	1	−.4145	−.0554	.3969	−.2854
x_3	−.2674	−.4145	1	−.0132	.3840	.6982
x_4	−.1933	−.0554	−.0132	1	−.0407	.3885
x_5	.3098	.3969	.3840	−.0407	1	.6074
y	.1643	−.2854	.6982	.3885	.6074	1

6.3.3.2 Regression Analyses

Our next step is to conduct a regression analysis. Using procedures described throughout the book, we obtain the values shown in Table 6.14. Notice that although the overall model is significant, only one of the regression coefficients attains significance. This pattern signals that collinearity might be inflating the standard errors. Looking more closely at the regression coefficients, we find further evidence that this is so. Notice that the only significant predictor is x_4, even though its zero-order correlation with the criterion is much smaller than the ones associated with x_3 and x_5. The final column in Table 6.14 shows why this occurs. The R^2 values were found by regressing each predictor on the others. Notice that only one predictor, x_4, is independent of the others. As a consequence, the variance it explains is unique and, in this case, significant.[14]

6.3.4 Quantifying Collinearity with the Variance Inflation Factor

As we have noted, collinearity inflates the standard errors. The diagonal entries of the inverted correlation matrix, known as the variance inflation factor (VIF), quantify this inflation.

$$\text{VIF}_k = \mathbf{R}_{ii} \qquad (6.31)$$

The square root of this statistic indicates how much the standard error of a regression coefficient is affected by the overlap among the predictors. We will illustrate the statistic using the standard error for our b_1 coefficient.

$$\text{VIF}x_1 = 3.45667$$

---

[14] The penultimate column in Table 6.14 (labeled Tolerance) is found as $1 - R^2$ (or, equivalently, as the inverse of the diagonal entry of the inverted correlation matrix). It indicates how independent each predictor is from all of the others.

Table 6.14 Regression analysis for hypothetical data set for collinearity diagnostics

Significance test of regression model

	SS	df	MS	F	p	R²
Regression	.361451	5	.072290	10.5750	.0062	.8981
Residual	.041016	6	.006836			
Total	.402467					

Regression coefficients and regression diagnostics

	b	se_b	t	p	Variance inflation factor	Tolerance	R²
b_0	-.43099	.14779	-2.91631	.0268			
b_1	.18854	.18394	1.02500	.3449	3.45667	.289296	.710704
b_2	-.13579	.19936	-.68111	.5212	5.18897	.192716	.807284
b_3	.44501	.27484	1.61917	.1565	6.12346	.163306	.836694
b_4	.35809	.11200	3.19712	.0187	1.16476	.858546	.141454
b_5	.93304	.70444	1.32450	.2335	6.16255	.162270	.837730

$(X'X)^{-1}$

3.195007	-2.108767	-2.186437	-3.181822	-1.201227	3.091053
-2.108767	4.949236	4.101974	5.975594	1.117111	-15.414616
-2.186437	4.101974	5.814028	6.989505	1.033642	-17.869307
-3.181822	5.975594	6.989505	11.049860	1.434812	-24.985599
-1.201227	1.117111	1.033642	1.434812	1.835158	-3.526030
3.091053	-15.414616	-17.869307	-24.985599	-3.526030	72.593129

$(R_x)^{-1}$

3.456670	3.238582	3.717588	.743768	-3.753402
3.238582	5.188972	4.915512	.777953	-4.918609
3.717588	4.915512	6.123464	.850936	-5.419291
.743768	.777953	.850936	1.164759	-.818465
-3.753402	-4.918609	-5.419291	-.818465	6.162552

If the predictors were orthogonal, then the standard error would involve only the deviation sum of squares (ss_x=.698425 from Table 6.12) and the mean square error (MS_{res}=.006836 from Table 6.14).

$$se_{b_1} = \sqrt{\frac{.006836}{.698425 * (1 - 0)}} = .098933$$

To derive the true standard error shown in Table 6.14, we can multiply this value by the square root of the VIF.

$$se_{b_1} = .098933 * \sqrt{3.456667} = .183937$$

Similar calculations yield the other standard errors shown in Table 6.14. In this manner, the diagonal entries of the inverted correlation matrix of predictors quantify how much collinearity inflates the standard error of each regression coefficient. Looking over our VIF entries, we see that the standard error associated with x_4 is hardly affected at all but that the standard errors associated with x_3 and x_5 are $\tilde{}\sqrt{6} = 2.5$ times greater than they would be if collinearity did not exist. This inflation explains why b_4 is significant but b_3 and b_5 are not.

6.3.5 Condition Index and Variance Proportion Decomposition

The VIF is not the only way to quantify collinearity. Another approach is to apply one of the decomposition techniques we learned in Chap. 5 to the matrix of predictors. A singular value decomposition (SVD) is a good choice (Belsley et al. 1980). As you learned in Chap. 5, SVD uses the eigenvalues and eigenvectors of a sum of squares matrix to decompose a rectangular matrix into three matrices [see Eq. (5.23) repeated below]:

$$\mathbf{A} = \mathbf{UDV}'$$ (6.32)

where $\mathbf{V}$ = eigenvectors of $\mathbf{A}'\mathbf{A}$, $\mathbf{D}$ = diagonal matrix of the square root of the eigenvalues of $\mathbf{A}'\mathbf{A}$ (called singular values), and $\mathbf{U} = \mathbf{AVD}^{-1}$. In this section, you will learn how SVD can be used to quantify collinearity.

6.3.5.1 Normalizing the Predictors

We begin by normalizing our predictors so that their variances can be more easily compared. To do so, we divide each column value by the square root of the sum of all squared column values.

$$n = \frac{x}{\sqrt{\Sigma x^2}} \qquad (6.33)$$

Following this transformation, each column vector has unit length (i.e., the sum of squares $= 1$ for each normalized variable). Table 6.12 shows the sum of squares for our hypothetical data, and Table 6.15 shows the predictors after they have been normalized.[15] The first entry for n_1 was found as follows:

$$n_1 = \frac{.52}{\sqrt{3.3775}} = .282947$$

Table 6.15 Normalized predictors (including intercept) for hypothetical data set for collinearity diagnostics

n_0	n_1	n_2	n_3	n_4	n_5
.288675	.282947	.400981	.073229	.316599	.239262
.288675	.103385	.181840	.407991	.403509	.348018
.288675	.440745	.438282	.183073	.291768	.377020
.288675	.435304	.102577	.277225	.273144	.282765
.288675	.337360	.023313	.339993	.415924	.224762
.288675	.250300	.289080	.146458	.006208	.232012
.288675	.228534	.326380	.319070	.291768	.326267
.288675	.141474	.410307	.397530	.173819	.340768
.288675	.380891	.191166	.198765	.043455	.268264
.288675	.348243	.279754	.360915	.142780	.348018
.288675	.092502	.335705	.115074	.490418	.203011
.288675	.043530	.130552	.371377	.161404	.195760

After normalizing all of the predictors (including the intercept), we create a sum of squares matrix. We will call this matrix $\mathbf{N'N}$ to remind ourselves that we are using normalized predictors, not raw ones. Although the diagonal elements of the sum of squares matrix all equal 1, this is not a correlation matrix and the off-diagonal elements are not standardized covariances.

$$\mathbf{N'N} = \begin{bmatrix} 1 & .89063 & .89776 & .92108 & .86914 & .97743 \\ .89063 & 1 & .78991 & .77299 & .73060 & .90028 \\ .89776 & .78991 & 1 & .75582 & .76822 & .91444 \\ .92108 & .77299 & .75582 & 1 & .79801 & .93187 \\ .86914 & .73060 & .76822 & .79801 & 1 & .84527 \\ .97743 & .90028 & .91444 & .93187 & .84527 & 1 \end{bmatrix}$$

---

[15] The normalization can also be performed using $\mathbf{N} = \mathbf{X}\left\{ \left[\sqrt{(\mathbf{I} * \mathbf{X'X})} \right]^{-1} \right\}$, where $\mathbf{I}$ is an $\mathbf{X} \times \mathbf{X}$ identity matrix.

Applying procedures we learned in Chap. 5, we then use the **QR** algorithm to compute the eigenvalues and eigenvectors of $\mathbf{N'N}$. These values, along with the inverse sum of squares matrix (to be discussed momentarily), are displayed in Table 6.16.

Table 6.16 Eigen pairs and $(\mathbf{N'N})^{-1}$ from normalized predictors for collinearity diagnostics

λ	5.264097	.286055	.233641	.187205	.024251	.004751
V	.431421	−.001896	−.050544	−.056529	−.882530	−.171064
	.394624	−.567833	−.284809	.604406	.124259	.244891
	.397967	−.329746	.659866	−.417167	.120188	.330155
	.402447	.336759	−.583259	−.425613	.176235	.415013
	.388292	.668772	.365898	.499534	.124901	.054295
	.432475	−.090365	−.082667	−.163673	.380259	−.791568
$(\mathbf{N'N})^{-1}$	38.340089	−13.425077	−16.244331	−21.072212	−6.703106	14.768506
	−13.425077	16.716046	16.168317	20.995319	3.307154	−39.072382
	−16.244331	16.168317	26.743949	28.659217	3.571122	−52.859300
	−21.072212	20.995319	28.659217	40.387237	4.418747	−65.882938
	−6.703106	3.307154	3.571122	4.418747	4.762052	−7.834040
	14.768506	−39.072382	−52.859300	−65.882938	−7.834040	138.093908

6.3.5.2   Condition Number

In Chap. 5, you learned that the sum of the eigenvalues equals the trace of the matrix. After normalizing our variables, the diagonal entries of $\mathbf{N'N} = 1$, so the trace of the matrix is equal to **p** (the number of predictors, including the intercept). If the predictors were orthogonal, there would be no variance to consolidate and each eigenvalue would $= 1$. Clearly this is not the case with our normalized matrix. Instead, the first eigenvalue is very large ($\lambda_1 = 5.2641$), indicating collinearity.

The condition number of a matrix quantifies the collinearity among the predictors. It is found by taking the square root of a ratio, with the largest eigenvalue in the numerator and the smallest eigenvalue in the denominator. If the predictors were independent, this ratio would $= 1$, so values > 1 signal collinearity.

$$\text{Condition } \# = \sqrt{\frac{\lambda_{max}}{\lambda_{min}}} \qquad (6.34)$$

Plugging in our values yields the condition number for our matrix of predictors.[16]

---

[16] Due to rounding error, the value calculated here differs from the true value (displayed in Table 6.17). This is because the matrix is ill conditioned, and small differences in decimal values produce large differences in calculated values.

$$\sqrt{\frac{5.2641}{.00475}} \cong 33.29011$$

Table 6.17 Condition index and variance proportion values for hypothetical data set for collinearity diagnostics

	λ	Condition index	Intercept	Variance proportions				
				x_1	x_2	x_3	x_4	x_5
1	5.26410	1	.00092	.00177	.00112	.00076	.00601	**.00026**
2	.28606	4.28980	.00000	.06743	.01421	.00982	**.32834**	.00021
3	.23364	4.74666	.00029	.02077	.06968	**.03605**	.12033	.00021
4	.18721	5.30277	.00045	.11674	.03476	.02396	.27991	.00104
5	.02425	14.73328	.83768	.03809	.02227	.03171	.13509	.04318
6	.00475	33.28801	.16066	.75520	.85795	.89770	.13031	.95511

Belsley et al. (1980) regard condition numbers greater than 30 to be a cause of concern (indicating that the largest eigenvalue is 900 times greater than the smallest eigenvalue), and our value exceeds this threshold. Table 6.17 shows that we can also compute a condition index for other eigenpairs by taking the square root of the largest eigenvalue divided by the next largest eigenvalue. For example, the second condition index value is found as follows:

$$\sqrt{\frac{5.2641}{.28606}} = 4.2898$$

6.3.5.3 Variance Proportion Decomposition

So far we have seen that the eigenvalues from a normalized matrix can be used to quantify the collinearity of a predictor matrix. We can also use the eigen pairs to pinpoint which variables contribute to collinearity. To begin, recall from earlier chapters that the diagonal elements of $(\mathbf{X'X})^{-1}$ are used to compute the standard errors and that these values can be reproduced from the eigenpairs of $\mathbf{X'X}$.

$$\left(\mathbf{X'X}\right)_{ii}^{-1} = \sum_{k=1}^{p} \frac{v_{jk}^2}{\lambda_k} \tag{6.35}$$

The same principle applies to $(\mathbf{N'N})^{-1}$. The bottom portion of Table 6.16 shows the inverse matrix, and the last diagonal entry representing x_5 can be found using the eigenpairs of $\mathbf{N'N}$.

$$\frac{.43248^2}{5.2641} + \frac{-.09036^2}{.28606} + \frac{-.08267^2}{.23364} + \frac{-.16367^2}{.18721} + \frac{-.38026^2}{.02425} + \frac{-.79157^2}{.00475} = 138.09391$$

If we then divide each eigenpair quotient by the sum of all terms [i.e., the diagonal value of $(\mathbf{N'N})^{-1}$], we find the proportion of a coefficient's variance that is associated with each singular value. This measure is known as the variance proportion score, and high scores indicate that the variance of the coefficient is strongly associated with a particular singular value.

$$VP = \frac{v_{jk}^2 / \lambda_k}{\sum_{k=1}^{P} v_{jk}^2 / \lambda_k} \qquad (6.36)$$

Table 6.17 displays the values. To help you with the calculations, the three values shown in bold font were calculated as follows:

$$\frac{.43248^2 / 5.2641}{138.09391} = .00026$$

$$\frac{.66877^2 / .28606}{4.76205} = .32834$$

$$\frac{-.58326^2 / .23364}{40.38724} = .03605$$

Because small singular values represent near dependencies, two (or more) predictors are said to be collinear when a high proportion of their variances is associated with a small singular value. As a guideline, Belsley et al. (1980) suggest that collinearity exits when a singular value with a condition index > 10 accounts for more than 50 % of the variance in at least two predictors. Using these criteria, our predictors suffer from collinearity. The bottom row of Table 6.17 shows that the final singular value has a condition index > 30 and accounts for more than 75 % of the variance in $x_1, x_2, x_3,$ and x_5.

6.3.6   Summary

Multiple regression is designed to identify the unique contribution a variable makes to the prediction of a criterion holding all other predictors constant. But if the predictors are too closely associated, the standard errors become inflated and imprecise. The question arises then, as to how much overlap is too much overlap. Unfortunately, the only standard for deciding whether collinearity is too excessive is whether there is a complete linear dependency. As long as the predictor matrix is invertible, a regression analysis can be performed.

At the same time, it is often desirable to minimize collinearity when it does arise. The first thing to do is check for a coding error. Perhaps a scale was inadvertently entered under two different names, or subscales were combined, and both variables (i.e., subscales and combined scales) were entered into the predictive equation. Barring errors like these, one can reconsider whether all overlapping variables are needed, increase sample size, or attempt to create combinations of variables using

Table 6.18 Hypothetical data set for collinearity diagnostics

Two predictors only			Add predictor correlated with aptitude				Add predictor correlated with hours spent studying			
Study time	Pretest	Grade	Study time	Pretest	SAT-Q	Grade	Study time	Pretest	Review session attendance	Grade
1	1	1	1	1	1	1	1	1	1	1
3	9	4	3	9	8	4	3	9	4	4
5	6	5	5	6	5	5	5	6	5	5
7	3	3	7	3	4	3	7	3	5	3
1	5	3	1	5	7	3	1	5	2	3
3	2	1	3	2	3	1	3	2	5	1
5	5	5	5	5	6	5	5	5	6	5
7	8	9	7	8	8	9	7	8	7	9
1	3	2	1	3	4	2	1	3	3	2
3	3	4	3	3	5	4	3	3	5	4
5	3	3	5	3	5	3	5	3	3	3
7	2	5	7	2	8	5	7	2	9	5
Regression coefficients			*Regression coefficients*				*Regression coefficients*			
	b	p		b	p			b	p	
b_0	$-.3725$	.6857	b_0	$-.8819$	.4049		b_0	$-.9024$	.3746	
b_1	.5183	.0108	b_1	.4389	.0379		b_1	.2612	.3393	
b_2	.4918	.0104	b_2	.3376	.1442		b_2	.4874	.0108	
			b_3	.2755	.3145		b_3	.3440	.2423	

various techniques. Generally, one of these remedies will reduce collinearity to a manageable degree.[17]

Ultimately, the biggest problem collinearity creates is apt to be conceptual rather than statistical. To appreciate the issues here, consider the data presented in Table 6.18. Looking only at the first three columns, we see that we are using hours spent studying and performance on a pretest to predict students' grades in a matrix algebra class. When we conduct a regression analysis using these variables, we find that each one uniquely explains variability in test performance, leading us to conclude that effort (study time) and ability (aptitude) predict success in math.

Now suppose that, for whatever reason, I want to show that study time is more important to success than is aptitude. The middle portion of Table 6.18 shows that all I have to do is add another predictor to the equation that is highly correlated ($r=.6924$) with aptitude (e.g., scores on the quantitative section of the SAT). When I do, I find that only time spent studying significantly predicts class performance. Of course, another investigator might wish to show that only aptitude matters. The final portion of Table 6.18 shows that this, too, can easily be accomplished by adding a predictor that is highly correlated ($r=.7986$) with hours spent studying (e.g., attendance at various study sessions).

---

[17] Ridge regression can also be used to mitigate the effects of collinearity. A discussion of this procedure can be found in Draper and Smith (1998, pp. 387–400).

Admittedly, this is an extreme example, but the larger point is valid. Unless the predictors are orthogonal, the contribution of a correlated predictor will reduce the impact of the variable with which it is correlated. More generally, we must remember that each predictor in a multiple regression analysis can only be understood in the context of the other predictors and that the more correlated the predictors are, the more we need to remind ourselves that this is true. So, ultimately, the biggest problem collinearity poses is that it might lead us to believe a particular variable is of little importance when its predictive utility is generally high.

6.3.7   R Code: Collinearity

```
x1=c(.52,.19,.81,.80,.62,.46,.42,.26,.70,.64,.17,.08)
x2=c(.86,.39,.94,.22,.05,.62,.70,.88,.41,.60,.72,.28)
x3=c(.14,.78,.35,.53,.65,.28,.61,.76,.38,.69,.22,.71)
x4=c(.51,.65,.47,.44,.67,.01,.47,.28,.07,.23,.79,.26)
x5=c(.33,.48,.52,.39,.31,.32,.45,.47,.37,.48,.28,.27)
y=c(.08,.60,.47,.39,.58,.11,.42,.37,.11,.38,.09,.20)

#Complete correlation matrix
V <-cbind(x1,x2,x3,x4,x5,y)
cor(V)

#Regression
model<-lm(y~x1+x2+x3+x4+x5)
summary(model)

#Variance Inflation Factor
library(car)                 #attach car package
vif(model)

#Condition Index and Variance Decomposition
X <-cbind(1,x1,x2,x3,x4,x5)
N <-X%*%(solve(sqrt(diag(1,ncol(X))*(t(X)%*%X))))
eigs <-eigen(t(N)%*%N);eigs
library(perturb)             #attach perturb package
colldiag(model)
```

6.4   Chapter Summary

1. Least squares estimation assumes that each set of observations (i.e., predictors and criterion) contributes equally to the value of the regression coefficients and standard errors. Yet some observations exert an undue influence, creating interpretive problems.
2. A predictor that is discrepant from other predictors is said to possess high leverage; a criterion that is discrepant from its fitted value is called an outlier.

Observations that possess high leverage and are outliers tend to be unduly influential.

3. Influential observations are identified by noting how various indices (e.g., fitted values, regression coefficients, standard errors) change when the observation is eliminated from the analysis. If the statistics change a great deal when an observation is excluded, the observation possesses high influence.

4. Various methods exist to efficiently calculate each observation's influence. Most of them use values from the hat matrix, an idempotent matrix that is used to compute the fitted values.

5. The most comprehensive assessment of an observation's influence is Cook's D, which measures how much the vector of regression coefficients changes when an observation is eliminated.

6. Influential observations often produce errors that are not normally distributed. A QQ probability plot can be used to visually inspect the normality of the residuals, and the Jarque-Bera test can be used to determine whether departures from normality are more severe than would be expected by chance.

7. Residuals that do not follow a normal distribution can sometimes be normalized by transforming the criterion. The Box-Cox procedure offers an efficient way to find the transformation that best achieves this goal.

8. It is normal for the predictors to be correlated in a multiple regression analysis, but if the correlations are too large, the predictors are termed "collinear," the design matrix become ill conditioned, and the standard errors become inflated. A statistic known as the variance inflation factor quantifies the magnitude of the inflation.

9. A singular value decomposition of a normalized design matrix can be used to judge the collinearity of the predictors.

10. Collinearity can be managed, but the biggest problem it creates is conceptual rather than statistical. Because each regression coefficient depends on the properties of the other predictors in the equation, the relative importance of a predictor can be minimized by including other variables with which it correlates.

Chapter 7
Errors and Residuals

In previous chapters, you learned that the errors in a linear regression model are assumed to be independent, and normally and identically distributed random variables with mean 0 and variance σ^2:

$$\varepsilon \sim NID\left(0, \sigma^2\right) \tag{7.1}$$

We discussed the normality assumption in Chap. 6, learning ways to determine whether the errors are normally distributed (e.g., QQ plot, Jarque-Bera test) and the steps we can take to normalize them if they are not (e.g., Box-Cox transformation). In this chapter, we will consider whether the errors are independent and identically distributed, learning ways to assess these assumptions and correct violations of them when they arise.

7.1 Errors and Their Assumed Distribution

A careful examination of the covariance matrix of errors Σ will clarify our task. For purposes of illustration, imagine that we conduct a study with only three values for x. The left-hand side of Table 7.1 presents the structure of the covariance matrix, and the right-hand side shows the expected values if we assume that the errors are independent and identically distributed. As you can see, the off-diagonal elements of the expected matrix are zero (because we assume that the errors derived from one fitted value are uncorrelated with the errors derived from all other fitted values), and the diagonal values of the expected matrix are identical (because we assume that the distribution of errors around each fitted value has the same variance). This is what it means to say that the errors are independent and identically distributed.

Electronic Supplementary Material: The online version of this chapter (doi: 10.1007/978-3-319-11734-8_7) contains supplementary material, which is available to authorized users

© Springer International Publishing Switzerland 2014
J.D. Brown, *Linear Models in Matrix Form*, DOI 10.1007/978-3-319-11734-8_7

Table 7.1 Covariance matrix of independent and identically distributed errors

Covariance matrix	Expected values
$Var(\varepsilon) = \Sigma = \begin{bmatrix} \varepsilon_1^2 & \varepsilon_1\varepsilon_2 & \varepsilon_1\varepsilon_3 \\ \varepsilon_2\varepsilon_1 & \varepsilon_2^2 & \varepsilon_2\varepsilon_3 \\ \varepsilon_3\varepsilon_1 & \varepsilon_3\varepsilon_2 & \varepsilon_3^2 \end{bmatrix}$	$E\{\Sigma\} = \begin{bmatrix} \sigma_\varepsilon^2 & 0 & 0 \\ 0 & \sigma_\varepsilon^2 & 0 \\ 0 & 0 & \sigma_\varepsilon^2 \end{bmatrix} = \sigma_\varepsilon^2 I$

Table 7.2 Two covariance matrices of errors that violate the assumptions of a linear model

Independent but not identically distributed	Identical but not independently distributed
$\Sigma = \begin{bmatrix} \sigma_1^2 & 0 & \cdots & 0 \\ 0 & \sigma_2^2 & \cdots & 0 \\ \vdots & \vdots & \ddots & \vdots \\ 0 & 0 & \cdots & \sigma_n^2 \end{bmatrix}$	$\Sigma = \begin{bmatrix} \sigma_\varepsilon^2 & \varepsilon_1\varepsilon_2 & \cdots & \varepsilon_1\varepsilon_n \\ \varepsilon_2\varepsilon_1 & \sigma_\varepsilon^2 & \cdots & \varepsilon_2\varepsilon_n \\ \vdots & \vdots & \ddots & \vdots \\ \varepsilon_n\varepsilon_1 & \varepsilon_n\varepsilon_2 & \cdots & \sigma_\varepsilon^2 \end{bmatrix}$

Keeping these points in mind, let's consider ways in which data can violate these assumptions. The first matrix in Table 7.2 shows the case where the errors are independent but not identically distributed. The matrix is still diagonal (i.e., with 0's on the off-diagonals), indicating that the errors are independently distributed, but the diagonal entries are no longer identical. Instead, the distributions around each error have their own variance. The next section of this chapter discusses this violation. The second matrix in Table 7.2 violates the independently distributed assumption. Here, the variances are identical (i.e., all diagonal entries are the same), but the off-diagonal entries are no longer 0. The final section of this chapter discusses this violation.

7.1.1   Why It Matters

The violations depicted in Table 7.2 do not influence the regression coefficients, but they do affect the precision of the standard errors (and, therefore, our confidence intervals and tests of statistical significance). Formally, we say that the regression estimates remain unbiased but become inefficient. To understand why the standard errors are affected, let's reexamine the formula we use to calculate the parameter covariance matrix, first discussed in Chap. 2:

$$C = \sigma^2 \left(X'X \right)^{-1} \tag{7.2}$$

As it turns out, Eq. (7.2) is a shortcut that works only when we assume that the errors are independent and identically distributed. The complete formula is as follows:

$$C = \left(X'X \right)^{-1} X'\Sigma X \left(X'X \right)^{-1} \tag{7.3}$$

or, equivalently,

$$\mathbf{C} = \sigma_{\varepsilon}^2 \left(\mathbf{X}' \mathbf{\Psi}^{-1} \mathbf{X} \right)^{-1} \tag{7.4}$$

where $\mathbf{\Psi}$ is a symmetric, $N \times N$ matrix of normalized residuals, with the normalized variances on the diagonals and their normalized covariances on the off-diagonals. As Table 7.3 shows, the normalization involves dividing each variance/covariance estimate by the overall variance estimate.

Table 7.3 Normalized covariance matrix of errors ($\mathbf{\Psi}$)

Normalized covariance matrix of errors ($\mathbf{\Psi}$)			

$$\mathbf{\Psi} = \begin{bmatrix} \dfrac{var_{\varepsilon 1}}{\sigma_{\varepsilon}^2} & \dfrac{cov_{\varepsilon 1,\varepsilon 2}}{\sigma_{\varepsilon}^2} & \cdots & \dfrac{cov_{\varepsilon 1,\,\varepsilon n}}{\sigma_{\varepsilon}^2} \\[2mm] \dfrac{cov_{\varepsilon 2,\varepsilon 1}}{\sigma_{\varepsilon}^2} & \dfrac{var_{\varepsilon 2}}{\sigma_{\varepsilon}^2} & \cdots & \dfrac{cov_{\varepsilon 2,\varepsilon n}}{\sigma_{\varepsilon}^2} \\[2mm] \vdots & \vdots & \ddots & \vdots \\[2mm] \dfrac{cov_{\varepsilon n,\varepsilon 1}}{\sigma_{\varepsilon}^2} & \dfrac{cov_{\varepsilon n,\varepsilon 2}}{\sigma_{\varepsilon}^2} & \cdots & \dfrac{var_{\varepsilon n}}{\sigma_{\varepsilon}^2} \end{bmatrix}$$

If you look carefully at the normalized matrix, you will notice something interesting. When the errors are independent and identically distributed, the normalized matrix becomes an identity matrix, allowing us to shorten the formula shown in Eq. (7.4) to the one shown in Eq. (7.2). But when the errors are not independent or identically distributed, the shortcut does not accurately produce the parameter covariance matrix, rendering our standard errors inefficient.

7.1.2   Errors and Residuals

The violations we are discussing pertain to the population errors ε, not the sample residuals e. To review, an error is a discrepancy between an observed value and its true population value, whereas a residual is the discrepancy between an observed value and its fitted value ($e = y - \hat{y}$). Because population errors are never known, we use the residuals to estimate them. But this estimation is just that—an estimation—and a covariance matrix of residuals is never as "perfect" as the covariance matrix of errors displayed in Table 7.1. In short, $\varepsilon \neq e$.

To reinforce your understanding of this point, let's reconsider an example we first encountered in Chap. 6. To demonstrate normally distributed residuals, I generated data using random values for x and e, and then set $y = 4 + .7x + e$. The data, originally shown in Table 6.6, are reproduced in Table 7.4, and an ordinary least squares (OLS) regression yields the values shown in Table 7.5. For present purposes, the critical value is the mean square error, $MS_{res} = .8921$.

Table 7.4 Illustration of independent and normally and identically distributed errors

x	y	$\hat{y}$	e	h
.5161	3.7293	4.3762	−.6469	.0834
.4688	3.3527	4.2784	−.9257	.0845
.0237	3.0408	3.3575	−.3167	.2921
.2626	4.2383	3.8519	.3864	.1363
.5866	4.1130	4.5220	−.4091	.0893
.2475	4.9917	3.8205	1.1712	.1431
.9705	6.4193	5.3163	1.1030	.2784
.4377	3.4769	4.2141	−.7372	.0874
.7817	3.4688	4.9256	−1.4568	.1522
.7573	6.1731	4.8753	1.2978	.1406
.0665	3.9843	3.4462	.5381	.2566
.9430	5.2553	5.2594	−.0041	.2560

Table 7.5 Regression output from data in Table 7.4

			Significance test of regression model				
	SS	df	MS	R^2	F	p	
Regression	4.7512	1	4.7512	.3475	5.3260	.0437	
Residual	8.9208	10	.8921				
Total	13.6720						

			Regression coefficients				
	b	se_b	t	p	$(\mathbf{X'X})^{-1}$		C	
b_0	3.3087	.5285	6.2599	.0001	.31316	−.45495	.27936	−.40585
b_1	2.0686	.8963	2.3078	.0437	−.45495	.90060	−.40585	.80341

The covariance matrix of residuals, calculated using Eq. (6.11) (reproduced below), yields the values shown in Table 7.6:

$$Var_{(e)} = (\mathbf{I} - \mathbf{H}) * MS_{res} \qquad (7.5)$$

Looking at the diagonal entries in Table 7.6, we can see that they all are close to the value for MS_{res}, though none match it exactly. We can also see that the off-diagonal entries are generally close to 0, though none is exactly 0. The $\mathbf{I} - \mathbf{H}$ matrix guarantees that these discrepancies will occur, even when the population errors are independent and identically distributed. Because the matrix is not diagonal, the residuals will never be completely independent; and because the hat values are not identical, the variances of the residuals won't be identical either.

In sum, actual residuals are never entirely independent and identically distributed. Nevertheless, when we calculate the parameter covariance matrix for this data set, we use the assumed matrix shown in Table 7.7, not the observed matrix shown in Table 7.6.

$$\mathbf{C} = \left(\mathbf{X'X}\right)^{-1} * MS_{res} = \begin{bmatrix} .27936 & −.40585 \\ −.40585 & .80341 \end{bmatrix}$$

Table 7.6 Covariance matrix of residuals from data in Table 7.4

.8176	−.0740	−.0701	−.0722	−.0750	−.0721	−.0784	−.0737	−.0768	−.0765	−.0705	−.0782
−.0740	.8166	−.0884	−.0814	−.0720	−.0819	−.0607	−.0763	−.0663	−.0670	−.0871	−.0615
−.0701	−.0884	.6314	−.1681	−.0428	−.1740	.1057	−.1004	.0326	.0232	−.2440	.0950
−.0722	−.0814	−.1681	.7704	−.0585	−.1245	.0163	−.0875	−.0205	−.0252	−.1598	.0110
−.0750	−.0720	−.0428	−.0585	.8123	−.0575	−.1048	−.0699	−.0924	−.0908	−.0456	−.1030
−.0721	−.0819	−.1740	−.1245	−.0575	.7643	.0220	−.0883	−.0171	−.0221	−.1651	.0163
−.0784	−.0607	.1057	.0163	−.1048	.0220	.6437	−.0491	−.1777	−.1686	.0896	−.2380
−.0737	−.0763	−.1004	−.0875	−.0699	−.0883	−.0491	.8140	−.0594	−.0607	−.0981	−.0506
−.0768	−.0663	.0326	−.0205	−.0924	−.0171	−.1777	−.0594	.7562	−.1303	.0231	−.1716
−.0765	−.0670	.0232	−.0252	−.0908	−.0221	−.1686	−.0607	−.1303	.7666	.0145	−.1630
−.0705	−.0871	−.2440	−.1598	−.0456	−.1651	.0896	−.0981	.0231	.0145	.6631	.0799
−.0782	−.0615	.0950	.0110	−.1030	.0163	−.2380	−.0506	−.1716	−.1630	.0799	.6637

Table 7.7 Assumed matrix used to calculate parameter covariance matrix for the data in Table 7.4

$$\Sigma = \begin{bmatrix} .8921 & 0 & 0 & 0 & 0 & 0 & 0 & 0 & 0 & 0 & 0 & 0 \\ 0 & .8921 & 0 & 0 & 0 & 0 & 0 & 0 & 0 & 0 & 0 & 0 \\ 0 & 0 & .8921 & 0 & 0 & 0 & 0 & 0 & 0 & 0 & 0 & 0 \\ 0 & 0 & 0 & .8921 & 0 & 0 & 0 & 0 & 0 & 0 & 0 & 0 \\ 0 & 0 & 0 & 0 & .8921 & 0 & 0 & 0 & 0 & 0 & 0 & 0 \\ 0 & 0 & 0 & 0 & 0 & .8921 & 0 & 0 & 0 & 0 & 0 & 0 \\ 0 & 0 & 0 & 0 & 0 & 0 & .8921 & 0 & 0 & 0 & 0 & 0 \\ 0 & 0 & 0 & 0 & 0 & 0 & 0 & .8921 & 0 & 0 & 0 & 0 \\ 0 & 0 & 0 & 0 & 0 & 0 & 0 & 0 & .8921 & 0 & 0 & 0 \\ 0 & 0 & 0 & 0 & 0 & 0 & 0 & 0 & 0 & .8921 & 0 & 0 \\ 0 & 0 & 0 & 0 & 0 & 0 & 0 & 0 & 0 & 0 & .8921 & 0 \\ 0 & 0 & 0 & 0 & 0 & 0 & 0 & 0 & 0 & 0 & 0 & .8921 \end{bmatrix}$$

7.1.3 Generalized Least Squares Estimation

A more formal way of integrating the points we've been making is to note that OLS estimation is a special case of a broader method known as generalized least squares (GLS) estimation. GLS estimation uses the following formulae to find the regression coefficients,

$$b_{gls} = \left(X'\Psi^{-1}X \right)^{-1} X'\Psi^{-1}y \tag{7.6}$$

and covariance matrix,

$$C_{gls} = \hat{s}_\varepsilon^2 \left(X'\Psi^{-1}X \right)^{-1} \tag{7.7}$$

When the errors are independent and identically distributed, $\Psi = I$ and Eqs. (7.6) and (7.7) reduce to the normal equations used by OLS estimation. This is why we say that OLS estimation is a special case of GLS estimation. In contrast, when the errors are not independent or identically distributed, we must estimate Ψ before we can use Eqs. (7.6) and (7.7). The remainder of this chapter examines this estimation process and discusses corrections that can be taken when the errors are not independent or identically distributed.

7.2   Heteroscedasticity

As you know, linear regression models assume that the errors are identically distributed across values of x (i.e., the diagonal elements of Σ are identical):

$$V\left(\varepsilon|x_i\right) = \sigma_\varepsilon^2 \tag{7.8}$$

When this assumption of constant error variance is upheld, we characterize the data as "homoscedastic" (roughly translated, "of equal variance"); when the assumption is violated, we describe the data as "heteroscedastic" (i.e., of unequal variance).[1]

Predicting some behavior from annual income often produces heteroscedasticity. Imagine we wish to predict how much money families set aside for college each year from their annual income. It seems likely that there will be little variability when annual income is low (if you don't have a lot of annual income, you can't save a lot of money for college), but a good deal of variability when annual income is high (wealthy families might decide to put aside a lot of money for college or figure they will come up with the money when the time comes, so they might not save anything). Figure 7.1 shows that these (hypothetical) differences reveal themselves in the distribution of residuals. Notice that the residuals are tightly clustered around the mean at low income levels, but not at high income levels. A funnel-shaped pattern like this is often a clue that heteroscedasticity exists.[2]

7.2.1   Small Sample Example

Analyzing a small sample example will help us learn more about heteroscedasticity. Consider the data in Table 7.8. Our predictor is age and our criterion is number of hours/night of sleep. The scatterplot shown in the left-hand side of Fig. 7.2 suggests

---

[1] In some textbooks, heteroscedasticity is spelled heteroskedasticity.

[2] The raw residuals are plotted against raw income levels in Fig. 7.1 for illustrative purposes. Ordinarily, we would plot a scaled residual against the standardized fitted values, as we learned to do in Chap. 6.

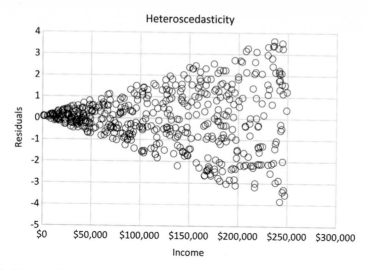

Fig. 7.1 *Heteroscedasticity.* In this hypothetical example, annual saving for college becomes more variable as income levels rise

two tendencies: teenagers generally sleep fewer hours than younger children and their sleep is characterized by greater variability (i.e., some teenagers sleep a lot and some hardly sleep at all). I've raised both kinds, so I'm pretty sure I know what I'm talking about!

Table 7.8 Small sample example of heteroscedasticity

x (age)	y (nightly hours of sleep)	$\hat{y}$	e	h	e^2	$\dfrac{e^2}{(1-h)^2}$	lel
8	11	10.8064	.1936	.2949	.0375	.0754	.1936
9	10.5	10.5249	−.0249	.2249	.0006	.0010	.0249
10	11.2	10.2435	.9565	.1690	.9149	1.3249	.9565
11	9.8	9.9620	−.1620	.1270	.0262	.0344	.1620
12	9.2	9.6805	−.4805	.0991	.2309	.2845	.4805
13	9.0	9.3991	−.3991	.0851	.1593	.1903	.3991
14	8.6	9.1176	−.5176	.0851	.2679	.3201	.5176
15	8.3	8.8361	−.5361	.0991	.2874	.3541	.5361
16	7.0	8.5547	−1.5547	.1270	2.4170	3.1716	1.5547
17	10.0	8.2732	1.7268	.1690	2.9819	4.3180	1.7268
18	10.5	7.9917	2.5083	.2249	6.2914	10.4733	2.5083
19	6.0	7.7103	−1.7103	.2949	2.9250	5.8828	1.7103

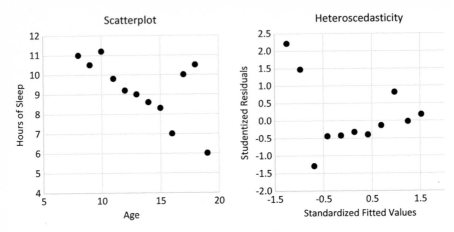

Fig. 7.2 Small sample example illustrating heteroscedasticity

7.2.1.1   Ordinary Regression Analysis

Our first step is to conduct an OLS regression. The results are shown in Table 7.9, and the slope of the regression line is significant, indicating that increases in age predict decreases in nightly hours of sleep.

Table 7.9 Regression analysis for hypothetical data set for heteroscedasticity

	SS	df	MS	R^2	F	p		
Significance test of regression model								
Regression	11.32911	1	11.32911	.4065	6.8495	.0257		
Residual	16.54006	10	1.65401					
Total	27.86917							
Regression coefficients								
	b	se_b	t	p	$(\mathbf{X'X})^{-1}$		$\mathbf{C}$	
b_0	13.05816	1.49861	8.7135	.0000	1.35781	−.09441	2.24582	−.15615
b_1	−.28147	.10755	−2.6172	.0257	−.09441	.00699	−.15615	.01157

7.2.1.2   Scatterplot of the Residuals

Plotting the studentized residuals against the standardized fitted values is a good way to assess heteroscedasticity. Following procedures described in Chap. 6, we calculate the relevant variables and construct the plot that appears in the right-hand side of Fig. 7.2. Looking the figure over, we see evidence of heteroscedasticity. The studentized residuals cluster tightly around the mean when the standardized fitted value of *y* is high, but become increasingly variable as the standardized fitted value of *y* decreases. Notice that these differences are not due to one or two unusual data points; rather, many of the residuals at the low end of the scale lie far from the regression line.

7.2.2 Detecting Heteroscedasticity

Visual inspection often reveals evidence of heteroscedasticity, but it is also desirable to have a less impressionist test of this violation. Several tests are available, and in this section you will learn two of them.

7.2.2.1 White's Test of Heteroscedasticity

To perform White's (1980) test, we take the squared residuals from our OLS analysis and regress them on our predictors, their squares, and all pairwise cross product terms. We then multiply the resultant coefficient of determination (i. e., R^2) by our sample size to form a value we refer to a χ^2 distribution with k degrees of freedom:

$$\chi^2_W = (R^2 * N); \ [df = k] \tag{7.9}$$

We have only one predictor in our example, so $k = 2$ (because we are predicting the squared residuals from the raw predictor and its square). The relevant calculations appear in Table 7.10, and the significant χ^2 value means we should reject the null hypothesis that the residuals have constant variance.

Table 7.10 White's test of heteroscedasticity

		x	x^2	e^2
	1	8	64	.0375
	1	9	81	.0006
	1	10	100	.9149
	1	11	121	.0262
	1	12	144	.2309
	1	13	169	.1593
	1	14	196	.2679
	1	15	225	.2874
	1	16	256	2.4170
	1	17	289	2.9819
	1	18	324	6.2914
	1	19	361	2.9250
	SS	R^2	χ^2	p
Regression	28.4655	.6918	8.3019	.0157
Residual	12.6801			
Total	41.1455			

7.2.2.2 Breusch-Pagan Test of Heteroscedasticity

White's test is easy to conduct when we have only a few predictors, but the calculations become laborious when we have several predictors that need to be squared and multiplied to create cross product terms.[3] The Breusch-Pagan test overcomes these limitations, although it performs best with a large sample size (Breusch and Pagan 1979).

As with White's test, we begin by saving the residuals from an OLS regression. We then regress the squared residuals on our predictors and compute SS^*_{reg}, with the asterisk indicating that the term comes from the auxiliary analysis, not the original one. Finally, we use this value along with the residual sum of squares from our original analysis to compute the test statistic which, with large samples, is distributed as χ^2 with 1 degree of freedom:

$$\chi^2_{BP} = \frac{SS^*_{reg}}{2} \div \left(\frac{SS_{res}}{N}\right)^2 \tag{7.10}$$

Table 7.9 provides the residuals from our original analysis, and Table 7.11 reports the remaining value needed to perform the Breusch-Pagan test with our data.

Table 7.11 Preliminary calculations for Breusch-Pagan test of heteroscedasticity

		x	e^2
	1	8	.0375
	1	9	.0006
	1	10	.9149
	1	11	.0262
	1	12	.2309
	1	13	.1593
	1	14	.2679
	1	15	.2874
	1	16	2.4170
	1	17	2.9819
	1	18	6.2914
	1	19	2.9250
	SS		
Regression	23.1529		

---

[3] To illustrate, if we had three original predictors, we would have nine predictors for White's test (i.e., three original predictors, three squared predictors, and three pairwise cross product terms). Note that we do not compute higher-order cross product terms, only pairwise ones, and we do not square dummy-coded variables (see Chap. 11). Finally, the general rule is that there will be $2k + \frac{(k)(k-1)}{2}$ predictors in the equation.

Plugging the relevant values into Eq. (7.10) confirms the presence of heteroscedasticity:

$$\chi^2_{BP} = \frac{23.1529}{2} \div \left(\frac{16.5401}{12}\right)^2 = 6.0934, \, p = .0136$$

7.2.2.3   Managing Heteroscedasticity

Having detected the presence of noncommon variance, our next step is to diminish its effect. Statistically, we have two approaches.[4] First, we can transform the data and use weighted least squares (WLS) estimation to find our regression coefficients and their standard errors. Doing so requires specifying the form of heteroscedasticity, estimating the variance of each residual, and recalculating a regression model with the transformed data. Alternatively, we can calculate a new covariance matrix to correct our standard errors without specifying the form of the heteroscedasticity or conducting a new regression analysis. The latter alternative is becoming increasingly popular and it is probably the one you will want to use, but it is instructive to work through the first alternative, so we will learn both methods.

7.2.3   Weighted Least Squares Estimation

WLS estimation is a special case of GLS estimation, used when the diagonal elements of the covariance matrix of residuals are unequal but the off-diagonal elements = 0. In other words, it is used when the errors are independently distributed but not identical.

The use of the term "weighted" calls attention to an important fact. As with all measures of variability, the variance of a residual represents its information value. The smaller the variance, the more precise is the estimate and the more informative is the observation about the population parameter. When we conduct an OLS regression, we assume that each observation is equally informative, so we give them equal weight. But when the variances of the residuals are heteroscedastic, observations with small residual variances provide more information than do observations with large residual variances.

---

[4] A third approach, which should always be considered before a statistical one, is to rethink our regression model to be sure we have included all relevant variables.

To take these informational differences into account, we weight each observation by the inverse of its corresponding residual variance estimate:

$$w = \frac{1}{\hat{s}_i^2} \tag{7.11}$$

In this manner, residuals with low variance receive more weight than residuals with high variance, and our regression analysis uses weighted least squares estimation rather than ordinary (equally weighted) least squares estimation.

7.2.3.1   Estimating the Variance of the Residuals

The trick is to derive the variance of each residual. Sometimes these values are known, but most of the time they need to be estimated from the data. The estimation technique we use depends on our intuitions about the form heteroscedasticity takes in our sample. If we are right, we improve the precision of our estimates; if we are wrong, we can wind up with an estimate that is less precise than our original, OLS one.

In many cases, the variances of the residuals are proportional to the size of a predictor variable, with large residuals associated with large predictors. When this occurs, we say that the errors enter the regression equation as a multiplicative function of the predictors rather than an additive one. To the extent that this is so, we can estimate the variance of the residuals using the following procedure:

1. Run an OLS regression and save the residuals.
2. Regress the absolute value of the residuals on $\mathbf{X}$ and calculate the fitted values from the regression coefficients. These fitted values estimate the standard deviation of the residuals (i.e., $\hat{y}_i = \hat{s}_i$).
3. Divide each observation (including the vector of leading 1's), by the absolute value of the estimated standard deviation of the residuals.
4. Perform an OLS regression on the transformed data, without including a term for the intercept (i.e., perform a regression through the origin).

Table 7.12 presents the values we need to illustrate the process. Before we perform the calculations, let's consider why we are regressing the absolute value of the residuals on our predictor. The residuals from the original analysis represent the portion of the variance in y that $\mathbf{X}$ cannot explain. When we regress these values on our predictors and examine the predicted values, we are asking "what portion of this unexplained variance can each observation explain?" Ideally, we would like to learn that each value explains (roughly) the same amount of this unexplained variance. We understand that this will not exactly be true, but that's the ideal given the assumption of identically distributed errors.

Table 7.12 Weighted least squares estimation using the absolute value of the residuals

	x	$|e|$	$\hat{y} = \hat{s}$	$\frac{1}{|\hat{s}|}$	$\frac{x}{|\hat{s}|}$	$\frac{y}{|\hat{s}|}$	$\hat{y}_w$	e_w
1	8	.1936	−.0962	10.3955	83.1638	114.3502	113.6977	.6525
1	9	.0249	.0845	11.8368	106.5312	124.2864	125.7015	−1.4150
1	10	.9565	.2652	3.7713	37.7130	42.2386	38.8515	3.3871
1	11	.1620	.4458	2.2430	24.6726	21.9811	22.3942	−.4131
1	12	.4805	.6265	1.5961	19.1535	14.6844	15.4290	−.7446
1	13	.3991	.8072	1.2389	16.1052	11.1497	11.5819	−.4321
1	14	.5176	.9879	1.0123	14.1719	8.7056	9.1420	−.4364
1	15	.5361	1.1685	.8558	12.8364	7.1028	7.4566	−.3538
1	16	1.5547	1.3492	.7412	11.8586	5.1882	6.2226	−1.0345
1	17	1.7268	1.5299	.6536	11.1118	6.5363	5.2801	1.2562
1	18	2.5083	1.7106	.5846	10.5227	6.1383	4.5367	1.6016
1	19	1.7103	1.8913	.5287	10.0462	3.1725	3.9353	−.7628

With that in mind, let's return to our estimation procedure. When we regress the absolute value of the residuals on our predictors, we obtain the following regression coefficients we need to calculate fitted values $(\hat{y})$:

$$b_0 = -1.5416$$

and

$$b_1 = .1807$$

As noted earlier, these fitted values estimate the standard deviation of each residual. We then divide our intercept vector of $1's$, our predictor $(\hat{x})$, and our original criterion (y) by the absolute value of each observation's residual standard deviation. These values are shown in Table 7.12.

7.2.3.2 Perform an OLS Regression on the Transformed Variables

If we then perform an OLS regression through the origin on the transformed data, we get the results shown in Table 7.13.[5] These values represent a weighted least squares solution, with the weight of each observation inversely proportional to its

Table 7.13 Weighted least squares regression analysis for data in Table 7.8

	b	se_b	t	p	$(\mathbf{X'X})^{-1}$		$\mathbf{C}$	
b_0	13.47870	.69436	19.4118	.0000	.23043	−.02569	.48213	−.05375
b_1	−.31768	.07803	−4.0712	.0022	−.02569	.00291	−.05375	.00609

---

[5] Because we have not included a true intercept, we do not interpret the R^2 value from the transformed data.

Table 7.14 Matrix formulae for weighted least squares estimation

Matrix	Regression coefficients $\mathbf{b}^*$	Parameter covariance matrix
$$\boldsymbol{\Sigma} = \begin{bmatrix} \hat{s}_1^2 & 0 & \cdots & 0 \\ 0 & \hat{s}_2^2 & \cdots & 0 \\ \vdots & \vdots & \ddots & \vdots \\ 0 & 0 & \cdots & \hat{s}_N^2 \end{bmatrix}$$	$(\mathbf{X}'\boldsymbol{\Sigma}^{-1}\mathbf{X})^{-1}\mathbf{X}'\boldsymbol{\Sigma}^{-1}\mathbf{y}$	$MS^*_{res}\,*$ $(\mathbf{X}'\boldsymbol{\Sigma}^{-1}\mathbf{X})^{-1}$
$$\boldsymbol{\Psi} = \begin{bmatrix} \hat{s}_1^2/\sigma^2 & 0 & \cdots & 0 \\ 0 & \hat{s}_2^2/\sigma^2 & \cdots & 0 \\ \vdots & \vdots & \ddots & \vdots \\ 0 & 0 & \cdots & \hat{s}_N^2/\sigma^2 \end{bmatrix}$$	$(\mathbf{X}'\boldsymbol{\Psi}^{-1}\mathbf{X})^{-1}\mathbf{X}'\boldsymbol{\Psi}^{-1}\mathbf{y}$	$MS^*_{res}\,*$ $\{\sigma^2(\mathbf{X}'\boldsymbol{\Psi}^{-1}\mathbf{X})^{-1}\}$
$$\boldsymbol{\Gamma} = \begin{bmatrix} \sqrt{\sigma^2/\hat{s}_1^2} & 0 & \cdots & 0 \\ 0 & \sqrt{\sigma^2/\hat{s}_2^2} & \cdots & 0 \\ \vdots & \vdots & \ddots & \vdots \\ 0 & 0 & \cdots & \sqrt{\sigma^2/\hat{s}_N^2} \end{bmatrix}$$	$(\mathbf{X}'\boldsymbol{\Gamma}\boldsymbol{\Gamma}\mathbf{X})^{-1}\mathbf{X}'\boldsymbol{\Gamma}\boldsymbol{\Gamma}\mathbf{y}$	$MS^*_{res}\,*$ $\{\sigma^2(\mathbf{X}'\boldsymbol{\Gamma}\boldsymbol{\Gamma}\mathbf{X})^{-1}\}$

estimated variance. With our data set, age remains a significant predictor of sleep after making this adjustment.

Finally, we can test the transformed residuals for heteroscedasticity to be certain we have eliminated it. The last column of Table 7.12 shows the relevant values and White's test confirms that heteroscedasticity has been reduced, $\chi^2 = .8179, p = .6643$.

7.2.3.3   Matrix Calculations for WLS Estimation

The calculations we have performed represent the easiest way to understand WLS estimation, but the analysis can also be performed using matrix formulae. Table 7.14 shows several, equivalent approaches. For all analyses,

$$MS^*_{res} = \frac{\mathbf{e}^{*'}\mathbf{X}\,\mathbf{e}^*}{(N-k-1)} \tag{7.12}$$

where $\mathbf{e}^* = \mathbf{y} - \mathbf{b}^*\mathbf{X}$ (i.e., the fitted values and residuals are found from the weighted regression coefficients).

I encourage you to work through the calculations to confirm their equivalence, as each offers a unique way of understanding WLS estimation. The final matrix $\boldsymbol{\Gamma}$ is of particular interest, as it is a transformation matrix created with the following properties:

$$\boldsymbol{\Gamma}'\boldsymbol{\Gamma} = \boldsymbol{\Psi}^{-1} \text{ and } \boldsymbol{\Gamma}\boldsymbol{\Psi}\boldsymbol{\Gamma}' = \mathbf{I} \tag{7.13}$$

Using Eq. (7.14), the transformation matrix can also be used to transform our original variables so that they can be analyzed using OLS estimation,

$$\mathbf{V}^* = \mathbf{\Gamma V} \tag{7.14}$$

where $\mathbf{V}$ refers to our original data matrix that includes a vector of leading 1's, $\mathbf{X}$ and $\mathbf{y}$.

7.2.4 Heteroscedasticity-Consistent Covariance Matrix

Because heteroscedasticity affects the standard errors but not the regression coefficients, we can address this violation by replacing our original covariance matrix with a heteroscedasticity-consistent covariance matrix (HCCM; White 1980). Earlier, we saw that Eq. (7.3) (reproduced below) can be used to find the parameter covariance matrix, where $\mathbf{\Sigma}$ is a diagonal matrix with σ^2 on the diagonals:

$$\mathbf{C} = \left(\mathbf{X'X}\right)^{-1} \mathbf{X'\Sigma X}\left(\mathbf{X'X}\right)^{-1}$$

Because $\mathbf{\Sigma}$ is a diagonal matrix, we can rewrite the equation as follows:

$$\mathbf{C} = (\mathbf{X'X})^{-1} \left(\mathbf{X'}\mathrm{diag}\left[\frac{\sigma^2}{1}\right]\mathbf{X}\right)(\mathbf{X'X})^{-1} \tag{7.15}$$

Building on this formula, the following equation can be used to find an HCCM developed by White and designated $\mathbf{HC_1}$ by Long and Ervin (2000):

$$\mathbf{HC_1} = \left(\mathbf{X'X}\right)^{-1} \left(\mathbf{X'}\mathrm{diag}\left[\frac{e_i^2}{1}\right]\mathbf{X}\right)\left(\mathbf{X'X}\right)^{-1} \tag{7.16}$$

Notice that the middle term is still a diagonal matrix, but the diagonal values now represent the square of each observation's residual rather than a common estimate of the variance.

The HCCM presented in Eq. (7.16) is accurate with large samples, but Long and Ervin recommend a slightly modified equation, designated $\mathbf{HC_3}$, when sample sizes are small:

$$\mathbf{HC_3} = \left(\mathbf{X'X}\right)^{-1} \left(\mathbf{X'}\mathrm{diag}\left[\frac{e_i^2}{(1-h_i)^2}\right]\mathbf{X}\right)\left(\mathbf{X'X}\right)^{-1} \tag{7.17}$$

The middle term is still a diagonal matrix, but the diagonal values now represent a unique estimate of the residual variance of each observation, computed by dividing its squared residual by $(1-h_i)^2$, where h_i is the hat value discussed in Chap. 6 [Eq. (6.4) repeated below]:

$$h_i = \mathbf{x_i'}\left(\mathbf{X'X}\right)^{-1}\mathbf{x_i}$$

If you look back to the penultimate column in Table 7.8, you will find the diagonal values needed to populate the matrix:

$$
HC_3 = \begin{bmatrix}
.0754 & 0 & 0 & 0 & 0 & 0 & 0 & 0 & 0 & 0 & 0 & 0 \\
0 & .0010 & 0 & 0 & 0 & 0 & 0 & 0 & 0 & 0 & 0 & 0 \\
0 & 0 & 1.3249 & 0 & 0 & 0 & 0 & 0 & 0 & 0 & 0 & 0 \\
0 & 0 & 0 & .0344 & 0 & 0 & 0 & 0 & 0 & 0 & 0 & 0 \\
0 & 0 & 0 & 0 & .2845 & 0 & 0 & 0 & 0 & 0 & 0 & 0 \\
0 & 0 & 0 & 0 & 0 & .1903 & 0 & 0 & 0 & 0 & 0 & 0 \\
0 & 0 & 0 & 0 & 0 & 0 & .3201 & 0 & 0 & 0 & 0 & 0 \\
0 & 0 & 0 & 0 & 0 & 0 & 0 & .3541 & 0 & 0 & 0 & 0 \\
0 & 0 & 0 & 0 & 0 & 0 & 0 & 0 & 3.1716 & 0 & 0 & 0 \\
0 & 0 & 0 & 0 & 0 & 0 & 0 & 0 & 0 & 4.3180 & 0 & 0 \\
0 & 0 & 0 & 0 & 0 & 0 & 0 & 0 & 0 & 0 & 10.4733 & 0 \\
0 & 0 & 0 & 0 & 0 & 0 & 0 & 0 & 0 & 0 & 0 & 5.8828
\end{bmatrix}
$$

If you then use this matrix in Eq. (7.17), you will derive a new covariance matrix that takes the heteroscedasticity of the errors into account:

$$
HC_3 = \begin{bmatrix}
2.95391 & -.26206 \\
-.26206 & .02362
\end{bmatrix}
$$

Taking the square root of the diagonal entries as our standard errors, we find that the regression slope, which was significant when homoscedasticity was assumed, is no longer significant when the correction is applied (see Table 7.15). Although this won't always occur, correcting our standard errors for heteroscedasticity usually produces larger standard errors and, therefore, more conservative tests of statistical significance.

Table 7.15 HC_3 corrected standard errors and tests of significance

	b	se_b	t	p
b_0	13.05816	1.71869	7.5977	.0000
b_1	−.28147	.15370	−1.8313	.0970

7.2.5   Summary

Table 7.16 summarizes some of the results from the analyses we have been conducting. As you can see, the **HC₃** analysis presents the most conservative test and the WLS analysis provides the most liberal. This will not always be the case but is true here because the method we used to estimate heteroscedasticity was very accurate. When we are less sure of the form that heteroscedasticity takes, the **HC₃** correction is recommended.

Table 7.16 Three ways of testing data with heteroscedasticity

	b_1	se_b	t	p
OLS	−.28147	.10755	−2.6172	.0257
WLS	−.31768	.07803	−4.0712	.0022
HC_3	−.28147	.15370	−1.8313	.0970

Regardless of which test we use, the question arises as to how serious a problem heteroscedasticity poses. Fox (2008) suggests that it will only be a problem when the largest residual is four times larger than the smallest residual. Our example far exceeded this threshold, but that's because I purposefully created a data set that presented a clear violation. Most violations will not be so extreme, so heteroscedasticity will rarely threaten the conclusions we draw from our data. That being said, statisticians are increasingly urging researchers to routinely use the **HC$_3$** correction whenever heteroscedasticity is suspected (Hayes and Li 2007).

7.2.6   R Code: Heteroscedasticity

```
x=c(8,9,10,11,12,13,14,15,16,17,18,19)
y=c(11,10.5,11.2,9.8,9.2,9,8.6,8.3,7,10,10.5,6)
hetero=lm(y~x)
summary(hetero)
ms.res <-sum(resid(hetero)^2)/((length(x)-2))

#White's Test of Heteroscedasticity
res.sqr <-resid(hetero)^2
white.test <-lm(res.sqr~x+I(x^2))
white.chi <-summary(white.test)$r.squared*length(x)
white.prob <- 1 - pchisq(white.chi, 2)
white <-cbind(white.chi, white.prob)
white

#Breusch Pagan Test of Heteroscedasticity
library(car)   #attach car package
BP <-ncvTest(hetero)
BP

#Weighted Least Squares
abs_e=abs(hetero$residuals) #Calculate absolute value of residuals
weights <-lm(abs_e~x)  #Regress absolute value of residuals on x
w.int =1/abs(fitted(weights)) #Create weighted variables
w.x =x/abs(fitted(weights))
w.y =y/abs(fitted(weights))
WLS <-lm(w.y~w.int+w.x-1) #regress weighted variables (no intercept)
summary(WLS)

#Shorter way to perform WLS using Weights commands
weight1=1/(weights$fitted^2)
WLS.1=lm(y~x,weights=weight1)
summary(WLS.1)

#Using Gamma as a Transformation Matrix for WLS
```

(continued)

7.2.6   R Code: Heteroscedasticity (continued)

```
V <-cbind(1,x,y)
gamma <-diag(sqrt((ms.res/fitted(weights)^2)))
vv <-gamma%*%V
vv.reg <-lm(vv[,3]~vv[,1:2]-1)
summary(vv.reg)

#HC_3 Heteroscedasticity Consistent Covariance Matrix
library(lmSupport) #attach lmSupport package
lm.correctSE(hetero, digits=6)
```

7.3   Autocorrelations

Linear regression models not only assume that the errors are identically distributed but also that they are independent. In terms of the covariance matrix of residuals, this assumption pertains to the off-diagonal entries, not the diagonal ones. Many circumstances can produce correlated errors. Suppose we measure task performance across several days of practice. If, as seems likely, performance on one day is associated with performance on the next, the residuals would no longer be independent. Similarly, crime statistics, weather patterns, and economic indicators are often measured at regular intervals, and observations at adjacent time periods tend to be more similar than observations with greater temporal distance. Spatial autocorrelation can also occur. As Tobler's first law of geography states, "Everything is related to everything else, but near things are more related than distant things."

As with heteroscedasticity, autocorrelations (as they are called) do not bias the regression coefficients, but they do affect the standard errors, thereby affecting tests of statistical significance and confidence intervals.[6] For this reason, it is just as important to detect and correct autocorrelations as it is to detect and correct noncommon variance estimates.

7.3.1   Mathematical Representation

Before proceeding to an analysis of data with autocorrelated errors, let's spend a moment looking at their mathematical representation.

---

[6] In some textbooks, autocorrelations are referred to as "serial correlations." There is no substantive difference between the terms, but I prefer "autocorrelation" because it underscores that the residuals are correlated with themselves.

7.3.1.1 Modeling Autocorrelated Errors

Equation (7.18) shows a simple model that we will examine throughout this section. The subscript t identifies a specific observation, and the subscript $t-1$ identifies its preceding observation. Thus, Eq. (7.18) indicates that the value of each error is determined by two factors: the size of the preceding error (ε_{t-1}) and a normally distributed random disturbance, designated u_t:

$$\varepsilon_t = \varepsilon_{t-1} + u_t \qquad (7.18)$$

Because the value of each error is influenced by the value of the previous error, the errors are no longer independent.

The model shown in Eq. (7.18) can be expanded in two ways. First, we can include a value that represents the weight of the preceding error. Equation (7.19) shows that we refer to the weight as ρ (rho). To illustrate, if $\rho=.50$, then each error would be found by first multiplying the preceding error by .50 and then adding a random disturbance term. The weight, which is known as the autoregressive parameter, represents the correlation between adjacent error terms. Like all correlations, it can assume values from -1 to $+1$:

$$\varepsilon_t = \rho\varepsilon_{t-1} + u_t \qquad (7.19)$$

Because it only considers the correlation between adjacent errors, the model depicted in Eq. (7.19) is known as a first-order autoregression or AR(1) process. This is not the only way to model autoregression, however. Equation (7.20) depicts an AR(2) process in which each error is influenced by the weighted values of the preceding error, the weighted error two observations back, and a random disturbance:

$$\varepsilon_t = \rho_1\varepsilon_{t-1} + \rho_2\varepsilon_{t-2} + u_t \qquad (7.20)$$

In this chapter, we will concern ourselves only with first-order lags, but the material we cover can readily be applied to longer lags

7.3.1.2 Building a Covariance Matrix of Errors in an AR(1) Model

Returning to an AR(1) model, recall that only adjacent errors directly influence subsequent errors. More distant errors do have an indirect effect, however. To better understand this point, let's imagine we have only four observations, shown in Table 7.17, with $\rho=.50$.

Table 7.17 Expanding a
first-order autoregressive
model

Observation #	u	ε
1	5	5
2	10	12.50
3	15	21.25
4	20	30.6250

Using our AR(1) formula $[\varepsilon_t = \rho\varepsilon_{t-1} + u_t]$, we can see how the last three error terms were computed:

$$\varepsilon_2 = \rho\varepsilon_1 + u_2 = (.5 * 5) + 10 = 12.50$$
$$\varepsilon_3 = \rho\varepsilon_2 + u_3 = (.5 * 12.50) + 15 = 21.25$$
$$\varepsilon_4 = \rho\varepsilon_3 + u_4 = (.5 * 21.25) + 20 = 30.6250$$

If we then take advantage of the fact that each successive term depends, in part, on the value of the preceding terms, we can derive the last two values using an expanded form:

$$\varepsilon_3 = \left(\rho^2 * \varepsilon_1\right) + (\rho * \varepsilon_2) + u_3$$
$$\varepsilon_4 = \left(\rho^3 * \varepsilon_1\right) + \left(\rho^2 * \varepsilon_2\right) + (\rho * \varepsilon_3) + u_4$$

Here we see that each error term is a linear combination of current disturbance and previous errors and that the weights diminish in magnitude as the absolute distance between a focal observation (t) and a preceding observation (s) increases (t):

$$\varepsilon_t = \sum_{s=0}^{\infty} \rho^s u_{t-s} \tag{7.21}$$

Exploiting this relation, we can construct a covariance matrix with the variances on the diagonals and the covariances on the off-diagonals, where T refers to the total number of observations rather than a particular observation:

$$\Sigma = \begin{bmatrix} 1 & \rho & \rho^2 & \rho^3 & \cdots & \rho^{T-1} \\ \rho & 1 & \rho & \rho^2 & \cdots & \rho^{T-2} \\ \rho^2 & \rho & 1 & \rho & \cdots & \rho^{T-3} \\ \rho^3 & \rho^2 & \rho & 1 & \cdots & \rho^{T-4} \\ \vdots & \vdots & \vdots & \vdots & \ddots & \vdots \\ \rho^{T-1} & \rho^{T-2} & \rho^{T-3} & \rho^{T-4} & \cdots & 1 \end{bmatrix} \tag{7.22}$$

7.3.2   Detecting Autocorrelations

Having discussed autocorrelations in abstract terms, we are ready for an example. Imagine a basketball fan wants to know whether crowd noise predicts a home team's performance. For 12 games, the fan records the median decibel level of crowd noise and the home team's field goal shooting percentage. Table 7.18 presents the (fabricated) data and a few summary statistics, and Table 7.19 presents the results from an OLS regression. Contrary to our fan's expectations, the coefficient relating crowd noise to performance is not significant in this analysis ($b = .0085, p = .0903$). A look at the scatterplot (see left-hand side of Fig. 7.3) confirms the lack of a consistent relation between crowd noise and shooting efficiency.

Table 7.18 Fictitious data demonstrating autocorrelation

Time	x (median decibels)	y (field goal percentage)	$\hat{y}$	e	h	Lagged residual differences
1	84	.5384	.56297	−.02457	.08537	
2	83	.5471	.55447	−.00737	.11585	.01721
3	81	.5663	.53745	.02885	.28659	.03621
4	83	.5677	.55447	.01323	.11585	−.01561
5	87	.6187	.58849	.03021	.21341	.01697
6	83	.6059	.55447	.05143	.11585	.02123
7	87	.6131	.58849	.02461	.21341	−.02683
8	87	.5944	.58849	.00591	.21341	−.01870
9	84	.5492	.56297	−.01377	.08537	−.01968
10	86	.5183	.57999	−.06169	.13415	−.04791
11	81	.5017	.53745	−.03575	.28659	.02594
12	86	.5689	.57999	−.01109	.13415	.02466

Table 7.19 Regression analysis for hypothetical data set for autocorrelation

	SS	df	MS	R^2	F	p		
			Significance test of regression model					
Regression	.00396	1	.00396	.2600	3.5138	.0903		
Residual	.01126	10	.00113					
Total	.01522							

	b	se_b	t	p	$(\mathbf{X'X})^{-1}$		C	
			Regression coefficients					
b_0	−.1516	.3829	−.3961	.7004	130.1829	−1.5427	.146583	−.001737
b_1	.0085	.0045	1.8745	.0903	−1.5427	.0183	−.001737	.000021

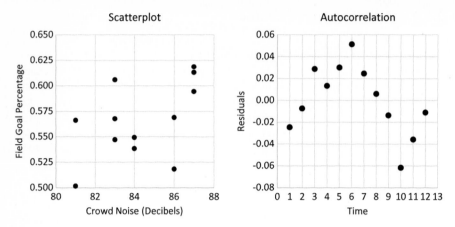

Fig. 7.3 Small sample example illustrating autocorrelation

7.3.2.1   Inspecting the Residuals

Many basketball aficionados believe in a phenomenon known as the "hot hand," in which a player or team "gets hot" and "goes on a roll." To the extent that this is so, performance on one occasion would predict performance on the next, and the residuals would be correlated. To examine whether the hot hand effect is operating in this situation, the fan decides to plot the residuals against the day the observations were made. The right-hand panel in Fig. 7.3 presents the data. Notice the undulating pattern of the residuals. If a residual at t is positive, we can predict with some certainty that the $t+1$ residual will also be positive; and if the residual at t is negative, we can predict with some certainty that the $t+1$ residual will be negative, too. We won't always be right, but it's apparent that knowledge of one residual does, in fact, provide knowledge of another. This fact violates the assumption that the residuals are independent.

7.3.2.2   Durbin-Watson Test

Autocorrelations won't always be as obvious as the pattern depicted in Fig. 7.3, so we will want to test for the pattern when we suspect its presence. The most common test is known as the Durbin-Watson test:

$$d = \frac{\sum_{t=2}^{t}(e_t - e_{t-1})^2}{\sum_{t=1}^{t} e_t^2} \tag{7.23}$$

To perform the test, we first conduct an OLS regression and save the residuals. We then find the numerator in Eq. (7.23) by subtracting from each residual the preceding residual in the sequence.[7] To illustrate, the first two difference scores shown in the last column of Table 7.18 were computed as follows:

$$(-.00737 - -.02457) = .01721$$
$$(.02885 - - .00737) = .03621$$

After finding all of the $N - 1$ difference scores, we sum their squared values and divide this sum by the sum of the squared residuals (i. e., SS_{res}) to find the Durbin-Watson statistic[8]:

$$d = \frac{.007624}{.011260} = .67707$$

7.3.2.3 Breusch-Godfrey (aka Lagrange Multiplier) Test of Autocorrelation

The expected value of the Durbin-Watson statistic equals 2 when the residuals are uncorrelated. The exact statistical significance of the Durbin-Watson test is determined by consulting tabled values that are presented in many statistical textbooks. As these tables are not always handy, it is useful to learn a self-contained test that can be performed with an ordinary spreadsheet. The Breusch-Godfrey test provides just such a test. Moreover, it can assess autocorrelations of various sizes [e.g., AR (2)], so it is more flexible than the Durbin-Watson test.

As was true with the Durbin-Watson test, the Breusch-Godfrey test begins by conducting a regression analysis and saving the residuals. We then conduct an auxiliary regression analysis using the residuals as our criterion. Our predictors are the original predictors (including the vector of leading 1's) and a lagged vector of residuals. Table 7.20 shows the data using our example. Looking over the last two columns, you can see that the values in the penultimate column lag one value behind the values in the final column of residuals.

Performing a regression analysis with these data yields an $R^2 = .4931$. We then multiply this value by our sample size (12) and refer the product (5.9173) to a χ^2 distribution with $df =$ to the number of lagged variables (in this case, 1).

---

[7] Notice that we cannot find the difference score for the first observation because there is no preceding error to subtract. Consequently, we will have $N - 1$ observations when performing this test.

[8] Spreadsheet functions can be used to perform these operations. To illustrate, if the vector of residuals lies in "e3:e14," the following command produces the Durbin-Watson statistic: SUMXMY2(e3:e13,e4:e14)/SUMSQ(e3:e14).

Observation #			e_{lagged}	e
1	1	84	0	−.02457
2	1	83	−.02457	−.00737
3	1	81	−.00737	.02885
4	1	83	.02885	.01323
5	1	87	.01323	.03021
6	1	83	.03021	.05143
7	1	87	.05143	.02461
8	1	87	.02461	.00591
9	1	84	.00591	−.01377
10	1	86	−.01377	−.06169
11	1	81	−.06169	−.03575
12	1	86	−.03575	−.01109

Table 7.20 Breusch-Godfrey test of autocorrelation

The probability level of our value given that no autocorrelation exists is small ($p=.0150$), so we reject the null hypothesis of independent residuals.[9]

7.3.2.4   Managing Autocorrelations

As with heteroscedasticity, two statistical procedures can be used to mitigate the impact of correlated error terms. The first involves GLS estimation and the second involves creating an autocorrelation-consistent covariance matrix. As before, we will learn each method, as each provides unique insights into the remediation of correlated error terms.

7.3.3   Generalized Least Squares Estimation for Managing Autocorrelation

Similar to WLS estimation, GLS begins by specifying a parameter, in this case, the autocorrelation parameter. Although there may be times when the parameter is known, it usually must be estimated from the data.

---

[9] Notice that we have entered a 0 for our first lagged residual. Alternatively, we can omit the observation entirely, performing the analysis on $t-1$ observations. In this case, we multiply R^2 by $t-1$. Using this method with our data, we find that $\chi^2 = 5.8925$, $p = .0152$.

7.3.3.1 Estimating the Autocorrelation Parameter

Estimation involves an iterative process that begins with an initial approximation. There are several ways to proceed, but the simplest is to regress e_{t-1} on e_t. Applied to our data set, our criterion consists of the residuals $e_2 \ldots e_{12}$ and our predictor consists of the residuals $e_1 \ldots e_{11}$ (along with a vector of leading 1's). The regression coefficient represents the estimate of the autocorrelation parameter. Performing the analysis with our data set yields an initial estimate:

$$\hat{\rho} = .63455$$

We then use this value as a starting point in a grid search method (see Chap. 3), seeking to maximize the log likelihood function of the AR(1) process. To refresh your memory, Eq. (7.24) reviews the log likelihood function for OLS regression [repeated from Eq. (3.21)]:

$$\ln L = -\frac{N}{2}\ln(2\pi) - \frac{N}{2}\ln(\sigma^2) - \left[\frac{(\mathbf{y} - \mathbf{Xb})^2}{2\sigma^2}\right] \tag{7.24}$$

When dealing with an AR(1) process, we simply add the estimated autocorrelation parameter to the log likelihood function:

$$\ln L = -\frac{N}{2}\ln(2\pi) - \frac{N}{2}\ln(\sigma^2) - \left[\frac{(\mathbf{y} - \mathbf{Xb})^2}{2\sigma^2}\right] + \left[.5\ln\left(1 - \rho^2\right)\right] \tag{7.25}$$

What we are looking for, then, is the value of ρ that will maximize the function. However, unlike OLS estimation, this function does not have a unique solution. So we set some standard, known as a stop value, to end our search. For example, we might decide to stop when the change in the likelihood function is $< .00000001$.

To begin our search, we transform our variables using a technique known as the Prais-Winsten transformation:[10]

$$\begin{cases} v_i^* = v_1 * \sqrt{1 - \hat{\rho}^2} & \text{for } v_i = 1 \\ v_i^* = v_t - (\hat{\rho} * v_{t-1}) & \text{for } v_i > 1 \end{cases} \tag{7.26}$$

Notice that this technique uses a different transformation for the first set of observations than for all of the other observations. This is because the first set of observations has no preceding term with which to create a lag, so we estimate it instead. Table 7.21 presents the transformed values, and the first two rows were created as follows.

---

[10] The procedure goes by a variety of other names, including *estimated generalized least squares*, the *Cochrane-Orcutt method*, and the *Yule-Walker method*. There are slight differences among these procedures, but they all estimate the magnitude of the autocorrelation parameter.

Table 7.21 Prais-Winsten transformation for GLS estimation and regression output

1*	x*	y*
.7729	64.9221	.4161
.3655	29.6978	.2055
.3655	28.3324	.2191
.3655	31.6015	.2084
.3655	34.3324	.2585
.3655	27.7942	.2133
.3655	34.3324	.2286
.3655	31.7942	.2054
.3655	28.7942	.1720
.3655	32.6978	.1698
.3655	26.4287	.1728
.3655	34.6015	.2505

Significance test of regression model

	SS	df	MS	F	p
Regression	.041998	1	.041998	69.15	.0000
Residual	.006073	10	.000607		
Total	.048071				

Intercept	x*	y*
$1 * \sqrt{1 - .63455^2} = .7729$	$84 * \sqrt{1 - .63455^2} = 64.9221$	$.5384 * \sqrt{1 - .63455^2} = .4161$
$1 - (.63455 * 1) = .3655$	$83 - (.63455 * 84) = 29.6978$	$.5471 - (.63455 * .5384) = .2055$

After transforming our variables, we perform an OLS regression through the origin using 1* and x* as predictors and y* as the criterion. The bottom portion of Table 7.21 shows that the residual sum of squares $= .006073$. Inserting this value into Eq. (7.25) yields the log likelihood function:

$$\ln L = -\frac{12}{2}\ln(2\pi) - \frac{12}{2}\ln\left(\frac{.006073}{12}\right) - \left(\frac{.006073}{2*.006073/12}\right) + \left[.5\ln\left(1 - .63455^2\right)\right] = 28.2475$$

If we continue trying out different values for ρ, we will find a value that maximizes the likelihood function. Table 7.22 shows the log likelihood function around a range of values for the autocorrelation parameter. As you can see, the largest likelihood function is found when $\rho=.66035$.

7.3.3.2 Transformation Method

Having found our estimate of ρ, we use it to transform our variables using Eq. (7.26) and perform another OLS regression through the origin. Table 7.23 presents the results. Notice that the regression coefficient is now significant, suggesting that, in

Table 7.22 Log likelihood functions using Prais-Winsten transformation and various estimates of the autocorrelation parameter

ρ	Log likelihood function
.65945	28.25593102
.65955	28.25593315
.65965	28.25593502
.65975	28.25593664
.65985	28.25593800
.65995	28.25593910
.66005	28.25593995
.66015	28.25594053
.66025	28.25594086
.66035	**28.25594094**
.66045	28.25594075
.66055	28.25594031
.66065	28.25593961
.66075	28.25593865
.66085	28.25593743
.66095	28.25593596
.66105	28.25593422
.66115	28.25593223
.66125	28.25592998
.66135	28.25592748

Table 7.23 GLS estimates following Prais-Winsten transformation using $\rho =.66035$

1*	x*	y*
.7382	62.0112	.3975
.3255	26.3378	.1839
.3255	25.0124	.1973
.3255	28.3615	.1857
.3255	31.0124	.2358
.3255	24.3142	.1886
.3255	31.0124	.2044
.3255	28.3142	.1808
.3255	25.3142	.1482
.3255	29.3378	.1478
.3255	22.9887	.1521
.3255	31.3615	.2305

Significance test of regression model

	SS	df	MS	F	p
Regression	.04363	1	.04363	72.4608	.0000
Residual	.00602	10	.00060		
Total	.04965				

Regression coefficients

	b	se_b	t	p	$(X'X)^{-1}$		C	
b_0	.03292	.2227	.1478	.8854	82.36430	−.96775	.049595	−.000583
b_1	.00627	.0026	2.3873	.0381	−.96775	.01145	−.000583	.000007

this fanciful example, crowd noise does predict the home team's performance once the correlated error terms are taken into account.

7.3.3.3 Matrix Method

As with WLS estimation, several matrix formulae produce identical results to the transformation method. The formulae appear in Table 7.14, and Table 7.24 shows the matrices needed to perform the analyses. As before, you should work through the calculations to solidify your understanding of GLS estimation.

Table 7.24 Matrices for generalized least squares estimation

Matrix

$$\Sigma = \begin{bmatrix} 1 & \rho & \rho^2 & \rho^3 & \cdots & \rho^{T-1} \\ \rho & 1 & \rho & \rho^2 & \cdots & \rho^{T-2} \\ \rho^2 & \rho & 1 & \rho & \cdots & \rho^{T-3} \\ \rho^3 & \rho^2 & \rho & 1 & \cdots & \rho^{T-4} \\ \vdots & \vdots & \vdots & \vdots & \ddots & \vdots \\ \rho^{T-1} & \rho^{T-2} & \rho^{T-3} & \rho^{T-4} & \cdots & 1 \end{bmatrix}$$

$$\Psi = \begin{bmatrix} 1/(1-\rho^2) & \rho/(1-\rho^2) & \rho^2/(1-\rho^2) & \rho^3/(1-\rho^2) & \cdots & \rho^{T-1}/(1-\rho^2) \\ \rho/(1-\rho^2) & 1/(1-\rho^2) & \rho/(1-\rho^2) & \rho^2/(1-\rho^2) & \cdots & \rho^{T-2}/(1-\rho^2) \\ \rho^2/(1-\rho^2) & \rho/(1-\rho^2) & 1/(1-\rho^2) & \rho\rho/(1-\rho^2) & \cdots & \rho^{T-3}/(1-\rho^2) \\ \rho^3/(1-\rho^2) & \rho^2/(1-\rho^2) & \rho/(1-\rho^2) & 1/(1-\rho^2) & \cdots & \rho^{T-4}/(1-\rho^2) \\ \vdots & \vdots & \vdots & \vdots & \ddots & \vdots \\ \rho^{T-1}/(1-\rho^2) & \rho^{T-2}/(1-\rho^2) & \rho^{T-3}/(1-\rho^2) & \rho^{T-4}/(1-\rho^2) & \cdots & 1/(1-\rho^2) \end{bmatrix}$$

$$\Gamma = \begin{bmatrix} \sqrt{(1-\rho^2)} & 0 & 0 & 0 & \cdots & 0 \\ -\rho & 1 & 0 & 0 & \cdots & 0 \\ 0 & -\rho & 1 & 0 & \cdots & 0 \\ 0 & 0 & -\rho & 1 & 0 & 0 \\ \vdots & \vdots & \vdots & \vdots & \ddots & \vdots \\ 0 & 0 & 0 & 0 & -\rho & 1 \end{bmatrix}$$

7.3.4 Autocorrelation-Consistent Covariance Matrix

GLS estimation is one way to accommodate the effects of autocorrelation, but just as we did with heteroscedastic errors, we can also handle autocorrelation by creating an autocorrelation-consistent covariance matrix (ACCM). Like an HCCM, an ACCM does not require us to estimate a parameter value (in this case, the autocorrelation parameter). In this sense, it is easier to conduct than GLS estimation.

The Newey-West (1987) method provides one approach to creating an ACCM. It begins by noting that since the influence of each observation weakens as the absolute distance between observations increases [see Eq. (7.21)], we can safely ignore the influence of observations that lie far from a focal observation. Imagine, for example, that we had a sample size of 250 observations. Although it is true that error 250 is influenced by error 2, the influence 248 observations away is so negligible that it can effectively be ignored.

7.3.4.1 Newey-West Procedure

Building on this fact, the Newey-West method constructs a covariance estimator $\hat{\mathbf{S}}$ using two matrices: a diagonal matrix $\mathbf{E}$ with the error terms on the diagonal and a matrix of lagged weights $\mathbf{L}$:

$$\hat{\mathbf{S}} = \mathbf{E}'\mathbf{L}\mathbf{E} \tag{7.27}$$

- The weights for $\mathbf{L}$ are determined using Equation (7.28),

$$1 - \frac{l}{(L+1)} \tag{7.28}$$

where l refers to the weight of a particular observation and L refers to the total number of lags being estimated; and
- the covariance estimator is then used to find a corrected covariance matrix:

$$\mathbf{C} = \left(\mathbf{X}'\mathbf{X}\right)^{-1}\mathbf{X}'\hat{\mathbf{S}}\,\mathbf{X}\left(\mathbf{X}'\mathbf{X}\right)^{-1} \tag{7.29}$$

Fortunately, it's not as complicated as it seems, so let's look at three examples using a 6-observation data set (see Table 7.25).[11] For our first example, we will assume a lag of four observations ($L=4$). With a lag of this size, observation 6, for example, is assumed to be affected by observations 2–5, but not observation 1. Using Eq. (7.28) yields the weights shown in the first lag matrix:

$$1 - \left(\frac{1}{4+1}\right) = .80; \ 1 - \left(\frac{2}{4+1}\right) = .60; \ 1 - \left(\frac{3}{4+1}\right) = .40; \ 1 - \left(\frac{4}{4+1}\right) = .20$$

---

[11] I have left the top half of each matrix empty to make it easier to see how the weights are constructed, but each matrix is symmetrical, so the top half is the transpose of the bottom half.

Table 7.25 Three lag matrices for a 6×6 matrix

$$L = 4 = \begin{bmatrix} 1 & \square & \square & \square & \square & \square \\ .80 & 1 & \square & \square & \square & \square \\ .60 & .80 & 1 & \square & \square & \square \\ .40 & .60 & .80 & 1 & \square & \square \\ .20 & .40 & .60 & .80 & 1 & \square \\ 0 & .20 & .40 & .60 & .80 & 1 \end{bmatrix}$$

$$L = 3 = \begin{bmatrix} 1 & \square & \square & \square & \square & \square \\ .75 & 1 & \square & \square & \square & \square \\ .50 & .75 & 1 & \square & \square & \square \\ .25 & .50 & .75 & 1 & \square & \square \\ 0 & .25 & .50 & .75 & 1 & \square \\ 0 & 0 & .25 & .50 & .75 & 1 \end{bmatrix}$$

$$L = 2 = \begin{bmatrix} 1 & \square & \square & \square & \square & \square \\ .6667 & 1 & \square & \square & \square & \square \\ .3333 & .6667 & 1 & \square & \square & \square \\ 0 & .3333 & .6667 & 1 & \square & \square \\ 0 & 0 & .3333 & .6667 & 1 & \square \\ 0 & 0 & 0 & .3333 & .6667 & 1 \end{bmatrix}$$

The middle matrix shows the weights when $L = 3$,

$$1 - \left(\frac{1}{3+1}\right) = .75; \ 1 - \left(\frac{2}{3+1}\right) = .50; \ 1 - \left(\frac{3}{3+1}\right) = .25$$

and the final matrix shows the weights when $L = 2$:

$$1 - \left(\frac{1}{2+1}\right) = .6667; \ 1 - \left(\frac{2}{2+1}\right) = .3333$$

The trick, then, is figure how long the lags should be. There is no hard and fast rule, but the following rule of thumb is useful:

$$L \geq N^{1/4} \tag{7.30}$$

With only six observations, the minimum lag would be 1.56, which we would round to 2. Generally, we will be dealing with much larger sample sizes so the lags will be greater.

Returning to our (fictitious) basketball example with $N = 12$, we'll assume that $L = 3$ to be conservative. We then have the following matrices (with **E** populated from the values in Table 7.18):

$$
\mathbf{E} =
\begin{bmatrix}
-.02457 & 0 & 0 & 0 & 0 & 0 & 0 & 0 & 0 & 0 & 0 & 0 \\
0 & -.00737 & 0 & 0 & 0 & 0 & 0 & 0 & 0 & 0 & 0 & 0 \\
0 & 0 & .02885 & 0 & 0 & 0 & 0 & 0 & 0 & 0 & 0 & 0 \\
0 & 0 & 0 & .01323 & 0 & 0 & 0 & 0 & 0 & 0 & 0 & 0 \\
0 & 0 & 0 & 0 & .03021 & 0 & 0 & 0 & 0 & 0 & 0 & 0 \\
0 & 0 & 0 & 0 & 0 & .05143 & 0 & 0 & 0 & 0 & 0 & 0 \\
0 & 0 & 0 & 0 & 0 & 0 & .02461 & 0 & 0 & 0 & 0 & 0 \\
0 & 0 & 0 & 0 & 0 & 0 & 0 & .00591 & 0 & 0 & 0 & 0 \\
0 & 0 & 0 & 0 & 0 & 0 & 0 & 0 & -.01377 & 0 & 0 & 0 \\
0 & 0 & 0 & 0 & 0 & 0 & 0 & 0 & 0 & -.06169 & 0 & 0 \\
0 & 0 & 0 & 0 & 0 & 0 & 0 & 0 & 0 & 0 & -.03575 & 0 \\
0 & 0 & 0 & 0 & 0 & 0 & 0 & 0 & 0 & 0 & 0 & -.01109
\end{bmatrix}
$$

and

$$
\mathbf{L} =
\begin{bmatrix}
1 & .75 & .50 & .25 & 0 & 0 & 0 & 0 & 0 & 0 & 0 & 0 \\
.75 & 1 & .75 & .50 & .25 & 0 & 0 & 0 & 0 & 0 & 0 & 0 \\
.50 & .75 & 1 & .75 & .50 & .25 & 0 & 0 & 0 & 0 & 0 & 0 \\
.25 & .50 & .75 & 1 & .75 & .50 & .25 & 0 & 0 & 0 & 0 & 0 \\
0 & .25 & .50 & .75 & 1 & .75 & .50 & .25 & 0 & 0 & 0 & 0 \\
0 & 0 & .25 & .50 & .75 & 1 & .75 & .50 & .25 & 0 & 0 & 0 \\
0 & 0 & 0 & .25 & .50 & .75 & 1 & .75 & .50 & .25 & 0 & 0 \\
0 & 0 & 0 & 0 & .25 & .50 & .75 & 1 & .75 & .50 & .25 & 0 \\
0 & 0 & 0 & 0 & 0 & .25 & .50 & .75 & 1 & .75 & .50 & .25 \\
0 & 0 & 0 & 0 & 0 & 0 & .25 & .50 & .75 & 1 & .75 & .50 \\
0 & 0 & 0 & 0 & 0 & 0 & 0 & .25 & .50 & .75 & 1 & .75 \\
0 & 0 & 0 & 0 & 0 & 0 & 0 & 0 & .25 & .50 & .75 & 1
\end{bmatrix}
$$

Performing the multiplication in Eqs. (7.27) and (7.29) gives us the corrected covariance matrix:

$$
\mathbf{C} =
\begin{bmatrix}
.030991 & -.000364 \\
-.000364 & .000004
\end{bmatrix}
$$

Taking the square root of the diagonal entries, we find our standard errors and use them to test the statistical significance of our regression coefficients (obtained from our initial analysis). Notice that the standard errors are quite a bit smaller than in the original analysis and that the regression coefficient relating crowd noise to performance is now statistically significant.

	b	se_b	t	p
b_0	$-.151642$	.176043	$-.8614$	.4092
b_1	.008507	.002072	4.1056	.0021

7.3.4.2 Autocorrelation- and Heteroscedasticity-Consistent Matrix

If you compare the Newey-West procedure with White's $\mathbf{HC}_1$ matrix [Eq. (7.16)], you will see that, when $L = 0$, the $\mathbf{HC}_1$ matrix is a special case of the Newey-West matrix. In this sense, the Newey-West procedure is able to accommodate both violations we have discussed in this chapter: heteroscedasticity and autocorrelation.

7.3.4.3 Summary

In this section, we have learned how to handle situations when our residuals are not independently distributed. Although statistical remedies are available, the most likely cause of this violation is a missing variable. Suppose, for example, that we find that sweater sales in Baltimore predict pneumonia deaths in Cleveland. More than likely, such an association would show evidence of autocorrelation, with seasonal or monthly variations influencing both variables. More generally, whenever temporal or spatial variations are present, we should first be certain we have specified our model correctly before seeking a statistical remedy.

7.3.5 R Code: Autocorrelations

```
x=c(84,83,81,83,87,83,87,87,84,86,81,86)
y=c(.5384,.5471,.5663,.5677,.6187,.6059,.6131,.5944,.5492,.5183,.5017,
.5689)
autoreg=lm(y~x)
summary(autoreg)

#Durbin Watson and Breusch Godfrey
library(lmtest)          #attach lmtest package
dwtest(autoreg)
bgtest(autoreg)

#Estimate Rho 1 iteration
e=autoreg$residuals
n <- length(e)
rho.1 <-lm(e[2:n]~e[1:(n-1)]); rho.1$coef[2]

#Generalized Least Squares using Maximum Likelihood Estimation
library(nlme)            #attach nlme package
gls1 <- gls(y~x, correlation = corAR1(), method = "ML")
summary(gls1)
gls1$fitted
gls1$residuals
vcov(gls1)

#Newey-West to Match Lag=3 Example Output is Corrected Covariance Matrix
library(sandwich)          #attach sandwich package
NeweyWest(autoreg, lag = 3, prewhite = FALSE)
```

7.4   Chapter Summary

1. Linear regression models assume that the errors are independent and identically distributed. In terms of a covariance matrix of errors, these assumptions maintain that the off-diagonal elements are zero and that the diagonal elements are identical. These assumptions pertain to the errors, not the residuals, as the covariance matrix of residuals never fully satisfies this ideal.

2. The regression coefficients are not impacted when the errors are not independent and identically distributed, but the standard errors are affected. Consequently, tests of statistical significance and confidence intervals are compromised.

3. Heteroscedasticity occurs when the errors are not identically distributed. Visually, the residuals show a greater scatter around some variables than around others. Often, heteroscedasticity takes the form of a funnel shape, in which the variability of the residuals becomes increasingly (or decreasingly) variable.

4. Heteroscedasticity can be tested using White's test or the Breusch-Pagan procedure. Both tests involve regressing the squared residuals from an ordinary least squares (OLS) regression on a set of predictors.

5. Heteroscedasticity can be managed using weighted least squares estimation, in which all variables (including the intercept) are divided by the estimated variance of each observation's residual. The transformed variables are then analyzed using OLS estimation.

6. Sometimes the variances of the residuals are known, but more commonly they need to be estimated from the data. Often this is accomplished by assuming that the variance of a residual is proportional to the magnitude of the predictor (i.e., that the variance of a residual is greater for large predictor values than for small predictor values).

7. Heteroscedasticity can also be managed by creating a heteroscedasticity-consistent covariance matrix. Several varieties are available, but one (termed HC_3) involves replacing the diagonal entries of the covariance matrix of residuals with a term that includes the squared residual in the numerator and $(1 - h_i)^2$ in the denominator.

8. Autocorrelations occur when the residuals are correlated rather than independent. Autocorrelation often arises when the predictors are associated in time or space.

9. In a first-order autoregression process, each error (ε_t) is influenced by a random disturbance (u_t), and the preceding error (ε_{t-1}), weighted by the autocorrelation parameter (ρ). Thus, $\varepsilon_t = \rho\varepsilon_{t-1} + u_t$.

10. The Durbin-Watson test and the Breusch-Godfrey test are used to identify the presence of autocorrelated errors.

11. Autocorrelations can be managed using generalized least squares estimation. After using an iterative technique to estimate the autocorrelation parameter, we use it to transform all of the variables (including the intercept). The transformed variables are then analyzed using OLS estimation.

12. Autocorrelations can also be managed by creating an autocorrelation-consistent covariance matrix. Using the Newey-West procedure, we first decide how many previous observations should be weighted in the prediction of a residual and then form a new covariance matrix to calculate our standard errors.

Chapter 8
Linearizing Transformations and Nonparametric Smoothers

In a linear regression model, the criterion is modeled as a linear combination of the weighted predictors and a disturbance term.

$$y = \beta_0 + \beta_1 x_1 + \beta_2 x_2 + \ldots + \beta_k x_k + \varepsilon \tag{8.1}$$

The model gives rise to two, related properties: linearity and additivity.

- The linearity property stipulates that the weight of each coefficient is constant across all levels of the variable with which it combines. If, for example, $b_1 = .5$, then holding all other predictors constant, we expect a .5 unit change in y with every 1 unit change in x_1, regardless of whether x_1 changes from 3 to 4, or from 16 to 17, or from 528 to 529.
- The additivity property stipulates that the total change in y is equal to the sum of the weighted changes in all of the predictors. If $b_1 = .5$ and $b_2 = .3$, then the fitted value of y will change by .8 units for every one unit change in x_1 and x_2.

In sum, a linear model describes an invariant rate of change in the fitted value of y for every one unit change in x. Given this property, you might assume that linear models can only accommodate linear relations. This is not so. Consider the relations depicted in Fig. 8.1. None of them is linear, yet all can be analyzed using a linear regression model. In the next three chapters, you will learn how to analyze nonlinear relations using various modifications of linear regression.

Electronic Supplementary Material: The online version of this chapter (doi: 10.1007/978-3-319-11734-8_8) contains supplementary material, which is available to authorized users

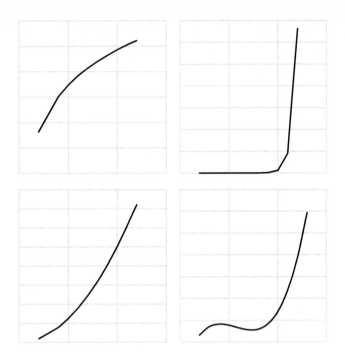

Fig. 8.1 Four nonlinear relations

8.1 Understanding Linearity

Let's begin by distinguishing two terms—*linear relation* and *linear model* (aka *linear function*). The first term describes the association between two (or more) variables, indicating that they form a straight line (or plane) when plotted on a graph; the second term specifies the mathematical operations that govern the prediction of a criterion. As first discussed in Chap. 2, a model is linear when (1) the criterion is the sum of the weighted coefficients and disturbance term and (2) each coefficient is of the first order and is not multiplied or divided by any other coefficient.[1]

These forms of linearity are conceptually independent. Consider the functions described in Table 8.1. Both functions in Column 1 produce a straight line when plotted on a graph (see left-hand side of Fig. 8.2), but only the first represents a linear model. The second is a nonlinear model because b_1 is multiplied by b_0. Conversely, both functions in Column 2 produce nonlinear relations (see right-hand side of Fig. 8.2), yet the first is linear and the second is not. Thus, knowing that a relation is linear does not guarantee it was produced by a linear function (and vice versa).

---

[1] Hereafter, the phrase "linear in the variables" will be used to describe a linear relation and the phrase "linear in the parameters" will be used to describe a linear model or function.

Table 8.1 Functions that vary along two dimensions of linearity

Linear function		Linear relation	
		Yes	**No**
	Yes	$\hat{y} = b_0 + b_1 x$	$\hat{y} = b_0 + b_1 \ln(x)$
	No	$\hat{y} = b_0 + b_0 b_1 x$	$\hat{y} = b_0 * x^{b_1}$

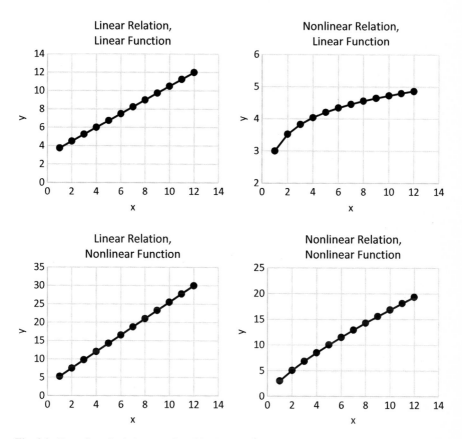

Fig. 8.2 Four plotted relations produced by the functions in Table 8.1

8.1.1 Partial Derivatives and Linear Functions

Partial derivatives distinguish linear functions from nonlinear ones. Recall from Chap. 3 that a partial derivative describes the instantaneous rate of change in a function when one of its input values change and all other values are held constant. When a function is linear, the partial derivatives are independent and have a unique, closed-form solution. When a function is nonlinear, one or more of the partial derivatives depends on the value of its own parameter or another parameter. Consequently, a unique, closed-form solution does not exist and iterative numerical techniques must be used to estimate the parameters.

Table 8.2 Partial derivatives of linear and nonlinear functions

	Equation	Partial derivatives	
Linear function	$y = \alpha + \beta x$	$\frac{\partial f}{\partial \alpha} = 1$	$\frac{\partial f}{\partial \beta} = x$
Nonlinear function	$y = \alpha \beta x$	$\frac{\partial f}{\partial \alpha} = \beta x$	$\frac{\partial f}{\partial \beta} = \alpha x$

To make this distinction less abstract, consider the functions shown in Table 8.2. The first function is linear, and the partial derivatives reflect this independence. The partial derivative with respect to α doesn't depend on β and the partial derivative with respect to β doesn't depend on α. The second function is nonlinear. Here, the parameters combine multiplicatively, and the partial derivative with respect to α depends on β and the partial derivative with respect to β depends on α. Because of these differences, the partial derivatives of the first equation have an analytical solution, but the partial derivatives of the second equation do not.

8.1.2   Assessing Linear Relations

Calculating the partial derivatives allows us to assess the linearity of a function when the data generating process is known. Unfortunately, this information is not always available. In many (perhaps most) situations, neither prior research nor theory describes a precise data generating process. Even more unfortunately, there is no surefire way to extract this information from the data. Consider again the scatterplots displayed in the first column of Fig. 8.2. Both figures show straight lines, but only the first was generated by a linear function. No statistical test would be able to divine this fact.

8.1.2.1   Using Scatterplots to Assess the Linearity of the Variables

Although scatterplots cannot be used to determine the linearity of the parameters, they can be used to assess the linearity of the variables. A scatterplot of the residuals from an ordinary least squares (OLS) regression is especially informative. Because OLS regression forces a linear relation on the data, the residuals represent the portion of y that cannot be explained by a linear function. Residuals that form a nonlinear pattern therefore suggest that the data are nonlinear in the variables.

To provide a context for this discussion, consider the data shown in Table 8.3. An OLS regression shows a very strong fit to the data ($R^2 > .97$), and the left-hand graph in Fig. 8.3 reveals a sharp linear trend.

By itself, the presence of a linear trend does not preclude the possibility of a nonlinear relation, however. In the present case, an examination of the residual plot in Fig. 8.3 shows strong evidence of curvature, suggesting the presence of a nonlinear component.

Table 8.3 Hypothetical data illustrating nonlinear relation

x	y	$\hat{y}$	e	h	$z\hat{y}$	Studentized residual
2	2.05	2.27765	−.22765	.29704	−1.53321	−1.46885
2	2.08	2.27765	−.19765	.29704	−1.53321	−1.27528
4	3.10	2.93562	.16438	.16196	−.92998	.97137
5	3.30	3.26461	.03539	.11923	−.62837	.20399
6	3.70	3.59360	.10640	.09304	−.32675	.60441
7	4.15	3.92258	.22742	.08339	−.02513	1.28501
8	4.50	4.25157	.24843	.09028	.27648	1.40905
9	4.60	4.58056	.01944	.11371	.57810	.11172
9	4.70	4.58056	.11944	.11371	.57810	.68635
11	5.05	5.23853	−.18853	.21020	1.18133	−1.14763
11	5.07	5.23853	−.16853	.21020	1.18133	−1.02589
11	5.10	5.23853	−.13853	.21020	1.18133	−.84327

Significance test of regression model						
	SS	df	MS	R^2	F	p
Regression	13.0871	1	13.0871	.9746	382.9987	.0000
Residual	.341701	10	.0342			
Total	13.4288					

Regression coefficients								
	b	se_b	t	p	$(\mathbf{X'X})^{-1}$		$\mathbf{C}$	
b_0	1.61968	.13048	12.4128	.0000	.49828	−.05858	.017026	−.002002
b_1	.32899	.01681	19.5704	.0000	−.05858	.00827	−.002002	.000283

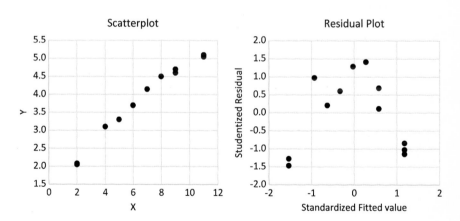

Fig. 8.3 Scatterplot and residual plot for hypothetical data from Table 8.3

8.1.2.2 Lack of Fit Test

Visual inspection of the residuals is one way to gauge nonlinearity. Another, less subjective approach is to conduct a lack of fit test. Use of the test requires that our predictor has at least one set of duplicate values. Such repetitions are termed "replicates," and looking over the data in Table 8.3, we see that we have three replicates (i.e., we have two values $= 2$, two values $= 9$, and three values $= 11$).

To perform the lack of fit test, we need to perform a second analysis that treats these replicates as categorical predictors rather than continuous ones. This approach is known as an analysis of variance, and we will discuss it in detail in Chaps. 11 and 12. Unlike a regression analysis, an analysis of variance does not assume that the predictors are related in a linear fashion. To illustrate, imagine we ask children to eat one of three desserts—cake, pie, or ice cream—and then indicate how much they enjoyed the treat. Obviously, cake, pie, and ice cream are not related in a linear way, but for statistical purposes, we could designate cake $= 1$, pie $= 2$, and ice cream $= 3$. The ordering of the groups is arbitrary (e.g., we could just have easily designated pie as the third group and cake as the second), so whatever variance is due to differences in the dessert is nonlinear in form.

Capitalizing on this fact, we form a test statistic comparing the unexplained variance from a linear regression analysis with the unexplained variance from a categorical analysis.

$$F = \frac{\left(SS_{res_{linear}} - SS_{res_{categorical}}\right) / \left(df_{linear} - df_{categorical}\right)}{SS_{res\_categorical} / df_{categorical}} \qquad (8.2)$$

The degrees of freedom for the categorical term is found by summing $(n-1)$ for each group of replicates. With our data set, $df = (2-1) + (2-1) + (3-1) = 4$.

Before we perform the calculations, let's look more closely at the formula itself. Because a linear model makes more assumptions than a categorical one, the unexplained variance from the linear analysis will always be greater than the unexplained variance from the categorical analysis. But if the difference between the two is small, then the fraction comprising the test statistic will be small and we will feel comfortable concluding that a linear model adequately fits the data. Conversely, if the unexplained variance from the linear model is much greater than the unexplained variance from the categorical one, our test statistic fraction will be large and we would conclude that a linear model does not adequately fit our data.

Keeping these points in mind, let's now consider how the test statistic is calculated. In Chap. 11 we will learn a more formal way to conduct an analysis of variance, but for the present purposes, we can take a shortcut by first computing the mean for each replicate. We then subtract the mean from the corresponding raw scores from which it is derived and square the differences. Finally, we sum the squares to get a measure of the squared deviations from each replicate's mean. This value is our residual sum of squares.

Table 8.4 presents the calculations. To illustrate, 11 has three replicates, and when we average the y scores for the three observations, we find that their average $= 5.0733$. We then subtract this value from each of the three scores and square the differences. When we do this for all three replicate groups and then sum the squares, we derive our residual sum of squares for the categorical predictors

Table 8.4 Calculations for a lack of fit test for data shown in Table 8.3

	x	y	$\bar{y}$	e	e^2
	2	2.05	2.065	−.0150	.00023
	2	2.08	2.065	.0150	.00023
	9	4.60	4.65	−.0500	.0025
	9	4.70	4.65	.0500	.0025
	11	5.05	5.0733	−.02333	.00054
	11	5.07	5.0733	−.00333	.00001
	11	5.10	5.0733	.02667	.00071
				Σ	**.006717**
	SS_{res}	df	MS_{res}	F (6,4)	p
Linear	.341701	10	.03417	33.2491	.00226
Categorical	.006717	4	.001679		

($SS_{res\_categorical}$=.006717). Table 8.3 provides the rest of the values needed to compute the test statistic.

$$F = \frac{(.341701 - .006717)/(10 - 4)}{.006717/4} = 33.2491$$

The F statistic is evaluated with the difference between the two degrees of freedom in the numerator (10 − 4) and the degrees of freedom for the categorical term in the denominator (4). In our data, we see a significant effect (p=.0023). Consequently, we fail to reject the null hypothesis of linearity and conclude that the data (probably) contain a nonlinear component.

8.1.3   Options for Analyzing Nonlinear Relations

Once we conclude that our data are nonlinear in the variables, we must decide how to proceed. Figure 8.4 presents our options. As you can see, one option is to use nonlinear regression methods, a topic not covered in this text. Another possibility is to add predictors, a topic we will consider in Chaps. 9 and 10. The other two possibilities—transform the variables and use nonparametric methods—comprise the remainder of this chapter.

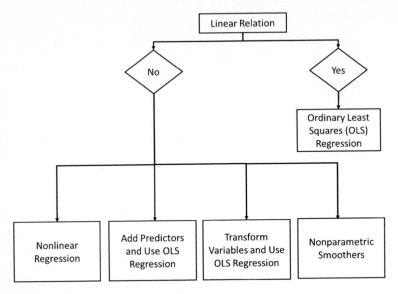

Fig. 8.4 Options for analyzing nonlinear relations

8.1.4   R Code: Assessing Nonlinearity

```
#Differentiation
#Linear Function
funct = expression(a+b*x, "a,b")
deriva1 <-D(funct,"a")
deriva1
derivb1 <-D(funct,"b")
derivb1

#Nonlinear Function
funct = expression(a*b*x, "a,b")
deriva1 <-D(funct,"a")
deriva1
derivb1 <-D(funct,"b")
derivb1

#Lack of Fit Test
x=c(2,2,4,5,6,7,8,9,9,11,11,11)
y=c(2.05,2.08,3.1,3.3,3.7,4.15,4.5,4.6,4.7,5.05,5.07,5.1)
pre.reduced <- lm(y ~ x)
anova(pre.reduced)
pre.full <- lm(y ~ factor(x))
anova(pre.full)
anova(pre.reduced, pre.full)
```

8.2   Transformations to Linearity

Many nonlinear relations can be transformed into ones that are at least approximately linear. In some cases these transformations involve altering the variables without changing the parameters; in other cases, the parameters are transformed. Before discussing these procedures, let's spend a moment discussing transformations more broadly.

8.2.1   Understanding Transformations

8.2.1.1   Linear Transformations and Nonlinear Transformations

Data transformations are of two varieties. Linear transformations extend or shrink the variables in a constant, uniform degree. If, for example, we add or subtract a constant from the scores in our data set, only the intercept in a regression equation will change. The same will be true if we multiply or divide the scores by a constant. For these reasons, a linear transformation will not linearize a nonlinear relation.

In contrast, nonlinear transformations are monotonic (i.e., the rank ordering of the values is preserved) but not uniform. For example, if we square our predictors, then the distance between large values will increase more than will the distance between smaller ones. Consequently, nonlinear transformations can linearize a nonlinear relation.

8.2.1.2   To Linearize or Not to Linearize?

Linear regression is less computationally intensive than is nonlinear regression, so transforming a nonlinear relation into a linear one was the usual approach to analyzing nonlinear relations before powerful computers became widely accessible. Many statisticians believe the transformation approach is no longer needed and recommend using nonlinear regression to model nonlinear relations; others disagree, noting that transformational methods are still preferred under some circumstances, in part because linear models are better developed and easier to interpret than are nonlinear models (Manning and Mullahy 2001; Xiao et al. 2011). Because there is disagreement on the matter, we will learn the transformational technique. Bear in mind, however, that not all nonlinear relations can be linearized, so nonlinear regression is always needed under some circumstances.[2]

---

[2] In some textbooks, nonlinear functions that can be linearized are called "intrinsically linear functions."

8.2.1.3   Selecting Variables to Transform

Having decided to transform our variables, we must decide which variables to transform. Transforming the criterion alters the distribution of the residuals. If the residuals are already normal, independent, and identically distributed (*NID*), it is better to achieve linearity by transforming the predictor than the criterion. Otherwise, the transformation could create problems in the residuals. On the other hand, if the residuals from an OLS regression are not *NID*, transforming the criterion might improve matters. In fact, we have already discussed such a transformation in Chap. 6 when we noted that the Box-Cox procedure can be used to find a transformation of *y* that will normalize the residuals.

8.2.1.4   Selecting a Linearizing Transformation

After deciding which variables to transform, we must pick a linearizing transformation. We have four options.

- Sometimes theory or previous research has identified a lawful, nonlinear relation between two or more variables. For example, the Michaelis-Menten equations describe the nonlinear rate of enzymatic reactions, and numerous linearizations of the model have been developed.
- When an established mathematical model is unavailable to guide us, we can use a rule of thumb known as the bulging rule (Mosteller and Tukey 1977). Table 8.5 presents the rules with λ indicating the power to which we raise *x* or *y*. For example, we might be able to linearize a negatively accelerated curve such as the one shown in the first panel by contracting values of $x\,(\lambda_x < 1)$ or stretching values of $y\,(\lambda_y > 1)$. Typically, λ is restricted to values that lend themselves to easy interpretation (e. g., -1, .5, 0, 2) with $\lambda = 0$ indicating a logarithmic transformation and $\lambda = 1$ indicating no transformation. The farther the transformation is from $\lambda = 1$, the stronger will be the contraction or stretching. To illustrate, a square root transformation ($\lambda = .5$) is less extreme than a logarithmic one ($\lambda = 0$).
- A third approach is to mathematically identify a transformation that best linearizes our data. The Box-Tidwell approach represents such a technique, and we will demonstrate its use later in this chapter.
- Finally, we can use trial and error until we stumble on an appropriate transformation. Clearly, this is an inefficient method, best used only as a last resort.

8.2.2   Logarithmic Model

We will begin our discussion of linearizing transformations by considering a logarithmic transformation that turns unit differences into ratios. To understand

Table 8.5 Power transformations for four nonlinear patterns

Description	Pattern	Suitable transformation of x	Suitable transformation of y
Large growth of y at low levels of x small growth of y at high levels of x		$x^{\lambda<1}$	$y^{\lambda>1}$
Steep decline of y at low levels of x small decline of y at high levels of x		$x^{\lambda<1}$	$y^{\lambda<1}$
Small decline of y at low levels of x large decline of y at high levels of x		$x^{\lambda>1}$	$y^{\lambda>1}$
Small growth of y at low levels of x large growth of y at high levels of x		$x^{\lambda>1}$	$y^{\lambda<1}$

After Mosteller and Tukey (1977)

their value, imagine we study the rate at which people master various tasks (commonly known as a learning curve). For our first investigation, we use an easy task—arithmetic—and measure how many problems high school students can solve in 10 min (y) as a function of the number of hours/week they study (x). The first two columns in Table 8.6 present the (fabricated) findings, and an OLS regression shows a strong linear trend ($b=.4858$, $p<.0001$). At the same time, the scatterplot displayed in Fig. 8.5 shows that performance tends to rise rapidly at first and then slow down, forming a nonlinear pattern similar to the negatively accelerated curve depicted in the first panel of Table 8.5. The residuals shown in the right-hand side of Fig. 8.5 provide further evidence for curvature, as does a significant lack of fit test, $F(7,3) = 12.37$, $p=.0317$. So even though the linear term in an OLS regression is significant, the data pattern is not entirely linear.

Table 8.6 Hypothetical data illustrating the linearization of a logarithmic function

Raw scores		After log transformation of x			
x	y	$\ln(x)$	$\hat{y}$	Residuals	h
2	2.99	.6931	2.91368	.07632	.25347
2	2.70	.6931	2.91368	−.21368	.25347
2	3.00	.6931	2.91368	.08632	.25347
3	4.02	1.0986	3.96194	.05806	.13887
4	4.71	1.3863	4.70568	.00432	.09548
6	5.83	1.7918	5.75393	.07607	.08777
7	6.28	1.9459	6.15246	.12754	.10124
8	6.27	2.0794	6.49768	−.22768	.12021
9	6.76	2.1972	6.80218	−.04218	.14257
10	7.04	2.3026	7.07457	−.03457	.16705
10	7.05	2.3026	7.07457	−.02457	.16705
12	7.66	2.4849	7.54593	.11407	.21937

Significance test after log transformation of x						
	SS	df	MS	R^2	F	p
Regression	35.1513	1	35.1513	.9957	2,300.3764	.0000
Residual	.1528	10	.0153			
Total	35.3041					

Regression coefficients after log transformation of x								
	b	se_b	t	p	$(\mathbf{X'X})^{-1}$		$\mathbf{C}$	
b_0	1.12169	.09528	11.7720	.0000	.59416	−.31166	.009079	−.004762
b_1	2.58531	.05390	47.9622	.0000	−.31166	.19014	−.004762	.002906

8.2.2.1 Log Transformation of x

Looking over the rules in Table 8.5, we see that a logarithmic transformation of x might linearize the data. Pursuing this approach yields the following regression equation, which is linear in the parameters but nonlinear in the variables.[3]

$$\hat{y} = a + b^*\ln(x) + v \qquad (8.3)$$

After applying the transformation, we conduct an OLS regression on the transformed data.[4] The results are shown in Table 8.6, and the scatterplot and residuals are shown in the bottom portion of Fig. 8.5. As expected, a log transformation of x linearizes the relation between study time and task performance, and a lack of fit test now produces a nonsignificant effect, $F(7, 3) = .6983$, $p = .6878$.

---

[3] I am using v to denote the error term because later we are going to use e to denote exponential functions.

[4] Log transformations cannot be made on negative numbers or zero. To accommodate this issue, some authors suggest adding a constant to all scores before using a log transformation. This practice is not easy to justify and should be avoided in favor of nonlinear regression models.

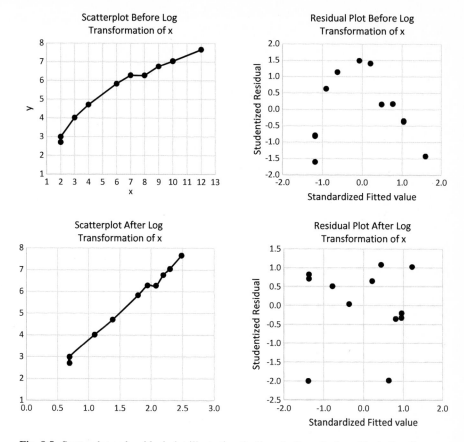

Fig. 8.5 Scatterplot and residual plot illustrating the linearization of a logarithmic function

8.2.2.2 Interpreting the Regression Coefficients and Fitted Values

Following a logarithmic transformation of a predictor, ratio changes in x predict unit changes in y. To illustrate this point, let's look at the mean expected difference in y when x doubles from 2 to 4 using the regression coefficients from the log transformed analysis.

$$\hat{y}\,|(x=2) = 1.12169 + \ln(2)^*2.58531 = 2.91368$$
$$\hat{y}\,|(x=4) = 1.12169 + \ln(4)^*2.58531 = 4.70568$$

Using simple subtraction, we can see that when x doubles, the expected increase in $y = 1.7920$. Equation (8.4) shows that we can also derive this value using the regression coefficient reported in Table 8.6.

$$\Delta \hat{y} = \ln\left(\frac{x_2}{x_1}\right)b \qquad\qquad (8.4)$$

Plugging in our values,

$$\Delta \hat{y} = \ln(2)^*2.58531 = 1.7920$$

Notably, this expected change in y occurs regardless of whether x doubles from 2 to 4, from 4 to 8, or from 6 to 12. In short, no matter where we start, a doubling in study time predicts a 1.7920 unit improvement in test performance. In a similar manner, we can predict the likely expected change in y for other ratio changes in x. For example, the expected difference in y when x triples

$$\ln(3)^*2.58531 = 2.8403$$

or quadruples.

$$\ln(4)^*2.58531 = 3.5840$$

You can verify these values by subtracting the relevant numbers shown in Table 8.6.

8.2.2.3   Summary

A log transformation of the predictor is appropriate when ratio changes in x predict unit changes in y. If you believe you will earn ~ 1.8 more points on your next math test if you double your study time, regardless of whether you double your study time from 2 to 4 h or from 6 to 12 h, you are assuming that a logarithmic function describes the association between study time and test performance.[5] Such a relation is commonly known as one of diminishing returns.

8.2.3   Exponential Model

The previous example dealt with a model that was nonlinear in the variables but linear in the parameters. Our next two examples involve models that possess both forms of nonlinearity. Our first example is an exponential function. Such functions are common in the behavioral and physical sciences. For example, compound interest follows an exponential growth function and radioactive isotopes decay according to an exponential decay function. With an exponential growth function,

---

[5] With a logged predictor, the intercept represents the fitted value of y when $x = 1$. Whether this value is meaningful will depend on the sample.

change is slow at first but then accelerates rapidly; with an exponential decay function, change is rapid at first and then levels off.

The following equation describes an exponential growth curve.[6]

$$y = ae^{bx} \tag{8.5}$$

Because the unknown terms combine multiplicatively, their derivatives are not independent. We can make them independent, however, by using a log transformation. As you might remember from elementary algebra, logarithms turn multiplication into addition and exponents into multipliers. Consequently, with one important stipulation, we can recast Eq. (8.5) as an additive model and use OLS to fit the data by predicting $\ln(y)$ from x.

$$\ln(y) = \ln(a) + bx \tag{8.6}$$

The stipulation is that we must treat the error term v in the original equation as a multiplicative term rather than an additive one,

$$y = ae^{bx}e^v \tag{8.7}$$

creating

$$\ln(y) = \ln(a) + bx + \ln(v). \tag{8.8}$$

In many cases this assumption will be reasonable. For example, with an exponential model, our ability to predict y from x will probably be less exact when y is large than when y is small. In this case, OLS regression on the transformed variables is appropriate and might even reduce heteroscedasticity. But if the error term is presumed to be additive on the original scale, an OLS regression on the transformed variables will not produce an accurate solution and might introduce heteroscedasticity rather than correct it. In this case, nonlinear methods must be used.

8.2.3.1   Example

To illustrate an exponential model, imagine we conduct another study, this time concerning the relation between hours of practice at typing and performance on a timed test. Learning to type is generally harder than learning to add or subtract, so we might expect a different learning curve than we saw in our earlier example. In particular, we might expect very little progress among those who practice infrequently and rapidly increasing progress among those who practice often. After all, once people take the time to learn the keyboard, muscle memory takes over and people become proficient at typing.

---

[6] We use a negative value for b to model an exponential decay curve.

The (phony) data appear in Table 8.7 and an OLS regression on the raw data reveals a strong linear trend ($b = 3.7922, p < .0001$), the nature of which is shown in Fig. 8.6. However, we also see evidence for a nonlinear trend, as performance improves little among students who seldom practice, then improves dramatically among those who practice frequently.[7] Moreover, the residuals from an OLS regression show strong evidence of curvature, and a significant lack of fit test provides additional evidence of nonlinearity, $F(7, 3) = 116.11$, $p < .0012$. Finally, notice that the residuals also show evidence of heteroscedasticity, as the variance increases with increasing values of y. A significant Breusch-Pagan test (see Chap. 7) confirms this violation, $\chi^2_{BP} = 6.27$, $p = .0123$, indicating that the error term is multiplicative, not additive.

Table 8.7 Hypothetical data illustrating the linearization of an exponential function

Raw scores		After log transformation of y				
x	y	$\ln(y)$	$\hat{y}$	residuals	h	$\hat{y}^* = e^{\hat{y}}$
2	5.92	1.77834	1.67637	.10196	.21031	5.34612
2	4.44	1.49065	1.67637	−.18572	.21031	5.34612
2	6.00	1.79176	1.67637	.11539	.21031	5.34612
3	6.94	1.93730	1.89561	.04169	.15759	6.65660
4	8.07	2.08815	2.11485	−.02669	.11892	8.28831
6	13.46	2.59972	2.55332	.04640	.08377	12.84969
7	17.62	2.86903	2.77256	.09648	.08729	15.99949
8	15.49	2.74019	2.99179	−.25160	.10486	19.92139
9	23.59	3.16082	3.21103	−.05021	.13650	24.80466
10	30.08	3.40386	3.43027	−.02641	.18219	30.88493
10	30.44	3.41576	3.43027	−.01451	.18219	30.88493
12	55.81	4.02195	3.86874	.15321	.31576	47.88215

Significance test after log transformation of y						
	SS	df	MS	R^2	F	p
Regression	6.83723	1	6.83723	.97681	421.2232	.0000
Residual	.16232	10	.01623			
Total	6.99955					

Regression coefficients after log transformation of y								
	b	se_b	t	p	$(\mathbf{X'X})^{-1}$		$\mathbf{C}$	
b_0	1.23790	.07622	16.24040	.0000	.35794	−.04394	.005810	−.000713
b_1	.21924	.01068	20.52372	.0000	−.04394	.00703	−.000713	.000114

---

[7] People commonly refer to tasks that are difficult to master as having a "steep learning curve," but this is incorrect. Easy tasks produce a steep learning curve, as was the case in our earlier example when performance rose quickly with practice, then leveled off. In contrast, difficult tasks produce shallow learning curves, like the function we are considering in this example.

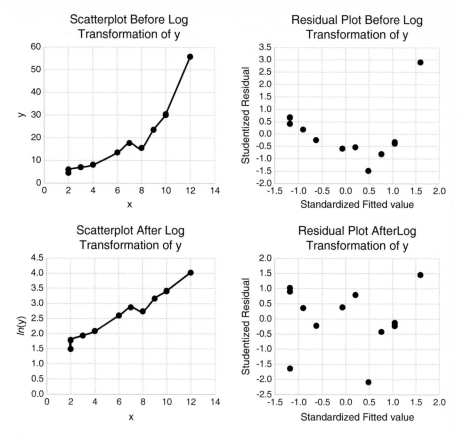

Fig. 8.6 Scatterplot and residual plot illustrating the linearization of an exponential function

8.2.3.2 Transforming the Criterion

After applying a logarithmic transformation to our criterion, we can use OLS regression to analyze our data. The pertinent data and results from this analysis are shown in Table 8.7 and appear graphically in the bottom portion of Fig. 8.6. Following the transformation, the relation is now (approximately) linear, with a nonsignificant lack of fit test, $F(7, 3) = .7721, p = .6511$. Heteroscedasticity is also eliminated, $\chi^2_{BP} = .0012, p = .9723$, indicating that the error term for the transformed equation is now appropriately additive.

8.2.3.3 Interpreting the Regression Coefficients and Fitted Values

To interpret our regression coefficients, we use Eq. (8.9) to return our fitted values to their original scale (see final column in Table 8.7).

$$\hat{y}^* = e^{\hat{y}} \tag{8.9}$$

When we do so, we find that a one unit change in x predicts an e^b ratio change in $\hat{y}^*$. To illustrate, consider what happens when x changes from 2 to 3.

$$\hat{y}^*|(x=2) = 5.34612$$
$$\hat{y}^*|(x=3) = 6.65660$$

The ratio of the two fitted values $= 1.24513$. This value can be derived from Eq. (8.10).

$$\left(\frac{\hat{y}_2^*}{\hat{y}_1^*}\right) = e^{(\Delta x * b)} = e^{(1*.21924)} = 1.2451 \tag{8.10}$$

So instead of saying "a one unit change in x predicts a b unit change in y" (as would normally be true with a linear model), we say "a one unit change in x predicts an e^b ratio change in y."

Applying this equation, we can find the ratio change in $\hat{y}^*$ for other changes in x. For example, a two unit change in x is associated with an $e^{(2 * .21924)} = 1.5503$ ratio change in $\hat{y}^*$ and a three unit change in x is associated with an $e^{(3 * .21924)} = 1.9304$ ratio change in $\hat{y}^*$.

With an exponential function, we can express changes in $\hat{y}^*$ as percentages. When b is small ($\sim <.15$), the regression coefficient approximates the expected percentage change. For example, if $b=.13$, a one unit increase in x predicts a $\sim 13\%$ increase in y; for larger values of b, percentages can be found by subtracting 1 from the results calculated using Eq. (8.10). Table 8.8 shows the percentage values for our data.

Table 8.8 Percentage changes in y as a function of unit changes in x: exponential function

Hourly change in study time	Expected ratio change in performance	Expected percentage change (%)
1	1.2451	24.5
2	1.5503	55.0
3	1.9304	93.0
4	2.4036	140.4
5	2.9927	199.3
6	3.7263	272.6
7	4.6397	364.0
8	5.7771	477.7
9	7.1932	619.3
10	8.9564	795.6

8.2.3.4 Summary

Exponential functions are used when unit changes in x predict ratio changes in y. In our example, practicing two additional hours yields greater benefits if you are increasing from 8 to 10 h than if you are increasing from 1 to 3 h. Why might this be the case? Perhaps 3 h/week is not enough to learn to type, so you might as well not bother practicing at all. More generally, if you believe that practice is disproportionately beneficial at high levels than at low levels, you are assuming that performance is an exponential function of effort.

8.2.4 Power Function

The final function we will consider in this section is a power function. Here, we raise the predictor to the value of a regression coefficient.

$$y = ax^b \tag{8.11}$$

With a power function, ratio changes in x predict ratio changes in y. Probably the best known example of a power function is Huxley's work on growth phenomena in animals. Across a variety of species, a power function describes how the expected size or weight of one physical characteristic can be predicted from the known size or weight of another physical characteristic. For example, using a power function, the mass of an animal's skeleton can be predicted from its body weight.

To keep things consistent with our previous examples, we will continue discussing practice time and test performance, this time looking at performance in a matrix algebra class. Matrix algebra is harder to learn than arithmetic, but it also does not require a substantial investment in time before any improvements appear (as was true with typing). So we might expect to find a third pattern in which ratio changes in x predict ratio changes in y.

The (phony) data appear in Table 8.9, and the results of an OLS regression on the raw data reveal a strong linear trend ($b = 1.18165$, $p < .0001$) that can be seen in Fig. 8.7. The scatterplot suggests a linear relation, but the residual plot shows evidence of curvature. A significant lack of fit test provides additional evidence of nonlinearity, $F(7, 3) = 135.34$, $p = .0009$. Finally, the residuals show evidence of heteroscedasticity, $\chi^2_{BP} = 5.63$, $p = .0177$, suggesting that they enter the model multiplicatively, not additively, as shown in Eq. (8.12).

$$y = ax^b e^v \tag{8.12}$$

8.2.4.1 Fitting the Regression Model

Taking the logarithm of both sides of Eq. (8.12) produces a model that is linear in the parameters but not in the variables.

Table 8.9 Hypothetical data illustrating the linearization of a power function

\multicolumn{2}{}{Raw scores}		\multicolumn{5}{}{Following log transformations of x and y}				
x	y	$\ln(x)$	$\ln(y)$	$\hat{y}$	e	h	$\hat{y}^* = e^{\hat{y}}$
2	3.02	.69315	1.10526	1.02795	.07731	.25347	2.7953
2	2.26	.69315	.81536	1.02795	−.21259	.25347	2.7953
2	3.06	.69315	1.11841	1.02795	.09046	.25347	2.7953
3	5.61	1.09861	1.72455	1.67072	.05383	.13887	5.3160
4	8.39	1.38629	2.12704	2.12678	.00027	.09548	8.3878
6	17.27	1.79176	2.84897	2.76955	.07942	.08777	15.9514
7	23.05	1.94591	3.13767	3.01392	.12375	.10124	20.3670
8	19.94	2.07944	2.99273	3.22560	−.23287	.12021	25.1687
9	29.06	2.19722	3.36936	3.41232	−.04295	.14257	30.3355
10	34.71	2.30259	3.54703	3.57934	−.03231	.16705	35.8499
10	35.13	2.30259	3.55906	3.57934	−.02029	.16705	35.8499
12	53.75	2.48491	3.98434	3.86837	.11597	.21937	47.8643

Significance test after log transformation of x and y

	SS	df	MS	R^2	F	p
Regression	13.21670	1	13.21670	.98842	853.50	.0000
Residual	.15485	10	.01549			
Total	13.37155					

Regression coefficients after log transformation of x and y

	b	se_b	t	p	\multicolumn{2}{}{$(\mathbf{X'X})^{-1}$}	\multicolumn{2}{}{C}
b_0	−.07087	.09592	−.7389	.4770	.59416	−.31166	.092007	−.048261
b_1	1.58527	.05426	29.2148	.0000	−.31166	.19014	−.048261	.029444

$$\ln(y) = \ln(a) + b * \ln(x) + \ln(v) \qquad (8.13)$$

Consequently, we can use OLS regression to analyze the transformed function.

The transformed values appear in Table 8.9 and the results of an OLS regression reveal a strong fit to the data. The bottom half of Fig. 8.7 shows that the log transformations linearized the data, an impression confirmed by a lack of fit test, $F = (7, 3) = .7009$, $p = .6854$; heteroscedasticity has also been eliminated, $\chi^2_{BP} = .1249$, $p = .7237$.

8.2.4.2 Interpreting the Regression Coefficients and Fitted Values

When both variables have been log transformed, ratio changes in x predict ratio changes in y:

$$\left(\frac{\hat{y}_2^*}{\hat{y}_1^*}\right) = \left(\frac{x_2}{x_1}\right)^b \qquad (8.14)$$

To illustrate, let's look at the mean expected difference in $\hat{y}^*$ when x doubles from 2 to 4.

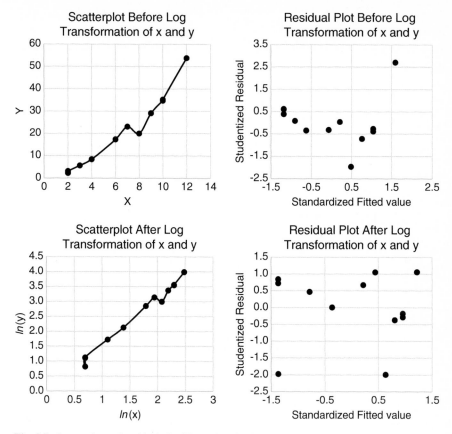

Fig. 8.7 Scatterplot and residual plot illustrating the linearization of a power function

$$\hat{y}^*|(x=2) = 2.7953$$
$$\hat{y}^*|(x=4) = 8.3878$$

The ratio change in $\hat{y}^* = 3.0007$. This value can also be found by calculating 2^b ($2^{1.58527} = 3.0007$). In our data, doubling your study time predicts a ~3-fold increase in test performance. If you triple your study time, you can expect a $3^{1.58527} = 5.70645$-fold increase in test performance. As before, this is true regardless of whether study time triples from 3 to 9 or from 4 to 12.

8.2.4.3 Summary

Power functions are appropriate when ratio changes in x predict ratio changes in y. If, regardless of how much you are studying now, you expect your performance to double when you double your study time, you are assuming that a power function explains the relation between study time and performance (and that the regression coefficient from the log transformed regression $= 1$).[8] When it comes to learning matrix algebra, it seems reasonable to assume that proportional gains in studying yield proportional gains in performance.

8.2.5   Box-Tidwell Transformation

In our previous examples, we have known the function that generated our nonlinear data. More commonly, the function underlying a nonlinear pattern is unknown. If all of the usual assumptions of OLS regression are met (i.e., the residuals are *NID*), we can attempt to linearize a nonlinear relation by using a power transformation of the predictor. Box and Tidwell (1962) developed an iterative technique to find the appropriate power function.

To illustrate the technique, imagine we conduct one last study to examine the association between study time and errors on a homework assignment in a math class. The first two columns in Table 8.10 present the (imaginary) raw data, and the

Table 8.10 First iteration of Box-Tidwell procedure for transforming predictor

		Initial analysis				First iteration			
x	y	x	$x[\ln(x)]$	y	x'	y	x'	$x'[\ln(x)]$	y
2	11.61	2	1.3863	11.61	.6017	11.61	.6017	.4170	11.61
2	11.58	2	1.3863	11.58	.6017	11.58	.6017	.4170	11.58
2	12.18	2	1.3863	12.18	.6017	12.18	.6017	.4170	12.18
3	7.35	3	3.2958	7.35	.4470	7.35	.4470	.4911	7.35
4	5.32	4	5.5452	5.32	.3620	5.32	.3620	.5019	5.32
6	4.26	6	10.7506	4.26	.2689	4.26	.2689	.4819	4.26
7	3.91	7	13.6214	3.91	.2402	3.91	.2402	.4674	3.91
8	2.89	8	16.6355	2.89	.2178	2.89	.2178	.4529	2.89
9	1.63	9	19.7750	1.63	.1998	1.63	.1998	.4390	1.63
10	1.59	10	23.0259	1.59	.1849	1.59	.1849	.4259	1.59
10	1.48	10	23.0259	1.48	.1849	1.48	.1849	.4259	1.48
12	1.71	12	29.8189	1.71	.1618	1.71	.1618	.4021	1.71
b_0	**12.2**	b_0	**21.2723**		b_0	**−2.5117**	b_0	**−.6945**	
b_1	**−1.07858**	b_1	**−6.2597**		b_1	**23.4869**	b_1	**23.4038**	
		b_2	**1.86912**				b_2	**−4.0209**	
		λ	**−.73295**				λ	**−.90415**	

---

[8] Economists use power functions to describe the elasticity of a commodity, defined as the expected percentage change in demand with a 1 % change in price. When b is < 2, the regression coefficient from the transformed analysis provides a good approximation of elasticity; when b is > 2, elasticity should be computed from the data: $E = (1.01^b * 100) - 100$.

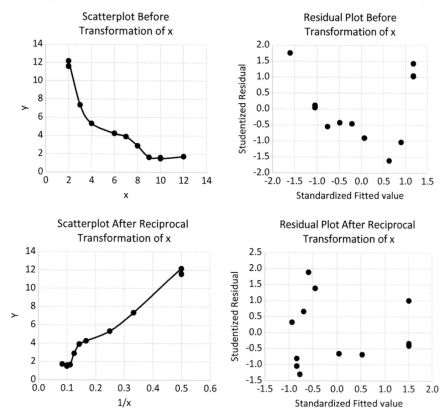

Fig. 8.8 Scatterplot and residual plot illustrating a Box-Tidwell (reciprocal) transformation of x

scatterplot shown in the first panel of Fig. 8.8 shows a clear nonlinear trend, with many errors when study time is low but very few errors once study time reaches 9 h/week. As with our earlier examples, the residuals also show evidence of curvature, and a lack of fit test produces a significant effect, $F(7, 3) = 51.70$, $p = .0040$; unlike our earlier examples, log transformations do not linearize the data. Consequently, we will use the Box-Tidwell procedure to find a power transformation of x (denoted λ) that will reduce (if not eliminate) nonlinearity.

To begin, we perform an OLS regression on the raw data, finding the regression coefficient relating study time to homework errors. The left-hand side of Table 8.10 shows that $b_1 = -1.07858$. We then perform a second analysis using x and $[x * \ln(x)]$ to predict y (see Table 8.10). Finally, we use Eq. (8.15) to find our first approximation of λ, with λ_n indicating that we are finding a new, updated value, λ_o indicating a previous value, b_{1*} indicating that the denominator comes from our first regression analysis, and b_2 indicating that our numerator comes from the second regression analysis.

Table 8.11 Five iterations of
the Box-Tidwell procedure
for transforming the predictor

Iteration	λ
0	−.732947315
1	−.904146720
2	−.923505612
3	−.925340152
4	−.925509867
5	−.925525531

$$\lambda_n = \lambda_o + \frac{b_2}{b_{1*}} \tag{8.15}$$

Plugging in our numbers yields the following value.[9]

$$\lambda_n = 1 + \frac{1.86912}{-1.07858} = -.73295$$

We then repeat the procedure forming a new predictor, $x' = x^\lambda$. Table 8.10 shows the data we use. The first value for x' was found using $x^\lambda = 2^{-.73295} = .6017$, and the first value for $x'[\ln(x)]$ was found using $[.6017 * \ln(2)] = .4170$. Notice that when computing this term, we use the updated value of x for x' but the original value of x when computing $[\ln(x)]$. After calculating the remaining values, we conduct two more regression analyses and compute an updated term for λ.

$$\lambda_n = -.73295 + \frac{-4.0209}{23.4869} = -.90415$$

We continue using this technique until changes in λ fall below some designated stop value (e.g., $<.000001$). Table 8.11 shows five iterations, but precision is not our goal, so two iterations usually suffice. As with the Box-Cox procedure (see Chap. 6), we are looking for the best interpretable transformation of x.[10]

A reciprocal transformation of x is appropriate when $\lambda \sim -.9$. Reciprocal transformations ($x^{-1} = 1/x$) exaggerate the differences between small values and attenuate the differences between large ones. Such a transformation makes sense with our (phony) data, because differences in error rates become miniscule once students study for a sufficient number of hours. Applying this transformation of x linearizes the data (see Fig. 8.8); produces a nonsignificant lack of fit test, $F(7, 3) = 4.22, p = .1319$; and reveals a strong effect of study time on performance, $b = 24.4645, p < .0001$.

---

[9] We are setting $\lambda_o = 1$ for our initial value (i.e., $x^1 = x$).

[10] $\lambda \sim 0$ indicates that a log transformation of the predictor is appropriate, and $\lambda \sim 1$ indicates that no transformation is needed.

8.2.6 Summary

In this section you have learned a variety of techniques that convert nonlinear relations into linear ones. In some cases, this transformation can be effected without changing the parameters of the regression model; in other cases the conversion requires reconfiguring the model itself. As the latter approach is appropriate only when the error term is assumed to be multiplicative, a careful examination of the residuals before and after any transformations have been made is required. For this reason, the diagnostic tools we discussed in Chaps. 6 and 7 should be applied whenever decisions regarding the transformation of variables are made.

8.2.7 R Code: Linear Transformations

```
#Log Transformation
x=c(2,2,2,3,4,6,7,8,9,10,10,12)
y=c(2.99,2.70,3.00,4.02,4.71,5.83,6.28,6.27,6.76,7.04,7.05,7.66)
log.mod <-lm(y~log(x))
summary(log.mod)


#Exponential Transformation
x=c(2,2,2,3,4,6,7,8,9,10,10,12)
y=c(5.92,4.44,6.00,6.94,8.07,13.46,17.62,15.49,23.59,30.08,30.44,
55.81)
exp.mod <-lm(log(y)~x)
summary(exp.mod)


#Percentile Changes with Unit changes
change <-seq(1,10,1)
percent <-100*((exp(change*exp.mod$coef[2]))-1)
percha <-rbind(paste(round(percent,2),"%"));percha


#Power Function
x=c(2,2,2,3,4,6,7,8,9,10,10,12)
y=c(3.02,2.26,3.06,5.61,8.39,17.27,23.05,19.94,29.06,34.71,35.13,
53.75)
pow.mod <-lm(log(y)~log(x))
summary(pow.mod)


#Box-Tidwell Transformation
x=c(2,2,2,3,4,6,7,8,9,10,10,12)
y=c(11.61,11.58,12.18,7.35,5.32,4.26,3.91,2.89,1.63,1.59,1.48,
1.71)


#Check Linearity Before Box-Tidwell Transformation
pre.reduced <- lm(y ~ x)
```

(continued)

8.2.7 R Code: Linear Transformations (continued)

```
pre.full <- lm(y ~ factor(x))
anova(pre.reduced, pre.full)

#Box-Tidwell
library(car)                   #attach car package
boxTidwell(y~x)

#Reciprocal Transformation
recip.mod <-lm(y~I(x^-1))
summary(recip.mod)

#Check Linearity After Box-Tidwell Transformation
x.recip=x^-1
post.reduced <- lm(y ~ x.recip)
post.full <- lm(y ~ factor(x.recip))
anova(post.reduced, post.full)
```

8.3 Nonparametric Smoothers

The transformations we have been studying are useful when the association between x and y is known or easily discerned. This is not always the case. Data are frequently noisy and a clearly defined relation between x and y is often obscured. Nonparametric smoothers can help clarify things. There are a large number of them, but they share two features: (1) they forgo estimating the parameters of a function (e.g., regression coefficients) in favor of identifying the function itself and (2) they smooth the data using a technique known as local fitting, in which multiple observations surrounding a focal value are combined to reduce noise. In this manner, nonparametric methods reveal the form of the association between x and y.

Nonparametric smoothers can be particularly useful when we are dealing with "big data," a popular term that refers to situations where the number of observations is excessively large and noisy (e.g., the number of people who GOOGLE the term big data in any day). In the examples we have covered in this chapter, I have purposefully made the nonlinear functions easy to see when plotting the data, but this is not always possible. To reinforce this point, consider the data on the left-hand side of Fig. 8.9. Looking at these data, it is difficult to discern any lawful pattern, but, in fact, there is one. After generating a list of 100 random x values between 0 and 1 and a standardized disturbance term v, I created these data using the following function.

$$y = \sin(10x) + v$$

It is very difficult to see the sine function through the noise unless you look at the right-hand figure. When we add the sine function without the disturbance term, the

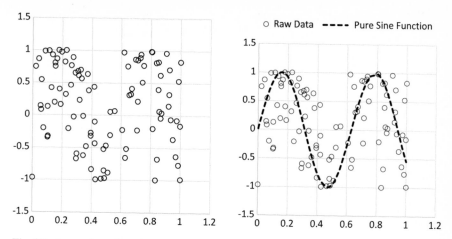

Fig. 8.9 Scatterplots of a noisy nonlinear function

pattern becomes apparent. Nonparametric scatterplot smoothers serve a similar function; they help us extract the big picture from big data.[11]

8.3.1 Understanding Nonparametric Regression

8.3.1.1 Three Regression Models

An easy way to understand nonparametric smoothers is to compare a nonparametric model with other regression models.

- Our first model is the familiar linear regression model. We don't specify a function in Eq. (8.16) because it is linear by default, and our interest centers around estimating the parameters (aka coefficients).

$$y = \beta x + \varepsilon \tag{8.16}$$

- Equation (8.17) shows a nonlinear regression model. Here, we assume that y is a weighted function of the parameters, but the function can take numerous forms (e.g., exponential, power). In this case, we first specify a nonlinear function before estimating its parameters.

$$y = f(\beta x) + \varepsilon \tag{8.17}$$

- Equation (8.18) shows a nonparametric model. Here, we assume that y is a function of x, but we are interested only in identifying that function, not

---

[11] With a single predictor, nonparametric smoothers are called scatterplot smoothers.

estimating the parameters. In fact, there aren't any parameters to estimate in Eq. (8.18), which is why it's called a nonparametric model.

$$y = f(x) + \varepsilon \tag{8.18}$$

8.3.1.2 Classifying Nonparametric Smoothers

Table 8.12 shows that nonparametric smoothers can be characterized along two dimensions. The first dimension is the estimation technique that is used to smooth the data; some smoothers compute averages and some calculate fitted values from an OLS regression. The second dimension is the weighting scheme; some smoothers use unweighted values and some use weighted values.

Table 8.12 2 × 2 Classification of nonparametric smoothers

		Weighting scheme	
		Unweighted	Weighted
Estimation technique	Averaging	Running average	Kernel regression[a]
	Fitted values using OLS regression	Running line	Locally weighted least squares regression (LOESS)

[a]The smoothing technique known as "kernel regression" involves weighted averages, not fitted values, so the term is somewhat of a misnomer.

8.3.1.3 Neighborhood Size

Regardless of which technique we use, we must decide how many observations will contribute to the smoothing function. The term *neighborhood size* is used when referring to this value, although other terms, such as *bandwidth*, *span*, and *window size*, are used as well. Two common approaches are *nearest neighbors* and *symmetric nearest neighbors*.

- *Nearest neighbors* uses the n closest values to a focal value, irrespective of whether these values are larger or smaller than the focal value. To illustrate, suppose we have the following sequence of values and wish to calculate the average of the four nearest neighbors for the focal value of 5.3.

$$1.0\ \ 2.4\ \ 2.8\ \ 3.9\ \ 4.7\ \ 5.3\ \ 7.6\ \ 8.1\ \ 8.3\ \ 9.9 \tag{8.19}$$

Looking over the numbers, we find the four closest values to 5.3 and compute the average.

$$\frac{2.8 + 3.9 + 4.7 + 5.3 + 7.6}{5} = 4.86$$

Notice that three of the values lie below the focal value and one lies above.

- *Symmetric nearest neighbors* uses $n/2$ values on either side of a focal value. With our example, we use two values lower than 5.3 and two values greater than 5.3. In our example, the average found using the symmetric nearest neighbors is quite a bit larger than the average found using nearest neighbors.

$$\frac{3.9 + 4.7 + 5.3 + 7.6 + 8.1}{5} = 5.92$$

8.3.1.4 Small Sample Example

All of the nonparametric techniques we will review involve numerous calculations, so performing them without a computer is impractical when sample sizes are large. Accordingly, we will use a small sample example to help us learn the techniques. Note, however, that the sample size is small only for illustrative purposes, and we ordinarily would be applying the techniques to a much larger number of observations.

Keeping that in mind, Table 8.13 shows a data set calculated using the same sine function as before [$y = \sin(10x) + v$], along with the results from four nonparametric smoothers. Figure 8.10 shows a scatterplot and the residuals from an OLS regression. Notice that the residuals look very much like the raw data, indicating that a linear trend is inadequate to describe the data. Notice also that the data do not

Table 8.13 Raw data and four nonparametric smoothers

v	x	y	Running average $(h = 1/4)$	Running line $(s = 1/3)$	Kernel regression $(h = .17044)$	LOESS $(f = 6)$
−.68452	.0931	.1177	.74955	1.16704	.27225	.94413
.41244	.1821	1.3814	.09500	.36970	−.01449	.34510
−.77306	.3598	−1.2141	−.12897	−1.22228	−.76228	−.88659
.19060	.3982	−.5542	−1.74667	−.56270	−.84111	−1.29944
−2.51631	.5012	−3.4717	−1.09303	−.81176	−.71345	−1.15602
.55026	.6481	.7468	−.42677	−.25050	−.00695	.39044
.56368	.7361	1.4446	.59290	.17844	.29000	.69600
−1.41086	.7914	−.4127	.42063	.48348	.38952	.46891
−.76646	.7938	.2300	.24230	.70884	.39263	.44954
−.00962	.8259	.9096	.93967	.29789	.42668	.61252
1.18895	.8912	1.6794	.63823	.50068	.46272	.48077
−1.03781	.9053	−.6743	.50255	.54446	.46642	.47601
	$sd = .280159$					

Note: $y = \sin(10x) + v$—angle in radians

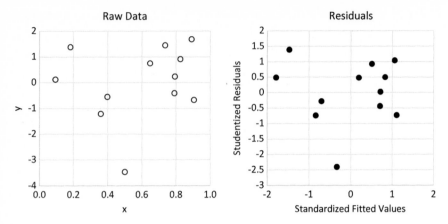

Fig. 8.10 Scatterplots of data from Table 8.13

readily suggest any other pattern, but seem to represent nothing but noise. Since we know how the data were generated, we know this is not so. But let's pretend we don't know, and see whether our smoothers can identify the pattern.

8.3.2 Running Average

The simplest, nonparametric smoother is a running average with symmetric nearest neighbors. For illustrative purposes, we will set the neighborhood size to equal 1/4 of the observations. Each estimated value of y will then be the average of three observations, one on each side of the focal value and the focal value itself.

We could work through all of the averaging by hand, but there's an easier way. If we create a smoother matrix $\mathbf{S}$, we can do the averaging using matrix multiplication.

$$\mathbf{a} = \mathbf{Sy} \tag{8.20}$$

The relevant values for $\mathbf{S}$ are shown below. First, notice that most of the values are weighted by a value of 1/3, which is 1/4 of our sample. This makes sense because we are averaging three values, so each receives equal weight. Now notice that a different weighting scheme is used at both ends of the distribution. Looking at the beginning of the distribution, we see that our first running average is computed using only two values. This is because we can't find a value below our first value, so we use only the focal value and the value that follows. The same pattern characterizes our final value, except here we use the last value and the preceding value.

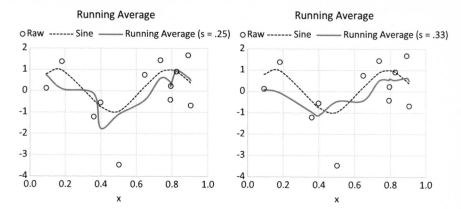

Fig. 8.11 Two running average smoothers for data from Table 8.13

$$
S_{.25} = \begin{bmatrix}
1/2 & 1/2 & 0 & 0 & 0 & 0 & 0 & 0 & 0 & 0 & 0 & 0 \\
1/3 & 1/3 & 1/3 & 0 & 0 & 0 & 0 & 0 & 0 & 0 & 0 & 0 \\
0 & 1/3 & 1/3 & 1/3 & 0 & 0 & 0 & 0 & 0 & 0 & 0 & 0 \\
0 & 0 & 1/3 & 1/3 & 1/3 & 0 & 0 & 0 & 0 & 0 & 0 & 0 \\
0 & 0 & 0 & 1/3 & 1/3 & 1/3 & 0 & 0 & 0 & 0 & 0 & 0 \\
0 & 0 & 0 & 0 & 1/3 & 1/3 & 1/3 & 0 & 0 & 0 & 0 & 0 \\
0 & 0 & 0 & 0 & 0 & 1/3 & 1/3 & 1/3 & 0 & 0 & 0 & 0 \\
0 & 0 & 0 & 0 & 0 & 0 & 1/3 & 1/3 & 1/3 & 0 & 0 & 0 \\
0 & 0 & 0 & 0 & 0 & 0 & 0 & 1/3 & 1/3 & 1/3 & 0 & 0 \\
0 & 0 & 0 & 0 & 0 & 0 & 0 & 0 & 1/3 & 1/3 & 1/3 & 0 \\
0 & 0 & 0 & 0 & 0 & 0 & 0 & 0 & 0 & 1/3 & 1/3 & 1/3 \\
0 & 0 & 0 & 0 & 0 & 0 & 0 & 0 & 0 & 0 & 1/2 & 1/2
\end{bmatrix}
$$

Table 8.13 shows the values obtained when we perform the multiplication, and Fig. 8.11 adds the smooth and the true sine function to the raw data. As you can see the running average does a pretty good job of identifying the undulating pattern of the data, although the pattern is far from smooth.

Using more values in our smoothing function will create a smoother pattern (and using fewer values will create a more jagged one). To illustrate, we'll form one more smoothing matrix, this time using 1/3 of the observations. The smoother matrix appears below, and its form is similar to our earlier matrix, with one exception. Our very last value is different than our first. Had there been an odd number of observations, this adjustment would not have been necessary. The right-hand side of Fig. 8.11 shows the running average obtained with the new smoother. It is less jagged than our earlier smoother, failing to show the underlying data pattern. In this case, the bandwidth is too large and we should choose the smaller bandwidth.

$$S_{.33} = \begin{bmatrix}
1/3 & 1/3 & 1/3 & 0 & 0 & 0 & 0 & 0 & 0 & 0 & 0 & 0 \\
1/4 & 1/4 & 1/4 & 1/4 & 0 & 0 & 0 & 0 & 0 & 0 & 0 & 0 \\
0 & 1/4 & 1/4 & 1/4 & 1/4 & 0 & 0 & 0 & 0 & 0 & 0 & 0 \\
0 & 0 & 1/4 & 1/4 & 1/4 & 1/4 & 0 & 0 & 0 & 0 & 0 & 0 \\
0 & 0 & 0 & 1/4 & 1/4 & 1/4 & 1/4 & 0 & 0 & 0 & 0 & 0 \\
0 & 0 & 0 & 0 & 1/4 & 1/4 & 1/4 & 1/4 & 0 & 0 & 0 & 0 \\
0 & 0 & 0 & 0 & 0 & 1/4 & 1/4 & 1/4 & 1/4 & 0 & 0 & 0 \\
0 & 0 & 0 & 0 & 0 & 0 & 1/4 & 1/4 & 1/4 & 1/4 & 0 & 0 \\
0 & 0 & 0 & 0 & 0 & 0 & 0 & 1/4 & 1/4 & 1/4 & 1/4 & 0 \\
0 & 0 & 0 & 0 & 0 & 0 & 0 & 0 & 1/4 & 1/4 & 1/4 & 1/4 \\
0 & 0 & 0 & 0 & 0 & 0 & 0 & 0 & 0 & 1/3 & 1/3 & 1/3 \\
0 & 0 & 0 & 0 & 0 & 0 & 0 & 0 & 0 & 0 & 1/2 & 1/2
\end{bmatrix}$$

Perhaps you are thinking that subjective impressions of smoothness are, well, too subjective to guide our choice of bandwidth. In Chap. 10 you will learn a more objective technique for selecting a bandwidth when we discuss another scatterplot smoother called a spline. For now, we will note only that scatterplot smoothers are used primarily to describe and understand our data and visual judgments are critical to their success. As Cohen, Cohen, West, and Aiken (2003) note, "Simple visual judgments by the analyst are normally sufficient to choose a reasonable window size that provides a good estimate of the shape of the distribution in the population" (p. 108).[12]

8.3.3 Running Line

Our next smoother is called a running line smoother. Here, we use symmetric nearest neighbors to find fitted values from an OLS regression. More observations are usually needed to compute a regression line than an average, so we will set our first window size for $s = 1/3$ and our second window size to $s = 1/2$. Table 8.14 shows the observations that are included for each fitted value, and Fig. 8.12 shows the smooth. As you can see, the first smoother does a good job of identifying the underlying pattern, but the second produces too much smoothing.

Because we don't need the regression coefficients, the easiest way to find the fitted values is to use the hat matrix, including only relevant values of x. To illustrate, the first hat matrix using $s = 1/3$ includes only the first five observations.

---

[12] See also Fox (2000, p. 23).

Table 8.14 Observations used in two running line smoothers with different bandwidths

For fitted value	$s = 1/3$ Observations included in the regression	$\hat{y}$	$s = 1/2$ Observations included in the regression	$\hat{y}$
1	1–5	1.167040	1–7	−.279368
2	1–5	.369704	1–7	−.263425
3	1–5	−1.222282	1–7	−.231593
4	2–6	−.562705	1–7	−.224715
5	3–7	−.811758	2–8	−.305070
6	4–8	−.250497	3–9	−.254043
7	5–9	.178437	4–10	.221065
8	6–10	.483477	5–11	.695757
9	7–11	.708842	6–12	.570399
10	8–12	.297894	6–12	.507120
11	8–12	.500675	6–12	.378394
12	8–12	.544461	6–12	.350598

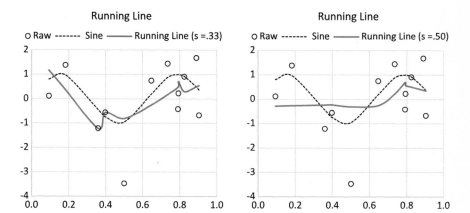

Fig. 8.12 Two running line smoothers for data from Table 8.13

$$\mathbf{H}[1:5] = \begin{bmatrix} .61482 & .44213 & .09731 & .02280 & -.17706 \\ .44213 & .34132 & .14006 & .09657 & -.02009 \\ .09731 & .14006 & .22542 & .24386 & .29334 \\ .02280 & .09657 & .24386 & .27569 & .36107 \\ -.17706 & -.02009 & .29334 & .36107 & .54274 \end{bmatrix}$$

Multiplying this matrix by the corresponding values in **y** produces five fitted values.

$$\hat{\mathbf{y}}\,[1{:}5] = \mathbf{H}[1{:}5]\mathbf{y}[1{:}5] = \begin{bmatrix} 1.16704 \\ .36970 \\ -1.22228 \\ -1.56630 \\ -2.48906 \end{bmatrix}$$

Even though we computed five fitted values, Table 8.14 shows that we only use the first three for our running line smoother. This is because we can find more symmetric nearest neighbors for observation 4 ($\mathbf{H}[2{:}6]$) and 5 ($\mathbf{H}[3{:}7]$). Similar calculations produce the rest of the fitted values.

8.3.4 Kernel Regression

In the previous examples, a fixed number of observations was specified and all observations included in the neighborhood received equal weight. An alternative approach is to include all observations but differentially weight them, with values lying close to a focal value receiving more weight than values that lie far from a focal value. Formally, this type of continuous weighting scheme is known as a kernel function, and its application with scatterplot smoothing is called kernel regression.[13] The procedures are somewhat complicated, but only arithmetic is involved.

8.3.4.1 Steps

1. Select a bandwidth value, h. A good rule of thumb, known as Silverman's rule, is to use the standard deviation of the predictors to find an initial value of h.

$$h = s_x * N^{-.2} \tag{8.21}$$

2. Create an $N \times N$ matrix of the predictors, then calculate u by centering each column value around its diagonal value and dividing by h.

$$u = \frac{\left(x_{ij} - x_{ii}\right)}{h} \tag{8.22}$$

3. Select a weighting scheme. Table 8.15 shows two common density functions, but several others are available (Buskirk et al. 2013).

---

[13] If you are thinking that this procedure sounds a lot like the Newey-West technique we learned in Chap. 7, you are absolutely right. The Newey-West technique uses a kernel known as the Bartlett kernel.

Table 8.15 Common kernel density functions, K

Kernel	Formula (K)	
Gaussian	$\dfrac{1}{\sqrt{2\pi}} * e^{-\dfrac{u^2}{2}}$	
Epanechnikov	$\left\{ \begin{array}{l} 3\left(1-\dfrac{u^2}{5}\right) \Big/ 4\sqrt{5} \\ \\ 0 \end{array} \right.$	for $u^2 < 5$ \\ otherwise

4. Compute the kernel smooth using the Nadaraya-Watson estimator.

$$\hat{\mu}\left(x^* | x, h\right) = \frac{\sum_{i=1}^{n} K(u)y}{\sum_{i=1}^{n} K(u)} \tag{8.23}$$

As is true with a regression coefficient, the numerator is a cross-product term and the denominator involves only the predictor. Unlike a regression coefficient, however, these are weighted sums, not estimates of variability.

8.3.4.2   Illustration

Let's work through the steps for our sample data set.

- Our first task is to find an initial value of h. Using the standard deviation shown in Table 8.13 ($s=.280159$), we derive the following value using Eq. (8.21).

$$h = .280159 * 12^{-.2} = .17044$$

- We then create a 12×12 matrix from our predictors and find u by centering each column entry around its diagonal value and dividing the difference by h [see Eq. (8.22)]. Table 8.16 shows the obtained values. To illustrate, the first three values in Column 1 were found as follows.

$$\frac{(x_{11} - x_{11})}{h} = \frac{(.0931 - .0931)}{.17044} = 0$$

$$\frac{(x_{21} - x_{11})}{h} = \frac{(.1821 - .0931)}{.17044} = .5222$$

and

$$\frac{(x_{31} - x_{11})}{h} = \frac{(.3598 - .0931)}{.17044} = 1.5648$$

The remaining values were found in a similar fashion. Notice that the matrix has 0 s along the main diagonals.

Table 8.16 Values for u using $h=.17044$

u_1	u_2	u_3	u_4	u_5	u_6	u_7	u_8	u_9	u_{10}	u_{11}	u_{12}
.0000	−.5222	−1.5648	−1.7901	−2.3944	−3.2563	−3.7726	−4.0971	−4.1112	−4.2995	−4.6826	−4.7653
.5222	.0000	−1.0426	−1.2679	−1.8722	−2.7341	−3.2504	−3.5749	−3.5890	−3.7773	−4.1604	−4.2432
1.5648	1.0426	.0000	−.2253	−.8296	−1.6915	−2.2078	−2.5323	−2.5464	−2.7347	−3.1178	−3.2006
1.7901	1.2679	.2253	.0000	−.6043	−1.4662	−1.9825	−2.3070	−2.3211	−2.5094	−2.8925	−2.9753
2.3944	1.8722	.8296	.6043	.0000	−.8619	−1.3782	−1.7027	−1.7167	−1.9051	−2.2882	−2.3709
3.2563	2.7341	1.6915	1.4662	.8619	.0000	−.5163	−.8408	−.8549	−1.0432	−1.4263	−1.5090
3.7726	3.2504	2.2078	1.9825	1.3782	.5163	.0000	−.3245	−.3385	−.5269	−.9100	−.9927
4.0971	3.5749	2.5323	2.3070	1.7027	.8408	.3245	.0000	−.0141	−.2024	−.5855	−.6683
4.1112	3.5890	2.5464	2.3211	1.7167	.8549	.3385	.0141	.0000	−.1883	−.5715	−.6542
4.2995	3.7773	2.7347	2.5094	1.9051	1.0432	.5269	.2024	.1883	.0000	−.3831	−.4659
4.6826	4.1604	3.1178	2.8925	2.2882	1.4263	.9100	.5855	.5715	.3831	.0000	−.0827
4.7653	4.2432	3.2006	2.9753	2.3709	1.5090	.9927	.6683	.6542	.4659	.0827	.0000

- Next, we weight our values using a kernel function. Table 8.17 presents the weights using a Gaussian kernel. To illustrate, the first three values in Column 1 were found as follows.

$$\frac{1}{\sqrt{2\pi}} * e^{-\frac{0^2}{2}} = .3989$$

$$\frac{1}{\sqrt{2\pi}} * e^{-\frac{.5222^2}{2}} = .3481$$

and

$$\frac{1}{\sqrt{2\pi}} * e^{-\frac{1.5648^2}{2}} = .1173$$

Notice that the weight of each observation decreases with distance from the diagonal value of .3989.

Table 8.17 Gaussian kernel weights (Ku_i)

	Ku_1	Ku_2	Ku_3	Ku_4	Ku_5	Ku_6	Ku_7	Ku_8	Ku_9	Ku_{10}	Ku_{11}	Ku_{12}
	.3989	.3481	.1173	.0804	.0227	.0020	.0003	.0001	.0001	.0000	.0000	.0000
	.3481	.3989	.2317	.1786	.0691	.0095	.0020	.0007	.0006	.0003	.0001	.0000
	.1173	.2317	.3989	.3889	.2828	.0954	.0349	.0162	.0156	.0095	.0031	.0024
	.0804	.1786	.3889	.3989	.3324	.1362	.0559	.0279	.0270	.0171	.0061	.0048
	.0227	.0691	.2828	.3324	.3989	.2752	.1543	.0936	.0914	.0650	.0291	.0240
	.0020	.0095	.0954	.1362	.2752	.3989	.3492	.2802	.2768	.2315	.1443	.1278
	.0003	.0020	.0349	.0559	.1543	.3492	.3989	.3785	.3767	.3472	.2637	.2437
	.0001	.0007	.0162	.0279	.0936	.2802	.3785	.3989	.3989	.3909	.3361	.3191
	.0001	.0006	.0156	.0270	.0914	.2768	.3767	.3989	.3989	.3919	.3388	.3221
	.0000	.0003	.0095	.0171	.0650	.2315	.3472	.3909	.3919	.3989	.3707	.3579
	.0000	.0001	.0031	.0061	.0291	.1443	.2637	.3361	.3388	.3707	.3989	.3976
	.0000	.0000	.0024	.0048	.0240	.1278	.2437	.3191	.3221	.3579	.3976	.3989
Σ	.9699	1.2397	1.5966	1.6541	1.8385	2.3269	2.6054	2.6410	2.6390	2.5811	2.2885	2.1983

- Finally, we compute our kernel-weighted average using the Nadaraya-Watson estimator [see Eq. (8.23)].

  – Multiply each column in Table 8.17 by the column of y values from Table 8.13 (see Table 8.18).
  – Sum the obtained values (see bottom row of Table 8.18).
  – Divide each sum by the sum of the column Ku_i values from the bottom row of Table 8.17.

Table 8.18 Calculations for Nadaraya-Watson estimator

	Ku_1y	Ku_2y	Ku_3y	Ku_4y	Ku_5y	Ku_6y	Ku_7y	Ku_8y	Ku_9y	$Ku_{10}y$	$Ku_{11}y$	$Ku_{12}y$
	.0470	.0410	.0138	.0095	.0027	.0002	.0000	.0000	.0000	.0000	.0000	.0000
	.4809	.5511	.3200	.2467	.0955	.0131	.0028	.0009	.0009	.0004	.0001	.0001
	−.1424	−.2813	−.4844	−.4722	−.3433	−.1158	−.0423	−.0196	−.0189	−.0115	−.0038	−.0029
	−.0445	−.0990	−.2156	−.2211	−.1842	−.0755	−.0310	−.0154	−.0150	−.0095	−.0034	−.0026
	−.0788	−.2400	−.9817	−1.1538	−1.3850	−.9553	−.5358	−.3250	−.3173	−.2256	−.1010	−.0833
	.0015	.0071	.0713	.1017	.2055	.2979	.2608	.2092	.2067	.1729	.1077	.0954
	.0005	.0029	.0504	.0808	.2229	.5044	.5763	.5468	.5442	.5016	.3809	.3521
	.0000	−.0003	−.0067	−.0115	−.0386	−.1156	−.1562	−.1646	−.1646	−.1613	−.1387	−.1317
	.0000	.0001	.0036	.0062	.0210	.0637	.0866	.0917	.0918	.0901	.0779	.0741
	.0000	.0003	.0086	.0156	.0591	.2106	.3159	.3555	.3565	.3629	.3372	.3256
	.0000	.0001	.0052	.0102	.0489	.2423	.4428	.5644	.5690	.6226	.6700	.6677
	.0000	.0000	−.0016	−.0032	−.0162	−.0862	−.1643	−.2152	−.2172	−.2413	−.2681	−.2690
Σ	.2641	−.0180	−1.2171	−1.3913	−1.3117	−.0162	.7556	1.0287	1.0362	1.1013	1.0589	1.0254

Performing the calculations yields the estimated values shown in Table 8.13 and displayed on the left-hand side of Fig. 8.13. Inspection of the smoother suggests that our initial value of h was too large and oversmoothed the data. The right-hand side of Fig. 8.13 shows the results when $h=.10$. This smoother more clearly reveals the underlying data pattern, underscoring the need to try several values and choose the most illuminating one.

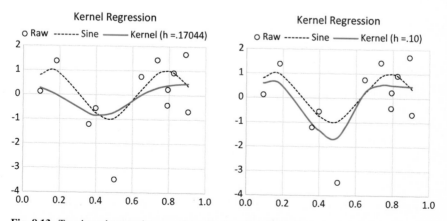

Fig. 8.13 Two kernel regression smoothers for data from Table 8.13

8.3.5 *Locally Weighted Regression*

The final smoother we will consider uses locally weighted regression to find fitted estimates of *y*. It is an unfortunate fact that it goes by two names, LOESS (local regression) and LOWESS (locally weighted scatterplot smoother). Although there are slight differences between LOESS and LOWESS, both involve selecting a portion of the data to be included in a regression analysis, weighting each selected value in accordance with its distance from the value being smoothed, and using weighted least squares estimation (see Chap. 7) to calculate a fitted value.

8.3.5.1 Steps

1. The first step is to select a fixed window width of nearest neighbors. As noted earlier, unlike symmetric nearest neighbors that uses $k/2$ values on either side of a focal value, nearest neighbors selects values without regard to whether they are higher or lower than the focal value. The bandwidth usually lies between 40 % and 60 % of the total sample, so we will chose 50 % to begin. With $N = 12$, this means we will be selecting the five values closest to our focal value, along with the focal value itself for a total of six values in each window.
2. Once we select our values, we calculate the absolute difference between each value and the focal value and then scale these absolute values by dividing each one by the maximum. In this manner, the value that lies farthest from our focal value receives a scaled value $= 1$. We will designate these scaled, absolute values, s.
3. Next, we apply a weighting function to the scaled values. Cleveland (1979) recommends a tricube function.

$$w = \left(1 - s^3\right)^3 \tag{8.24}$$

Notice that now the value closest to our focal value receives a weight of 1.

4. Finally, we perform a weighted least squares regression using only the six scaled values. Recall from Chap. 7 that with a weighted regression, we multiply each of our original values by the square root of the weight and find the fitted values from an OLS regression through the origin.

8.3.5.2 Illustration

Table 8.19 illustrates the steps involved in computing a LOESS smoother for the seventh observation (in bold font). Notice that the seventh observation presents an unusual case, as four of the five nearest neighbors lie above the seventh value and only one lies below. Other than that, the steps involved are straightforward.

Table 8.19 Illustrative calculations for LOESS procedure for seventh observation

x	y	$|x_i - x|$	s	w	$1 * \sqrt{w}$	$x * \sqrt{w}$	$y * \sqrt{w}$
.0931	.1177	.64300					
.1821	1.3814	.55400					
.3598	−1.2141	.37630					
.3982	−.5542	.33790					
.5012	−3.4717	.23490					
.6481	.7468	.08800	.56738	.54605	.73895	.47891	.55185
.7361	**1.4446**	0	0	1	1	.73610	1.44460
.7914	−.4127	.05530	.35654	.87009	.93279	.73821	−.38496
.7938	.2300	.05770	.37202	.85336	.92377	.73329	.21247
.8259	.9096	.08980	.57898	.52344	.72349	.59753	.65809
.8912	1.6794	.15510	1	0	0	0	0
.9053	−.6743	.16920				$\hat{y}$	.69600

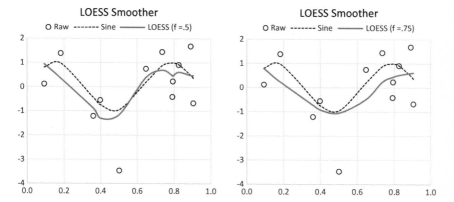

Fig. 8.14 Two LOESS smoothers for data from Table 8.13

- Calculate the absolute difference between the focal observation and the five nearest neighbors.
- Scale these differences by dividing each by the maximum.
- Weight the scaled values using a tricube function.
- Multiply the intercept, the predictor, and the criterion by the square root of the weighted values.
- Conduct an OLS regression through the origin (i.e., don't add a column vector of leading 1's) and find the fitted value using the hat matrix and weighted criterion.

The left-hand side of Fig. 8.14 plots the fitted values. As you can see, they do an excellent job of representing the function that generated the data. The right-hand side of Fig. 8.14 shows another LOESS smooth using a larger bandwidth. In this case, the curve is a bit too smooth and the first one is more informative.

8.3.6 Extensions and Applications

Four scatterplot smoothers have been described in this section. The calculations are manageable with a small data set, but computers are needed with more observations. Moreover, although the techniques can be extended to situations involving multiple predictors, the calculations become increasingly difficult and the interpretation increasingly murky (a state of affairs known as the "curse of dimensionality").

Although they are primarily used to identify the form of the function relating x and y, nonparametric smoothers can also be used for interpolation (i.e., to find fitted values corresponding to values included with the range of predictors but not represented among the predictors) and parametric estimation (i.e., one can test a parametric model based on the visual depiction of the nonparametric function). In some cases, confidence limits around the fitted values can be calculated and inferential statistics regarding goodness of fit can be performed. We will cover these issues in more detail in Chap. 10 when we discuss penalized cubic splines.

8.3.7 R Code: Nonparametric Smoothers

```
#Sine Function
x=c(.0931,.1821,.3598,.3982,.5012,.6481,.7361,.7914,.7938,.8259,.8912,
.9053)
y=c(.1177,1.3814,-1.2141,-.5542,-3.4717,.7468,1.4446,-.4127,.2300,.9096,
1.6794,-.6743)


pure=c(0.802217374,0.968862012,-0.440726043,-0.744914947,-0.955451368,
0.196527112,0.880925705,0.998199438,0.996472533,0.91909516,0.49059981,
0.363272501)


#Running Average (h=.33)
library(caTools)                    #(attach caTools package)
run.aver <-runmean(y,3)
run.aver


#Running Line (s = 1/3)
run.line33 <-supsmu(x,y,span=.33)
run.line33


#Kernel Regression (h=.17044)
library(np)                     #(attach np package)
kernel.reg <-npreg(y~x,bws=.17044,ckertype="gaussian")
fitted(kernel.reg)


#LOESS (f=.5)
reg.lo <- lowess(y ~ x,f=.5,iter=0)
reg.lo
```

8.4 Chapter Summary

1. A model is linear in the parameters when the criterion is an additive function of weighted predictors and a disturbance term, and the regression weights are of the first power and are not multiplied or divided by any other value.
2. A model is linear in the variables when it forms a straight line or plane when plotted. A model that is linear in the variables might or might not be linear in the parameters.
3. If our predictor has at least one set of duplicate values, a lack of fit test can be used to identify the presence of a nonlinear relation. The test compares the residual sum of squares from an OLS regression to the residual sum of squares from a categorical analysis that treats the duplicate values as nonlinear factors.
4. Many nonlinear relations can be linearized by transforming the predictors and/or the criterion.
5. Logarithmic functions with additive errors can be converted to a linear form by regressing the criterion on the log of the predictors. In this case, ratio changes in x predict unit changes in y.
6. Exponential functions with multiplicative errors can be converted to a linear form by regressing the log of the criterion on the raw predictors. In this case, unit changes in x predict ratio changes in y.
7. Power functions with multiplicative errors can be converted to a linear form by regressing the log of the criterion on the log of the predictors. In this case, ratio changes in x predict ratio changes in y.
8. The Box-Tidwell procedure can be used to find a linearizing transformation of x.
9. Nonparametric methods describe the functional relation between predictors and a criterion by using subsets of the data to calculate central tendencies. With a single predictor, nonparametric methods are called scatterplot smoothers.
10. A running average is the simplest scatterplot smoother. Here, the average value of y is computed from a moving window of predictor values. The window is constructed using symmetric nearest neighbors, with $k/2$ values on either side of a focal observation.
11. A running line is a fitted value from an unweighted subset of symmetric nearest neighbors.
12. With kernel regression, a smoother parameter is chosen and each predictor is weighted in accordance with its distance from a focal value being smoothed. The nearer the predictor is to the value being smoothed, the more weight it receives. Smoothed estimates of y are found using the Nadaraya-Watson formula, with a weighted xy cross-product term in the numerator and the sum of the weighted predictors in the denominator.
13. With locally weighted regression (LOESS/LOWESS) k predictors closest to a focal value are chosen without regard to whether they are equally distributed around the focal value. The predictors are scaled and weighted, and the fitted value from a weighted least squares regression comprises the smoothed value.

Chapter 9
Cross-Product Terms and Interactions

In a linear regression model with multiple predictors, the regression coefficients represent the unique contribution of each variable to the prediction of a criterion holding all other variables constant. Because the coefficients are statistically independent, a one-unit change in x_i predicts a b_i change in y across all levels of x_k. This *conditional invariance* (as it is called) is demanded by the form of an ordinary linear regression model:

$$y = \beta_0 + \beta_1 x_1 + \beta_2 x_2 + \ldots + \beta_k x_k + \varepsilon \tag{9.1}$$

9.1 Understanding Interactions

One can easily imagine situations, however, in which the expected change in y associated with a one-unit change in one predictor depends on the value of another predictor. For example, stressful life events predict well-being, but this is especially true among people who lack social support (i.e., a friend or loved one they can trust in times of trouble). In this case, the magnitude of the expected change in well-being following a one-unit increase in stress depends on whether social support is low or high. When the predictive effect of one variable changes with the values of another variable, we say that the two variables interact to predict a criterion or that one predictor moderates the effects of another.

Electronic Supplementary Material: The online version of this chapter (doi: 10.1007/978-3-319-11734-8_9) contains supplementary material, which is available to authorized users

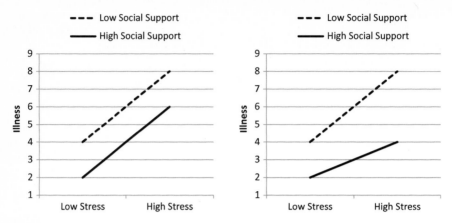

Fig. 9.1 Examples of no interaction (*Panel A*) and an interaction (*Panel B*) involving social support, stress, and illness

9.1.1   Depicting an Interaction

Visually, an interaction can be detected by examining the slope of the regression lines relating one predictor to a criterion across values of another predictor. If the slopes are parallel, there is no interaction; if the slopes are not parallel, there is. Figure 9.1 displays both possibilities, using the relation between stress and illness as a function of social support. In Panel A, the regression lines are parallel, indicating that the simple slope of stress is constant across levels of social support.[1] In this case, there is no interaction. Panel B shows a different pattern. Here, the simple slopes are not parallel, signifying the presence of an interaction. The interaction indicates that the negative effects of stress are less severe when social support is high than when it is low. Interactions can take several forms, so they won't always look like the one displayed in Panel B; but whenever we see nonparallel regression lines, we know that an interaction exists.

9.1.2   Modeling Interactions with Cross-Product Terms

To model an interaction, we expand our regression equation so that the magnitude of a regression coefficient can change across levels of another predictor. To effect this modification, we add a new term to our equation, formed by multiplying two (or more) predictors to create a cross-product term. It will be a bit easier to illustrate this point if we name our predictors x and z rather than x_1 and x_2, so let's rewrite our

---

[1] The term "simple slope" is used to describe the slope of a regression line at a particular value of another predictor. We will denote it b_s to distinguish it from the unstandardized regression coefficient b.

regression equation using this new nomenclature, adding a cross-product term we will designate xz:

$$\hat{y} = b_0 + b_1x + b_2z + b_3xz \qquad (9.2)$$

In a moment, we will see that the regression coefficient associated with the cross-product term (b_3) tests whether our two predictors are conditionally invariant. Before we describe this test, let's examine why we multiply our predictors to model an interaction. We can best see the logic here by rearranging some of the terms in our regression equation:

$$\hat{y} = b_0 + (b_1 + \boldsymbol{b_3z})x + b_2z \qquad (9.3)$$

The highlighted term in the middle portion of the equation indicates that the slope of the regression of y on x now depends on the value of z. Similarly, we can recast the equation to show that a one-unit change in z now depends on values of x:

$$\hat{y} = b_0 + (b_2 + \boldsymbol{b_3x})z + b_1x \qquad (9.4)$$

9.1.2.1 Cross-Product Terms and Partial Derivatives

Considering that x and z combine multiplicatively in Eq. (9.2), you might think that adding a cross-product term to a regression equation creates a nonlinear model. This is not so. In a linear model, the *parameters* cannot combine multiplicatively, but the *predictors* can. To reinforce this point, recall from Chap. 8 that all partial derivatives are independent in a linear model. This property applies to Eq. (9.2):

$$\frac{\partial y}{\partial b_1} = x; \quad \frac{\partial y}{\partial b_2} = z; \quad \frac{\partial y}{\partial b_3} = xz \qquad (9.5)$$

In short, Eq. (9.2) is nonlinear in the variables, but linear in the parameters.

9.1.2.2 Numerical Example of Cross-Product Terms

Rather than continuing to discuss interactions in abstract terms, let's consider how the introduction of a cross-product term influences our fitted values using the following equation as an illustration:

$$\hat{y} = 4.0 + 2.5x + 1.25z + 3.0xz$$

Now, we will plug in various values for x and z. The left-hand column in Table 9.1 shows what happens when we hold z constant and vary x. Several points are of interest.

- First, notice that when $z = -2$, a one-unit change in x is associated with a -3.50 unit change in y, but when $z = -1$, a one-unit change in x is associated with a

Table 9.1 Illustration of simple slopes with a cross-product term

	Simple slopes of x at z					Simple slopes of z at x			
x	z	$\hat{y}$	b_s	Δb_s	x	z	$\hat{y}$	b_s	Δb_s
1	−2	−2.0			−2	1	−5.75		
2	−2	−5.5	−3.50		−2	2	−10.5	−4.75	
1	−1	2.25		3.0	−1	1	−0.25		3.0
2	−1	1.75	−0.50		−1	2	−2.0	−1.75	
1	0	6.5		3.0	0	1	5.25		3.0
2	0	9.0	2.50		0	2	6.50	1.25	
1	1	10.75		3.0	1	1	10.75		3.0
2	1	16.25	5.50		1	2	15.0	4.25	
1	2	15.0		3.0	2	1	16.25		3.0
2	2	23.5	8.50		2	2	23.5	7.25	

Fig. 9.2 Simple slope of x at five levels of z

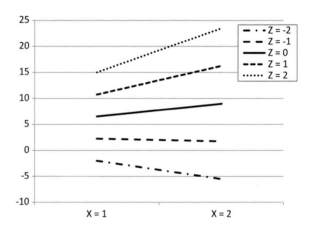

−.50 unit change in y. Thus, these simple slopes are not conditionally invariant. We can see this effect graphically in Fig. 9.2, which plots the simple slopes of x at each level of z. Clearly, these lines are not parallel, indicating the presence of an interaction.

- Second, notice that the difference between the two simple slopes $= 3.0$. It is no coincidence that this value equals the regression coefficient for the cross-product term (i.e., $b_3 = 3.0$), as this is precisely what the cross-product coefficient represents. It quantifies how much the simple slope of x changes with a one-unit change in z.

- Third, notice that this change in slope is uniform across all values of z. No matter which value we start with, the simple slope of x changes by 3.0 units with every one-unit change in z.

- The right-hand side of Table 9.1 reveals a complementary pattern when we hold x constant and vary z. When $x = -2$, a one-unit change in z is associated with a -4.75 unit change in y, but when $x = -1$, a one-unit change in z is associated with a -1.75 unit change in y. As before, these simple slopes are not conditionally invariant.

- Note, however, that the simple slope of z changes $b_3 = 3.0$ units with every one-unit increase in x. Because the cross-product coefficient is symmetrical, this will always be true. Just as the simple slope of x changes b_3 units with a one-unit change in z, so, too, does the simple slope of z change b_3 units with a one-unit change in x.
- Table 9.1 reveals one more point of interest. Notice that, when $z = 0$, the simple slope of $x = 2.50$ and that this value is the same as b_1 in our regression equation. Similarly, when $x = 0$, the simple slope of $z = 1.25$, which equals b_2 in our regression equation. These are not coincidences either. When a regression equation contains a cross-product term, the lower-order effects (as they are called) represent the expected change in y with every one-unit change in x_i when all $x_k = 0$.

It is important to fully understand the meaning of a lower-order coefficient when a cross-product term is included in a regression equation. Without a cross-product term in the equation, b_i represents the expected change in y with a one-unit change in x_i at every level of x_k; with a cross-product term in the equation, b_i represents the expected change in y with a one-unit change in x_i *only* when $x_k = 0$. In short, the lower-order terms in a cross-product model represent simple slopes, not general ones. Unfortunately, not all researchers are aware of this fact. Instead of recognizing that lower-order coefficients represent a conditional effect of x_i when $x_k = 0$, they erroneously assume that these coefficients represent an invariant effect of x_i across all levels of x_k.

9.1.2.3 Mean-Centering the Predictors Before Computing the Cross-Product Term

Even when lower-order coefficients are interpreted correctly, they are not necessarily of interest. Many variables do not have a true zero point, so knowing the simple slope when a variable equals zero is not always informative. To mitigate the problems involved in interpreting lower-order coefficients, many statisticians recommend centering predictors before computing a cross-product term. Doing so isn't necessary, but it does clarify the meaning of the lower-order terms.[2] Mean-centering a variable means nothing more than computing its deviate score (observed score − mean). If we then use these deviate scores in our regression analysis, each predictor's mean will equal 0, and the value of our lower-order regression coefficients will represent the simple slope of one predictor at the mean of another predictor.

---

[2] Some textbooks encourage researchers to center the predictors to minimize collinearity, but there is no statistical basis to this recommendation. The only reason to center the predictors is to facilitate the interpretation of lower-order effects.

Table 9.2 Small sample example using deviate scores and a cross-product term to model an interaction

	Original variables			Variables used in the analysis				
	Weather forecast (x)	Time to spend outdoors (z)	Excitement for the weekend (y)		dev_x	dev_z	xz	y
	1	3	2	1	−4.0833	−2.00	8.1667	2
	9	7	9	1	3.9167	2.00	7.8333	9
	5	1	6	1	−.0833	−4.00	.3333	6
	2	6	1	1	−3.0833	1.00	−3.0833	1
	5	5	5	1	−.0833	0.00	0.00	5
	6	6	6	1	.9167	1.00	.9167	6
	4	9	2	1	−1.0833	4.00	−4.3333	2
	7	1	3	1	1.9167	−4.00	−7.6667	3
	6	1	5	1	.9167	−4.00	−3.6667	5
	8	5	8	1	2.9167	0.00	0.00	8
	2	7	2	1	−3.0833	2.00	−6.1667	2
	6	9	6	1	.9167	4.00	3.6667	6
Mean	5.0833	5.00	4.5833		0.00	0.00	−.3333	4.5833
Standard deviation	2.4664	2.9233	2.5746		2.4664	2.9233	5.0421	2.5746

9.1.3   Testing Cross-Product Terms

Now that we understand the logic behind a cross-product term, let's consider a complete example. Suppose we are interested in whether the weekend weather forecast predicts people's moods. To address this issue, we code the weekend weather forecast for 12 cities (x) using a 9-point scale (1 = lousy; 9 = glorious). We then call up one person from each city and ask the respondent how many hours they expect to spend outside that weekend (z) and how excited they are for the upcoming weekend (y). Table 9.2 presents the (imaginary) raw data.

To analyze these data using a cross-product term, we will first create deviate scores. These values appear to the right of the original scores in Table 9.2. Four points are noteworthy.

- First, although both deviate variables now have a mean of zero, the cross-product term does not. This is because the cross-product term is formed by multiplying the deviate predictors, but is not, itself, centered.
- Notice also that the standard deviations of the deviate scores match the standard deviations of the original scores. So although we have altered the mean, we have not altered the variance (as would be true if we had standardized the scores).
- Third, we do not center the criterion.
- Finally, we still include a column of leading 1's to model the intercept.

9.1.3.1 Regression Model

Table 9.3 shows the results of a regression analysis in which y was regressed on all predictors. As you can see, the overall model is significant and, in combination, the predictors explain ~87 % of the variance in the criterion. However, with a cross-product term in the equation, our interest normally centers on the cross-product term's regression coefficient, b_3. Table 9.3 shows that the coefficient is statistically significant, indicating that the regression line relating x to y is not uniform across levels of z (and, equivalently, that the regression line relating z to y is not uniform across levels of x).

Using Eq. (9.6), we can calculate the squared semipartial correlation of the cross-product term or, equivalently, how much R^2 changes when the cross-product term is subtracted from the regression model:

$$\Delta R^2 = \frac{(b/SE_b)^2 * MS_e}{SS_y} \tag{9.6}$$

Plugging in values from Table 9.3, we see that the cross-product term explains ~15 % of the total variance:

$$\Delta R^2 = \frac{(.1985/.0662)^2 * 1.1738}{72.9167} = .1447$$

9.1.3.2 Simple Slopes

The value of the cross-product coefficient tells us how much the simple slope of each variable changes across levels of the other. In our case, we see that the simple slope of x changes by .1985 units with every one-unit increase in z (and that the

Table 9.3 Regression model for small sample example with cross-product term

			Significance test of regression model				
	SS	df	MS	R^2	F	p	
Regression	63.5266	3	21.1755	.8712	18.0408	.0006	
Residual	9.3901	8	1.1738				
Total	72.9167						
			Regression coefficients				
	b	se_b	ΔR^2	t	p		
b_0	4.6495	.3135		14.8296	.0000		
b_1	.8379	.1338	.6315	6.2636	.0002		
b_2	−.0090	.1135	.0001	−.0790	.9390		
b_3	.1985	.0662	.1447	2.9978	.0171		

$(\mathbf{X'X})^{-1}$				C			
.083748	−.000332	−.000361	.001245	.098301	−.000389	−.000424	.001462
−.000332	.015247	.000926	−.000995	−.000389	.017897	.001087	−.001168
−.000361	.000926	.010979	−.001082	−.000424	.001087	.012887	−.001271
.001245	−.000995	−.001082	.003736	.001462	−.001168	−.001271	.004386

simple slope of z changes by .1985 units with every one-unit increase in x). We can verify this is true by calculating the simple slopes of x at various levels of z^3:

$$b_s @ z = b_1 + zb_3 \qquad (9.7)$$

For simplicity, let's calculate the simple slope of x when $z = 2$ and when $z = 4$:

$$b_s @ z = 2 = .8379 + 2(.1985) = 1.2349$$
$$b_s @ z = 4 = .8379 + 4(.1985) = 1.6319$$

Unsurprisingly, the difference between the two simple slopes divided by the difference between the z values $= .1985$, which is the value associated with b_3. We get the same result when we calculate the simple slope change in z at every one-unit change in x:

$$b_s @ x = b_2 + zb_3 \qquad (9.8)$$
$$b_s @ x = 2 = -.0090 + 2(.1985) = .3880$$
$$b_s @ x = 4 = -.0090 + 4(.1985) = .7850$$

9.1.3.3 Lower-Order Effects

Now let's examine the lower-order effects shown in Table 9.3. With a cross-product term in the regression equation, b_1 represents the simple slope of x when $z = 0$ and b_2 represents the simple slope of z when $x = 0$. Because both variables have been centered around their mean, the mean of each variable is 0 and the lower-order coefficients indicate the simple slopes at the mean of each variable. In this case, we can see that x is a significant predictor of y at the mean of z, but z is not a significant predictor of y at the mean of x.

Mean-centering also allows us to interpret the lower-order coefficients as the average simple slope across all observed values. To illustrate, if we compute the simple slopes for each observed (deviate) score of x, we find that the average simple slope of $x = b_1$; similarly, if we compute the simple slopes for each observed (deviate) score of z, we find that the average simple slope of $z = b_2$. These calculations appear in Table 9.4.

Finally, lower-order effects are sometimes called "main effects," a term derived from a factorial analysis of variance (to be discussed in Chap. 12). There is, however, an important difference in the way the two terms are calculated. Because they give each score equal weight, lower-order effects in the presence of a cross-product term represent weighted averages. In contrast, the main effects in a factorial analysis of variance are ordinarily computed as unweighted averages, collapsing across cell size differences. We will have more to say about these differences in

---

[3] This method of calculating simple slopes differs from the one we used earlier. Previously, we generated fitted values from our overall equation and then found the simple slopes by subtracting one fitted value from another; here, we are directly calculating simple slopes by using a subset of terms from our original regression. The two approaches yield identical results.

Table 9.4 Mean-centered lower-order effects as weighted averages of simple slopes

| b_1 as average slope | | b_2 as average slope | |
dev_z	Simple slope of x	dev_x	Simple slope of z
−2	.4409	−4.0833	−.8196
2	1.2350	3.9167	.7686
−4	.0438	−.0833	−.0255
1	1.0365	−3.0833	−.6211
0	.8379	−.0833	−.0255
1	1.0365	.9167	.1730
4	1.6321	−1.0833	−.2240
−4	.0438	1.9167	.3715
−4	.0438	.9167	.1730
0	.8379	2.9167	.5701
2	1.2350	−3.0833	−.6211
4	1.6321	.9167	.1730
Average	**.8379**	**Average**	**−.0090**

Chap. 12; for now, we will simply note that it is preferable to refer to lower-order effects as lower-order effects, not main effects.

9.1.4   R Code: Testing a Cross-Product Term

```
x=c(1,9,5,2,5,6,4,7,6,8,2,6)
z=c(3,7,1,6,5,6,9,1,1,5,7,9)
y=c(2,9,6,1,5,6,2,3,5,8,2,6)

#Center Variables
dx=scale(x, center = TRUE, scale = FALSE)
dz=scale(z, center = TRUE, scale = FALSE)

#Regression with all three terms
mod <-lm(y~dx*dz)     #asterisk includes lower order terms
summary(mod)

#View model matrix and covariance matrix
X <-model.matrix(mod)
X
covar <-vcov(mod)
covar

#Simple slopes as Weighted Averages
simp.x <-mod$coef[2]+(dz*mod$coef[4])
simp.z <-mod$coef[3]+(dx*mod$coef[4])
simple <-cbind(simp.x,simp.z)
simple
mean(simple[,1])
mean(simple[,2])
```

9.2   Probing an Interaction

Knowing that a cross-product term is a significant predictor of a criterion tells us little about the nature of the interaction. To unravel its meaning, we need to examine its form. One common technique calculates and tests the simple slopes of one predictor at three values of another predictor—one standard deviation below the mean, at the mean, and one standard deviation above the mean. The following steps are involved:

1. Calculate predicted values of y at each of the 9 (3 × 3) locations.
2. Plot the predicted values and examine the form of the interaction.
3. Compute simple slopes at each location, calculate their standard errors, and test their significance against the null hypothesis that each slope equals 0.

9.2.1   Calculating Predicted Values

It is easy to calculate predicted values using simple algebra, but we will save time later if we learn how to do so using matrix algebra. The first step is to compute a matrix of weights, with each column representing one of the 9 (3 levels of x×3 levels of z) combinations. The weights for this matrix, which we will designate **P**, are shown in the left-hand side of Table 9.5. As you can see, we enter a "1" to model the intercept and then insert the relevant values for x, z, and xz. Using the standard deviations from Table 9.3 produces the nine entries in our **P** matrix.[4]

If we then multiply $\mathbf{P'b}$, we generate a column vector that displays the predicted values for all nine combinations (see Table 9.6).

Table 9.5 Coefficient weights for a **P** matrix of predicted values

		1 sd. below the mean of x			Mean of x			1 sd. above the mean of x		
Weight		$x_L z_L$	$x_L z_M$	$x_L z_H$	$x_M z_L$	$x_M z_M$	$x_M z_H$	$x_H z_L$	$x_H z_M$	$x_H z_H$
b_0	1	1	1	1	1	1	1	1	1	1
b_1	x	−2.4664	−2.4664	−2.4664	0	0	0	2.4664	2.4664	2.4664
b_2	z	−2.9233	0	2.9233	−2.9233	0	2.9233	−2.9233	0	2.9233
b_3	xz	7.2101	0	−7.2101	0	0	0	−7.2101	0	7.2101

---

[4] The subscripts indicate whether the value is one standard deviation below the mean (L), at the mean (M), or one standard deviation above the mean (H).

Table 9.6 Predicted values
at various combinations of
low, medium, and high scores
on two predictors

Cell combination	Predicted value
$x_L z_L$	4.0404
$x_L z_M$	2.5828
$x_L z_H$	1.1252
$x_M z_L$	4.6757
$x_M z_M$	4.6495
$x_M z_H$	4.6233
$x_H z_L$	5.3111
$x_H z_M$	6.7162
$x_H z_H$	8.1214

9.2.2   Plotting Predicted Values

Our next step is to plot the predicted values to observe the form of the interaction. The left-hand panel in Fig. 9.3 reveals the nature of the interaction when we plot the values of x at three levels of z. Looking at the figure, it appears that the weekend weather forecast has very little effect on mood when people do not expect to be able to get outside, a moderately positive effect on mood when people expect to be outside a bit, and a large positive effect on mood when people expect to spend a lot of time outdoors.

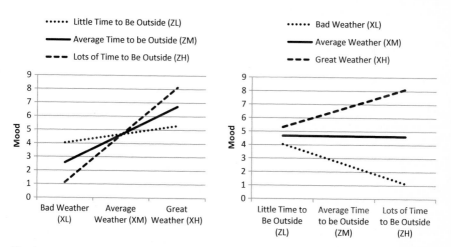

Fig. 9.3 Simple slope of x at three levels of z (*left-hand panel*) and simple slope of z at three levels of x (*right-hand panel*)

9.2.3   Testing Simple Slopes

Observing a trend is suggestive, but oftentimes we'd like to test the statistical significance of our simple slopes. Doing so involves dividing each simple slope by its standard error.

9.2.3.1   Computing Simple Slopes

To compute the simple slopes of the regression lines we see depicted in Fig. 9.3, we construct a new matrix we will designate $\mathbf{S}$.[5] The weights, which appear in Table 9.7, were generated using the same rules we used in Chap. 3 to find the first-order partial derivative of a function. This is no coincidence, because simple slopes *are* first-order partial derivatives [i.e., they represent the (near) instantaneous rate of change at a particular point on a curve holding other variables constant]:

- Enter a "0" for all effects that do not include the simple slope (because the derivative of a constant equals 0).
- Enter a "1" for the simple slope of interest (because the derivative of a derivative equals 1).
- Multiply all effects that include the simple slope by "1."

Table 9.7 Rules for constructing a matrix of weights for computing simple slopes, **S**

	Simple slope of x @ z	Simple slope of z @ x
b_0	0	0
b_1 (x)	1	0
b_2 (z)	0	1
b_3 (xz)	$1 * z$	$1 * x$

Table 9.8 shows the results when we enter values from our data set. For example, to model the simple slope of x when z is one standard deviation below the mean, we enter a "0" for the intercept (which will always be true), a "1" for x, and "0" for z. Finally, we multiply 1 * the value of z at which we seek the simple slope of x. Since the standard deviation of $z = 2.9233$, we want the value for -2.9233 (i.e., one standard deviation below the mean of 0). The remaining values in Table 9.8 were found in a comparable way. We will be making great use of these procedures throughout this text, so take the time now to be sure you understand how the values were generated.

Table 9.8 **S** matrix of weights for computing simple slopes

	x @ z_L	x @ z_M	x @ z_H	z @ x_L	z @ x_M	z @ x_H
b_0	0	0	0	0	0	0
$b_1(x)$	1	1	1	0	0	0
$b_2(z)$	0	0	0	1	1	1
$b_3(xz)$	−2.9233	0	2.9233	−2.4664	0	2.4664

---

[5] The designation of this matrix as **S** (for simple slopes) is arbitrary, and the matrix should not be confused with a sum of squares and cross-product matrix discussed in Chap. 1 or the smoother matrix discussed in Chap. 8.

If we then multiply $\mathbf{S'b}$, we create a column vector of simple slopes (see Table 9.9). Notice that the simple slope of x @ z_M equals b_1 from our original regression analysis and that the simple slope of z @ x_M equals b_2 from our original regression analysis. This is because the lower-order coefficients in our original analysis are simple slopes of one variable when the other variable equals zero (which, after centering our variables, now equals the mean).

Table 9.9 Simple slopes at three levels

Term	s_b
x @ z_L	.2576
x @ z_M	.8379
x @ z_H	1.4183
z @ x_L	−.4986
z @ x_M	−.0090
z @ x_H	.4807

9.2.3.2   Calculating Standard Errors of the Simple Slopes

Our next task is to calculate the standard errors of the simple slopes. The algebraic formula for doing so is shown below, where c = entries from the parameter covariance matrix and z refers to the level of z for which we seek the simple slope of x:

$$se_{b_s} = \sqrt{c_{ii} + 2zc_{ij} + z^2 c_{jj}} \tag{9.9}$$

To illustrate, we will calculate the standard error for the simple slope of x when z is one standard deviation below the mean:

$$se_{b_s} = \sqrt{c_{22} + 2zc_{23} + z^2 c_{33}}$$

Inserting the relevant values from Table 9.3 produces the standard error:

$$se_{b_s} = \sqrt{.017897 + (2 * -2.9233 * -.001168) + \left(-2.9233^2 * .004386\right)}$$
$$= \sqrt{.0622} = .2494$$

We could continue using this formula to compute each of our standard errors, but it's easier to use our $\mathbf{S}$ matrix. If we calculate $\mathbf{S'CS}$, the standard errors of the simple slopes can be found by taking the square root of the diagonal elements of the product matrix (see highlighted portion of Table 9.10).

Table 9.10 Parameter covariance matrix for simple slopes

	x @ z_L	x @ z_M	x @ z_H	z @ x_L	z @ x_M	z @ x_H
x @ z_L	.062205	.021312	−.019581	.039304	.004801	−.029702
x @ z_M	.021312	.017897	.014481	.003968	.001087	−.001795
x @ z_H	−.019581	.014481	.048544	−.031367	−.002627	.026112
z @ x_L	.039304	.003968	−.031367	.045834	.016021	−.013793
z @ x_M	.004801	.001087	−.002627	.016021	.012887	.009753
z @ x_H	−.029702	−.001795	.026112	−.013793	.009753	.033299

Standard errors are found by taking the square root of the diagonal entries

9.2.3.3   Testing the Statistical Significance of Simple Slopes

If we then divide each simple slope by its corresponding standard error, we derive a
t-statistic we can use to test each slope's statistical significance. We can also compute
confidence intervals by applying Eq. (2.31). The first three rows in Table 9.11
provide the statistical tests for the regression lines shown in the left-hand side of
Fig. 9.3. The probability values confirm our earlier impression: the weather forecast
has little effect on people who do not expect to be outdoors (b_s=.2576, p=.3319),
but a sizable effect among people who expect to spend some time outdoors
(b_s=.8379, p=.0002) or a lot of time outdoors (b_s = 1.4183, p = .0002). To characterize these findings, we might say that the weekend weather forecast is only
important when people expect to spend at least some time outdoors. Note, also,
that the probability levels for the last two simple slopes are approximately equal
(both ps = .0002), suggesting that it might not matter whether one is expecting to
spend a lot of time outdoors or only a little time outdoors. We will have more to say
about this (apparent, but illusory) equivalence in a moment.

Table 9.11 Simple slope tests of significance

Term	b_s	se_b	t	p	$CI-$	$CI+$
x @ z_L	.2576	.2494	1.0328	.3319	−.3176	.8327
x @ z_M	.8379	.1338	6.2636	.0002	.5294	1.1464
x @ z_H	1.4183	.2203	6.4372	.0002	.9102	1.9264
z @ x_L	−.4986	.2141	−2.3291	.0482	−.9923	−.0049
z @ x_M	−.0090	.1135	−.0790	.9390	−.2707	.2528
z @ x_H	.4807	.1825	2.6342	.0300	.0599	.9015

Although interactions are symmetrical, their interpretation is not. We can appreciate this point by considering the simple slopes of z at various levels of x (see the
last three rows of Table 9.11 and the right-hand side of Fig. 9.3). Here, we see that
time available to spend outdoors negatively predicts excitement for the weekend
when the weather forecast is bad (b_s=−.4986, p= .0482), has no effect on
excitement for the weekend when the weather forecast is average
(b_s =−.0090, p= .9390), and has a positive effect on excitement for the weekend

when the weather forecast is good ($b_s=.4807$, $p=.0300$). In short, whether spending time outside is good or bad depends entirely on whether the weather is expected to be good or bad: If the weather forecast is good, more time outside is good; if the weather forecast is bad, more time outside is bad.

9.2.3.4   Comparing Simple Slopes

Earlier, we noted that the probability level for the simple slope of x @ z_M is nearly equivalent to the probability level for the simple slope of x @ z_H ($p=.0002$ for both simple slopes). Given this equivalence, we might reasonably expect that if we compared the two slopes, we would find that they are not statistically different from each other. We can test the difference between any two simple slopes using the following formula, where s refers to an entry in our $\mathbf{S'CS}$ matrix shown in Table 9.10:

$$t = \frac{b_{s1} - b_{s2}}{\sqrt{s_{ii} + s_{jj} - 2s_{ij}}} \tag{9.10}$$

Inserting our values, we learn something interesting:

$$t = \frac{.8379 - 1.4183}{\sqrt{.017897 + .048544 - (2*.014481)}} = \frac{-.5804}{.1936} = -2.9978,\ p = .0171$$

Why is this interesting? Even though the two simple slopes are "equally probable," the simple slope of x @ z_M is significantly smaller than the simple slope of x @ z_H. If you look back at our original regression coefficients in Table 9.3, you will understand why. Notice that the t and p values for the xz product term match the absolute values of the t and p values from our test of slope differences. This is because a cross-product coefficient in the original table is a test of the difference between *any* two simple slopes. So no matter which simple slopes we compare, the statistical significance of their difference will always equal the probability level associated with our cross-product term.

To reinforce this point, let's test whether the simple slope of z @ x_L differs from the simple slope of z @ x_H. Notice that this is now a test of the simple slopes in the right-hand panel in Fig. 9.3 and that the slopes are now farther apart than they were in the preceding comparison. Nonetheless, the test of differences yields the same (absolute) t value as the original cross-product term, and this will be true for any two values we select, regardless of whether they are very close together or very far apart:

$$t = \frac{-.4986 - .4807}{\sqrt{.045834 + .033299 - (2*-.013793)}} = \frac{-.9793}{.3267} = -2.9978,\ p = .0171$$

9.2.4   *Characterizing an Interaction*

By definition, nonparallel lines eventually cross, so simple slopes always cross whenever a significant interaction is present. Whether they cross within the range of the data we have collected, however, is not guaranteed. If they do, the interaction is said to be disordinal; if they do not, the interaction is termed ordinal.

The following formula can be used to calculate the crossing point (*cp*) for the simple slopes of *x*:

$$cp_x = \frac{-b_2}{b_3} \tag{9.11}$$

Inserting values from Table 9.3 yields the crossing point:

$$cp_x = \frac{.0090}{.1985} = .0453$$

This value indicates that the simple slopes of *x* cross when $dev_x = .0453$. Glancing back to Table 9.2, we can see that the deviate values for *x* range from -4.0833 to $+3.9167$. A crossing point of .0453 falls well within this range, indicating that our interaction is disordinal.

We compute the crossing point of the simple slopes of *z* in a comparable manner,

$$cp_z = \frac{-b_1}{b_3} \tag{9.12}$$

and inserting values from Table 9.3 yields the crossing point:

$$cp_z = \frac{-.8379}{.1985} = -4.2207$$

This values indicates that the simple slopes of *z* cross when $dev_z = -4.2207$. Table 9.2 shows that this value lies outside the range of scores we observed (dev_z range $= -4$ to $+2$), so we conclude that the interaction is ordinal.

If you're confused, you're not alone. What we see here is that the same interaction can be classified as disordinal or ordinal depending on how it is plotted. For this reason, I recommend describing an interaction as taking one of three forms:

1. If one of the simple slopes is not significant, we have an "only for" interaction (e.g., weekend weather is a significant predictor of mood *only* if you are going to be outside at least some of the time).
2. If all simple slopes are significant in the same direction, we have an "especially for" interaction (e.g., weekend weather is a significant predictor of mood, *especially* when you are going to be spending a lot of time outdoors).
3. If at least two of the simple slopes are significant but of opposite sign, we have a crossover interaction (e.g., time outdoors negatively predicts mood when the weather is bad, but positively predicts mood when the weather is good).

Applying these rules, we would classify our interaction as an "only for" interaction when describing how time spent outdoors moderates the effects of weather
and a crossover interaction when describing how weather moderates the effects of
time spent outdoors.

9.2.5   R Code: Predicted Values and Simple Slopes

```
x=c(1,9,5,2,5,6,4,7,6,8,2,6)
z=c(3,7,1,6,5,6,9,1,1,5,7,9)
y=c(2,9,6,1,5,6,2,3,5,8,2,6)
dx=scale(x, center = TRUE, scale = FALSE)
dz=scale(z, center = TRUE, scale = FALSE)
mod <-lm(y~dx*dz)    #asterisk includes lower order terms

#Construct P Matrix for Predicted Values
p0 <-rep(1,9);p1 <-c(rep(-sd(x),3),rep(0,3),rep(sd(x),3))
p2 <-rep(c(-sd(z),0,sd(z)),3);p3 <-p1*p2
P <-round(rbind(p0,p1,p2,p3),digits=5);P
pred.val <-t(P)%*%coef(mod)
dimnames(pred.val)=list(c("lo.x/lo.z","lo.x/med.z","lo.x/hi.
z","med.x/lo.z","med.x/med.z","med.x/hi.z","hi.x/lo.z","hi.x/med.
z","hi.x/hi.z"))
pred.val

#Plot Predicted Values 1
byrow <-rbind(c(pred.val[1:3]),c(pred.val[4:6]),c(pred.val[7:9]))
matplot((byrow), main = "Simple Slopes of Weather at Three Levels of Free
Time", type="l",ylab = "Mood", xlab = "Weather",lwd=2)
legend("topleft",legend=c("Low Free Time","Average Free Time","Lots
of Free Time"),
lty=1,lwd=2,pch=21,col=c("black","red","darkgreen"),
ncol=1,bty="n",cex=0.8,
text.col=c("black","red","darkgreen"),
inset=0.01)
dev.new()
#Plot Predicted Values 2
bycol <-cbind(c(pred.val[1:3]),c(pred.val[4:6]),c(pred.val[7:9]))
matplot((bycol), main = "Simple Slopes of Free Time at Three Levels of
Weather", type="l",ylab = "Mood", xlab = "Free Time")
legend("topleft",legend=c("Lousy Weather","Average Weather","Great
Weather"),
lty=1,lwd=2,pch=21,col=c("black","red","darkgreen"),
ncol=1,bty="n",cex=0.8,
text.col=c("black","red","darkgreen"),
inset=0.01)
```

(continued)

9.2.5   R Code: Predicted Values and Simple Slopes (continued)

```
#Construct S Matrix for Simple Slopes and Simple Standard Errors
s0 <-rep(0,6);s1 <-c(rep(1,3),rep(0,3));s2 <-c(rep(0,3),rep(1,3))
s3 <-c(-sd(z),0,sd(z),-sd(x),0,sd(x))
S <-round(rbind(s0,s1,s2,s3),digits=5);S
simp.slope <-t(S)%*%coef(mod)
simp.cov <-t(S)%*%vcov(mod)%*%S
simp.err <-sqrt(diag(simp.cov))
simples <-simp.slope/sqrt(diag(simp.cov))
tvalues <-2*pt(-abs(simples),df=(length(x)-nrow(S)))
crit <-abs(qt(0.025,(length(x)-nrow(S))))
CI.low <-simp.slope-(crit*simp.err)
CI.high <-simp.slope+(crit*simp.err)
simp.table<-round(matrix(c(simp.slope,simp.err,simples,tvalues,
CI.low,CI.high),nrow=6,ncol=6),digits=5)
dimnames(simp.table)=list(c("x@z.low","x@z.med","x@z.high","z@x.
low","z@x.med","z@x.high"), c("slope","stderr","t","p","CI.low",
"CI.high"))
simp.table
```

9.2.6   Johnson-Neyman Technique

Before concluding this section, we will examine an alternative approach to probing an interaction called the Johnson-Neyman technique. Instead of computing the significance of a simple slope of x at selected values of z, this approach identifies the range of z values for which the simple slope of x will be significant (and likewise for the simple slope of z). Calculating these "regions of significance" isn't difficult, but there are quite a few steps. Before describing them, let's review the formula we use to test the significance of a simple slope [see Eq. (9.9)], using the simple slope of x @ z for illustrative purposes

$$t = \frac{b_1 + b_3 z}{\sqrt{c_{11} + 2zc_{13} + z^2 c_{33}}}$$

Using the Johnson-Neyman technique, we solve the equation to find values of z that return a significant value of t. Once we do, we have identified the range of values of z for which the simple slope of x will be significant. The following steps solve the equation:

- Cross multiply to set up equality:

$$t * \sqrt{c_{11} + 2zc_{13} + z^2 c_{33}} = b_1 + b_3 z$$

- Square all terms to eliminate square root sign:

$$\left[t^2 * \left(c_{11} + 2zc_{13} + z^2 c_{33}\right)\right] = (b_1 + b_3 z)(b_1 + b_3 z)$$

- Multiply and rearrange terms to $= 0$:

$$\left[t^2 * \left(c_{11} + 2zc_{13} + z^2 c_{33}\right)\right] - \left(b_1^2 + 2b_1 b_3 z + b_3^2 z^2\right) = 0$$

- Collect like terms using quadratic equation: $Az^2 + Bz + C$:

$$\begin{aligned} A &= \left(t^2 * c_{33}\right) - b_3^2 \\ B &= 2 * \left[\left(t^2 * c_{13}\right) - (b_1 * b_3)\right] \\ C &= \left(t^2 * c_{11}\right) - b_1^2 \end{aligned} \tag{9.13}$$

- Find the critical t value for our sample size and desired level of significance. In our example, with 8 degrees of freedom and a desired .05 level of significance, the critical value of $t = 2.306$.
- Calculate terms and insert returned values into the quadratic formula to find the boundaries of the regions of significance (rs):

$$rs = \frac{-B \pm \sqrt{B^2 - 4AC}}{2A} \tag{9.14}$$

Inserting values from Table 9.3 into the various equations produces the range of z for which the simple slope of x is significant:

$$\begin{aligned} A &= \left(2.306^2 * .004386\right) - .1985^2 = -.0161 \\ B &= 2 * \left[\left(2.306^2 * -.001168\right) - (.8379 * .1985)\right] = -.3451 \\ C &= \left(2.306^2 * .017897\right) - .8379^2 = -.6069 \end{aligned}$$

and

$$rs = \frac{-(-.3451) \pm \sqrt{-.3451^2 - 4(-.0161)(-.6069)}}{2(-.0161)}$$

Solving the equation yields the lower (-19.5019) and upper boundary (-1.9329) of the region of significance.[6] These values indicate that the simple slope of x is significant when dev_z is less than -19.5019 (which falls well outside the data we observed), or greater than -1.9329, but not when dev_z falls between these two

---

[6] Due to rounding error, these values differ a bit from the ones you will find if you use the **R** code that accompanies this section.

values. Because 0 does not fall between these two values, these calculations are consistent with the fact that b_1 was significant in our original regression analysis.

We can perform similar calculations to find the regions of significance for the simple slope of z:

$$A = \left(2.306^2 * .004386\right) - .1985^2 = -.0161$$
$$B = 2 * \left[\left(2.306^2 * -.001271\right) - (-.0090 * .1985)\right] = -.0099$$
$$C = \left(2.306^2 * .012887\right) - .0090^2 = .0684$$

and

$$rs = \frac{-(-.0099) \pm \sqrt{-.0099^2 - 4(-.0161)(.0684)}}{2(-.0161)}$$

Here, we find that the simple slope of z will be significant when dev_x is less than -2.3914 or greater than 1.7765. Because 0 falls between these two values, these calculations match our finding that b_2 was not significant in our original regression analysis.

9.2.7　R Code: Johnson-Neyman Regions of Significance

```
#Johnson-Neyman Regions of Significance
x=c(1,9,5,2,5,6,4,7,6,8,2,6)
z=c(3,7,1,6,5,6,9,1,1,5,7,9)
y=c(2,9,6,1,5,6,2,3,5,8,2,6)
dx=scale(x, center = TRUE, scale = FALSE);
dz=scale(z, center = TRUE, scale = FALSE)

#Function for 2-way regions --enter 1 for x; 2 for z
JN <-function(simple){
    mod <-lm(y~dx*dz)
      coef <-mod$coef[2:4]
      cov <-vcov(mod)[2:4,2:4]
    df <-length(y)-4
      t.crit <-abs(qt(.025,df))
      A <-(t.crit^2*cov[3,3])-coef[3]^2
      B <-2*((t.crit^2*cov[simple,3]-(coef[simple]*coef[3])))
      C <-(t.crit^2*cov[simple,simple])-coef[simple]^2
      lower <-(-B+sqrt(B^2-4*A*C))/(2*A)
      upper <-(-B-sqrt(B^2-4*A*C))/(2*A)
      return(cbind(lower,upper))
}
JN(1)
JN(2)
```

9.3 Higher-Order Interactions

So far we have learned how to test an interaction using only two predictors, but higher-order interactions can also be tested. To illustrate, suppose a researcher decides to expand upon our study of weekend weather and time outdoors as predictors of upcoming mood. The researcher believes that the effects we reported apply only to cities that generally have bad weather (e.g., Seattle), not ones that usually have good weather (e.g., San Diego). After all, if the weather is normally good, one bad weekend might not affect one's mood very much. To test her ideas, the researcher repeats our earlier study, adding another variable, typical weather (w), coded on a 1–9 scale. Equation (9.15) shows the regression model, and Table 9.12 displays the (bogus) data. The last two rows of the table provide the mean and standard deviation of each variable:

$$\hat{y} = b_0 + b_1 x + b_2 z + b_3 w + b_4 xz + b_5 xw + b_6 zw + b_7 xzw \tag{9.15}$$

Notice that now, there are four cross-product terms: three two-way interactions and one three-way interaction. The two-way interactions are formed by multiplying two of the deviate scores, and the three-way interaction is found by multiplying all three deviate scores. As before, we do not center the criterion or the cross-product terms after forming them.

9.3.1 Testing the Regression Equation

Table 9.13 shows the results of a regression in which all predictors are entered at once. As you can see, collectively, our seven predictors account for a significant amount of the variance in our criterion. Because lower-order coefficients represent conditional slopes, we concern ourselves only with the three-way interaction represented by b_7. Supporting the researcher's intuition, the interaction is statistically significant, uniquely explaining ~10 % of the variance in our criterion.

9.3.2 Probing a Three-Variable Interaction

Having found a significant three-variable interaction, we must now probe its form. Insofar as the researcher predicted that the xz interaction we observed earlier depends on a city's usual weather, we will focus on the simple slopes of x and z when w is low and high.

Table 9.12 Small sample example with three predictors and four cross-product terms

x (weather forecast)	z (time to spend outside)	w (months of nice weather)	y	dev_x	dev_z	dev_w	xz	xw	zw	xzw
1	3	7	3	-3.6667	-2.00	2.5833	7.3333	-9.4722	-5.1667	18.9444
8	7	4	7	3.3333	2.00	-.4167	6.6667	-1.3889	-.8333	-2.7778
2	1	3	8	-2.6667	-4.00	-1.4167	10.6667	3.7778	5.6667	-15.1111
2	6	1	3	-2.6667	1.00	-3.4167	-2.6667	9.1111	-3.4167	9.1111
5	6	8	4	.3333	1.00	3.5833	.3333	1.1944	3.5833	1.1944
6	6	8	5	1.3333	1.00	3.5833	1.3333	4.7778	3.5833	4.7778
5	7	5	4	.3333	2.00	.5833	.6667	.1944	1.1667	.3889
7	2	1	2	2.3333	-3.00	-3.4167	-7.00	-7.9722	10.2500	23.9167
4	3	8	5	-.6667	-2.00	3.5833	1.3333	-2.3889	-7.1667	4.7778
8	6	1	9	3.3333	1.00	-3.4167	3.3333	-11.3889	-3.4167	-11.3889
3	5	5	3	-1.6667	0.00	.5833	0.00	-.9722	0.00	0.00
5	8	2	6	.3333	3.00	-2.4167	1.00	-.8056	-7.2500	-2.4167
Mean 4.6667	5.00	4.4167	4.9167	0.00	0.00	0.00	1.9167	-1.2778	-.2500	2.6181
SD 2.3484	2.2156	2.8431	2.1933	2.3484	2.2156	2.8431	4.6734	5.9863	5.3778	11.0803

Table 9.13 Regression analysis for higher-order interactions example

	SS	df	MS	R^2	F	p	
			Significance test of regression model				
Regression	50.788	7	7.2554	.9598	13.6310	.0119	
Residuals	2.1291	4	.5323				
Total	52.917						

	b	se_b	ΔR^2	t	p	
			Regression coefficients			
b_0	4.8317	.2863		16.8776	.0001	
b_1	.5620	.1601	.1239	3.5093	.0247	
b_2	−.3182	.1637	.0380	−1.9434	.1239	
b_3	−.1317	.0858	.0237	−1.5346	.1997	
b_4	.1978	.0743	.0714	2.6642	.0561	
b_5	.0485	.0549	.0079	.8836	.4268	
b_6	−.1278	.0600	.0457	−2.1319	.1000	
b_7	−.1009	.0316	.1028	−3.1966	.0330	

Covariance matrix

.081956	−.004627	−.009665	.005888	−.013429	−.000656	−.001539	−.004997
−.004627	.025645	−.016258	−.000399	.003208	.006005	−.005667	.001809
−.009665	−.016258	.026807	−.000779	.002242	−.004601	.006611	.000436
.005888	−.000399	−.000779	.007367	−.002323	−.000824	.000155	−.000936
−.013429	.003208	.002242	−.002323	.005514	.001163	.000397	.001698
−.000656	.006005	−.004601	−.000824	.001163	.003018	−.001676	.000712
−.001539	−.005667	.006611	.000155	.000397	−.001676	.003594	−.000177
−.004997	.001809	.000436	−.000936	.001698	.000712	−.000177	.000996

9.3.2.1   Calculating and Plotting Predicted Values

Our first step is to calculate and plot predicted values, again using values one standard deviation below and above the mean. Table 9.14 shows the **P** matrix we use, and the predicted values generated by multiplying **P′b** are displayed in Fig. 9.4.[7]

---

[7] Table 9.14 does not report the predicted values when the variables are at their means. These values were derived in Fig. 9.5 by simple arithmetic $\left(\text{mean} = \dfrac{(\text{low} + \text{high})}{2} \right)$.

Table 9.14 **P** matrix for calculating predicted values of x and z at two levels of w

			w_L				w_H	
	x_Lz_L	x_Lz_H	x_Hz_L	x_Hz_H	x_Lz_L	x_Lz_H	x_Hz_L	x_Hz_H
b_0	1	1	1	1	1	1	1	1
$b_1(x)$	−2.3484	−2.3484	2.3484	2.3484	−2.3484	−2.3484	2.3484	2.3484
$b_2(z)$	−2.2156	2.2156	−2.2156	2.2156	−2.2156	2.2156	−2.2156	2.2156
$b_3(w)$	−2.8431	−2.8431	−2.8431	−2.8431	2.8431	2.8431	2.8431	2.8431
$b_4(xz)$	5.2033	−5.2033	−5.2033	5.2033	5.2033	−5.2033	−5.2033	5.2033
$b_5(xw)$	6.6769	6.6769	−6.6769	−6.6769	−6.6769	−6.6769	6.6769	6.6769
$b_6(zw)$	6.2994	−6.2994	6.2994	−6.2994	−6.2994	6.2994	−6.2994	6.2994
$b_7(xzw)$	−14.7936	14.7936	14.7936	−14.7936	14.7936	−14.7936	−14.7936	14.7936

Looking over the figure, it seems that our researcher was right: weather forecast and time outdoors interact in cities with generally bad weather (left-hand panel in Fig. 9.4), but not in cities with generally good weather (right-hand panel in Fig. 9.4).

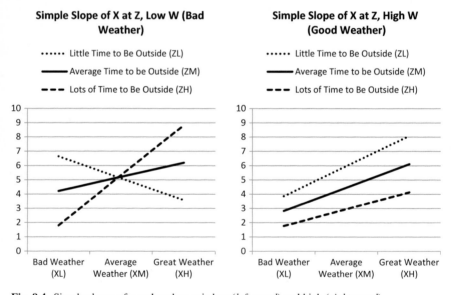

Fig. 9.4 Simple slopes of x and z when w is low (*left panel*) and high (*right panel*)

9.3.2.2 Calculating and Testing Simple Slopes

Our next step is to calculate simple slopes and test their significance using their associated standard errors. As before, we differentiate our function and create an **S** matrix using the following rules:

- Enter a "0" for all effects that do not include the simple slope.
- Enter a "1" for the simple slope of interest.
- Multiply all effects that include the simple slope by "1."

Table 9.15 Matrix weights for calculating simple slopes of x, z, and xz at w

	Weights			Illustration using simple slopes of x and z @ w_L						
	xz	$x @ z$	$z @ x$	xz	$x @ z_L$	$x @ z_M$	$x @ z_H$	$z @ x_L$	$z @ x_M$	$z @ x_H$
b_0	0	0	0	0	0	0	0	0	0	0
$b_1(x)$	0	1	0	0	1	1	1	0	0	0
$b_2(z)$	0	0	1	0	0	0	0	1	1	1
$b_3(w)$	0	0	0	0	0	0	0	0	0	0
$b_4(xz)$	1	z	x	1	−2.2156	0	2.2156	−2.3484	0	2.3484
$b_5(xw)$	0	w	0	0	−2.8431	−2.8431	−2.8431	0	0	0
$b_6(zw)$	0	0	w	0	0	0	0	−2.8431	−2.8431	−2.8431
$b_7(xzw)$	w	zw	xw	−2.8431	6.2994	0	−6.2994	6.67689	0	−6.6769

Applying these rules generates the **S** matrix shown in the left-hand portion of Table 9.15. The right-hand portion of the table illustrates their application when w is one standard deviation below its mean. Because the researcher had predicted that the xz interaction would be significant when w was low but not when w was high, the first column represents the simple xz cross-product term.

After computing the simple slopes using **S′b** and finding the standard errors by taking the square root of the diagonal entries of **S′CS**, we construct t-tests. In accordance with predictions, the top portion of Table 9.16 shows that the simple xz interaction is significant when w is low ($b_s = .4847$, $p = .0015$). Additional tests showed that the simple slope of weather is negative for people who can't get outside very much ($b_s = −.6500$, $p = .0230$), but positive for people who expect to get

Table 9.16 Simple slopes of x, z, and xz at two levels of w

	Simple slopes @ w_L					
	sb	se_b	t	p	CI_{low}	CI_{high}
xz	.4847	.0626	7.7490	.0015	.3110	.6584
$x @ z_L$	−.6500	.1812	−3.5875	.0230	−1.1530	−.1469
$x @ z_M$	.4240	.1261	3.3630	.0282	.0739	.7740
$x @ z_H$	1.4979	.1933	7.7478	.0015	.9611	2.0347
$z @ x_L$	−1.0931	.2172	−5.0332	.0073	−1.6961	−.4901
$z @ x_M$	.0452	.1351	.3344	.7548	−.3300	.4204
$z @ x_H$	1.1835	.1803	6.5627	.0028	.6828	1.6842
	Simple slopes @ w_H					
	sb	se_b	t	p	CI_{low}	CI_{high}
xz	−.0890	.1524	−.5842	.5904	−.5121	.3341
$x @ z_L$	.8972	.3479	2.5790	.0614	−.0687	1.8632
$x @ z_M$	.7000	.2901	2.4125	.0734	−.1056	1.5056
$x @ z_H$	.5027	.5247	.9581	.3923	−.9542	1.9596
$z @ x_L$	−.4725	.4545	−1.0395	.3573	−1.7345	.7895
$z @ x_M$	−.6816	.3057	−2.2295	.0897	−1.5303	.1672
$z @ x_H$	−.8907	.4863	−1.8316	.1410	−2.2408	.4595

outside a bit ($b_s=.4240$, $p=.0282$), or a lot ($b_s = 1.4979$, $p=.0015$). These findings support the claim that weekend weather and time to spend outdoors matter a lot when you live in a gloomy climate with only a few nice weekends a year. In fact, the situation is so dire that people who cannot get outside on a nice weekend feel sad.[8]

The bottom portion of Table 9.16 shows the simple slopes for cities that generally have good weather.[9] As you can see, none of the simple slopes (including the xz interaction) is significant. Thus, when the weather is generally good, weekend weather and time outdoors have very little effect on mood.

9.3.2.3   Comparing Simple Slopes

Earlier, you learned how to test whether two simple slopes are significantly different. With a two-way interaction, there really wasn't much reason to do so because all comparisons have the same probability value as the cross-product term. This is not true with a three-way interaction, so it will be useful to review the steps we take to compare simple slopes. To illustrate, Table 9.17 presents an **S** matrix we will use to compare the simple slope of x at various combinations of w and z.

Table 9.17   **S** matrix for comparing simple slopes of x at various combinations of z and w

	$z_L w_L$ Little time to spend outside in a bad weather city	$z_L w_H$ Little time to spend outside in a good weather city	$z_H w_L$ Lots of time to spend outside in a bad weather city	$z_H w_H$ Lots of time to spend outside in a good weather city
b_0	0	0	0	0
$b_1(x)$	1	1	1	1
$b_2(z)$	0	0	0	0
$b_3(w)$	0	0	0	0
$b_4(xz)$	−2.2156	−2.2156	2.2156	2.2156
$b_5(xw)$	−2.8431	2.8431	−2.8431	2.8431
$b_6(zw)$	0	0	0	0
$b_7(xzw)$	6.2994	−6.2994	−6.2994	6.2994

After calculating simple slopes (**S′b**) and their associated standard errors by taking the square root of the diagonal entries of **S′CS**,

$$\mathbf{S'b} = \begin{bmatrix} -.6500 \\ .8972 \\ 1.4979 \\ .5027 \end{bmatrix} \quad \mathbf{S'CS} = \begin{bmatrix} .032825 & .000074 & -.003314 & .021850 \\ .000074 & .121033 & .005582 & -.029820 \\ -.003314 & .005582 & .037377 & -.022503 \\ .021850 & -.029820 & -.022503 & .275345 \end{bmatrix}$$

---

[8] Living in Seattle, I can attest to the [fabricated] effect!

[9] I leave it as an exercise for you to construct an **S** matrix to derive the values shown in the bottom portion of Table 9.16.

Table 9.18 Comparing simple slopes of x at various combinations of z and w

Group Comparisons		t	p
1 v 2	Little time to spend outside in a bad weather city vs. little time to spend outside in a good weather city	−3.9463	.0169
1 v 3	Little time to spend outside in a bad weather city vs. lots of time to spend outside in a bad weather city	−7.7490	.0015
1 v 4	Little time to spend outside in a bad weather city vs. lots of time to spend outside in a good weather city	−2.2414	.0885
2 v 3	Little time to spend outside in a good weather city vs. lots of time to spend outside in a bad weather city	−1.5653	.1926
2 v 4	Little time to spend outside in a good weather city vs. lots of time to spend outside in a good weather city	.5842	.5904
3 v 4	Lots of time to spend outside in a bad weather city vs. lots of time to spend outside in a good weather city	1.6639	.1715

we use the following formula, first presented in Eq. (9.10), to compare any two simple slopes, where s refers to values in $\mathbf{S'CS}$:

$$t = \frac{b_{s1} - b_{s2}}{\sqrt{s_{ii} + s_{jj} - 2s_{ij}}}$$

To illustrate, we will compare the simple slope of x when one has little time to spend outside in a city with generally bad weather ($x @ z_L w_L$) vs. the simple slope of x when one has little time to spend outside in a city with generally good weather ($x @ z_L w_H$):

$$t = \frac{-.6500 - .8972}{\sqrt{.032825 + .121033 - (2 * .000074)}} = \frac{-1.5472}{.3921} = -3.9463, \ p = \ .0169$$

The comparison is significant, supporting the following conclusion: when people have little time to spend outside, the simple slope of x is significantly different for cities with bad weather ($b_s = -.6500$) than for cities with good weather ($b_s = .8972$), $t(8) = 3.9463, p = .0169$. Table 9.18 presents the remaining comparisons. It is a good idea to test your understanding of how the terms were calculated and practice interpreting the [sham] effects in your own words.

9.3.2.4 Other Ways to Probe a Three-Way Interaction

In our example, we probed the simple slopes of x and z at two levels of w. But we could have probed the simple slopes of z and w at two levels of x or the simple slopes of x and w at two levels of z. Only the weights we use to construct the $\mathbf{S}$ matrix will change with these variations. Table 9.19 presents these weights, and it is a good idea to carefully study them because we will be computing and testing simple slopes throughout the rest of this book.

Table 9.19 S matrix weights for calculating simple slopes

	zw	$z \, @ \, w$	$w \, @ \, z$	xw	$x \, @ \, w$	$w \, @ \, x$
b_0	0	0	0	0	0	0
$b_1(x)$	0	0	0	0	1	0
$b_2(z)$	0	1	0	0	0	0
$b_3(w)$	0	0	0	0	0	1
$b_4(xz)$	0	x	0	0	z	0
$b_5(xw)$	0	0	x	1	w	x
$b_6(zw)$	1	w	z	0	0	z
$b_7(xzw)$	x	xw	xz	z	zw	xz

9.3.3 R Code: Three-Way Interaction

```
x=c(1,8,2,2,5,6,5,7,4,8,3,5)
z=c(3,7,1,6,6,6,7,2,3,6,5,8)
w=c(7,4,3,1,8,8,5,1,8,1,5,2)
y=c(3,7,8,3,4,5,4,2,5,9,3,6)
dx=scale(x, center = TRUE, scale = FALSE)
dz=scale(z, center = TRUE, scale = FALSE)
dw=scale(w, center = TRUE, scale = FALSE)
mod <-lm(y~dx*dz*dw)

#Function for Predicted Values
pp <-function(a,b,c){
p0 <-rep(1,8)
p1<-c(rep(c(rep(-sd(a),2),rep(sd(a),2)),2))
p2 <-c(rep(c(-sd(c),sd(c)),4))
p3 <-c(rep(-sd(b),4),rep(sd(b),4))
p4 <-p1*p2
p5 <-p1*p3
p6 <-p2*p3
p7 <-p1*p2*p3
P <-round(rbind(p0,p1,p2,p3,p4,p5,p6,p7),digits=5)
pred.val <-t(P)%*%coef(mod)
dimnames(pred.val)=list(c("lo.x,lo.z,lo.w","lo.x,hi.z,lo.w","hi.x,
lo.z,lo.w","hi.x,hi.z,lo.w","lo.x,lo.z,hi.w","lo.x,hi.z,hi.w","hi.
x,lo.z,hi.w","hi.x,hi.z,hi.w"))
return(list(P,pred.val))
}
predicted <-pp(x,w,z)
predicted

#Function for Simple Slopes Tests
ss <-function(a,b,c,d){
s0 <-rep(0,7)
```

(continued)

9.3.3 R Code: Three-Way Interaction (continued)

```
s1 <-c(0,rep(1,3),rep(0,3));s2 <-c(0,rep(0,3),rep(1,3))
s3 <-rep(0,7);s4 <-c(1,-sd(b),0,sd(b),-sd(a),0,sd(a))
s5 <-c(0,rep(d*sd(c),3),rep(0,3));s6 <-c(0,rep(0,3),rep(d*sd(c),3))
s7 <-c(d*sd(c),-sd(b)*d*sd(c),0,sd(b)*d*sd(c),-sd(a)*d*sd(c),0,
sd(a)*d*sd(c))
S <-round(rbind(s0,s1,s2,s3,s4,s5,s6,s7),digits=5)
simp.slope <-t(S)%*%coef(mod)
simp.err <-sqrt(diag(t(S)%*%vcov(mod)%*%S))
ttests <-simp.slope/simp.err
pvalues <-2*pt(-abs(ttests),df=(length(x)-nrow(S)))
crit <-abs(qt(0.025, df=(length(x)-nrow(S))))
CI.low <-simp.slope-(crit*simp.err)
CI.high <-simp.slope+(crit*simp.err)
simp.table<-round(matrix(c(simp.slope,simp.err,ttests,pvalues,
CI.low,CI.high),nrow=7,ncol=6),digits=5)
dimnames(simp.table)=list(c("xz@low","x@z.low","x@z.med","x@z.
high","z@x.low","z@x.med","z@x.high"), c("slope","stderr","t","p",
"CI.low","CI.high"))
return(list(S,simp.table))
}


#Simple Slopes Tests (first two variables define simple slopes at third
#variable; specify -1 for @ low and 1 for @ high
simple <-ss(x,z,w,-1)
simple

#Compare any two simple slopes
simptest =function(a,b,c,d,e,f) #use 0 for effect of interest, 1 or -1
#for high or low
{
s1<-c(0, 1, 0, 0, b*sd(z), c*sd(w), 0, b*sd(z)*c*sd(w))
s2<-c(0, 0, 1, 0, a*sd(x), 0, c*sd(w), a*sd(x)*c*sd(w))
s3<-c(0, 0, 0, 1, 0, b*sd(z), c*sd(w), a*sd(x)*b*sd(z))
if (a == 0) {slope1 <- s1}
else if(b==0) {slope1<-s2}
else slope1 <-s3
s4<-c(0, 1, 0, 0, e*sd(z), f*sd(w), 0, e*sd(z)*f*sd(w))
s5<-c(0, 0, 1, 0, d*sd(x), 0, f*sd(w), d*sd(x)*f*sd(w))
s6<-c(0, 0, 0, 1, 0, e*sd(z), f*sd(w), d*sd(x)*e*sd(z))
if (d == 0) {slope2 <- s4}
else if(e==0) {slope2<-s5}
else slope2 <-s6
S= cbind(slope1,slope2)
SB<-t(S)%*%coef(mod);
```

(continued)

9.3.3   R Code: Three-Way Interaction (continued)

```
sb<-SB[1]-SB[2]
SC<-t(S)%*%vcov(mod)%*%S
seb<-sqrt(SC[1]+SC[4]-(2*SC[2]))
ttest <-sb/seb
p <-2*pt(-abs(ttest),df=(length(x)-nrow(S)))
slopetest <-cbind(ttest,p)
return(list(S,slopetest))
}

S<-simptest(0, -1, -1,  0, -1, 1) #simple slope of X at Z_low, W_low vs.
#simple slope of X at Z_low, W_high
S
```

9.3.4   Recentering Variables to Calculate Simple Slopes

Differentiating a function to create an **S** matrix helps us identify the terms that determine the shape and statistical significance of each simple slope. There is, however, another way to calculate simple slopes. It is less informative than constructing an **S** matrix, but a bit easier to perform, especially if we are using statistical software to analyze our data. The trick is to remember that when a cross-product term is added to a regression equation, the lower-order coefficients represent the simple slopes of one variable when other variables equal 0. When we centered each variable around its mean, we made 0 equal each variable's mean. But we could make 0 equal something else. For example, if 0 actually represented 1 standard deviation below z's mean, then b_1 in the original analysis would give us the simple slope of x at z_L.

To illustrate, we will calculate the simple slopes of x and z when w is one standard deviation below its mean. To begin, we *add* the standard deviation of w to every dev_w score: recentered Score = [(Raw Score − Mean) + Standard Deviation]. Table 9.20 shows the transformed data, with the key column highlighted. Notice that I have changed the name to lo_w (to indicate that the values have been recentered) and that the mean of lo_w now equals 2.8431 instead of 0. Finally, it might seem odd that we *add* the standard deviation from the mean-centered values to find the simple slope when w is low, but you can understand the logic if you examine the variable's new mean. Because this variable's mean is now one standard deviation above 0, 0 represents one standard deviation below the mean.

If we now recalculate our cross-product terms using our recoded variable, and then perform a normal regression analysis using the modified variables, we get the

Table 9.20 Using recentering to find the simple slopes of x and z when w is one standard deviation below its mean

	x (weather forecast)	z (time to spend outside)	w (months of nice weather)	y	dev_x	dev_z	lo_w	xz	xw	zw	xzw
	1	3	7	3	−3.6667	−2.00	**5.4264**	7.3333	−19.8969	−10.8529	39.7938
	8	7	4	7	3.3333	2.00	**2.4264**	6.6667	8.0881	4.8529	16.1762
	2	1	3	8	−2.6667	−4.00	**1.4264**	10.6667	−3.8038	−5.7057	15.2153
	2	6	1	3	−2.6667	1.00	**−.5736**	−2.6667	1.5295	−.5736	1.5295
	5	6	8	4	.3333	1.00	**6.4264**	.3333	2.1421	6.4264	2.1421
	6	6	8	5	1.3333	1.00	**6.4264**	1.3333	8.5686	6.4264	8.5686
	5	7	5	4	.3333	2.00	**3.4264**	.6667	1.1421	6.8529	2.2843
	7	2	1	2	2.3333	−3.00	**−.5736**	−7.0000	−1.3383	1.7207	4.0150
	4	3	8	5	−.6667	−2.00	**6.4264**	1.3333	−4.2843	−12.8529	8.5686
	8	6	1	9	3.3333	1.00	**−.5736**	3.3333	−1.9119	−.5736	−1.9119
	3	5	5	3	−1.6667	.00	**3.4264**	.0000	−5.7107	.00	.00
	5	8	2	6	.3333	3.00	**.4264**	1.0000	.1421	1.2793	.4264
Mean	**4.6667**	**5.00**	**4.4167**	**4.9167**	**.00**	**.00**	**2.8431**	**1.9167**	**−1.2778**	**−.2500**	**8.0673**
SD	**2.3484**	**2.2156**	**2.8431**	**2.1933**	**2.3484**	**2.2156**	**2.8431**	**4.6734**	**7.3501**	**6.5585**	**11.5785**

Table 9.21 Simple slopes of x, z, and xz when w is one standard deviation below its mean

	b	se_b	t	ΔR^2	p
b_0	5.2062	.3287	15.8401		.0001
$b_1(x)$	.4240	.1261	3.3630	.1138	.0282
$b_2(z)$	.0452	.1351	.3344	.0011	.7549
$b_3(w)$	−.1317	.0858	−1.5346	.0237	.1997
$b_4(xz)$	.4847	.0626	7.7490	.6040	.0015
$b_5(xw)$	.0485	.0549	.8836	.0079	.4268
$b_6(zw)$	−.1278	.0600	−2.1319	.0457	.1000
$b_7(xzw)$	−.1009	.0316	−3.1966	.1028	.0330

regression coefficients displayed in Table 9.21. We interpret only coefficients that do *not* involve the recentered variable, as these coefficients represent simple slopes when w is one standard deviation below its mean. Notice that the coefficient for the xz cross-product term is identical to the value reported in the top row of Table 9.16 (both coefficients = .4847). By recentering the data, the xz cross-product term in this analysis represents the simple slope of xz when w is one standard deviation below its mean.

Suppose we then want to find the simple slope of x when z is one standard deviation above its mean and w is one standard deviation below its mean (i.e., simple slope of $x @ z_H w_L$). In this case, we *subtract* the standard deviation of z from dev_z, add the standard deviation of w to dev_w, and perform a new analysis with both recentered variables. Table 9.22 shows the transformed data, and Table 9.23 shows the regression coefficients.

As before, we interpret only effects that do *not* involve recentered variables. In this case, we concern ourselves only with the simple slope of x. If you look back at Table 9.16, you will find that the simple slope of $x @ z_H w_L$ is identical to the simple slope of x in Table 9.23 (both coefficients = 1.4979). By recentering both variables, the regression coefficient for x now represents the simple slope of $x @ z_H w_L$.

Table 9.22 Using recentering to find the simple slope of x when z is one standard deviation above its mean and w is one standard deviation below its mean

x (weather forecast)	z (time to spend outside)	w (months of nice weather)	y	dev_x	hi_z	lo_w	xz	xw	zw	xzw
1	3	7	3	−3.6667	−4.2156	5.4264	15.4572	−19.8969	−22.8757	83.8775
8	7	4	7	3.3333	−.2156	2.4264	−.7187	8.0881	−.5231	−1.7438
2	1	3	8	−2.6667	−6.2156	1.4264	16.5749	−3.8038	−8.8661	23.6430
2	6	1	3	−2.6667	−1.2156	−.5736	3.2416	1.5295	.6972	−1.8593
5	6	8	4	.3333	−1.2156	6.4264	−.4052	2.1421	−7.8120	−2.6040
6	6	8	5	1.3333	−1.2156	6.4264	−1.6208	8.5686	−7.8120	−10.4160
5	7	5	4	.3333	−.2156	3.4264	−.0719	1.1421	−.7387	−.2462
7	2	1	2	2.3333	−5.2156	−.5736	−12.1697	−1.3383	2.9915	6.9802
4	3	8	5	−.6667	−4.2156	6.4264	2.8104	−4.2843	−27.0913	18.0608
8	6	1	9	3.3333	−1.2156	−.5736	−4.0520	−1.9119	.6972	2.3241
3	5	5	3	−1.6667	−2.2156	3.4264	3.6927	−5.7107	−7.5916	12.6527
5	8	2	6	.3333	.7844	.4264	.2615	.1421	.3345	.1115
Mean	**5.00**	**4.4167**	**4.9167**	**.00**	**−2.2156**	**2.8431**	**1.9167**	**−1.2778**	**−6.5492**	**10.8984**
4.6667										
SD	**2.2156**	**2.8431**	**2.1933**	**2.3484**	**2.2156**	**2.8431**	**7.8016**	**7.3501**	**9.5902**	**24.9206**
2.3484										

Table 9.23 Simple slope of x @ $z_H w_L$

	b	se_b	t	ΔR^2	p
b_0	5.3063	.4354	12.1862		.0003
$b_1(x)$	1.4979	.1933	7.7478	.6038	.0015
$b_2(z)$	.0452	.1351	.3344	.0011	.7549
$b_3(w)$	−.4149	.1603	−2.5883	.0674	.0608
$b_4(xz)$	.4847	.0626	7.7490	.6040	.0015
$b_5(xw)$	−.1750	.1052	−1.6639	.0278	.1715
$b_6(zw)$	−.1278	.0600	−2.1319	.0457	.1000
$b_7(xzw)$	−.1009	.0316	−3.1966	.1028	.0330

9.3.5   R Code: Three-Way Interaction Using Recentering

```
x=c(1,8,2,2,5,6,5,7,4,8,3,5)
z=c(3,7,1,6,6,6,7,2,3,6,5,8)
w=c(7,4,3,1,8,8,5,1,8,1,5,2)
y=c(3,7,8,3,4,5,4,2,5,9,3,6)
dx=scale(x, center = TRUE, scale = FALSE)
dz=scale(z, center = TRUE, scale = FALSE)
dw=scale(w, center = TRUE, scale = FALSE)
mod <-lm(y~dx*dz*dw)

#Simple Slopes by Recentering Variables
lo.x <-dx+sd(x)
hi.x <-dx-sd(x)
lo.z <-dz+sd(z)
hi.z <-dz-sd(z)
lo.w <-dw+sd(w)
hi.w <-dw-sd(w)

#Simple slopes of x and z at low w - - (interpret only effects that do NOT
#include recentered variable)
simple.1 <-lm(y~dx*dz*lo.w)
summary(simple.1)

#Simple slope of x for high z and low w - - (interpret only effects that do
#NOT include recentered variable)
simple.2 <-lm(y~dx*hi.z*lo.w)
summary(simple.2)
```

9.4   Effect Size and Statistical Power

In theory, we could keep adding variables to our regression equation and continue exploring four-variable, five-variable, and even six-variable interactions. In actuality, this is rarely done. There are two problems. First, it becomes difficult to interpret interactions that involve more than three variables. Second, detecting a significant multivariable interaction isn't very likely. In the final section of this chapter, we will consider the power to detect cross-product terms in a multiple regression analysis.

9.4.1   Effect Size

The small sample examples I create are contrived. Because I believe it is easier for students to understand statistics when the findings are "statistically significant," I create examples where the key effects will be strong and interpretable. But doing so with such a small sample requires constructing an unrealistic data set.

To appreciate this fact, we need to quantify the strength of an effect by computing its effect size. Effect sizes are calculated in different ways depending on the particular effect of interest, but when we are testing the unique contribution of a variable in a regression analysis, we commonly use a measure developed by Cohen (1988) called f^2. This statistic quantities the proportion of variance explained by a single predictor to the total amount of unexplained variance:

$$f^2 = \frac{\Delta r^2}{1 - R^2_{all}} \tag{9.16}$$

You might recognize from Chap. 4 that the numerator represents the squared semipartial correlation, and the denominator is $1 -$ the coefficient of determination. If we desire, the value of the effect size can be converted to a squared partial correlation:

$$\text{partial } r^2 = \frac{f^2}{1 + f^2} \tag{9.17}$$

To compute the effect size for the three-way interaction we discussed earlier, we take the overall R^2 value from Table 9.13 (.9598) and the Δr^2 value from Table 9.23 (.1028) and perform the calculations:

$$f^2 = \frac{.1028}{1 - .9598} = 2.5545$$

An effect size of this magnitude is extremely large. Values of .10 are more common, so this effect is more than 25 * more powerful than average. In short, it is unrealistically large and you shouldn't believe for a second that the effects we have been discussing are representative of real-world effects.

9.4.2   Statistical Power

A discussion of effect sizes leads naturally to a discussion of statistical power. The power of statistical tests refers to the probability that we will correctly reject a null hypothesis. For example, if it is the case that weekend weather and time outdoors interact only in cities with lousy weather, the statistical power of a test refers to the likelihood we would find a significant three-way interaction in our sample. Clearly, we can't expect to find the effect every time we run our study, so there will always be some error.

Statistical power depends on four interrelated factors: (a) effect size, (b) sample size, (c) our adopted level of statistical significance [designated alpha (α) and generally set at .05], and (d) our tolerance for failing to detect a true population effect [designated beta (β) and generally set at .80]. Power, which equals $1 - \beta$, represents the likelihood of finding a significant sample effect given that a true effect exists in the population.

Because statistical power is affected by sample size, we can use knowledge of the other three parameters to determine how many subjects we need in order to achieve a desired level of power. The following formula can be used to determine how many subjects are needed to achieve 80 % power with $\alpha = .05$ (two-tailed) for the 1df cross-product terms we have been testing (in the formula, k refers to the total number of predictors in the regression equation, including the cross-product term):

$$N = \frac{7.85}{f^2} + (k - 1) \tag{9.18}$$

To illustrate its use, suppose that based on previous research, we believe a three-way interaction is likely to have a small effect size of .03. There are seven predictors in our regression equation (x, z, w, xz, xw, zw, and xzw), so $k = 7$. Performing the calculations yields the sample size needed to detect the effect:

$$N = \frac{7.85}{.03} + (7 - 1) = 267.6667$$

After rounding, we learn that 268 subjects are needed to detect the effect 80 % of the time. Now, suppose you have reason to believe that the effect size is 10 times larger (i.e., $f^2 = .30$). In this case, we would need only 33 subjects to detect the effect 80 % of the time:[10]

$$N = \frac{7.85}{.30} + (7 - 1) = 32.1667$$

If the effect size were as large as the one I contrived, you would need only 10 subjects:

$$N = \frac{7.85}{2.5545} + (7 - 1) = 9.0731$$

---

[10] It's a good idea to round up for these calculations.

Obviously, it pays to be working with large effect sizes or have access to lots of subjects. Insofar as it is difficult to control our effect sizes (unless we are making up our data!), we need to be sure we have lots of subjects to achieve a satisfactory level of power. This is especially important when testing cross-product terms, as their effect sizes tend to be rather low (McClelland and Judd 1993).

Finally, considering the difficulties involved in detecting them, you might be wondering whether searching for interactions is worth the costs it entails. My answer is an unqualified "yes." Almost all effects occur only within a limited range of conditions. Interactions help us identify those conditions, thereby providing clues to an effect's causes and consequences. But interpreting them accurately takes time and practice. As we have seen, the same interaction that yields one conclusion when viewed from one vantage point can yield a different conclusion when viewed from another angle. Being able to extract the meaning from an interaction is a valuable skill, well worth the effort it takes to cultivate.

9.4.3   R Code: Effect Size of Three-Way Cross-Product Term

```
x=c(1,8,2,2,5,6,5,7,4,8,3,5)
z=c(3,7,1,6,6,6,7,2,3,6,5,8)
w=c(7,4,3,1,8,8,5,1,8,1,5,2)
y=c(3,7,8,3,4,5,4,2,5,9,3,6)
dx=scale(x, center = TRUE, scale = FALSE)
dz=scale(z, center = TRUE, scale = FALSE)
dw=scale(w, center = TRUE, scale = FALSE)
mod <-lm(y~dx*dz*dw)

#Calculate Effect Size for Highest Order Cross-Product Term
r.2 <-summary(mod)$r.squared
ss.y <-var(y)*(length(y)-1)
ms.e <-(sum(resid(mod)^2)/(length(y)-nrow(summary(mod)$coef)))
r.cha <-((((mod$coef[8]/sqrt(vcov(mod)[8,8]))^2)*ms.e)/ss.y)
f.2 <-r.cha/(1-r.2)
f.2

#Sample Size Needed for 80% Power  - enter effect size and k
sampsize <-function(f,k){
N = 7.85/f+(k-1)
}
sample <-sampsize(.03,7)
sample
```

9.5   Chapter Summary

1. Two variables interact when the slope relating one of the predictors to the criterion changes across levels of another predictor. In a multiple regression analysis, interactions can be tested by including a cross-product term, formed by multiplying two (or more) predictors. If the cross-product term is significant, the slope relating a predictor to the criterion varies significantly across levels of another predictor.

2. Adding a cross-product term to a regression equation transforms the lower-order coefficients into simple slopes, such that each lower-order coefficient represents the slope of a line relating a predictor to a criterion only when other variables equal 0. To render these lower-order coefficients more interpretable, many researchers use mean-centered (deviate) scores as predictors.

3. To probe the form of an interaction, researchers commonly calculate simple slopes for values one standard deviation below the mean, at the mean, and one standard deviation above the mean. The nature of an interaction is revealed by visually examining the simple slopes and testing their statistical significance.

4. Simple slopes represent first-order partial derivatives, and they can be calculated using rules of differentiation first described in Chap. 3. Alternatively, we can recalculate our predictors so that a score of zero equals a particular value of interest (e.g., one standard deviation below the mean).

5. Regions of significance that identify the range of values for which a simple slope will be statistically significant can be calculated using the Johnson-Neyman technique.

6. Cross-product terms can test higher-order interactions. In theory, we can add as many variables as we like, but it is difficult to find and interpret interactions involving more than three variables.

7. An effect size quantifies the strength of an observed effect, and the power of a statistical test refers to the likelihood that the test will correctly reject a null hypothesis. The greater the effect size, the more powerful the statistical test. The effect sizes of cross-product terms tend to be low, so sample sizes need to be large to have a test powerful enough to detect them.

Chapter 10
Polynomial Regression

Linear regression normally fits a straight line through a set of observations. Not all relations are characterized by a straight line, however. Suppose we wish to determine how much children enjoy a dessert as a function of the amount of sugar we use in its preparation. Surely children like a sugary treat, but only up to a point. If we put in too much sugar, they will stop enjoying it. The same is true for just about anything else we can think of (which is why we have phrases like "too much of a good thing" and "everything in moderation"). Formally, these phrases describe curvilinear relations. Figure 10.1 displays two common curvilinear relations, but other forms are possible.

We have already seen that linear regression models can accommodate nonlinear relations by transforming variables (Chap. 8) or by adding cross-product terms (Chap. 9). In this chapter we will discuss another approach to accommodating nonlinear relations: polynomial regression.

Consider the following, third-order polynomial, named for the highest power term in the equation.

$$y = \beta_0 + \beta_1 x + \beta_2 x^2 + \beta_3 x^3 + \varepsilon \tag{10.1}$$

Looking over the equation, we see that it satisfies our properties of linearity (see Chap. 8). None of the regression coefficients is raised to a power other than 1 or is multiplied or divided by another coefficient, and the equation is solved by summing the weighted terms and the disturbance. Notice, however, that not all of the predictors are of the first order. This is not a problem; only the regression coefficients (i.e., the parameters) must be of the first order in a linear model.

Electronic Supplementary Material: The online version of this chapter (doi: 10.1007/978-3-319-11734-8_10) contains supplementary material, which is available to authorized users

© Springer International Publishing Switzerland 2014
J.D. Brown, *Linear Models in Matrix Form*, DOI 10.1007/978-3-319-11734-8_10

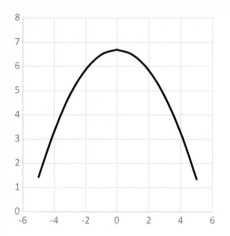 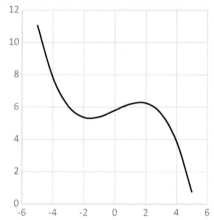

Fig. 10.1 Two curvilinear relations

Including power terms in a regression equation allows us to determine whether various nonlinear functions (e.g., quadratic, cubic) describe our data. We can do so using linear regression only if lower-order terms are also included in the regression equation. This is because each power term can be thought of as a cross-product term [e.g., $x^2 = (x * x)$], so before we examine the contribution of a higher-order term, we must first take into account its lower-order term. In this sense, we can think of polynomial regression as a subset of the models we covered in Chap. 9.

10.1 Simple Polynomial Regression

To make our discussion more concrete, imagine that a behavioral scientist is interested in testing the relation between temperature and violent crimes. To address the issue, she records the maximum temperature for 12 summer days and the number of violent crimes committed in the city in which she resides.

10.1.1 Testing the Linear Component

The (pretend) data appear in Table 10.1, and a scatterplot (with residuals) from a simple linear regression is shown in Fig. 10.2.

The scatterplot shows that temperature and crime are not related in a linear fashion. Instead, crime rises as temperatures increase up to a point and then decreases as temperatures climb even higher. More formally, the relation between temperature and crime appears to be curvilinear, best represented by a second-order,

Table 10.1 Linear and polynomial regression

		Simple linear regression			Polynomial regression			
	x	y	$\hat{y}$	e	dev_x	dev_{x^2}	$\hat{y}$	e
	36	3	5.1228	−2.1228	−2.75	7.5625	3.2158	−.2158
	36	2	5.1228	−3.1228	−2.75	7.5625	3.2158	−1.2158
	37	4	5.0175	−1.0175	−1.75	3.0625	4.7994	−.7994
	37	7	5.0175	1.9825	−1.75	3.0625	4.7994	2.2006
	38	8	4.9123	3.0877	−0.75	.5625	5.7858	2.2142
	39	8	4.8070	3.1930	0.25	.0625	6.1749	1.8251
	39	6	4.8070	1.1930	0.25	.0625	6.1749	−.1749
	39	6	4.8070	1.1930	0.25	.0625	6.1749	−.1749
	40	5	4.7018	.2982	1.25	1.5625	5.9668	−.9668
	39	3	4.8070	−1.8070	0.25	0.0625	6.1749	−3.1749
	42	4	4.4912	−.4912	3.25	10.5625	3.7587	.2413
	43	2	4.3860	−2.3860	4.25	18.0625	1.7588	.2412
Mean	**38.75**	**4.8333**			**.0000**	**4.3542**		
SD	**2.1794**	**2.1672**			**2.1794**	**5.6103**		
SS	**52.25**	**51.6667**			**52.25**	**346.2292**		
	b	se_b	t	p	b	se_b	t	p
b_0	8.9123	12.1343	.7345	.4795	6.1336	.6679	9.1829	.0000
b_1	−.1053	.3127	−.3366	.7433	.2398	.2653	.9039	.3896
b_2					−.2986	.1031	−2.8974	.0177

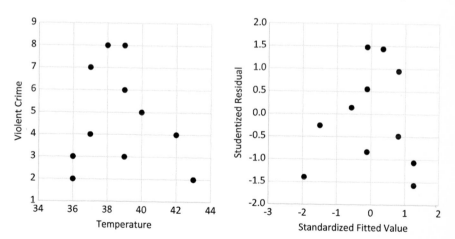

Fig. 10.2 Scatterplot and residuals relating temperature to violent crime: simple linear regression

quadratic, polynomial. The residuals also show a nonlinear pattern, confirming that a linear fit to the data is inadequate. The lack of a significant linear effect in Table 10.1 provides additional evidence that this is so.[1]

---

[1] A lack of fit test (see Chap. 8) does not produce a significant effect ($p = .2529$). This is usually true with a curvilinear relation, so the test is not very useful in this case.

10.1.2   Adding a Quadratic Term

Having determined that our (fictitious) relation between temperature and crime is nonlinear, we will perform a polynomial regression analysis. With our data, we first create deviate scores for our predictor and then square them to find a second order polynomial. The right-hand side of Table 10.1 shows the calculations. Notice that we square the deviate score, not the raw score.

10.1.2.1   Hierarchical Regression

We use hierarchical regression to analyze the data. The term "hierarchical" means that lower-order terms are entered before higher-order ones. In our case, we enter the linear term (dev_x) to the predictive equation before adding our power term (dev_{x^2}). The bottom right-hand portion of Table 10.1 shows that the quadratic term makes a significant contribution to the prediction of our criterion, uniquely accounting for ~48 % of the total variance.[2] Our sample regression equation thus becomes $\hat{y} = 6.1336 + .2398x - .2986x^2$.

10.1.2.2   Predicted Values

To understand the nature of the quadratic effect, we can calculate predicted values by constructing a **P** matrix with scores one standard deviation below the mean, at the mean, and above the mean as weights (see Chap. 9). The left-hand side of Table 10.2 displays the relevant values. Notice that the weight for the power term is found by squaring the standard deviation of x, not by using the standard deviation of the power term itself.

Table 10.2   Matrices to calculate predicted values and simple slopes for polynomial regression

P Matrix for predicted values				S Matrix for simple slopes			
Weights	x_L	x_M	x_H	Weights	x_L	x_M	x_H
1	1	1	1	0	0	0	0
x	−2.1794	0	2.1794	1	1	1	1
x^2	4.7500	0	4.7500	2x	−4.3589	0	4.3589

When we multiply **P′b**, we compute our predicted values.[3] These values ($x_L = 4.1925, x_M = 6.1336, x_H = 5.2378$) are displayed in Fig. 10.3.[4]

---

[2] Equation (9.6) was used to find the increment in R^2 associated with the regression coefficient.

[3] Although we do not interpret the linear coefficient from the overall equation, we use it when computing the predicted values.

[4] Figures 10.3 and 10.4 have been smoothed for illustrative purposes. The topic of smoothing polynomial functions will be discussed later in this chapter.

Fig. 10.3 Polynomial
regression with a
quadratic term

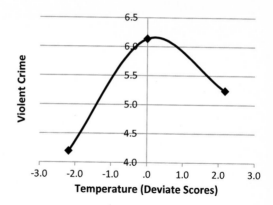

10.1.2.3   Calculating and Testing Simple Slopes

The pattern shown in Fig. 10.3 confirms that violent crime initially increases as the temperature rises, levels off just past the mean, and then declines as temperatures climb even higher. Testing these apparent relations involves computing and testing the simple slopes by differentiating the function to find its first derivative. Before describing the specific procedures, recall that we can rearrange the terms of a regression equation with a cross-product term to highlight the regression of y on x at different values of z [see Eqs. (9.3) and (9.4)].

$$\hat{y} = b_0 + b_1 x + b_2 z + b_3 xz$$

becomes

$$\hat{y} = b_0 + (b_1 + \boldsymbol{b_3} z) x + b_2 z$$

Now let's do the same thing using a power term.

$$\hat{y} = b_0 + b_1 x + b_2 x^2$$

becomes

$$\hat{y} = b_0 + (b_1 + \boldsymbol{b_2} x) x$$

Do you see the difference here? Instead of finding the change in x at different values of z, we are now looking for the change in x at different values of x. The solution is to find the slope of a tangent line at a point along the curve. As first discussed in Chap. 3, a tangent line touches the curve at a single point without crossing it. Figure 10.4 shows a tangent line that touches the curve at $x=0$. If we find the slope of this regression line, we have found the simple slope of x when $x=0$.

Fig. 10.4 Polynomial
regression with a tangent
line to represent a simple
slope

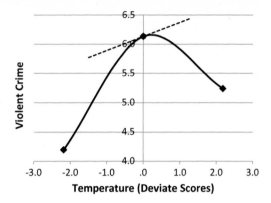

From Chap. 3, we know that we find the tangent line by differentiating a function to find its derivative. Our function appears below, and our task is to find the first partial derivative of the function with respect to x.

$$\hat{y} = b_0 + b_1 x + b_2 x^2$$

Implementing the rules we learned in Chap. 3, we disregard all terms that do not involve x, change x to 1 when it appears by itself, and apply the power rule when x appears with an exponent.

$$\frac{\partial \hat{y}}{\partial x} = b_1 + 2b_2 x \tag{10.2}$$

We then translate this formula into weights in an **S** matrix (see right-hand side of Table 10.2) and plug in values for x that are one standard deviation below the mean, at the mean, and above the mean.

Next, we find the simple slopes by calculating $\mathbf{S'b}$ and their standard errors by taking the square root of the diagonal elements of $\mathbf{S'CS}$. Finally, we can construct t-tests and create confidence intervals if we desire. These values appear in Table 10.3. It may be seen that the simple slope of x when x is one standard deviation below the mean is significantly positive, the simple slope of x when $x = 0$ is not significant, and the simple slope of x when x is one standard deviation above the mean is significantly negative. These values confirm the researcher's hypothesis that temperature and violent crime are related in a curvilinear fashion.

Table 10.3 Simple slopes of x for a polynomial regression

	s_b	se_b	t	p	$CL-$	$CI+$
x_L	1.5415	.6158	2.5032	.0337	.1484	2.9346
x_M	.2398	.2653	.9039	.3896	−.3604	.8400
x_H	−1.0619	.4065	−2.6125	.0282	−1.9814	−.1424

Notice also that the simple slope of x when $x=0$ is identical to b_1 in our overall regression equation. As is true with a cross-product term, when a power term is added to a regression equation, the lower-order coefficients represent simple slopes at $x=0$. Because we have centered the predictor around its mean, zero represents an average temperature. But we do not interpret the lower-order coefficients in a hierarchical analysis because we are not interested in whether the linear component explains variance that the quadratic component does not; we are interested only in whether the quadratic component explains variance that the linear component does not.

10.1.2.4 Finding the Maximum Point

We know from Chap. 3 that the maximum or minimum point on a curve occurs when the simple slope is 0 (i.e., the tangent line is horizontal).[5] Mathematically, it is found by setting the first partial derivative equal to 0.

$$b_1 + 2b_2x = 0$$

Rearranging terms,

$$b_1 = -2b_2x$$

and solving for x produces Eq. (10.3).

$$mv_x = \frac{-b_1}{2b_2} \tag{10.3}$$

Inserting our data yields the maximum value.

$$mv_x = \frac{-.2398}{(2 * -.2986)} = .4015$$

When $dev_x=.4015$ (just a bit higher than the mean), the tangent line is horizontal, the simple slope is 0, and the maximum is found. In terms of raw temperature values, violent crime (in this fictitious example) is at its maximum when the temperature is ~39.15° Celsius.

10.1.3 Testing Other Polynomials

After we test a quadratic model, we test for a cubic function by adding x^3 to the regression equation. In our example, this term does not significantly increase the prediction of our criterion, so we stop and decide that the quadratic model is the highest polynomial function for these data.

---

[5] To be a maximum, the second partial derivative must be negative. This is the case in our example.

You might have noticed that I added "for these data." It's important to underscore that we are only modeling the data we have collected. Most (if not all) linear relations eventually level off or decline, but we are interested only in whether, within the range of the data we have collected, the functional relation is linear or curvilinear in form. For this reason, you should not speculate about the nature of relations that lie outside the range of your data. It is conceivable that violence returns when temperatures climb higher than the ones we recorded.

One other thing. Suppose we have two predictors that are highly correlated. If we then find that their cross-product increases the prediction of y, we might actually have uncovered a curvilinear relation rather than an interactive one. This possibility provides another reason why we want to choose predictors that are somewhat independent (see Chap. 6).

10.1.4   R Code: Cubic Polynomial

```
x=c(36,36,37,37,38,39,39,39,40,39,42,43)
y=c(3,2,4,7,8,8,6,6,5,3,4,2)

#Plot function
plot(x, y, cex = 1, col ="black", ,ylab = "Violent Crimes", xlab =
"Temperature", main = "Scatterplot")

#Regression
dx=scale(x, center = TRUE, scale = FALSE)
summary(linear <-lm(y~x))
summary(quad <-lm(y~dx+I(dx^2)))

#Predicted Values
p0 <-rep(1,3);p1 <-c(-sd(x),0,sd(x));p2 <-c(sd(x)^2,0,sd(x)^2)
P.mat <-round(rbind(p0,p1,p2),digits=5)
P <-round(rbind(p0,p1,p2),digits=5)
pred.val <-t(P)%*%coef(quad)
dimnames(pred.val)=list(c("lo.x","med.x","hio.x"))
pred.val

#Plotting Predicted Values
dev.new()
plot(pred.val,type="l",ylab = "Violent Crime", xlab = "Temperature",
main = "Predicted Values")

#Simple Slopes and Simple Standard Errors
s0 <-rep(0,3);s1 <-rep(1,3);s2 <-c(-2*sd(x),0,2*sd(x))
S <-round(rbind(s0,s1,s2),digits=5)
simp.slope <-t(S)%*%coef(quad)
simp.cov <-t(S)%*%vcov(quad)%*%S
simp.err <-sqrt(diag(simp.cov))
simples <-simp.slope/sqrt(diag(simp.cov))
```

(continued)

10.1.4   R Code: Cubic Polynomial (continued)

```
tvalues <-2*pt(-abs(simples),df=(length(x)-nrow(S)))
crit <-abs(qt(0.025,(length(x)-nrow(S))))
CI.low <-simp.slope-(crit*simp.err)
CI.high <-simp.slope+(crit*simp.err)
simp.table<-round(matrix(c(simp.slope,simp.err,simples,tvalues,
CI.low,CI.high),nrow=3,ncol=6),digits=5)
dimnames(simp.table)=list(c("x@z.low","x@z.med","x@z.high"),
c("slope", "stderr", "t","p","CI.low","CI.high"))
simp.table

#Maximum Point
maximum <-(-quad$coef[2]/(2*quad$coef[3]));maximum
```

10.2   Polynomial Interactions

In Chap. 9, we discussed the construction and interpretation of cross-product terms and interactions. We can also use cross-product terms to model polynomial inter-actions. Suppose after reading our study of temperature and aggression another researcher hypothesizes that these variables have a curvilinear relation in crowded cities, but not in cities with low population density. After all, if you don't live near anyone, being hot and bothered is unlikely to foster aggression. To test this hypothesis, the researcher repeats our study adding a new variable: population density (in 100,000 increments). The researcher's model is shown below, and Table 10.4 presents the (imaginary) data.

$$\hat{y} = b_0 + b_1 x + b_2 x^2 + b_3 z + b_4 xz + b_5 x^2 z \qquad (10.4)$$

10.2.1   Regression Equations

After creating deviate scores, we calculate our power and cross-product terms. We then conduct a hierarchical regression analysis, entering the linear predictors (dev_x, dev_z, dev_{xz}) before adding the quadratic terms (dev_{x^2} and $dev_{x^2 z}$). The top half of Table 10.5 shows that the model is not significant before we add the power terms, but is significant once they are included.

The fact that the latter model is significant but the former is not does not prove that including the power terms significantly improves the prediction of our crite-rion. To make this determination we need to test whether the model with the power terms is significantly better than the model without them. This determination involves testing the difference between the two models, designating the linear model as the reduced model (since it has less terms).

Table 10.4 Small sample example for polynomial interaction

	x (heat)	z (density)	y (crime)	dev_x	dev_{x^2}	dev_z	dev_{xz}	dev_{x^2z}
	37	2	4	−.50	.25	−3.0833	1.5417	−.7708
	35	1	2	−2.50	6.25	−4.0833	10.2083	−25.5208
	38	8	8	.50	.25	2.9167	1.4583	.7292
	38	3	2	.50	.25	−2.0833	−1.0417	−.5208
	35	6	1	−2.50	6.25	.9167	−2.2917	5.7292
	40	9	3	2.50	6.25	3.9167	9.7917	24.4792
	36	7	6	−1.50	2.25	1.9167	−2.8750	4.3125
	39	4	5	1.50	2.25	−1.0833	−1.6250	−2.4375
	40	3	4	2.50	6.25	−2.0833	−5.2083	−13.0208
	39	8	8	1.50	2.25	2.9167	4.3750	6.5625
	36	2	5	−1.50	2.25	−3.0833	4.6250	−6.9375
	37	8	9	−.50	.25	2.9167	−1.4583	.7292
Mean	37.5000	5.0833	4.7500	.0000	2.9167	.0000	1.4583	−.5556
s	1.7838	2.8749	2.5981	1.7838	2.6054	2.8749	4.9184	11.9349

Table 10.5 Regression equation for power interaction example before adding power terms (*top panel*) and after adding power terms (*bottom panel*)

	SS	df	MS	R^2	F	p
	Regression equation before adding power terms					
	SS	df	MS	R^2	F	p
Regression	22.3014	3	7.4338	.3004	1.1448	.3882
Residual	51.9486	8	6.4936			
Total	74.2500					
	Regression equation after adding power terms					
	SS	df	MS	R^2	F	p
Regression	60.4911	5	12.0982	.8147	5.2758	.0334
Residual	13.7589	6	2.2931			
Total	74.2500					

$$F = \frac{\left(R^2_{full} - R^2_{reduced}\right)/\left(k_{full} - k_{reduced}\right)}{\left(1 - R^2_{full}\right)/\left(N - k_{full} - 1\right)} \tag{10.5}$$

After plugging in our values, we find that adding the power terms does, in fact, significantly improve the prediction of our criterion.

$$F(2,6) = \frac{(.8147 - .3004)/(5 - 3)}{(1 - .8147)/(12 - 5 - 1)} = 8.3269, p = .0186$$

10.2.2 Testing the Regression Coefficients

Next, we examine the individual coefficients. Concerning ourselves only with the two power terms, b_2 and b_5, we see that both are significant (see Table 10.6).

Table 10.6 Regression coefficients for a polynomial interaction

	b	se_b	ΔR^2	t	p	$CI-$	$CI+$
b_0	6.3289	.6741		9.3886	.0001	4.6794	7.9783
$b_1(x)$	.3055	.2783	.0372	1.0978	.3144	−.3754	.9864
$b_2(x^2)$	−.5937	.1831	.3248	−3.2431	.0176	−1.0416	−.1457
$b_3(z)$	.9195	.2511	.4141	3.6616	.0106	.3050	1.5339
$b_4(xz)$	.0399	.0974	.0052	.4100	.6960	−.1984	.2782
$b_5(x^2z)$	−.1699	.0625	.2284	−2.7195	.0347	−.3228	−.0170

10.2.3 Probing a Polynomial Interaction

Having found a significant polynomial interaction, we turn our attention toward probing its form. As before, we will begin by constructing a **P** matrix and computing **P′b** to generate predicted values. The weights are shown in Table 10.7, and the predicted values are displayed in Fig. 10.5. Confirming the researcher's intuitions, the relation between temperature and crime appears to be relatively flat for cities of low density, moderately curvilinear for cities of moderate density, and substantially curvilinear for cities of high density.

Table 10.7 **P** Matrix for predicted values for a polynomial interaction

	z_L			z_M			z_H		
	x_L	x_M	x_H	x_L	x_M	x_H	x_L	x_M	x_H
b_0	1	1	1	1	1	1	1	1	1
$b_1(x)$	−1.7838	0	1.7838	−1.7838	0	1.7838	−1.7838	0	1.7838
$b_2(x^2)$	3.1818	0	3.1818	3.1818	0	3.1818	3.1818	0	3.1818
$b_3(z)$	−2.8749	−2.8749	−2.8749	0	0	0	2.8749	2.8749	2.8749
$b_4(xz)$	5.1282	0	−5.1282	0	0	0	−5.1282	0	5.1282
$b_5(x^2z)$	−9.1475	0	−9.1475	0	0	0	9.1475	0	9.1475

10.2.3.1 Testing the Linear and Quadratic Coefficients

Testing the significance of the relations shown in Fig. 10.5 involves testing the significance of the linear and quadratic terms at three levels of z. Table 10.8 shows the **S** matrix.

Fig. 10.5 Predicted values
for power term interaction

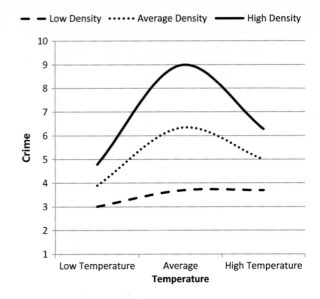

Table 10.8 S Matrix for testing linear and quadratic terms at three levels of z

	Linear (x)				Quadratic (x^2)			
	Weights	z_L	z_M	z_H	Weights	z_L	z_M	z_H
b_0	**0**	0	0	0	**0**	0	0	0
$b_1(x)$	**1**	1	1	1	**0**	0	0	0
$b_2(x^2)$	**0**	0	0	0	**1**	1	1	1
$b_3(z)$	**0**	0	0	0	**0**	0	0	0
$b_4(xz)$	z	−2.8749	0	2.8749	**0**	0	0	0
$b_5(x^2 z)$	**0**	0	0	0	z	−2.8749	0	2.8749

If we then calculate $\mathbf{S'b}$ to find the coefficients and take the square root of the
diagonal elements of $\mathbf{S'CS}$ to find the standard errors, we can construct t-tests to test
the significance of each coefficient (see Table 10.9). Confirming our impression of
the relations displayed in Fig. 10.5, we find that the quadratic term is negligible for
sparsely populated cities, moderate in size for cities of average density, and large
for crowded cities.

Table 10.9 Significance tests for linear and quadratic terms at three levels of z

		s_b	se_b	t	p	$CI-$	$CI+$
Linear	z_L	.1907	.3733	.5108	.6277	−.7228	1.1041
	z_M	.3055	.2783	1.0978	.3144	−.3754	.9864
	z_H	.4203	.4151	1.0126	.3503	−.5954	1.4360
Quadratic	z_L	−.1052	.2464	−.4268	.6844	−.7081	.4978
	z_M	−.5937	.1831	−3.2431	.0176	−1.0416	−.1457
	z_H	−1.0822	.2662	−4.0659	.0066	−1.7334	−.4309

10.2.3.2 Simple Slopes

Our final step is to calculate simple slopes. You might have assumed that we just did that, but calculating the linear and quadratic coefficients is not the same as calculating the simple slopes. Recall that a simple slope refers to a tangent line that touches a curve at one point only. There is no bend to a tangent line; it is always linear. Consequently, only the linear coefficients shown in Table 10.9 provide information about simple slopes. These coefficients represent the simple slopes at three levels of z when $x = 0$. If we are interested in calculating the simple slopes at other levels of x, we need to use the weighting scheme shown in the second column of Table 10.10. The rest of the table shows the relevant values for our example.

Table 10.10 S matrix weights for calculating simple slopes of x at various values of x and z

		z_L			z_M			z_H		
	Weights	x_L	x_M	x_H	x_L	x_M	x_H	x_L	x_M	x_H
b_0	0	0	0	0	0	0	0	0	0	0
$b_1(x)$	1	1	1	1	1	1	1	1	1	1
$b_2(x^2)$	$2x$	-3.5675	0	3.5675	-3.5675	0	3.5675	-3.5675	0	3.5675
$b_3(z)$	0	0	0	0	0	0	0	0	0	0
$b_4(xz)$	z	-2.8749	-2.8749	-2.8749	0	0	0	2.8749	2.8749	2.8749
$b_5(x^2z)$	$2xz$	10.2564	0	-10.2564	0	0	0	-10.2564	0	10.2564

If we then calculate $S'b$ to find the coefficients and take the square root of the diagonal elements of $S'CS$ to find the standard errors, we can construct t-tests to test the simple slope at each of the 9 combinations shown in Table 10.11.

Table 10.11 Significance tests for simple slopes at nine locations along three curves

		s_b	se_b	t	p	$CI-$	$CI+$
z_L	x_L	.5658	.8921	.6343	.5493	-1.6171	2.7488
	x_M	.1907	.3733	.5108	.6277	$-.7228$	1.1041
	x_H	$-.1845$	1.0140	$-.1819$	.8616	-2.6656	2.2967
z_M	x_L	2.4234	.7258	3.3387	.0156	.6473	4.1995
	x_M	.3055	.2783	1.0978	.3144	$-.3754$	.9864
	x_H	-1.8124	.6935	-2.6133	.0399	-3.5094	$-.1154$
z_H	x_L	4.2809	1.1529	3.7132	.0099	1.4599	7.1019
	x_M	.4203	.4151	1.0126	.3503	$-.5954$	1.4360
	x_H	-3.4403	.9048	-3.8025	.0089	-5.6542	-1.2265

Here we see that none of the simple slopes is significant at z_L, but the simple slopes of x at x_L and x_H are significant at z_M and z_H. To be clear about what these coefficients tell us, let's consider what happens when z is one standard deviation above its mean (i.e., a city with lots of people in a small area). When temperatures start out at a relatively comfortable level (i.e., when x is one standard deviation below its mean), a one unit increase in temperature is associated with a 4.2809 unit increase in crime; in contrast, when it's already very hot (i.e., when x is one standard deviation above its mean), a one unit increase in temperature is associated with a 3.4403 unit decrease in crime.

10.2.3.3 Maximum Value

We can find the maximum/minimum value of the curve at any level of z using the coefficients from our original polynomial regression.

$$\frac{-(b_1 + b_4 z)}{2 * (b_2 + b_5 z)} \tag{10.6}$$

To illustrate, we will find the maximum value when z is one standard deviation below its mean.

$$mv_L = \frac{-[.3055 + (.0399 * -2.8749)]}{2 * [-.5937 + (-.1699 * -2.8749)]} = .9067$$

Performing the rest of the calculations yields the maximum values when z is at its mean ($mv_M = .2573$) and one standard deviation above its mean ($mv_H = .1942$). Thus, in a crowded city, crime is at its maximum when $dev_x = .1942$ (~37° Celsius).

10.2.4 R Code: Polynomial Interaction

```
x=c(37,35,38,38,35,40,36,39,40,39,36,37)
z=c(2,1,8,3,6,9,7,4,3,8,2,8)
y=c(4,2,8,2,1,3,6,5,4,8,5,9)
dx=scale(x, center = TRUE, scale = FALSE)
dz=scale(z, center = TRUE, scale = FALSE)

#Compare Linear and Polynomial Model
summary(lin.reg <-lm(y~dx*dz))
summary(poly.reg <-lm(y~dx+I(dx^2)+dz+dx*dz+I(dx^2)*dz))
anova(lin.reg,poly.reg)

#Predicted values
p0 <-rep(1,9);p1 <-rep(c(-sd(x),0,sd(x)),3);p2 <-rep(c(sd(x)^2,0,
sd(x)^2),3)
p3 <-c(rep(-sd(z),3),rep(0,3),rep(sd(z),3));p4 <-p1*p3;p5 <-p2*p3
P <-rbind(p0,p1,p2,p3,p4,p5)
preds <-matrix(t(P)%*%coef(poly.reg),nrow=3,ncol=3)
```

(continued)

10.2.4   R Code: Polynomial Interaction (continued)

```
dimnames(preds)=list(c("low.z","med.z","high.z"),
c("low.x","med.x","high.x"))
preds

#Simple Polynomial Coefficients
q0 <-rep(0,6);q1 <-c(rep(1,3),rep(0,3));q2 <-c(rep(0,3),rep(1,3))
q3 <-rep(0,6);q4 <-c(-sd(z),0,sd(z),rep(0,3));q5 <-c(rep(0,3),-sd(z),
0,sd(z))
Q <-round(rbind(q0,q1,q2,q3,q4,q5),digits=5)

simp.slope <-t(Q)%*%coef(poly.reg)
simp.cov <-t(Q)%*%vcov(poly.reg)%*%Q
simp.err <-sqrt(diag(simp.cov))
simples <-simp.slope/sqrt(diag(simp.cov))
tvalues <-2*pt(-abs(simples),df=(length(x)-nrow(Q)))
crit <-abs(qt(0.025, df=(length(x)-nrow(Q))))
CI.low <-simp.slope-(crit*simp.err)
CI.high <-simp.slope+(crit*simp.err)
simp.poly<-round(matrix(c(simp.slope,simp.err,simples,tvalues,
CI.low,CI.high),nrow=6,ncol=6),digits=5)
dimnames(simp.poly)=list(c("linear@z.low","linear@z.med","linear@z.
high","quad@x.low","quad@x.med","quad@x.high"),c("slope","stderr",
"t","p","CI.low","CI.high"))
simp.poly

#Simple Slopes
s0 <-rep(0,9);s1 <-rep(1,9);s2 <-rep(c(-sd(x)*2,0,sd(x)*2),3)
s3 <-rep(0,9);s4 <-c(rep(-sd(z),3),rep(0,3),rep(sd(z),3));s5 <-s2*s4
S <-round(rbind(s0,s1,s2,s3,s4,s5),digits=5)

simp.slope <-t(S)%*%coef(poly.reg)
simp.cov <-t(S)%*%vcov(poly.reg)%*%S
simp.err <-sqrt(diag(simp.cov))
simples <-simp.slope/sqrt(diag(simp.cov))
tvalues <-2*pt(-abs(simples),df=(length(x)-nrow(S)))
crit <-abs(qt(0.025, df=(length(x)-nrow(S))))
CI.low <-simp.slope-(crit*simp.err)
CI.high <-simp.slope+(crit*simp.err)
simp.slop<-round(matrix(c(simp.slope,simp.err,simples,tvalues,
CI.low,CI.high),nrow=9,ncol=6),digits=5)
dimnames(simp.slop)=list(c("x.lo@z.low","x.med@z.low","x.hi@z.lo",
"x.lo@z.med","x.med@z.med","x.hi@z.med","x.lo@z.hi","x.med@z.hi",
"x.hi@z.med"),c("slope","stderr", "t","p","CI.low","CI.high"))
simp.slop
```

(continued)

10.2.4   R Code: Polynomial Interaction (continued)

```
#maximum value
max<-(-(poly.reg$coef[2]+poly.reg$coef[5]*-sd(z))/((2*(poly.reg
$coef[3]+poly.reg$coef[6]*-sd(z))))))
max
```

10.3   Piecewise Polynomials

To this point we have used polynomials to perform a global fit to the data. In this context, the word "global" means that the power term applies to all data points rather than only some. For example, in a quadratic model, we square all terms; in a cubic model, we raise all predictors to the third power; and so on.

10.3.1   Regression Splines

An alternative approach uses "local" fitting, applying a power term to only some of the observations. This approach goes by several names, including a piecewise polynomial, truncated power series, or a regression spline.[6] Regardless of the name, these models involve local fitting in which different polynomial terms are applied to different segments of the data.

Figure 10.6 presents a schematic depiction of two piecewise polynomials. The first panel illustrates a quadratic model with one knot. A "knot" refers to the point at

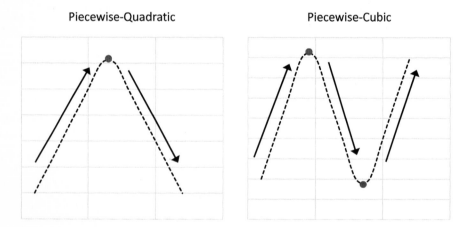

Piecewise-Quadratic   Piecewise-Cubic

Fig. 10.6 Schematic depiction of piecewise polynomials

---

[6] A spline is a thin strip of wood, flexible enough to be bent. Historically, engineers and builders used splines to model a curved connection between two points. My guess is that computer imaging has rendered this practice obsolete.

which the relation between x and y changes in a nonlinear way. In our first example, increases in x predict increases in y below the knot, but increases in x predict decreases in y above the knot. The second panel depicts a cubic model. Here we have two knots and three linear equations, two positive and one negative.

10.3.1.1 Piecewise Cubic Regression Spline

To clarify matters, let's use an example that is familiar to many students and all too familiar to way too many professors! Suppose we track the amount of time students spend studying throughout an academic term. Table 10.12 presents the (hypothetical) data and Fig. 10.7 presents a scatterplot and the studentized residuals from a linear fit.[7] The scatterplot indicates two points at which the relation appears to change direction (filled-in markers) and two points when exams are given (large markers). The pattern is clear: study time increases as an exam approaches, declines rapidly after the exam, and gradually picks up again as another exam nears.

Table 10.12 Small sample example predicting study time during an academic term

| x | y | | | | Cubic polynomial | |
Weeks	Study hours/week	dev_x	dev_{x^2}	dev_{x^3}	$\hat{y}$	e
1	1.0	−5.5	30.25	−166.3750	.42857	.5714
2	1.5	−4.5	20.25	−91.1250	2.27872	−.7787
3	1.6	−3.5	12.25	−42.8750	3.23996	−1.6400
4	5.0	−2.5	6.25	−15.6250	3.53362	1.4664
5	6.0	−1.5	2.25	−3.3750	3.38102	2.6190
6	2.0	−.5	.25	−.1250	3.00350	−1.0035
7	1.4	.5	.25	.1250	2.62238	−1.2224
8	1.25	1.5	2.25	3.3750	2.45899	−1.2090
9	3.0	2.5	6.25	15.6250	2.73467	.2653
10	4.0	3.5	12.25	42.8750	3.67073	.3293
11	7.2	4.5	20.25	91.1250	5.48851	1.7115
12	7.3	5.5	30.25	166.3750	8.40934	−1.1093

10.3.1.2 Global Fitting of a Cubic Polynomial

One way to analyze our data is to use a cubic polynomial of the form shown in Eq. (10.1). Following procedures described earlier, we would first create deviate scores and then create two power terms (i.e., dev_{x^2} and dev_{x^3}). We then include all lower-order terms before testing the significance of the power terms. The right-hand side of Table 10.12 shows the relevant vectors and Table 10.13 presents

---

[7] We will pretend that the academic term is 12 weeks long in order to maintain consistency with other examples used in this book.

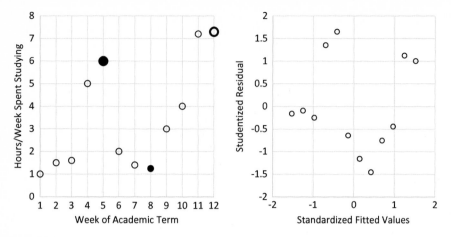

Fig. 10.7 Hours/week spent studying across weeks of an academic term. For the scatterplot, *larger markers* indicate exams; *filled-in* markers indicate knots

Table 10.13 Regression analysis using global fitting of a cubic polynomial

	SS	df	MS	R^2	F	p
Regression	41.7324	3	13.9108	.6660	5.3162	.0262
Residual	20.9333	8	2.6167			
Total	62.6656					
	b	se_b	t	p		
b_0	2.7996	.7046	3.9732	.0041		
b_1	−.3903	.3469	−1.1254	.2931		
b_2	.0535	.0443	1.2090	.2612		
b_3	.0369	.0150	2.4543	.0397		

a summary of the analysis. Note the probability level of the cubic term (b_3), indicating a significant cubic trend.

10.3.1.3   Local Fitting with a Piecewise Polynomial

So far, all we have done is follow the procedures we learned in the first part of the chapter using global fitting. Now let's try local fitting. Looking at the data, we see that study time shows a steady rise from week 1 to week 5, a steep decline from week 5 to week 8, and a second rise from week 8 to week 12. So we have two knots: one at week 5 and one at week 8. Coincidentally (not!), the first exam is given at week 5 and the second exam is given at week 12.

To model these changes locally, we create two new vectors that yield a value of 0 for scores that fall below or match the knot (k), and scores of a cubed difference score [$(x − k)^3$] for values that fall above the knot.

$$
\begin{cases}
0 & \dots \quad (x \le k) \\
(x - k)^3 & \dots \quad (x > k)
\end{cases}
\tag{10.7}
$$

Table 10.14 presents these calculations for $k_1 = 5$ and $k_2 = 8$. When we use these terms (along with a vector of leading $1's$) to predict study time, we get the results shown in Table 10.15. The overall fit of the model is good, and both piecewise terms b_4 and b_5 are significant.[8]

Table 10.14 Truncated power basis for a cubic piecewise polynomial with two knots

x	x^2	x^3	$(x - k_1)_1^3$	$(x - k_2)_1^3$	$\hat{y}$	e
1	1	1	0	0	1.02729	−.02729
2	4	8	0	0	1.15161	.34839
3	9	27	0	0	2.64711	−1.04711
4	16	64	0	0	4.28282	.71718
5	25	125	0	0	4.82782	1.17218
6	36	216	1	0	3.53983	−1.53983
7	49	343	8	0	1.63135	−.23135
8	64	512	27	0	.80356	.44644
9	81	729	64	1	2.21013	.78987
10	100	1,000	125	8	4.81457	−.81457
11	121	1,331	216	27	7.03288	.16712
12	144	1,728	343	64	7.28103	.01897

Table 10.15 Regression analysis for a cubic piecewise polynomial with two knots

	SS	df	MS	R^2	F	p
Regression	55.6191	5	11.1238	.8876	9.4718	.0082
Residual	7.0465	6	1.1744			
Total	62.6656					
	b	se_b	t	p		
b_0	3.5051	3.4317	1.0214	.3465		
b_1	−4.1892	3.6186	−1.1577	.2910		
b_2	1.9165	1.0634	1.8022	.1216		
b_3	−.2052	.0922	−2.2251	.0677		
b_4	.4887	.1598	3.0590	.0223		
b_5	−.5475	.1598	−3.4273	.0140		

---

[8] Along with a column vector of leading 1's, the first 5 columns of Table 10.14 form a truncated power basis, so called because the values before the knots are truncated (i.e., $= 0$).

Fig. 10.8 Global cubic, piecewise (local) cubic, and raw data

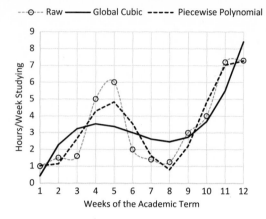

Figure 10.8 compares the fitted values from the global and local model against the raw scores. Looking over the figure, it appears that the local fit hews closer to the data than the global fit. If we test the two models using Eq. (10.5), we find that the piecewise polynomial model does, in fact, fit the data better ($R^2=.8896$) than the global cubic model ($R^2=.6660$), $F_{(2,6)}=5.9122, p=.0381$.

10.3.2 Natural Cubic Splines

Piecewise polynomials often provide a good fit to the data, but they are not without problems. First, the function can oscillate when the predictor is unevenly spaced; second, with increases in the number of knots or degrees, the predictors become highly correlated, creating collinearity (see Chap. 6). These problems can be minimized by using an equivalent, but better conditioned basis function than a truncated power basis. Several are available, including a B-spline basis (De Boors 1972) that is implemented in most statistical packages.

In this section, you will learn to create a cubic spline using a basis function described by Gu (2002) and Wood (2006). A cubic spline is comprised of many cubic polynomial equations joined to form a continuous function.[9] Figure 10.9 shows a simple example with four control points, two interior knots, and the following properties.

- Each curve segment passes through its control point (i.e., the $x-y$ coordinate).
- The curve segments have the same slope where they join (i.e., their first derivatives are constrained to be equal).

---

[9] Cubic splines are similar to the nonparametric smoothers we discussed in Chap. 8, and many textbooks discuss them together. Because splines are not entirely nonparametric, I have chosen to distinguish them from nonparametric smoothers.

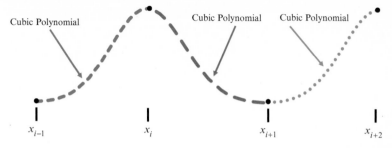

Fig. 10.9 Schematic representation of a cubic spline with four control points and two interior knots

- The curvature of the curve segments is equal where they join (i.e., their second derivatives are constrained to be equal).

  If we further constrain the curves to be linear at their ends, we produce a *natural* cubic spline, which is the variety we will learn.

10.3.2.1 Basis Function for a Natural Cubic Spline

Finding the coefficients for each segment of a natural cubic spline involves solving a system of linear equations subject to the restrictions just described. Each segment is a separate cubic polynomial, so we have four coefficients for each segment.

$$y = a + bx + cx^2 + dx^3 \tag{10.8}$$

Although there are methods for solving these equations directly, a less intuitive but computationally advantageous solution is to apply Eq. (10.9) after transforming the predictor so that all values lie within a range of 0 to 1 (i.e., $x \in [0,1]$).[10]

$$
\begin{cases}
b_1(x) = 1, \\
b_2(x) = x, \\
b_{i+2} = R(x_i, k_i) \text{ for } i = 1 \ldots q - 2 \\[1ex]
\text{where } R(x_i, k_i) = \\[1ex]
\dfrac{\left[(k_i - .5)^2 - \dfrac{1}{12} \right]\left[(x_i - .5)^2 - \dfrac{1}{12} \right]}{4} - \dfrac{\left[(|x_i - k_i| - .5)^4 - .5(|x_i - k_i| - .5)^2 + \dfrac{7}{240} \right]}{24}
\end{cases}
$$

$$\tag{10.9}$$

---

[10] It is only coincidental that one of the terms in Eq. (10.9) uses a denominator that matches our sample size (i.e., 1/12). This value is used regardless of how many observations we have.

It looks complicated, but it's just arithmetic, so let's work through the calculations, beginning with the transformation of our predictor to unit size.[11]

- Subtract the minimum value from all values to transform the smallest value to 0.
- Divide all subtracted values by the maximum value to transform the largest value to 1.

Table 10.16 shows the result of these calculations. The knots associated with weeks 5 and 8 are now represented by $k_1=.3636$ and $k_2=.6364$, respectively.

Table 10.16 Basis function for natural cubic spline with two knots

x	$x =$ Unit scaling	$k_1=.3636$	$k_2=.6364$	$\hat{y}$	e
1	0	−.003540	−.003540	.37271	.62729
2	.0909	−.001610	−.002532	1.92936	−.42936
3	.1818	.000177	−.001462	3.21859	−1.61859
4	.2727	.001617	−.000330	3.99114	1.00886
5	.3636	.002437	.000797	4.01759	1.98241
6	.4545	.002420	.001782	3.24063	−1.24063
7	.5455	.001782	.002420	2.23209	−.83209
8	.6364	.000797	.002437	1.73593	−.48593
9	.7273	−.000330	.001617	2.32724	.67276
10	.8182	−.001462	.000177	3.84636	.15364
11	.9091	−.002532	−.001610	5.96473	1.23527
12	1	−.003540	−.003540	8.37364	−1.07364

Our next step is to compute the two basis functions. To illustrate, the first two values in the third column of Table 10.16 were computed as follows:

$$\frac{\left[(.3636-.5)^2-\tfrac{1}{12}\right]\left[(0-.5)^2-\tfrac{1}{12}\right]}{4}$$
$$-\frac{\left[(|0-.3636|-.5)^4-.5(|0-.3636|-.5)^2+\tfrac{7}{240}\right]}{24}=-.003540$$

and

$$\frac{\left[(.3636-.5)^2-\tfrac{1}{12}\right]\left[(.0909-.5)^2-\tfrac{1}{12}\right]}{4}$$
$$-\frac{\left[(|.0909-.3636|-.5)^4-.5(|.0909-.3636|-.5)^2+\tfrac{7}{240}\right]}{24}=-.001610$$

[11] The transformation to unit length is made "without loss of generality," meaning it does not alter the relation between the predictor and the criterion.

10.3.2.2 OLS Using Basis Function to Predict Criterion

If we perform an OLS regression using the basis function (along with a column of leading 1's) to predict y, we obtain the fitted values shown in Table 10.16. Figure 10.10 compares the fit of a piecewise polynomial and a natural cubic spline. As you can see, they are quite similar, except only the natural cubic spline is linear at the ends.[12]

Fig. 10.10 Comparing the fit of a piecewise polynomial and a natural cubic spline

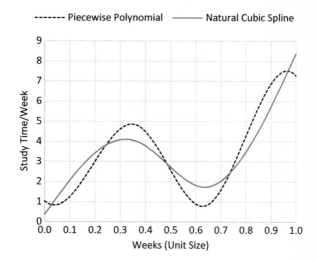

10.3.3 R Code: Unpenalized Regression Splines

```
x=seq(1:12)
y=c(1.0,1.5,1.6,5.0,6.0,2.0,1.4,1.25,3.0,4.0,7.2,7.3)

#Plot
plot(x, y, cex = 1, col ="gray", main = "Scatterplot")
lines(x = x, y = y, lwd = 1, col = "red")

#Linear model
lin.reg <-lm(y~x)
summary(lin.reg)

#Global Cubic
dx=scale(x, center = TRUE, scale = FALSE)
summary(cubic <-lm(y~dx+I(dx^2)+I(dx^3)))
```

(continued)

---

[12] These figures were created using 100 data points to create a smoother curve.

10.3.3   R Code: Unpenalized Regression Splines (continued)

```
#Piecewise cubic spline using two knots
k1=5
k2=8
dum1 <-ifelse(x <= k1, 0, (x-k1))^3
dum2 <-ifelse(x <= k2, 0,(x-k2))^3
summary(piece <-lm(y~x+I(x^2)+I(x^3)+dum1+dum2))

#Compare Fit of Cubic and Piecewise Polynomial
anova(cubic,piece)

##Natural Cubic Spline Without Penalty
x <-seq(1:12)
x<-x-min(x);x<-x/max(x)
y=c(1,1.5,1.6,5,6,2,1.4,1.25,3,4,7.2,7.3)
kn <- c(x[5],x[8])

#Cubic Spline Function
basis<-function(x,z) {
 ((z-0.5)^2-1/12)*((x-0.5)^2-1/12)/4-
 ((abs(x-z)-0.5)^4-(abs(x-z)-0.5)^2/2+7/240)/24
}

# Model Matrix for Unpenalized Natural Cubic Spline
spline.X<-function(x,kn){
 q<-length(kn)+2
 n<-length(x)
 X<-matrix(1,n,q)
 X[,2]<-x
 X[,3:q]<-outer(x,kn,FUN=basis)
 X
}

X<-spline.X(x,kn)
nat.spline<-lm(y~X-1)
fitted(nat.spline)
```

10.3.4   Penalized Natural Cubic Spline

Regression splines avoid the ill-conditioning of a piecewise polynomial, but they leave us with another problem: How many knots should we use and where should they go? This was not an issue in our contrived example because we knew when the exams were given, but this will not always be the case. When it's not, we have two choices. First, we can create an interpolating spline that passes through all of the data points. An interpolating spline will fit the data perfectly, but will be extremely wiggly. Alternatively, we can attempt to balance fit and curvature by selecting a subset of knots and adding a term that penalizes excessive wiggliness. Equation (10.10) formalizes this compromise between fit and wiggliness.

$$S(f) = \sum_{i=1}^{n} [\mathbf{y} - \mathbf{X}\boldsymbol{\beta}]^2 + \lambda \int_{x_{min}}^{x_{max}} \left[f''(x) \right]^2 dx \qquad (10.10)$$

The first part of Eq. (10.10) represents the sum of the squared residuals and the second part penalizes curvature. In this equation, the penalty is the integrated squared second derivative of the regression function. Why use this penalty? Because the second derivative of a straight line equals 0, any departure from a straight line (in either direction) represents curvature, and the squared departures quantify a function's wiggliness.[13]

The smoothing coefficient (λ) controls the weight of the penalty. When $\lambda = 0$, there is no penalty and we will have a wiggly spline that minimizes the residual sum of squares; as $\lambda \rightarrow \infty$, we will have a minimally wiggly function that passes between the data points much like a least squares line (though it need not be linear). What we are looking for, then, is a smoother value that falls between these extremes, providing the best, not-too-wiggly, fit.

10.3.4.1 Select Knots

Our first step is to select a set of knots. Absent any information about their likely location, it is customary to use quantiles (with an equal number of observations in each interval) or knots that are equally far apart (even if the number of observations within each interval varies). Because our raw predictor values are all one unit apart, these approaches are identical with our data set. For our example, we will use quantiles, although the small sample size means there will be very few observations within each interval. There is no hard-and-fast rule, but with 12 observations, 8 knots seems sufficient. Our first quantile is found as 1/9=.1111, our second is found as 2/9=.2222, and so on through 8/9=.8888.

10.3.4.2 Create a Basis Function

Our next step is to use Eq. (10.9) to create a basis function with the quantiles. Table 10.17 presents the result of these calculations. To illustrate, the first two values in the third column were computed as follows:

---

[13] Eilers and Marx (1996) describe another common penalty function.

Table 10.17 Basis function for penalized natural cubic spline with eight quantile knots

	x Unit scaling				Quantiles				
x		.111111	.222222	.333333	.444444	.555556	.666667	.777778	.888889
1	0	.003812	−.000113	−.002984	−.004495	−.004495	−.002984	−.000113	.003812
2	.090909	.002799	.000717	−.001183	−.002473	−.002875	−.002264	−.000668	.001732
3	.181818	.001513	.001299	.000452	−.000533	−.001253	−.001459	−.001055	−.000095
4	.272727	.000086	.001342	.001694	.001182	.000310	−.000546	−.001166	−.001480
5	.363636	−.001195	.000875	.002252	.002458	.001685	.000429	−.000964	−.002301
6	.454545	−.002109	.000189	.002045	.003015	.002676	.001354	−.000480	−.002506
7	.545455	−.002506	−.000480	.001354	.002676	.003015	.002045	.000189	−.002109
8	.636364	−.002301	−.000964	.000429	.001685	.002458	.002252	.000875	−.001195
9	.727273	−.001480	−.001166	−.000546	.000310	.001182	.001694	.001342	.000086
10	.818182	−.000095	−.001055	−.001459	−.001253	−.000533	.000452	.001299	.001513
11	.909091	.001732	−.000668	−.002264	−.002875	−.002473	−.001183	.000717	.002799
12	1	.003812	−.000113	−.002984	−.004495	−.004495	−.002984	−.000113	.003812

$$\frac{\left[(.1111-.5)^2-\frac{1}{12}\right]\left[(0-.5)^2-\frac{1}{12}\right]}{4}-\frac{\left[(|0-.1111|-.5)^4-.5(|0-.1111|-.5)^2+\frac{7}{240}\right]}{24}$$

$$=.003812$$

and

$$\frac{\left[(.1111-.5)^2-\frac{1}{12}\right]\left[(.0909-.5)^2-\frac{1}{12}\right]}{4}$$

$$-\frac{\left[(|.0909-.1111|-.5)^4-.5(|.0909-.1111|-.5)^2+\frac{7}{240}\right]}{24}=.002799$$

If you conduct an OLS regression using this basis function, you will generate *unpenalized* fitted values, similar to what we did before using only two knots.

10.3.4.3 Incorporating the Penalty

Our next step is to incorporate a penalty into our design matrix. The integrated squared second derivative can be written as follows, with **T** representing a matrix to be described momentarily (Ruppert et al. 2003)[14]:

---

[14] In Eq. (10.11), **T** is known as a smoother matrix and is commonly designated **S**. The designation is arbitrary, but we are using **T** because we designated our matrix of simple slopes as **S**.

$$\int_{x_{min}}^{x_{max}} \left[f''(x) \right]^2 dx = \boldsymbol{\beta}' \mathbf{T} \boldsymbol{\beta} \tag{10.11}$$

Substituting for the relevant portion in Eq. (10.10) gives us

$$S(f) = \sum_{i=1}^{n} [\mathbf{y} - \mathbf{X}\boldsymbol{\beta}]^2 + \lambda \boldsymbol{\beta}' \mathbf{T} \boldsymbol{\beta} \tag{10.12}$$

and solving for $\boldsymbol{\beta}$ produces augmented normal equations:

$$\boldsymbol{\beta} = \left(\mathbf{X}'\mathbf{X} + \lambda \mathbf{T} \right)^{-1} \mathbf{X}'\mathbf{y} \tag{10.13}$$

and an augmented hat matrix:

$$\mathbf{H} = \mathbf{X} \left(\mathbf{X}'\mathbf{X} + \lambda \mathbf{T} \right)^{-1} \mathbf{X}' \tag{10.14}$$

Looking over these equations, we see that we incorporate a weighted penalty by including it as part of our design matrix. Doing so involves three steps.

- First, substitute the quantile knot values for the predictors, and use Eq. (10.9) to create another knot matrix $\mathbf{T}$. The top portion of Table 10.18 presents the matrix, and the first value was found as follows:

$$\frac{\left[(.1111 - .5)^2 - \frac{1}{12} \right] \left[(.1111 - .5)^2 - \frac{1}{12} \right]}{4}$$

$$- \frac{\left[(|.1111 - .1111| - .5)^4 - .5(|.1111 - .1111| - .5)^2 + \frac{7}{240} \right]}{24} = .00254$$

- Using procedures described in Chap. 5, perform an eigen decomposition of $\mathbf{T}$ to find its square root:

$$\sqrt{\mathbf{T}} = \mathbf{V}\mathbf{D}\mathbf{V}' \tag{10.15}$$

where $\mathbf{V}$ is a matrix of the eigenvectors of $\mathbf{T}$ and $\mathbf{D}$ is a diagonal matrix of the square root of the eigenvalues of $\mathbf{T}$. Table 10.18 shows the relevant values using our data set.[15]

- Create an augmented matrix, inserting $\sqrt{\mathbf{T}}$ and two rows and columns of zeros as shown in Table 10.19. Notice that I have included a column for the criterion that also includes zeros.

---

[15] Finding the eigenpairs of $\mathbf{T}$ using the procedures described in Chap. 5 requires many iterations, so you might want to save yourself some time by using a statistical package for these calculations.

Table 10.18 Appended knot matrix and eigen decomposition

	.11111	.22222	.33333	.44444	.55556	.66667	.77778	.88889
.11111	.00254	.00088	−.00080	−.00203	−.00251	−.00209	−.00077	.00130
.22222	.00088	.00140	.00107	.00027	−.00054	−.00107	−.00114	−.00077
.33333	−.00080	.00107	.00216	.00210	.00126	.00010	−.00107	−.00209
.44444	−.00203	.00027	.00210	.00300	.00259	.00126	−.00054	−.00251
.55556	−.00251	−.00054	.00126	.00259	.00300	.00210	.00027	−.00203
.66667	−.00209	−.00107	.00010	.00126	.00210	.00216	.00107	−.00080
.77778	−.00077	−.00114	−.00107	−.00054	.00027	.00107	.00140	.00088
.88889	.00130	−.00077	−.00209	−.00251	−.00203	−.00080	.00088	.00254

$$\sqrt{T}=VDV'$$

.028297	.008306	−.003691	−.016264	−.024108	−.023090	−.009772	.010264
.008306	.025059	.014513	.000858	−.008952	−.009385	−.014364	−.010402
−.003691	.014513	.032358	.021042	.009164	−.003775	−.005928	−.019509
−.016264	.000858	.021042	.036099	.019845	.010229	−.006752	−.021636
−.024108	−.008952	.009164	.019845	.034794	.018494	−.000166	−.017789
−.023090	−.009385	−.003775	.010229	.018494	.028255	.014309	−.004651
−.009772	−.014364	−.005928	−.006752	−.000166	.014309	.027962	.011276
.010264	−.010402	−.019509	−.021636	−.017789	−.004651	.011276	.031197

Table 10.19 Form of the augmented matrix for a penalized smoothing spline

1	x	x_{b1}	x_{b2}	x_{b3}	x_{b4}	x_{b5}	x_{b6}	x_{b7}	x_{b8}	y
1	x	x_{b1}	x_{b2}	x_{b3}	x_{b4}	x_{b5}	x_{b6}	x_{b7}	x_{b8}	y
1	x	x_{b1}	x_{b2}	x_{b3}	x_{b4}	x_{b5}	x_{b6}	x_{b7}	x_{b8}	y
1	x	x_{b1}	x_{b2}	x_{b3}	x_{b4}	x_{b5}	x_{b6}	x_{b7}	x_{b8}	y
1	x	x_{b1}	x_{b2}	x_{b3}	x_{b4}	x_{b5}	x_{b6}	x_{b7}	x_{b8}	y
1	x	x_{b1}	x_{b2}	x_{b3}	x_{b4}	x_{b5}	x_{b6}	x_{b7}	x_{b8}	y
1	x	x_{b1}	x_{b2}	x_{b3}	x_{b4}	x_{b5}	x_{b6}	x_{b7}	x_{b8}	y
1	x	x_{b1}	x_{b2}	x_{b3}	x_{b4}	x_{b5}	x_{b6}	x_{b7}	x_{b8}	y
1	x	x_{b1}	x_{b2}	x_{b3}	x_{b4}	x_{b5}	x_{b6}	x_{b7}	x_{b8}	y
1	x	x_{b1}	x_{b2}	x_{b3}	x_{b4}	x_{b5}	x_{b6}	x_{b7}	x_{b8}	y
1	x	x_{b1}	x_{b2}	x_{b3}	x_{b4}	x_{b5}	x_{b6}	x_{b7}	x_{b8}	y
1	x	x_{b1}	x_{b2}	x_{b3}	x_{b4}	x_{b5}	x_{b6}	x_{b7}	x_{b8}	y
0	0	0	0	0	0	0	0	0	0	0
0	0	0	0	0	0	0	0	0	0	0
0	0	$\sqrt{T}$	$\sqrt{T}$	$\sqrt{T}$	$\sqrt{T}$	$\sqrt{T}$	$\sqrt{T}$	$\sqrt{T}$	$\sqrt{T}$	0
0	0	$\sqrt{T}$	$\sqrt{T}$	$\sqrt{T}$	$\sqrt{T}$	$\sqrt{T}$	$\sqrt{T}$	$\sqrt{T}$	$\sqrt{T}$	0
0	0	$\sqrt{T}$	$\sqrt{T}$	$\sqrt{T}$	$\sqrt{T}$	$\sqrt{T}$	$\sqrt{T}$	$\sqrt{T}$	$\sqrt{T}$	0
0	0	$\sqrt{T}$	$\sqrt{T}$	$\sqrt{T}$	$\sqrt{T}$	$\sqrt{T}$	$\sqrt{T}$	$\sqrt{T}$	$\sqrt{T}$	0
0	0	$\sqrt{T}$	$\sqrt{T}$	$\sqrt{T}$	$\sqrt{T}$	$\sqrt{T}$	$\sqrt{T}$	$\sqrt{T}$	$\sqrt{T}$	0
0	0	$\sqrt{T}$	$\sqrt{T}$	$\sqrt{T}$	$\sqrt{T}$	$\sqrt{T}$	$\sqrt{T}$	$\sqrt{T}$	$\sqrt{T}$	0
0	0	$\sqrt{T}$	$\sqrt{T}$	$\sqrt{T}$	$\sqrt{T}$	$\sqrt{T}$	$\sqrt{T}$	$\sqrt{T}$	$\sqrt{T}$	0

Note: x_{bi} denotes a basis function, and $\sqrt{T}$ denotes the square root of the appended knot matrix

10.3.4.4 Calculating the Smoothing Parameter

The final step is to weight the penalty matrix by the smoothing parameter, λ. As noted earlier, if the smoothing parameter is very small, then curvature is not penalized and a wiggly function that (nearly) interpolates the data will provide the best fit. On the other hand, if the smoothing parameter is large, then curvature is penalized heavily, resulting in a relatively flat line of best fit (that could be linear or nonlinear). Figure 10.11 shows examples of each kind. The top half shows a linear fit. When $\lambda = 0$, the plotted function is very wiggly; when the smoothing parameter is increased ($\lambda = 0.5$), the function is flattened out. The bottom half of the figure shows a similar pattern using a quadratic function. As before, the magnitude of the smoothing parameter controls the wiggliness of the line, not its form.

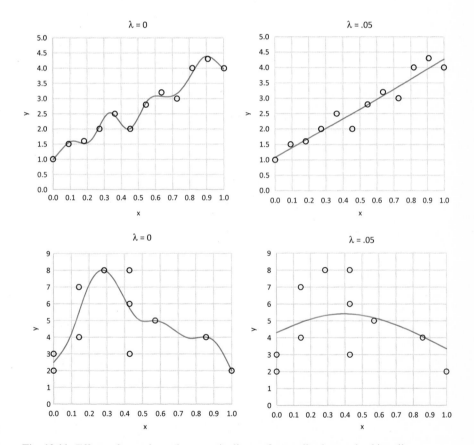

Fig. 10.11 Effects of smoother values on wiggliness of a penalized natural cubic spline

The following steps can be used to find a value of λ that best balances fit and smoothness:

- Create a small starting value for λ (e.g., 1.00 E − 8) and increase it incrementally [e. g., $(\lambda_{new} = \lambda_{old} + \lambda_{old}^{1.5})$] after cycling through the remaining steps.
  - Multiply $\sqrt{T} * \sqrt{\lambda}$ to complete the design matrix.
  - Perform an OLS regression using the design matrix, and calculate the residual sum of squares and hat values (h_i) from the hat matrix.
  - Compute the generalized cross-validation term.

$$\text{GCV} = (N * SS_{res})/\left(N - \sum h_i\right)^2 \qquad (10.16)$$

Similar to the DFFIT statistic described in Chap. 6, this term represents the results from a series of "leave one observation out" analyses, quantifying the averaged squared difference between the omitted value and its fitted value. Small values signify a good fit, so we perform a grid-search looking for the smoother value λ that minimizes the term.

Table 10.20 shows the results from 20 smoother values. As you can see, the smallest GCV value is found when $\lambda = .000022168$.

Table 10.20 Generalized cross-validation term, smoothing parameter, and penalized natural cubic spline: example 1

	λ	GCV	$\hat{y}$
1	.000000577	2.709330598	1.048477
2	.000000865	2.644660469	1.238514
3	.000001297	2.556719041	2.083817
4	.000001946	2.442410940	4.807188
5	.000002919	2.303005798	5.601235
6	.000004379	2.147369542	2.457529
7	.000006568	1.993367637	1.163116
8	.000009853	1.864280784	1.549287
9	.000014779	1.779688145	2.551352
10	**.000022168**	**1.746171002**	4.554544
11	.000033253	1.755721124	6.740053
12	.000049879	1.793659612	7.454887
13	.000074818	1.849238001	
14	.000112227	1.920584570	
15	.000168341	2.011652669	
16	.000252512	2.125307061	
17	.000252512	2.125307061	
18	.000378768	2.258648482	
19	.000568151	2.404439496	
20	.000852227	2.557884776	

Table 10.21 Augmented matrix for penalized cubic spline: example 1

1	.00000	−.000113	−.002984	−.004495	−.004495	−.002984	−.000113	.003812	1.0
1	.09091	.000717	−.001183	−.002473	−.002875	−.002264	−.000668	.001732	1.5
1	.18182	.001299	.000452	−.000533	−.001253	−.001459	−.001055	−.000095	1.6
1	.27273	.001342	.001694	.001182	.000310	−.000546	−.001166	−.001480	5.0
1	.36364	.000875	.002252	.002458	.001685	.000429	−.000964	−.002301	6.0
1	.45455	.000189	.002045	.003015	.002676	.001354	−.000480	−.002506	2.0
1	.54545	−.000480	.001354	.002676	.003015	.002045	.000189	−.002109	1.4
1	.63636	−.000964	.000429	.001685	.002458	.002252	.000875	−.001195	1.3
1	.72727	−.001166	−.000546	.000310	.001182	.001694	.001342	.000086	3.0
1	.81818	−.001055	−.001459	−.001253	−.000533	.000452	.001299	.001513	4.0
1	.90909	−.000668	−.002264	−.002875	−.002473	−.001183	.000717	.002799	7.2
1	1.00000	−.000113	−.002984	−.004495	−.004495	−.002984	−.000113	.003812	7.3
0	0	0	0	0	0	0	0	0	0
0	0	0	0	0	0	0	0	0	0
0	0	.0000482	−.0000349	−.0000835	−.0001029	−.0000929	−.0000496	.0000325	0
0	0	.0001237	.0000659	.0000071	−.0000315	−.0000513	−.0000538	−.0000496	0
0	0	.0000659	.0001454	.0000930	.0000338	−.0000127	−.0000513	−.0000929	0
0	0	.0000071	.0000930	.0001648	.0001045	.0000338	−.0000315	−.0001029	0
0	0	−.0000315	.0000338	.0001045	.0001648	.0000930	.0000071	−.0000835	0
0	0	−.0000513	−.0000127	.0000338	.0000930	.0001454	.0000659	−.0000349	0
0	0	−.0000538	−.0000513	−.0000315	.0000071	.0000659	.0001237	.0000482	0
0	0	−.0000496	−.0000929	−.0001029	−.0000835	−.0000349	.0000482	.0001520	0

Multiplying $\sqrt{T}$ by $\sqrt{\lambda}$ yields the augmented matrix shown in Table 10.21, and performing a regression analysis with this matrix yields the fitted values shown in Table 10.20. Because the smoother value lies very close to 0, the function nearly interpolates the data.

We can quantify the strength of the association shown in Table 10.20 by forming a test statistic known as "deviance explained" (DE).

$$DE = SS_{regression}/SS_{total} \qquad (10.17)$$

This statistic is comparable to the coefficient of determination (R^2) in a linear regression analysis and represents the proportion of the total variance in y that can be explained by the penalized cubic spline. In our example, this statistic is extremely large (DE=.9745), indicating an excellent fit.

10.3.4.5   A Second Example

In our first example, the predictors were unique and evenly spaced, making it easy to compute the sample quantiles. This will not always be the case, so we will finish this section with a second example that requires us to compute the quantiles using a different method. To help us out, imagine I had also asked students "How concerned are you with getting good grades," and we now want to examine how this new variable predicts our criterion (i.e., hours/week spent studying). The new predictor appears in Table 10.22 and unlike our first predictor, (1) the values are not evenly spaced, (2) the values are not sorted in ascending order, and (3) there are duplicates.[16] To accommodate these features, we need to modify the quantiles for our sample.

Table 10.22 Basis function for a second data set with quantile knots

	x	x Unit scaling	Sample quantiles							
			.014815	.050370	.164444	.268148	.423704	.524444	.670222	.813630
1	1.45	.32000	−.002421	−.001740	.000297	.001665	.002016	.001340	−.000077	−.001406
1	2.91	.70933	−.001989	−.001889	−.001463	−.000852	.000420	.001260	.001867	.000877
1	.30	.01333	.007225	.005896	.001826	−.001249	−.004025	−.004388	−.002814	.000901
1	3.35	.82667	.001305	.000827	−.000525	−.001317	−.001476	−.000948	.000346	.001470
1	.35	.02667	.006733	.005547	.001830	−.001050	−.003726	−.004126	−.002716	.000746
1	.75	.13333	.002877	.002666	.001725	.000444	−.001371	−.002031	−.001882	−.000390
1	1.10	.22667	−.000101	.000254	.001184	.001387	.000513	−.000255	−.001032	−.001093
1	.25	.00000	.007716	.006245	.001822	−.001448	−.004325	−.004650	−.002912	.001057
1	2.15	.50667	−.004380	−.003650	−.001332	.000630	.002762	.003099	.001741	−.000810
1	2.25	.53333	−.004334	−.003652	−.001471	.000415	.002585	.003086	.001925	−.000616
1	3.35	.82667	.001305	.000827	−.000525	−.001317	−.001476	−.000948	.000346	.001470
1	4	1.00000	.007716	.006245	.001822	−.001448	−.004325	−.004650	−.002912	.001057

---

[16] The (fabricated) data in Table 10.22 were collected by asking students to place a mark on a line with endpoints (0 = not at all 4 = very) to indicate how concerned they were about getting good grades. The tabled values represent distance from 0.

There are several ways to calculate sample quantiles, and the method we will learn is the default method in **R**.

- After ensuring that all values lie within [0,1], sort the data in ascending order using only unique values, and record their rank order (see Table 10.23).

Table 10.23 Calculations for sample quantiles

Rank	Unique, ordered values	Desired quantile	h	Calculation of sample quantiles using Eq. (10.19)
1	.00	.11111	2.11111	$.01333 + \{.11111 * [(.02667) - .01333]\} = .014815$
2	.01333	.22222	3.22222	$.02667 + \{.22222 * [(.13333) - .02667]\} = .050370$
3	.02667	.33333	4.33333	$.13333 + \{.33333 * [(.22667) - .13333]\} = .164444$
4	.13333	.44444	5.44444	$.22667 + \{.44444 * [(.32000) - .22667]\} = .268148$
5	.22667	.55556	6.55556	$.32000 + \{.55556 * [(.50667) - .50667]\} = .423704$
6	.32000	.66667	7.66667	$.50667 + \{.66667 * [(.53333) - .50667]\} = .524444$
7	.50667	.77778	8.77778	$.53333 + \{.77778 * [(.70933) - .53333]\} = .670222$
8	.53333	.88889	9.88889	$.70933 + \{.88889 * [(.82667) - .70933]\} = .813630$
9	.70933			
10	.82667			
11	1.00			

- Perform the following calculations, where $q =$ the quantile, and designate the integer (j) and the remainder (g).

$$h = [(N - 1)q] + 1 \qquad (10.18)$$

To illustrate, with 11 unique values in our data set, the following calculations produce h for our first value:

$$h = [(11 - 1) * .11111] + 1 = 2.11111$$

and

$$j = 2$$
$$g = .11111$$

- Locate the predictor value associated with j and ($j + 1$) and use the following formula to calculate the sample quantile:

$$q_s = j + \{g * [(j + 1) - j]\} \qquad (10.19)$$

Notice that j refers to the predictor value associated with a particular ordinal position. Using our sample, $j = 2$, refers to our second, sorted predictor (.01333), and $j + 1$ refers to our third, sorted predictor (.02667). Table 10.23 presents the complete calculations.

After computing the sample quantiles, we use them to compute the basis function and an appended knot matrix and its eigen decomposition in the manner described earlier. Finally, we create an augmented design matrix and perform OLS regression using a grid-search method to find a smoother value that minimizes the GCV. If you follow those steps, you will find that $\lambda=.0218416$ and $GCV=4.29011$. Table 10.24 shows the pattern. As you can see, the form of this smoothing spline is quite different from the one we found in our original example and does not fit the data as well (*deviance explained*$=.5173$). Apparently, students have to care a lot about getting good grades before they study hard!

Table 10.24 Generalized cross-validation term, smoothing parameter, and penalized natural cubic spline: example 2

	λ	GCV	$\hat{y}$
1	.00056815	5.93436146	2.323445
2	.00085223	5.70021776	4.222038
3	.00127834	5.45261218	2.329945
4	.00191751	5.20332861	5.208435
5	.00287627	4.96503609	2.318033
6	.00431440	4.74923697	2.230435
7	.00647160	4.56587209	2.216508
8	.00970740	4.42368620	2.341798
9	.01456110	4.33021362	2.961791
10	**.02184164**	**4.29011151**	3.094437
11	.03276247	4.30183034	5.208435
12	.04914370	4.35530204	6.794699
13	.07371555	4.43384697	
14	.11057332	4.51979577	
15	.16585998	4.59979617	
16	.24878998	4.66679353	
17	.37318497	4.71898650	
18	.55977745	4.75766759	
19	.83966617	4.78536911	
20	1.25949926	4.80474924	

10.3.5 R Code: Penalized Natural Cubic Splines

```
#Penalized Splines (Returned values approximate textbook values)
x=seq(1:12)
y=c(1,1.5,1.6,5,6,2,1.4,1.25,3,4,7.2,7.3)
spl.x <-smooth.spline(x,y,spar=.25);fitted(spl.x);spl.x$lambda

z=c(1.45,2.91,.3,3.35,.35,.75,1.10,.25,2.15,2.25,3.35,4)
spl.z <-smooth.spline(z,y,spar=.875);fitted(spl.z);spl.z$lambda

#Create split plots
old.par <- par(mfrow=c(1, 2))
plot (x,fitted(spl.x))
plot(z,fitted(spl.z))
par(old.par)
```

10.4 Chapter Summary

1. Power terms can be used in a linear regression model to analyze curvilinear relations. Two common polynomial models are a quadratic model (x^2) and a cubic model (x^3).

2. When conducting a polynomial regression, all lower-order terms should be entered into the predictive equation before adding higher-order ones. It isn't necessary, but it is also a good idea to center a predictor before computing any of its higher-order terms.

3. The simple slopes in a polynomial regression are calculated by differentiating the function to find its first derivative.

4. Interactions involving polynomial terms can be modeled by creating cross-product terms. Simple effects involving lower-order terms (e.g., linear and quadratic components) can then be tested for statistical significance.

5. Polynomial regression constitutes a global fit, as all observations are raised to a given power (e.g., all terms in a quadratic model are squared). Piecewise polynomials use local fitting (e.g., only some observations are squared). The value above which the observations are altered is known as a knot.

6. Regression splines are piecewise polynomials with relatively few knots, chosen by the researcher based on theory or observation. They can be computed using a truncated power series or an equivalent basis function with greater numerical stability.

7. A cubic spline is comprised of sections of cubic polynomials, constrained to be continuous in value, and equal in their first and second derivatives. A natural cubic spline is further constrained to be linear at the ends.

8. Interpolating splines use as many knots as there are observations. Such splines fit the data perfectly, but are very wiggly.

9. Penalized regression splines use a large subset of knots, but penalize wiggliness to create a smoother function. The weight of the penalty is controlled by the smoothing parameter, λ.

10. Generalized cross-validation is used to find a smoothing parameter that best balances fit and wiggliness.

Chapter 11
Categorical Predictors

The predictors in a linear regression model are usually continuous, not categorical. A categorical variable has clearly defined levels, and the differences between levels are qualitative (either-or) rather than quantitative (more or less). Examples include subject variables (male, female), affiliations (Democrat, Republican, Libertarian), or an experimental treatment (control group vs. experimental group). Historically, analysis of variance (ANOVA) was used to analyze data with categorical predictors, but multiple regression can be used as well. This is because ANOVA and multiple regression are subsets of a broader approach, known as the general linear model. Of the two, multiple regression is the more flexible. Whereas ANOVA can only be used with categorical predictors, multiple regression can be used with continuous and categorical predictors.

The two approaches differ in their typical application. ANOVA is most often used with experimental research conducted under controlled laboratory conditions, and multiple regression is more commonly used in field settings with naturally occurring variables. These distinctions are somewhat arbitrary, because multiple regression can analyze experimentally manipulated or naturally occurring categorical variables.

The biggest practical difference between ANOVA and multiple regression is that a coding scheme is needed to model the categorical predictors when using multiple regression, such that each of the J groups needs a unique code on a set of $J - 1$ vectors. Creating these codes has pros and cons. On the one hand, constructing a coding scheme offers more flexibility and forces us to think carefully about the statistical hypotheses we wish to test; on the other hand, constructing a coding scheme takes time, especially as the number of groups increases. Considering these issues, I recommend that you learn the coding method and then decide whether you want to use ANOVA or multiple regression when you analyze data with categorical predictors.

Electronic Supplementary Material: The online version of this chapter (doi: 10.1007/978-3-319-11734-8_11) contains supplementary material, which is available to authorized users

11.1   Coding Schemes

Imagine a fitness trainer claims to have invented a painless way to get in shape
called the Ripomatic 450. Unlike traditional exercise that requires exertion and
sweat, the Ripomatic 450 builds muscle tone effortlessly while you sleep. The
trainer conducts an experiment to test his invention. He randomly assigns 12 subjects to 1 of 3 groups—(1) a no exercise control condition, (2) a traditional weight
training condition, or (3) the Ripomatic 450—with 4 subjects in each group.[1] After
several weeks, he measures muscle tone using a 1–9 scale. Table 11.1 presents the
(supposed) data, along with three coding schemes we can use to analyze them, and
Fig. 11.1 presents the means.

11.1.1   Analysis of Variance

Before learning how to use multiple regression to analyze the data in Table 11.1,
let's analyze them using ANOVA. As you might know, an ANOVA partitions the

Table 11.1 Small sample example using coding schemes for categorical predictors with equal
cell sizes

Subject	Group	y	Orthogonal contrast codes				Dummy codes (group 1 as reference)		Effect codes (group 3 as base)	
			Whole numbers		Fractions					
1	Control	2	−1	−1	−.5	−.3333	0	0	1	0
2	Control	3	−1	−1	−.5	−.3333	0	0	1	0
3	Control	4	−1	−1	−.5	−.3333	0	0	1	0
4	Control	3	−1	−1	−.5	−.3333	0	0	1	0
5	Weights	6	0	2	0	.6667	1	0	0	1
6	Weights	2	0	2	0	.6667	1	0	0	1
7	Weights	7	0	2	0	.6667	1	0	0	1
8	Weights	9	0	2	0	.6667	1	0	0	1
9	Ripomatic 450	7	1	−1	.5	−.3333	0	1	−1	−1
10	Ripomatic 450	6	1	−1	.5	−.3333	0	1	−1	−1
11	Ripomatic 450	6	1	−1	.5	−.3333	0	1	−1	−1
12	Ripomatic 450	7	1	−1	.5	−.3333	0	1	−1	−1

---

[1] Notice that the number of subjects in each group is equal in this example. Later in this chapter, we
will learn how to accommodate unbalanced cell sizes.

Fig. 11.1 Group means from small sample example in Table 11.1

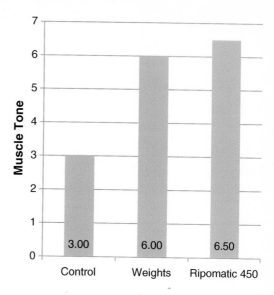

total variance in a criterion into two parts—a between-group sum of squares (SS_{bg}) and a residual sum of squares (SS_{res}):[2]

- The between-group component is found by computing the deviation of each group mean $(\overline{Y}_j)$ from the grand mean $(\overline{Y})$, multiplying the squared deviations by the number of subjects in each group (n_j), and summing the weighted squared deviations:

$$SS_{bg} = \sum_j n_j (\overline{Y}_j - \overline{Y})^2 \tag{11.1}$$

The mean muscle tone in our example is 5.1667, and plugging in the remaining values yields our between-group sum of squares:

$$SS_{bg} = 4(3 - 5.1667)^2 + 4(6 - 5.1667)^2 + 4(6.5 - 5.1667)^2 = 28.6667$$

- The residual sum of squares is found by summing the squared deviations of each score (y_{ij}) from its group mean $(\overline{Y}_j)$:

$$SS_{res} = \sum_j \sum_i (y_{ij} - \overline{Y}_j)^2 \tag{11.2}$$

---

[2] The residual sum of squares is sometimes called the within-subject sum of squares.

Plugging in our values produces our residual sum of squares:

$$
\begin{aligned}
SS_{res} = &\left[(2-3)^2 + (3-3)^2 + (4-3)^2 + (3-3)^2 \right] \\
&+ \left[(6-6)^2 + (2-6)^2 + (7-6)^2 + (9-6)^2 \right] \\
&+ \left[(7-6.5)^2 + (6-6.5)^2 + (6-6.5)^2 + (7-6.5)^2 \right] = 29.00
\end{aligned}
$$

- We can derive the total sum of squares by summing the previous components or by summing the squared deviation of each score from the grand mean:

$$
SS_{tot} = \sum_j \sum_i \left(y_{ij} - \overline{Y} \right)^2 \tag{11.3}
$$

   In our example, the total sum of squares $= 28.6667 + 29.00 = 56.6667$.
- Finally, we can form an F ratio and test the significance of the between-group effect:

$$
F = \frac{SS_{bg}/df_{bg}}{SS_{res}/df_{res}} \tag{11.4}
$$

The degrees of freedom for the between-group effect is $J - 1$, and the degrees of freedom for the residual effect is $J(n_j - 1)$.

Table 11.2 summarizes the calculations from our example in a form used by many statistical packages. As you can see, the between-group effect is significant, indicating that the variability among the three groups is unlikely to be due to chance alone. At this point, it would be customary to perform post hoc comparisons to pinpoint the source of this variability.

Table 11.2 Analysis of variance summary table for small sample example in Table 11.1

	SS	df	MS	F	p
Groups	28.6667	2	14.3333	4.4483	.0454
Residual	29.00	9	3.2222		
Total	57.6667				

11.1.2 Overview of Coding Schemes

Now that we know how to analyze categorical data using ANOVA, let's look at how multiple regression can produce the same results. As indicated earlier, we begin by constructing a coding scheme. Before describing them in detail, let's consider why we need them at all. Suppose we conduct a study with only two groups, assigning a "1" to group 1 and a "2" to group 2. If we then use this categorical designation to predict some criterion, the regression sum of squares will represent a 1 *df* test of group differences.

Now suppose we have three groups in our study, as in our exercise example. It might seem that all we need to do is assign a "3" to the third group, but this is not so. As we have seen, with categorical data, we model the variability among the groups by subtracting each group's mean from the overall mean. Consequently, each of the $J = 3$ groups needs a unique code on $J - 1$ vectors. When we then regress our criterion on these vectors, the test of our regression model yields a 2 *df* test of group differences, just as with an ANOVA.

Table 11.3 describes three common coding schemes: orthogonal contrast codes, dummy codes, and effect codes. With three groups, each coding scheme is comprised of two vectors, with each vector representing a 1 *df* contrast among group means. A contrast can specify a simple comparison between two groups (e.g., compare the average of group 1 vs. the average of group 2) or designate a contrast involving multiple groups (e.g., contrast the average of group 1 vs. the combined average of groups 2 and 3). Table 11.3 shows that coding schemes vary along this dimension (i.e., some can handle multiple-group comparisons, whereas others can accommodate only a single comparison). Coding schemes can also be differentiated along two other dimensions: do the column entries sum to zero and do the cross-product terms sum to zero. The importance of these dimensions will be described momentarily.

Table 11.3   Comparisons among three coding schemes

Coding dimension	Orthogonal contrast codes	Dummy codes	Effect codes
Can handle multiple-group contrasts	Yes	No	Yes
Column entries sum to zero	Yes	No	Yes
Cross products sum to zero	Yes	Yes	No

11.1.3   Orthogonal Contrast Codes

We will first discuss orthogonal contrast codes because this is the scheme most commonly associated with ANOVA. Table 11.4 illustrates this coding scheme for a three-group design, using whole numbers for clarity. The entries specify the weight we assign to each group mean. The first vector compares group 1 against group 3, and the second vector compares group 2 against the average of groups 1 and 3. Notice, then, that orthogonal contrast codes can handle simple comparisons and multiple contrasts.

Table 11.4 Orthogonal
contrast codes for a three-
group design

Group	Vector 1	Vector 2	Cross product
Group 1	−1	−1	1
Group 2	0	2	0
Group 3	1	−1	−1
Σ	0	0	0

Notice also that the entries within each column sum to zero and the sum of the cross product equals zero. Collectively, these features create orthogonal contrast codes. When we have the same number of subjects in each group, we can express these conditions using Eq. (11.5):[3]

$$\Sigma c_{ij} = \Sigma c_{ij} c_{kj} = 0 \qquad (11.5)$$

Orthogonal contrast codes produce uncorrelated regression coefficients. This independence appears as a diagonal covariance matrix (i.e., variances on the diagonals and 0's on the off diagonals) with two related consequences. First, each regression coefficient is calculated independently of the other coefficients; second, in combination, the regression coefficients generated from orthogonal contrasts account for all of the between-group variance in our criterion.

It is useful (though not essential) to convert orthogonal contrast codes to fractions by dividing each whole number by the number of groups being compared.[4] There are two groups being compared in the first contrast, so the contrast weights become −.5, 0, and .5; there are three groups being compared in our second contrast, so the weights become −.3333, .6667, and−.3333. If we then regress y on these vectors in the usual manner (i.e., including a vector of leading 1's), we get the output shown in Table 11.5. The top portion matches the output you get using ANOVA. As before, we find a significant group effect, indicating that the variability among our three groups is unlikely to be due to chance alone. The R^2 value (which ordinarily would not appear in an ANOVA table) shows that, collectively, our vectors explain ~50 % of the variance in muscle tone.

---

[3] Technically, two *vectors* are orthogonal if their cross products sum to zero, regardless of whether their individual column values sum to zero. However, two *contrasts* are orthogonal only if both criteria are met.

[4] The advantage of using fractional orthogonal codes is that the regression coefficients represent differences between group means.

Table 11.5 Regression output using orthogonal contrast coding

			Significance test of regression model			
	SS	df	MS	R^2	F	p
Regression	28.6667	2	14.3333	.4971	4.4483	.0454
Residual	29.00	9	3.2222			
Total	57.6667					

			Regression coefficients			
	b	se_b	SS	t	p	Explanation
b_0	5.1667	.5182		9.9706	.0000	Unweighted grand mean
b_1	3.50	1.2693	24.50	2.7574	.0222	Group 1 vs. group 3
b_2	1.25	1.0992	4.1667	1.1371	.2848	Group 2 vs. average of groups 1 and 3

Covariance matrix		
.2685	.0000	.0000
.0000	1.6111	.0000
.0000	.0000	1.2083

11.1.3.1 Interpreting the Regression Coefficients

Now let's look at the regression coefficients themselves. The intercept represents the *unweighted* grand mean, found by first calculating the mean of each group and then averaging the group means:[5]

$$M_u = \frac{\Sigma \bar{y}_j}{J} \qquad (11.6)$$

Plugging in our values yields the value for b_0 shown in Table 11.5:

$$M_u = \frac{(3 + 6 + 6.5)}{3} = 5.1667$$

The b_1 coefficient ($b_1 = 3.50$) represents the difference between the mean of group 3 (Ripomatic 450) and the mean of group 1 (control). When we divide the coefficient by its standard error, we get the t value for this two-group comparison. As you can see, the effect is significant in our example, indicating that the

---

[5] We will define the weighted grand mean in the last section of this chapter. For now, we note that the unweighted and weighted grand means are the same with equal group sizes. Finally, you might find it helpful to think of the unweighted mean as the *equally weighted* mean because we are giving equal weight to each group. (Since the weights are equal, it's the same as giving no weight at all, which is why it's called unweighted.)

Ripomatic 450 is significantly different from the control condition. Applying Eq. (11.7), we can also calculate the coefficient's sum of squares:

$$SS_{b_i} = \left(\frac{b_i}{se_{b_i}}\right)^2 * MS_{res} \tag{11.7}$$

Plugging in our numbers produces the sum of squares associated with b_1:

$$SS_{b_1} = \left(\frac{3.5}{1.2693}\right)^2 * 3.2222 = 24.50$$

The b_2 coefficient ($b_2 = 1.25$) represents the difference between the mean of group 2 (weights) and the average of groups 1 and 3. Dividing this coefficient by its standard error yields the t value for this contrast, which, in our example, is not significant. The sum of squares associated with this coefficient can also be found using Eq. (11.7):

$$SS_{b_2} = \left(\frac{1.25}{1.0992}\right)^2 * 3.2222 = 4.1667$$

11.1.3.2 Orthogonal Contrasts Reproduce the Regression Sum of Squares

Looking at the covariance matrix, we see that all of the off-diagonal elements $= 0$. This is because these contrasts are orthogonal. To understand why this is important, let's add up the SS values of our two regression coefficients:

$$24.50 + 4.1667 = 28.6667$$

Notice that the total equals the SS_{reg} entry in the ANOVA table shown in Table 11.2. This is no coincidence. With three groups, the 2 df between-groups term in an ANOVA represents the sum of two orthogonal contrasts.

To reinforce this point, let's consider a different set of orthogonal contrasts shown in Table 11.6. The first vector compares group 1 vs. group 2, and the second vector contrasts group 3 against the average of groups 1 and 2.

Table 11.6 Second set of orthogonal contrast codes for a three-group design

Group	Vector 1	Vector 2	Cross product
Group 1	−.5	−.3333	−.16667
Group 2	.5	−.3333	.16667
Group 3	0	.6667	0
Σ	0	0	0

When we repeat our regression analysis substituting these codes, we get the output shown in Table 11.7. Looking things over, we see that tests of our overall regression model, b_0 coefficient, and covariance matrix remain unchanged, but our two regression coefficients are not the same as before. Now, b_1 represents the difference between the mean of group 2 (weights) and the mean of group 1 (control), and b_2 represents the difference between the mean of group 3 (Ripomatic 450) and the average of groups 1 and 2. Nevertheless, if we add the sum of squares associated with these tests, we still obtain SS_{reg} (18.00 + 10.6667 = 28.6667). This is because the contrasts are orthogonal and, in combination, represent all of the explained variance in our criterion.

Table 11.7 Regression coefficients and covariance matrix from second set of orthogonal contrast codes

			Significance test of regression model				
	SS	df	MS	R^2	F	p	
Regression	28.6667	2	14.3333	.4971	4.4483	.0454	
Residual	29.00	9	3.2222				
Total	57.6667						
			Regression coefficients				
	b	se_b	SS	t	p	Explanation	
b_0	5.1666	.5182		9.9706	.0000	Grand mean	
b_1	3.00	1.2693	18.00	2.3635	.0424	Group 1 vs. group 2	
b_2	2.00	1.0992	10.6667	1.8194	.1022	Group 3 vs. average of groups 1 and 2	

Covariance matrix		
.2685	.0000	.0000
.0000	1.6111	.0000
.0000	.0000	1.2083

11.1.3.3   Alternative Formula for Computing 1 *df* Contrasts

The sum of squares associated with a 1 *df* contrast can also be directly calculated from the group means:

$$SS_{contrast} = \frac{\Sigma(c_i\bar{y})^2 * m_h}{\Sigma c_i^2} \qquad (11.8)$$

where c_i = the contrast coefficient, $\bar{y}$ = the mean involved in the comparison, and m_h = the harmonic mean. The harmonic mean is found using Eq. (11.9):

$$m_h = \frac{j_c}{\Sigma\left(\frac{1}{n_j}\right)} \qquad (11.9)$$

where $j_c =$ the number of groups involved in the comparison and $n_j =$ the sample size of each group. When the number of subjects in each group is equal (as is true in our example), computing the harmonic mean is the same as taking the size of any group; but when the sample sizes are unequal, as they will be later in the chapter, an adjustment is needed. For clarity, we will compute the term now and then use Eq. (11.8) to calculate the regression sum of squares for the comparisons in Table 11.7:

$$m_h = \frac{3}{\frac{1}{4} + \frac{1}{4} + \frac{1}{4}} = \frac{3}{.75} = 4$$

- Compare group 1 vs. group 2:

$$SS_{contrast} = \frac{[(-.5 * 3) + (.5 * 6)]^2 * 4}{(-.5^2 + . - 5^2)} = \frac{1.5^2 * 4}{.5} = 18.00$$

- Compare group 3 vs. groups 1 and 2:

$$SS_{contrast} = \frac{[(-.3333 * 3) + (-.3333 * 6) + (.6667 * 6.5)]^2 * 4}{(-.3333^2 + . - 3333^2 + .6667^2)} = \frac{1.3333^2 * 4}{.6667} = 10.6667$$

These values match the ones we found using multiple regression. If you like, you can apply the formula to derive the values from Table 11.5.

11.1.3.4   Fitted Values Equal the Group Mean

When we use a set of codes to define our groups, everyone in the group receives the same code. So our expectation is that everyone's score will equal the group mean and that all variations from the group mean represent error. To better appreciate this point, let's compute our fitted values in the usual manner by multiplying **Xb**. Using the regression coefficients from Table 11.7 produces the fitted values shown in Table 11.8. As you can see, each fitted value equals its group mean, which is why the residual sum of squares in an ANOVA is found by summing the squared deviations of each score from its associated group mean [see Eq. (11.3)]. Because the fitted values *are* the group means, we are simply substituting the group mean for the fitted value in our usual equation:

$$\left[SS_{res} = \sum (y - \hat{y})^2 \right]$$

Moreover, this will be true no matter which coding scheme we use to analyze our data.

Table 11.8 Fitted values for a three-group design

Subject	Group	$\hat{y}$
1	Control	3.0
2	Control	3.0
3	Control	3.0
4	Control	3.0
5	Weights	6.0
6	Weights	6.0
7	Weights	6.0
8	Weights	6.0
9	Ripomatic 450	6.5
10	Ripomatic 450	6.5
11	Ripomatic 450	6.5
12	Ripomatic 450	6.5

11.1.4 Dummy Codes

Table 11.9 shows our next set of codes, called dummy codes. With dummy coding, we designate one of our groups to be a reference group. This group receives a score of 0 on all vectors. It is customary to use the control condition as the reference group, so I have selected group 1 to be the reference group in our example. Each of the other groups then receives a score of 1 or 0 on the remaining vectors, such that each column sum equals 1 and the cross-product sum equals 0. Notice, then, that even though the cross-product sum equals 0, these codes are *not* orthogonal because the columns do not sum to 0.

Table 11.9 Dummy codes for a three-group design

Group	Vector 1	Vector 2	Cross product
Group 1	0	0	0
Group 2	1	0	0
Group 3	0	1	0
Σ	1	1	0

Table 11.1 shows how this coding scheme can be implemented with our data. If we regress y on these vectors in the usual manner, we get the results shown in Table 11.10.

11.1.4.1 Test of the Regression Model

The test of our overall regression model is identical to the one we found using contrast codes. Because all coding schemes produce the same fitted values, this equivalence will occur no matter which coding scheme we use.

Table 11.10 Regression output using dummy coding

			Significance test of regression model			
	SS	df	MS	R^2	F	p
Regression	28.6667	2	14.3333	.4971	4.4483	.0454
Residual	29.00	9	3.2222			
Total	57.6667					
			Regression coefficients			
	b	se_b	SS	t	p	Explanation
b_0	3.0	.8975		3.3425	.0086	Mean of reference group (group 1)
b_1	3.0	1.2693	18.00	2.3635	.0424	Group 1 vs. group 2
b_2	3.5	1.2693	24.50	2.7574	.0222	Group 1 vs. group 3

Covariance matrix		
.8056	−.8056	−.8056
−.8056	1.6111	.8056
−.8056	.8056	1.6111

11.1.4.2   Interpreting the Regression Coefficients

Our regression coefficients, however, are not the same as the ones we observed using orthogonal contrast codes. Now, b_0 represents the mean of our reference group (group 1), b_1 represents the difference between groups 2 and 1 ($6 - 3.0 = 3.0$), and b_2 represents the difference between groups 3 and 1 ($6.5 - 3.0 = 3.5$). The t-tests associated with these coefficients indicate whether the means being compared are significantly different from each other. In our example, we see that the control group differs from each of the two experimental conditions.

Finally, the off-diagonal entries of the covariance matrix do not equal 0. This is because dummy coding does not produce orthogonal comparisons. As a consequence, adding the coefficient sum of squares does not reproduce SS_{reg}:

$$18.00 + 24.50 \neq 28.6667$$

11.1.5   Effect Codes

Table 11.11 shows our final set of codes, called effect codes. With effect coding, we designate one of our groups to be a base group and assign this group a score of -1 on both vectors. In our example, I have set group 3 (the Ripomatic 450) to be the base group. Each of the other groups then receives a score of 1 or 0 on the remaining vectors, such that each column sum equals 0 and the cross-product sum equals 1. Because the cross products do not sum to 0, these codes are not orthogonal.

The final columns in Table 11.1 illustrate the use of this coding scheme with our data set, and regressing y on these vectors produces the results shown in Table 11.12.

Table 11.11 Effect codes for a three-group design

Group	Vector 1	Vector 2	Cross product
Group 1	1	0	0
Group 2	0	1	0
Group 3	−1	−1	1
Σ	0	0	1

Table 11.12 Regression output using effect coding

			Significance test of regression model				
	SS	df	MS	R^2	F	p	
Regression	28.6667	2	14.3333	.4971	4.4483	.0454	
Residual	29.00	9	3.2222				
Total	57.6667						

				Regression coefficients			
	b	se_b	SS	t	p	Explanation	
b_0	5.1667	.5182		9.9707	.0000	Unweighted grand mean	
b_1	−2.1667	.7328	28.1667	2.9566	.0160	Group 1 − grand mean [=compare group 1 vs. average of groups 2 and 3]	
b_2	.8333	.7328	4.1667	1.1371	.2848	Group 2 − grand mean [=compare group 2 vs. average of groups 1 and 3]	

Covariance matrix		
.2685	.0000	.0000
.0000	.5370	−.2685
.0000	−.2685	.5370

11.1.5.1 Interpreting the Regression Coefficients

As before, the test of our regression model is identical to the one we found when using contrast codes and dummy codes, but the regression coefficients differ. Here, our intercept represents the unweighted grand mean ($b_0 = 5.1667$), b_1 is the difference between the mean of group 1 and the unweighted grand mean, and b_2 is the difference between the mean of group 2 and the unweighted grand mean. Thus, with effect codes, each group receiving a "1" is compared against the unweighted grand mean. In some cases, we can think of these effects as "treatment effects," because they indicate how far a mean is from the overall average.

Effect codes also compare the average of one group vs. the average of all other groups. Why? Since the unweighted grand mean is formed by averaging the group means, each coefficient represents a comparison between the group receiving "1" and all other groups combined. So, the t-test for b_1 is a test of group 1 vs. groups 2 and 3, and the t-test for b_2 is a test of group 2 vs. groups 1 and 3. In our example, the first comparison shows that the control condition is significantly different from the two experimental conditions.

We can best appreciate this point by reexamining our three group means. Looking back to Fig. 11.1, we find control condition ($M = 3.00$), weights ($M = 6.00$), and Ripomatic 450 ($M = 6.50$). Now, let's multiply these values by contrast weights that represent a group 1 vs. groups 2 and 3 pattern:

$$(3.00 * .6667) + (6.00 * -.3333) + (6.50 * -.3333) = -2.1667$$

When we do, we find that the sum equals our b_1 coefficient. That's because the b_1 coefficient describes the slope of the line connecting the mean of the control condition to the average of the other two groups. We find the same pattern for b_2[6]:

$$(3.00 * -.3333) + (6.00 * .6667) + (6.50 * -.3333) = .8333$$

It's important to note that a base group in effect coding is not the same as a reference group in dummy coding. With dummy coding, the reference group is compared against each of the other groups, one at a time; with effect coding, the group with a "1" is tested against all other groups at once, and the base group (with -1 for both vectors) is never compared individually against any other group.

11.1.5.2   Covariance Matrix

Glancing at the covariance matrix in Table 11.12, we see that not all of the off-diagonal entries equal zero. This is because effect codes do not create orthogonal comparisons. Consequently, if we sum the regression coefficient sum of squares, our total does not equal SS_{reg}:

$$28.1667 + 4.1667 \neq 28.6667$$

11.1.5.3   Effect Contrast Vectors Must Be Considered as a Unit

If you examine the coefficients for our first effect vector $(1, 0, -1)$, you might assume that its associated regression coefficient would provide a simple comparison between groups 1 and 3. Similarly, you might assume that our second effect vector $(0, 1, -1)$ provides a simple comparison between groups 2 and 3. These assumptions would be true if the contrast vectors were orthogonal, but they are not. Recall from Chap. 4 that when two predictors are not orthogonal, the regression coefficients are characterized by interdependence (i.e., the value of a regression coefficient depends on both vectors, not just the one with which it is associated). In the present case, even though the codes seem to indicate a two-group comparison, the regression coefficients actually produce a contrast in which one group is

---

[6] Notice that the t value for b_2 in Table 11.12 matches the t value for b_2 in Table 11.5. This is because both coefficients are tests of group 2 vs. groups 1 and 3.

contrasted against the average of the other two. Only with orthogonal vectors can we interpret the contrasts in one vector independent of the codes in other vectors.

11.1.6 Summary

In this section, we have discussed three coding schemes. They all produce the same overall regression results, but each offers different information regarding comparisons among the groups. So which one should you use? The answer is it depends on your specific hypotheses:

- Use orthogonal contrast codes when you want to test independent hypotheses that reproduce the regression sum of squares.
- Use dummy codes when you want to compare one group against each of the others.
- Use effect codes when you want to test one group vs. the average of all of the other groups.

11.1.7 R Code: Coding Schemes

```
grp <-c(1,1,1,1,2,2,2,2,3,3,3,3)
y <-c(2,3,4,3,6,2,7,9,7,6,6,7)

#ANOVA
anova.mod <-aov(y~factor(grp))
summary(anova.mod)

#Create Chart
grpmeans <- tapply(y, grp, mean)
barplot(grpmeans, main="Fitness by Condition",
col=c("black","gray","black"), density=c(20,15,10), angle=c(30,20,0),
xlab= "Conditions", names = c("Control","Weights","Ripomatic 450"),
ylim = c(0, 7), ylab = "Fitness")

#Create coding scheme function for a balanced 3-grp design
codes.3 <-function(y,a,b,c,d,e,f){
n=length(y)/3
c1 <-c(rep(a,4),rep(b,4),rep(c,4))
c2 <-c(rep(d,4),rep(e,4),rep(f,4))
mod <-lm(y~c1+c2)
}

#Enter criterion and coding scheme
summary(ortho1 <-codes.3(y, -.5, 0, .5, -1/3, 2/3, -1/3))
summary(dummy <-codes.3(y, 0, 1, 0, 0, 0, 1))
summary(effect <-codes.3(y, 1, 0, -1, 0, 1, -1))
```

11.2   Creating Orthogonal Contrast Codes

Each coding scheme has advantages, but orthogonal contrast codes are the most flexible. They allow us to test single mean comparisons (as with dummy codes), one group vs. all other group contrasts (as with effect codes), and any other hypothesis of interest.[7] In order to take advantage of this flexibility, however, our vectors must be orthogonal. The question arises, then, as to how to create orthogonal contrast codes. We know we can check to see if they are orthogonal by verifying that the columns and cross products sum to 0, but is there a way other than trial and error to ensure that these conditions are met?

The answer is "yes." In fact, we can use several approaches. Before describing them, let's imagine another fitness expert is interested in the effects of exercise on muscle flexibility. She randomly assigns subjects to one of four exercise conditions—aerobics, weights, Pilates, and yoga—measures flexibility 4 weeks later, and then wishes to use a set of orthogonal contrasts to analyze the data.

11.2.1   Helmert Contrasts

Table 11.13 shows one set of orthogonal contrasts, using whole numbers for clarity. The contrasts, formally known as Helmert contrasts, were generated by following a simple rule of thumb: Start with the broadest possible contrast (1 group vs. the other 3) and then narrow your comparison by omitting the group with the largest weight from subsequent comparisons. In terms of our example, the first vector contrasts yoga against the other three forms of exercise, the second compares Pilates against weights and aerobics, and the third compares weights vs. aerobics. We could easily change the groups involved in these comparisons by rearranging the order of the groups (i.e., we could list yoga first and weights last), but once we settle on an order, we must retain it across all three contrasts.

Table 11.13 Helmert contrasts for a four-group design with cross-product terms

Group	Contrast 1	Contrast 2	Contrast 3	Cross product_12	Cross product_13	Cross product_23
Aerobics	-1	-1	-1	1	1	1
Weights	-1	-1	1	1	-1	-1
Pilates	-1	2	0	-2	0	0
Yoga	3	0	0	0	0	0
Σ	0	0	0	0	0	0

---

[7] In Chap. 12, we will learn how effect codes can be modified to test a variety of contrasts.

11.2.2   Gram-Schmidt Orthogonalization

The rule of thumb approach is convenient, but it doesn't necessarily test hypotheses of interest, which, after all, is the entire point of using orthogonal contrasts. For this reason, it is useful to learn another approach known as the Gram-Schmidt orthogonalization. The approach works best when we first use a theory to devise one or more orthogonal contrasts and then use the Gram-Schmidt procedure to generate the remaining contrasts.

11.2.2.1   Generating the Coefficients

To illustrate, let's assume our researcher first wishes to test whether the combination of Pilates and yoga yields greater flexibility than the combination of aerobics and weights. The coefficients for this test appear in Table 11.14. Suppose further that the researcher believes that yoga is superior to Pilates; the second set of contrast coefficients in Table 11.14 tests this contrast. Notice that these contrasts are orthogonal (i.e., each sums to zero, and their cross product sums to zero).

Table 11.14 Gram-Schmidt orthogonalization technique

Group	Contrast 1	Contrast 2	Seed 3	Contrast 3
Aerobics	−.25	0	1	−.5
Weights	−.25	0	2	.5
Pilates	.25	−.5	3	0
Yoga	.25	.5	4	0
Σ	0	0	10	0

In order to test both of these hypotheses in a single analysis, the researcher must find a third set of codes that is orthogonal to the other two. This is where the Gram-Schmidt procedure comes in:

$$v_k = c_k - \sum_{j=1}^{k-1} \frac{v_j' c_k v_j}{v_j' v_j} \qquad k = 1, \ldots, n \qquad (11.10)$$

Starting with an arbitrary seed vector (c_3), we solve Eq. (11.10) for our third orthogonal vector (v_3):

$$v_3 = c_3 - \frac{v_1' c_3 v_1}{v_1' v_1} - \frac{v_2' c_3 v_2}{v_2' v_2}$$

To illustrate, we will create a seed vector

$$c_3 = \begin{bmatrix} 1 \\ 2 \\ 3 \\ 4 \end{bmatrix}$$

and plug in the remaining values to solve the equation

$$v_3 = \begin{bmatrix} 1 \\ 2 \\ 3 \\ 4 \end{bmatrix} - \frac{\begin{bmatrix} -.25 \\ -.25 \\ .25 \\ .25 \end{bmatrix} \begin{bmatrix} 1 & 2 & 3 & 4 \end{bmatrix} \begin{bmatrix} -.25 \\ -.25 \\ .25 \\ .25 \end{bmatrix}}{\begin{bmatrix} -.25 & -.25 & .25 & .25 \end{bmatrix} \begin{bmatrix} -.25 \\ -.25 \\ .25 \\ .25 \end{bmatrix}} - \frac{\begin{bmatrix} 0 \\ 0 \\ -.5 \\ .5 \end{bmatrix} \begin{bmatrix} 1 & 2 & 3 & 4 \end{bmatrix} \begin{bmatrix} 0 \\ 0 \\ -.5 \\ .5 \end{bmatrix}}{\begin{bmatrix} 0 & 0 & -.50 & .50 \end{bmatrix} \begin{bmatrix} 0 \\ 0 \\ -.5 \\ .5 \end{bmatrix}} = \begin{bmatrix} 2 \\ 3 \\ 2.5 \\ 2.5 \end{bmatrix}$$

If we then express the vector in deviate form, we find our third orthogonal contrast code:

$$v_3 = \begin{bmatrix} -.5 \\ .5 \\ 0 \\ 0 \end{bmatrix}$$

In terms of our example, this vector contrasts weight training vs. aerobics.

11.2.2.2   Using the Coefficients in a Regression Analysis

Now let's go ahead and use these vectors in a regression analysis. Table 11.15 presents some (fictitious) data, along with the orthogonal contrast codes we have created, and Fig. 11.2 presents the means for each group.

	Subject	Exercise	y	c_1	c_2	c_3
Table 11.15 Orthogonal coding using Gram-Schmidt orthogonalization	1	Aerobics	6	−.25	0	−.5
	2	Aerobics	6	−.25	0	−.5
	3	Aerobics	2	−.25	0	−.5
	4	Weights	1	−.25	0	.5
	5	Weights	2	−.25	0	.5
	6	Weights	2	−.25	0	.5
	7	Pilates	7	.25	−.5	0
	8	Pilates	7	.25	−.5	0
	9	Pilates	8	.25	−.5	0
	10	Yoga	8	.25	.5	0
	11	Yoga	4	.25	.5	0
	12	Yoga	6	.25	.5	0

Fig. 11.2 Group means
for data in Table 11.15

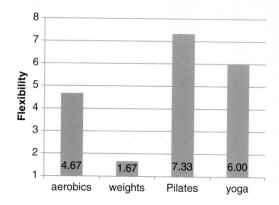

If we then perform a multiple regression analysis using these contrast codes as predictors and flexibility as our criterion, we get the results shown in Table 11.16. The following findings are noteworthy:

- The overall regression model is significant, indicating that group differences in flexibility are unlikely to be due to chance alone.
- The intercept represents the unweighted grand mean.
- The first regression coefficient is significant, indicating that yoga and Pilates predict greater flexibility than aerobics and weights.

Table 11.16 Regression output using orthogonal coding

	SS	df	MS	R^2	F	p	
			Significance test of regression model				
Regression	52.9167	3	17.6389	.7257	7.0556	.0123	
Residual	20.00	8	2.50				
Total	72.9167						

	b	se_b	SS	t	p	Explanation
			Regression coefficients			
b_0	4.9167	.4564		10.7719	.0000	Unweighted grand mean
b_1	7.00	1.8257	36.75	3.8341	.0050	Average of groups 1 and 2 vs. Average of groups 3 and 4
b_2	−1.3333	1.2910	2.6667	1.0328	.3319	Group 3 vs. group 4
b_3	−3.00	1.2910	13.50	2.3238	.0486	Group 1 vs. group 2

Covariance matrix			
.2083	.0000	.0000	.0000
.0000	3.3333	.0000	.0000
.0000	.0000	1.6667	.0000
.0000	.0000	.0000	1.6667

- The second regression coefficient is not significant, indicating that yoga does not predict greater flexibility than Pilates.
- The third regression coefficient is significant, indicating that aerobics predicts greater flexibility than weight training.
- The covariance matrix is diagonal (confirming that the contrasts are orthogonal). Consequently, summing the contrast sum of squares reproduces the regression sum of squares:

$$36.75 + 2.6667 + 13.50 = 52.9167$$

11.2.3   Polynomial Terms in a Trend Analysis

Earlier we noted that we ordinarily cannot assign a meaningful numerical value to categorical data because the ordering of groups is arbitrary. Trend analysis constitutes an exception. Consider research designed to test the effectiveness of a new medicine. Suppose a researcher randomly assigns subjects to receive either 10, 15, 20, or 25 mg of the medicine. In this case, the groups have a meaningful numerical order and can be analyzed with a set of orthogonal polynomial contrast codes. Three vectors are needed, each representing one of the three possible polynomial trends (e.g., linear, quadratic, and cubic) in a four-group design.

Recall from Chap. 10 that polynomial terms are powers of an initial linear predictor. For example, starting with x, we can test a quadratic function by adding x^2, a cubic function by adding x^3, and so on. A similar logic can be applied to categorical predictors with an interpretable numerical order. These predictors can be of two varieties: equally spaced intervals (e.g., high school class: freshman, sophomore, junior, senior) or unequally spaced intervals (e.g., ages: 10, 20, 40, and 65).

Most statistics textbooks provide tables of polynomial coefficients. These tables are convenient, but they are useful only when the groups are equally spaced. Moreover, they provide no insight into how the terms are calculated. Since one of the reasons I am writing this book is to demystify such calculations, let's spend a moment learning how to calculate the coefficients themselves, using a four-group design with unequal spacing.

Imagine we are predicting stiffness as a function of how many days/week a person exercises. As just noted, we will use unequal spacing in our example (e.g., 1, 2, 5, and 7 days/week), but the procedure works just as well with even spacing:

- Create three seed vectors: (1) The first represents the true numerical scores assigned to the groups, (2) the second is the square of each of the first vector values, and (3) the third is the cube of each of the first vector values.
- To create the linear coefficients, calculate deviate scores for the first vector by subtracting its mean from each value.
- To create the quadratic coefficients, regress the second seed variable on a constant (1's are fine) and the linear term you just calculated. The residuals are the quadratic term.

- To create the cubic term, regress the third seed variable on a constant and the linear and quadratic terms. The residuals are the cubic term.

Table 11.17 illustrates the calculations, and Table 11.18 applies the contrast codes to a (hypothetical) data set. The means appear in Fig. 11.3, and Table 11.19 shows the results from a regression analysis. The following findings are noteworthy:

Table 11.17 Creating orthogonal polynomials for a four-group design

Seed	Seed^2	Seed^3	Linear: subtract mean from all scores of seed 1	Quadratic: regress Seed^2 on a vector of 1's and the linear term. The residuals are the quadratic term	Cubic: regress Seed^3 on a vector of 1's, the linear term, and the quadratic term. The residuals are the cubic term
1	1	1	−2.75	3.09890	−5.08475
2	4	8	−1.75	−1.84615	8.13559
5	25	125	1.25	−4.68132	−5.08475
7	49	343	3.25	3.42857	2.03390

- The overall regression model is significant, indicating that group differences in stiffness are unlikely to be due to chance alone.
- The intercept represents the unweighted grand mean.
- The first regression coefficient represents the linear trend, which is not significant.
- The second regression coefficient represents the quadratic trend, which is significant. Looking over Fig. 11.3, we see that people who exercise infrequently (1 day/week) and those who exercise constantly (7 days/week) experience more stiffness than do those who exercise a moderate amount.
- The third regression coefficient, which represents the cubic trend, is not significant.

Table 11.18 Trend analysis using orthogonal polynomials with unequal spacing

Subject	Days/week of exercise	y	c_1	c_2	c_3
1	1	2	−2.75	3.0989	−5.0847
2	1	1	−2.75	3.0989	−5.0847
3	1	3	−2.75	3.0989	−5.0847
4	2	3	−1.75	−1.8462	8.13559
5	2	8	−1.75	−1.8462	8.13559
6	2	7	−1.75	−1.8462	8.13559
7	5	5	1.25	−4.6813	−5.0847
8	5	8	1.25	−4.6813	−5.0847
9	5	7	1.25	−4.6813	−5.0847
10	7	2	3.25	3.42857	2.0339
11	7	4	3.25	3.42857	2.0339
12	7	3	3.25	3.42857	2.0339

Fig. 11.3 Group means
from orthogonal
polynomials with unequal
spacing

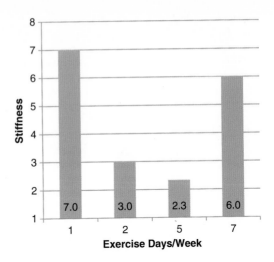

Table 11.19 Regression output using orthogonal polynomials with unequal spacing

				Significance test of regression model			
	SS	*df*	*MS*	R^2	*F*	*p*	
Regression	46.25	3	15.4167	.6711	5.4412	.0247	
Residual	22.6667	8	2.8333				
Total	68.917						
				Regression coefficients			
	b	se_b	*SS*	*t*	*p*	Explanation	
b_0	4.5833	.4859		9.4324	.0000	Unweighted grand mean	
b_1	−.0916	.2037	.5723	.4494	.6650	Linear	
b_2	.5527	.1422	42.7850	3.8859	.0046	Quadratic	
b_3	−.0889	.0880	2.8927	1.0104	.3419	Cubic	
	Covariance matrix						
.2361	.0000	.0000	.0000				
.0000	.0415	.0000	.0000				
.0000	.0000	.0202	.0000				
.0000	.0000	.0000	.0077				

- The covariance matrix is diagonal (confirming that the contrasts are orthogonal).
  Consequently, summing the contrast sum of squares reproduces the regression
  sum of squares:

$$.5723 + 42.7850 + 2.8920 = 46.25$$

11.2.4   R Code: Creating Orthogonal Contrasts

```
#Gram-Schmidt Orthogonalization
v1 <-rbind(-.25,-.25,.25,.25)
v2 <-rbind(0,0,-.5,.5)
c3 <-rbind(1,2,3,4)    #Seed vector
v3<-c3-(v1%*%t(c3)%*%v1)/as.vector(t(v1)%*%v1)-
(v2%*%t(c3)%*%v2)/as.vector(t(v2)%*%v2)
v3 <-v3-mean(v3)
codes <-cbind(v1,v2,v3);codes


#Regression Analysis Using Orthogonal Codes
y=c(6,6,2,1,2,2,7,7,8,8,4,6)
orth1 <-c(rep(v1[1],3),rep(v1[2],3),rep(v1[3],3),rep(v1[4],3))
orth2 <-c(rep(v2[1],3),rep(v2[2],3),rep(v2[3],3),rep(v2[4],3))
orth3 <-c(rep(v3[1],3),rep(v3[2],3),rep(v3[3],3),rep(v3[4],3))
gram.reg <-lm(y~orth1+orth2+orth3)
summary(gram.reg)


#Create Orthogonal Polynomial Coefficients
seed1 <-rbind(1,2,5,7)
seed2 <-seed1^2
seed3 <-seed1^3
line <-seed1-mean(seed1)
quad.reg <-lm(seed2~seed1+line)
quad <-resid(quad.reg)
cubic.reg <-lm(seed3~line+quad)
cube <-resid(cubic.reg)


#Regression Analysis Using Orthogonal Polynomial Coefficients
y=c(7,8,6,6,1,2,4,1,2,7,5,6)
poly1 <-c(rep(line[1],3),rep(line[2],3),rep(line[3],3),rep(line[4],3))
poly2 <-c(rep(quad[1],3),rep(quad[2],3),rep(quad[3],3),rep(quad[4],3))
poly3 <-c(rep(cube[1],3),rep(cube[2],3),rep(cube[3],3),rep(cube[4],3))
poly.reg <-lm(y~poly1+poly2+poly3)
summary(poly.reg)
```

11.3   Contrast Codes with Unbalanced Designs

The procedures we have been discussing are appropriate when the number of subjects in each group is the same, but adjustments need to be made to orthogonal contrast codes and effect codes when cell sizes are unequal. Dummy codes do not need to be adjusted.

To help us understand the issues involved, let's concoct another study. Imagine a researcher wishes to determine which form of exercise is associated with more physician visits for injury or accident. The researcher selects four forms of exercise for study (swimming, jogging, rowing, and cycling) and then records exercise-related visits to a physician. The (phony) data appear in Table 11.20, and it is apparent that the cell sizes are uneven.

Table 11.20 Small sample example for analysis with unequal group sizes

Group	y	Group	n_j	Mean
Swimming	2	Swimming	2	1.5
Swimming	1	Jogging	5	7.2
Jogging	8	Rowing	3	4.0
Jogging	9	Cycling	2	6.5
Jogging	8			
Jogging	6			
Jogging	5			
Rowing	3			
Rowing	6			
Rowing	3			
Cycling	6			
Cycling	7			

Depending on why they have come about, we have two ways of dealing with unequal cell sizes. If we have randomly assigned subjects to conditions and, by chance, the cell sizes are unequal, we use an unweighted means analysis that compensates for the cell size differences. This is the method most commonly used with ANOVA because, in an experimental setting, cell size differences are usually accidental, not meaningful.

On the other hand, there might be instances in which different cell sizes have occurred because of some meaningful event, such as differential attrition rates in an experiment, or because of true differences in the population. For example, if many more people choose one form of exercise over another, then a survey study would most likely want to take those differences into account rather than ignore them. In this case, we use a weighted means analysis.

11.3.1 Analysis with Unweighted Means

We will begin by considering the case where cell size differences are unintentional rather than meaningful, which is the default setting for an ANOVA in most statistical packages. To illustrate the process, we will use the Helmert procedure to generate a set of orthogonal contrast weights. These weights are displayed in Table 11.21, using whole numbers for clarity. Our first contrast compares swimming against the other three forms of exercise, our second contrast compares jogging against rowing and cycling, and our third contrast compares rowing vs. cycling.

Table 11.21 Unweighted contrast coefficients with unbalanced designs

Subject	GRP	c_1	c_2	c_3	y
1	Swimming	3	0	0	2
2	Swimming	3	0	0	1
3	Jogging	−1	2	0	8
4	Jogging	−1	2	0	9
5	Jogging	−1	2	0	8
6	Jogging	−1	2	0	6
7	Jogging	−1	2	0	5
8	Rowing	−1	−1	−1	3
9	Rowing	−1	−1	−1	6
10	Rowing	−1	−1	−1	3
11	Cycling	−1	−1	1	6
12	Cycling	−1	−1	1	7

If we regress y on these vectors in the usual manner, we get the output shown in Table 11.22.

Table 11.22 Regression output using unweighted contrast coding with an unbalanced design

			Significance test of regression model				
	SS	df	MS	R^2	F	p	
Regression	54.8667	3	18.2889	.7550	8.2197	.0079	
Residual	17.80	8	2.2250				
Total	72.667						

		Regression coefficients				
	b	se_b	SS	t	p	Explanation
b_0	4.80	.4618		10.3948	.0000	Unweighted grand mean
b_1	−1.10	.2924	31.4892	3.7620	.0055	Swimming vs. other sports
b_2	.65	.3177	9.3122	2.0458	.0750	Jogging vs. rowing and cycling
b_3	1.25	.6808	7.50	1.8360	.1037	Rowing vs. cycling

Covariance matrix			
.2132	.0216	−.0402	.0464
.0216	.0855	.0134	−.0155
−.0402	.0134	.1009	−.0309
.0464	−.0155	−.0309	.4635

We know from previous analyses that the test of our overall regression equation does not vary across coding schemes, so we will not spend time interpreting it. Instead, let's look at our covariance matrix. Even though we used orthogonal predictors, we can see that we do not have a diagonal covariance matrix. This is because the cell sizes are unequal. Formally, we can say that a set of contrasts is orthogonal only if it satisfies the requirements in Eq. (11.11):

$$\Sigma n_j c_j = \Sigma n_j c_{ij} c_{kj} = 0 \qquad (11.11)$$

The equation looks complicated, but all we've done is add each group's sample size (n_j) to Eq. (11.5). We didn't need to do that earlier because each group had the same number of observations, but we do need to make this adjustment when the cell sizes are uneven.

11.3.1.1   Interpreting the Regression Coefficients

Turning now to the regression coefficients, we see that our first contrast is significant, but our second and third contrasts are not. We might also note that when we add up the sum of the squares associated with each coefficient, we do not reproduce our regression sum of squares:

$$31.4892 + 9.3122 + 7.50 \neq 54.8667$$

Finally, notice that the intercept represents the unweighted grand mean [see Eq. (11.6)]:

$$\frac{(1.5 + 7.2 + 4 + 6.5)}{4} = 4.80$$

11.3.1.2   Calculating the Sum of Squares from the Group Means

The sum of squares for an unweighted contrast (SS_{cu}) can also be calculated using the group means:

$$SS_{cu} = \frac{\left[\Sigma (\bar{y} c_i)^2 \right] * cm_h}{\Sigma c_i^2} \qquad (11.12)$$

where $\bar{y}$ refers to a group mean, c_i refers to the contrast coefficients, and cm_h refers to the harmonic mean of the contrast:

$$cm_h = \frac{1}{\Sigma \left[\frac{1}{n} * \frac{c^2}{\Sigma c^2} \right]} \qquad (11.13)$$

To see how the formula works, we will apply it to our first contrast in which we compared swimming against the other forms of exercise. We begin by calculating the harmonic mean of the contrast:

$$cm_h = \frac{1}{\left(\frac{1}{2} * \frac{3^2}{12}\right) + \left(\frac{1}{5} * \frac{-1^2}{12}\right) + \left(\frac{1}{3} * \frac{-1^2}{12}\right) + \left(\frac{1}{2} * \frac{-1^2}{12}\right)} = 2.1687$$

We then compute the sum of squares as follows, matching the value shown in Table 11.22:

$$SS_{cu} = \frac{[(1.5 * 3) + (7.2 * -1) + (4 * -1) + (6.5 * -1)]^2 * 2.1687}{12} = 31.4892$$

Similar calculations produce our second contrast:

$$cm_h = \frac{1}{(0) + \left(\frac{1}{5} * \frac{2^2}{6}\right) + \left(\frac{1}{3} * \frac{-1^2}{6}\right) + \left(\frac{1}{2} * \frac{-1^2}{6}\right)} = 3.6735$$

and

$$SS_{cu} = \frac{[(0) + (7.2 * 2) + (4 * -1) + (6.5 * -1)]^2 * 3.6736}{6} = 9.3122$$

I will leave it to you to calculate the sum of squares for our third contrast.

11.3.2   Weighted Means Analysis

In most situations, the unweighted approach is the one to use when analyzing categorical data with uneven cell sizes. When cell size differences are meaningful, however, you might wish to incorporate them rather than ignore them by using a weighted means analysis.

11.3.2.1   Calculating Orthogonal Coefficients

To perform a weighted means analysis, we need to adjust our contrast coefficients by their cell sizes. There are several ways to do this, but I prefer a slight modification of the method of successive residuals that we learned when calculating orthogonal polynomial coefficients. Here, we will use the following terminology:

u_i = unweighted coefficients we are using as seeds
w_i = weighted coefficients we wish to calculate

Table 11.23 Unweighted and weighted orthogonal contrast coefficients with unbalanced design

Subject	Group	u_1	u_2	u_3	w_1	w_2	w_3
1	Swimming	3	0	0	3.3333	0	0
2	Swimming	3	0	0	3.3333	0	0
3	Jogging	−1	2	0	−0.6667	1.5	0
4	Jogging	−1	2	0	−0.6667	1.5	0
5	Jogging	−1	2	0	−0.6667	1.5	0
6	Jogging	−1	2	0	−0.6667	1.5	0
7	Jogging	−1	2	0	−0.6667	1.5	0
8	Rowing	−1	−1	−1	−0.6667	−1.5	−0.8
9	Rowing	−1	−1	−1	−0.6667	−1.5	−0.8
10	Rowing	−1	−1	−1	−0.6667	−1.5	−0.8
11	Cycling	−1	−1	1	−0.6667	−1.5	1.2
12	Cycling	−1	−1	1	−0.6667	−1.5	1.2

Table 11.23 reproduces Table 11.21 using this new terminology. We calculate the weighted coefficients using the same procedures we used to calculate orthogonal polynomial coefficients in Table 11.17:

- To calculate w_1, create deviate scores of u_1.
- To calculate w_2, regress u_2 on leading 1's and w_1. The residuals $= w_2$.
- To calculate w_3, regress u_3 on leading 1's, w_1, and w_2. The residuals $= w_3$.

11.3.2.2 Weighted Regression Analysis

Regressing y on the adjusted coefficients produces the results shows in Table 11.24. Five points merit attention:

- The overall test of the regression model remains unchanged. As noted throughout this chapter, this will be true no matter which coding scheme we use.
- The intercept now equals the weighted grand mean, not the unweighted grand mean. The weighted mean is found using Eq. (11.14):

$$M_w = \frac{\Sigma Y}{N} \tag{11.14}$$

- The covariance matrix is now diagonal, indicating that our coefficients are orthogonal.
- Because the coefficients are orthogonal, summing the contrast sum of squares reproduces the regression sum of squares:

$$35.2667 + 12.10 + 7.50 = 54.8667$$

Table 11.24 Regression output using weighted contrast coding with an unbalanced design

			Significance test of regression model			
	SS	df	MS	R^2	F	p
Regression	54.8667	3	18.2889	.7550	8.2197	.0079
Residual	17.80	8	2.2250			
Total	72.667					

			Regression coefficients			
	b	se_b	SS	t	p	Explanation
b_0	5.3330	.4306		12.3849	.0000	Weighted grand mean
b_1	−1.15	.2889	35.2667	3.9812	.0041	Swimming vs. other sports
b_2	.7333	.3145	12.10	2.3320	.0480	Jogging vs. rowing and cycling
b_3	−1.25	.6808	7.50	1.8360	.1037	Rowing vs. cycling

Covariance matrix			
.1854	.0000	.0000	.0000
.0000	.0834	.0000	.0000
.0000	.0000	.0989	.0000
.0000	.0000	.0000	.4635

- The significance levels of some of our coefficients have changed. A comparison involving only two means will be the same no matter which coding scheme we use (which is why dummy codes require no adjustment for unequal cell sizes), but a contrast involving more than two means will vary with changes in the cell sizes. In our example, we see that our second contrast (jogging vs. rowing and cycling), which was not significant with the unweighted means analysis, is now significant. The change occurs because jogging has a disproportionate number of subjects in our example. With an unweighted means analysis, this disparity is neutralized; with a weighted means analysis, it is preserved. As I said before, if the group size differences are interpretable, we might want to use the weighted means analysis to incorporate these disparities.

11.3.2.3   Sum of Squares with Weighted Analysis

If we desire, we can calculate the sum of squares for a weighted contrast directly from the means. Here, we ignore the group labels when computing our terms, focusing only on the observations that go into creating the contrast:

$$SS_{cw} = \Sigma \left[\left(\bar{y}_{ci} - \bar{Y}_C \right)^{2} * n_{ci} \right] \quad (11.15)$$

where $\bar{y}_{ci}$ refers to the mean of each group involved in the contrast, $\bar{Y}_C$ refers to the mean of the entire contrast, and n_{ci} refers to each contrast group's cell size.

To see how the formula works, we will compute our first weighted contrast, comparing group 1 to the average of groups 2, 3, and 4. We'll begin by computing three means:

$$\text{Group 1} = \frac{(2+1)}{2} = 1.5$$

$$\text{Groups (2, 3, 4)} = \frac{(8+9+8+6+5+3+6+3+6+7)}{10} = 6.1$$

$$\text{Groups (1, 2, 3, 4)} = \frac{(2+1+8+9+8+6+5+3+6+3+6+7)}{12} = 5.3333$$

We then find our contrast:

$$SS_{cw} = \left[(1.5 - 5.3333)^2 * 2\right] + \left[(6.1 - 5.3333)^2 * 10\right] = 35.2667$$

Our second contrast compares group 2 vs. groups 3 and 4:

$$\text{Group 2} = \frac{(8+9+8+6+5)}{5} = 7.2$$

$$\text{Groups (3, 4)} = \frac{(3+6+3+6+7)}{5} = 5.0$$

$$\text{Groups (2, 3, 4)} = \frac{(8+9+8+6+5+3+6+3+6+7)}{10} = 6.1$$

$$SS_{cw} = \left[(7.2 - 6.1)^2 * 5\right] + \left[(5.0 - 6.1)^2 * 5\right] = 12.10$$

You can compute the sum of the squares for the third contrast if you desire.

11.3.2.4   Weighted Polynomial Trends and Effect Codes

As noted earlier, dummy codes do not need to be adjusted for unbalanced sample sizes, but polynomial coefficients and effect codes do. The procedures described in Sect. 11.3.2.1 can be used to adjust polynomial codes, and the easiest way to create weighted effect codes is simply to divide each effect code by its group sample size. The **R** code that accompanies this chapter performs the latter analysis.

11.3.3   R Code: Unbalanced Designs

```
#Unbalanced Designs with Orthogonal Contrast Codes
g <-c(1,1,2,2,2,2,2,3,3,3,4,4)
y <-c(2,1,8,9,8,6,5,3,6,3,6,7)
c1 <-c(3,3,rep(-1,10))
c2 <-c(0,0,rep(2,5),rep(-1,5))
c3 <-c(rep(0,7),rep(-1,3),rep(1,2))

#Unweighted Regression
unweighted.reg <-lm(y~c1+c2+c3)
summary(unweighted.reg)
vcov(unweighted.reg)
anova(unweighted.reg)

#Weighted Regression
linear <-c1-mean(c1)
quad.reg <-lm(c2~c1+linear)
quadratic <-resid(quad.reg)
cubic.reg <-lm(c3~linear+quadratic)
cubic <-resid(cubic.reg)
wgts <-cbind(linear,quadratic,cubic)
wgts
weighted.reg <-lm(y~wgts)
summary(weighted.reg)
vcov(weighted.reg)
anova(weighted.reg)

#Weighted Effect Codes With Last Group as Base Group
u1 <-c(1,1,0,0,0,0,0,0,0,0,-1,-1)
u2 <-c(0,0,1,1,1,1,1,0,0,0,-1,-1)
u3 <-c(0,0,0,0,0,0,0,1,1,1,-1,-1)
gsize <-c(rep(2,2),rep(5,5),rep(3,3),rep(2,2))
e1 <-u1/gsize
e2 <-u2/gsize
e3 <-u3/gsize
wgt.eff <-lm(y~e1+e2+e3)
summary(wgt.eff)
```

11.4   Chapter Summary

1. Multiple regression is commonly used with continuous predictors, but it can also be used with categorical ones. Doing so requires constructing a set of vectors that uniquely identifies each of the J groups. The number of vectors will always be 1 less than the number of groups.
2. Orthogonal contrast codes partition the between-group variance into an independent set of $J - 1$ contrasts. This coding scheme can accommodate two-group comparisons and multiple-group contrasts.
3. Dummy codes are used when a single group is compared with one other group.
4. Effect codes compare one group against the average of all other groups.
5. Orthogonal contrast codes can be computed using the Gram-Schmidt technique or by saving the residuals from a series of regression analyses using seeded vectors.
6. A trend analysis using orthogonal polynomial terms can be used when a grouping variable reflects a naturally increasing or decreasing category. The differences can be evenly spaced or unevenly spaced.
7. Modifications to orthogonal contrast coefficients and effect codes are required when cell sizes are unequal or disproportionate. The nature of the modification depends on whether the imbalance is accidental or meaningful. We use unweighted means when the imbalance is accidental and weighted means when it is meaningful.

Chapter 12
Factorial Designs

In Chap. 11, we used multiple regression to analyze data with a single categorical predictor. Yet multiple regression, like analysis of variance (ANOVA), can also be used with designs that combine two or more categorical predictors. In most cases, the categorical variables are crossed to form a factorial design. In this chapter, you will learn how to use multiple regression to analyze and interpret factorial designs.

12.1 Basics of Factorial Designs

The easiest way to understand the mechanics of a factorial design is to compare it to a single factor design. Imagine that after reading Chap. 11, a researcher decides to conduct another exercise study to follow-up on the (fictitious) ones we conducted. The researcher randomly assigns 12 subjects to one of four conditions: a control (no exercise) condition, weightlifting, cycling, or a cross-training condition in which subjects lift weights and cycle. Several weeks later, the researcher measures muscle tone. The (imaginary) data, along with the coding scheme we will use are shown in Table 12.1; the means are displayed in Fig. 12.1.

12.1.1 Regression Analysis of a One-Way Design

We have used dummy codes to create our 3 vectors, designating the control condition as the reference group. If we then regress our criterion on these coded vectors in the usual manner, we get the output displayed in Table 12.2. By now, interpreting the output should be familiar.

Electronic Supplementary Material: The online version of this chapter (doi: 10.1007/978-3-319-11734-8_12) contains supplementary material, which is available to authorized users

© Springer International Publishing Switzerland 2014

J.D. Brown, *Linear Models in Matrix Form*, DOI 10.1007/978-3-319-11734-8_12

Table 12.1 Four-group experiment as a single factor design

Subject	Group	y	Dummy code 1	Dummy code 2	Dummy code 3
1	Control	3	0	0	0
2	Control	4	0	0	0
3	Control	3	0	0	0
4	Weightlifting	5	1	0	0
5	Weightlifting	3	1	0	0
6	Weightlifting	4	1	0	0
7	Cycling	5	0	1	0
8	Cycling	6	0	1	0
9	Cycling	4	0	1	0
10	Cross-training	9	0	0	1
11	Cross-training	9	0	0	1
12	Cross-training	8	0	0	1

Fig. 12.1 Group means from small sample example of one-way, four-group design

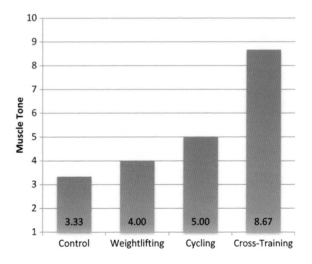

- Our overall regression model is significant, indicating that the variation among the four conditions is unlikely to be due to chance alone.
- The weightlifting group does not differ from the control condition, but the cycling and cross-training groups do.
- The covariance matrix is not diagonal (because dummy codes are not orthogonal).

12.1.2   Recasting the Data as a Factorial Design

It is perfectly acceptable to analyze the data as we have done, but Table 12.3 shows another way to organize our data, known as a 2 × 2 (read "2 by 2") factorial design. The first factor is cycling, and it has two levels (no or yes); the second factor is

Table 12.2 Regression output from a one-way, four-group design

	SS	df	MS	R^2	F	p	
			Significance test of regression model				
Regression	50.9167	3	16.9722	.9052	25.4583	.0002	
Residual	5.3333	8	.6667				
Total	56.25						

	b	se_b	SS	t	p	Explanation
			Regression coefficients			
b_0	3.3333	.4714		7.0711	.0001	Mean of control group
b_1	.6667	.6667	.6667	1.00	.3466	Control vs. weightlifting
b_2	1.6667	.6667	4.1667	2.50	.0369	Control vs. cycling
b_3	5.3333	.6667	42.6667	8.00	.0000	Control vs. cross-training

Covariance matrix			
.2222	−.2222	−.2222	−.2222
−.2222	.4444	.2222	.2222
−.2222	.2222	.4444	.2222
−.2222	.2222	.2222	.4444

Table 12.3 Recasting a one-way, four-group design as a 2×2 factorial design

		Weightlifting	
		No	Yes
Cycling	No	Control (group 1)	Weightlifting (group 2)
	Yes	Cycling (group 3)	Cross-training (group 4)

weightlifting, and it also has two levels (no or yes). The control condition is represented by the absence of both forms of exercise; the cross-training condition is represented by the presence of both forms of exercise; and the other two groups represent a single form of exercise. Figure 12.2 shows the means using this organizational scheme.

12.1.3   Properties of a Factorial Design

In a factorial design, two or more factors are crossed such that all possible combinations of all factors are represented.[1] The number of combinations (or cells) is found by multiplying the levels of each factor. For example, in a 3×4 factorial design, there are two factors: the first factor has 3 levels and the

---

[1] To be precise, the preceding definition defines a *complete* factorial design. Partial (or fractional) factorial designs, which will not be considered here, occur when only some combinations are represented.

Fig. 12.2 Means for a 2×2
factorial design

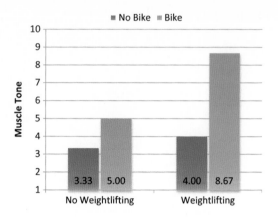

second factor has 4 levels, yielding 12 possible combinations. A $2 \times 3 \times 5$ design
has 3 factors: the first factor has 2 levels, the second factor has 3 levels, and the third
factor has 5 levels. Crossing all 5 factors yields 30 combinations.

Factorial designs possess several desirable features. First, by allowing us to
examine multiple factors in a single design, they are efficient. Factorial designs
are also very informative. In fact, their chief advantage is their ability to partition
the total amount of explained variance (i.e., SS_{reg}) into components, helping us to
better understand our data. Before discussing this partitioning, you should be aware
that the following presentation applies only to balanced designs. As shown in
Table 12.4, a balanced design is one in which the cell frequencies are equal or
proportional. The variance in unbalanced factorial designs can also be partitioned,
but the components do not always equal SS_{reg}. For this reason, we will start with a
balanced design and consider unbalanced designs later in this chapter.

Table 12.4 Cell sizes in three 2×2 factorial designs

	Balanced (equal)			Balanced (proportional)			Unbalanced	
	B1	B2		B1	B2		B1	B2
A1	3	3	A1	3	4	A1	3	4
A2	3	3	A2	6	8	A2	2	5

12.1.4   Sources of Variance in a Balanced Factorial Design

The explained variance (SS_{reg}) in a factorial design can be partitioned into two
components: main effects and interactions.

1. Main effects examine the overall impact of a variable without taking other
   variables into account. Sometimes we say main effects "collapse" across other
   variables because they are represented by the marginal means. A marginal mean
   is the unweighted average of two (or more) cells in a factorial design. Table 12.5
   presents the data from our exercise study, with the marginal means identified.

2. Interactions test whether the effects of one variable vary across levels of another variable. If, for example, the effects of weightlifting depend on whether one is also cycling, we would say that weightlifting and cycling interact to predict muscle tone. Conversely, if the effects of weightlifting do not depend on whether one is also cycling, there is no interaction.

Table 12.5 Cell means and marginal means for a 2 × 2 factorial design

		Weightlifting		
		No	Yes	Marginal mean
Cycling	No	3.3333	4.00	3.6667
	Yes	5.00	8.6667	6.8333
	Marginal mean	4.1667	6.3333	5.25

12.1.4.1 Coding with Factorial Designs

In a 2 × 2 factorial design, there are two main effects. The first main effect in our example examines whether subjects who rode a bike developed better muscle tone than those who did not. Here we are comparing the row marginal means in Table 12.5. The first column of contrast weights in Table 12.6 shows that we can test these means by comparing Groups 1 and 2 vs. Groups 3 and 4. Our second main effect is the main effect of weightlifting. Here we are comparing the column marginal means in Table 12.5. The second column of contrast weights in Table 12.6 shows that we test these means by comparing Groups 1 and 3 vs. Groups 2 and 4. The third column of contrast weights in Table 12.6 shows a cross-product term, formed by multiplying contrast 1 and contrast 2. As in Chap. 9, the cross-product term will allow us to test the interaction between the two factors.

Table 12.6 Contrast coding for a 2 × 2 factorial design

Subject		y	Contrast 1 (main effect of A)	Contrast 2 (main effect of B)	Contrast 3 (A × B interaction)
1	Control	3	−.5	−.5	.25
2	Control	4	−.5	−.5	.25
3	Control	3	−.5	−.5	.25
4	Weights	5	−.5	.5	−.25
5	Weights	3	−.5	.5	−.25
6	Weights	4	−.5	.5	−.25
7	Cycling	5	.5	−.5	−.25
8	Cycling	6	.5	−.5	−.25
9	Cycling	4	.5	−.5	−.25
10	Cross-training	9	.5	.5	.25
11	Cross-training	9	.5	.5	.25
12	Cross-training	8	.5	.5	.25

Table 12.7 Regression output from a 2×2 factorial design

	SS	df	MS	R^2	F	p	
			Significance test of regression model				
Regression	50.9167	3	16.9722	.9052	25.4583	.0002	
Residual	5.3333	8	.6667				
Total	56.25						

	b	se_b	SS	t	p	Explanation
			Regression coefficients			
b_0	5.25	.2357		22.2739	.0000	Unweighted grand mean
b_1	3.1667	.4714	30.0833	6.7175	.0001	Cycling vs. no cycling
b_2	2.1667	.4714	14.0833	4.5962	.0018	Weights vs. no weights
b_3	3.00	.9428	6.75	3.1820	.0130	Interaction

Covariance matrix			
.0556	0	0	0
0	.2222	0	0
0	0	.2222	0
0	0	0	.8889

12.1.4.2   Main Effect Sum of Squares

Table 12.7 displays the output from a regression analysis that uses the coding scheme shown in Table 12.6. The test of the regression equation is identical to the one we obtained earlier using dummy coding, but the regression coefficients have changed. The intercept shows the unweighted grand mean, and the value of the first regression coefficient ($b_1 = 3.1667$) corresponds to the difference in the average muscle tone of subjects who cycled ($M = 6.8333$) vs. those who did not ($M = 3.6667$).[2] Notice that these values match the marginal row means in Table 12.5. A test of this coefficient represents the main effect of cycling. With a balanced 2×2 design, we can use Eq. (11.7) to compute the sum of squares associated with this main effect.[3]

---

[2] With categorical predictors, the slope of the regression line equals the difference between two means.

[3] If we had used dummy coding to create our grouping vectors, the b_1 coefficient would represent a simple effect rather than a main effect. For example, if we had assigned 0's to those who did not lift weights and 1's to those who did, the b_1 coefficient would represent the simple effect of cycling when no weights were lifted rather than the main effect of cycling across weightlifting conditions. Unfortunately, many researchers are unaware of this fact, leading them to erroneously interpret their dummy-coded coefficients as main effects rather than simple effects. To avoid confusion, you should not use dummy coding with a factorial design.

$$SS_A = \left(\frac{3.1667}{.4714}\right)^2 * .6667 = 30.0833$$

The b_2 coefficient (2.1667) corresponds to the difference in the average muscle tone of subjects who lifted weights ($M = 6.3333$) vs. those who did not ($M = 4.1667$). As before, these means match the column marginal means in Table 12.5, and a test of this coefficient represents a main effect of weightlifting.

$$SS_B = \left(\frac{2.1667}{.4714}\right)^2 * .6667 = 14.0833$$

12.1.4.3   Interaction

Our final source of variance is the A × B interaction. In Chap. 9, you learned how to model an interaction using a cross-product term, formed by multiplying two (or more) deviate vectors. The same procedure is used with categorical variables, except we don't bother creating deviate values with a balanced design because our contrasts already sum to 0. Our third contrast vector in Table 12.6 was formed by multiplying vector 1 and vector 2. Table 12.7 shows that its associated regression coefficient ($b_3 = 3.00$) is significant and its sum of squares is found as follows:

$$SS_{AB} = \left(\frac{3.00}{.9428}\right)^2 * .6667 = 6.75$$

12.1.4.4   Reproducing SS_{reg} in a Balanced Factorial Design

In a moment, we will probe the nature of the interaction in detail. Before we do, let's consider our three sources of variance taken as a whole. First, notice that the covariance matrix is diagonal, indicating that our three contrast vectors are orthogonal. We should not be surprised, therefore, to find that when we add up the sum of squares of each term, we reconstruct our regression sum of squares.

$$30.0833 + 14.08333 + 6.75 = 50.9167$$

Thus, in a balanced, two factor factorial design, the regression sum of squares equals the sum of the main effects and interaction sum of squares.

$$SS_{reg} = SS_A + SS_B + SS_{AB} \tag{12.1}$$

12.1.4.5   Summary

In this section you learned how to analyze a balanced factorial design using multiple regression with coded vectors. If we had used ANOVA instead of multiple regression, we would have gotten output similar to that shown in Table 12.8.

Table 12.8 ANOVA table
for a 2 × 2 factorial design

	SS	df	MS	F	p
A	30.0833	1	30.0833	45.1250	.0001
B	14.0833	1	14.0833	21.1250	.0018
AB	6.75	1	6.75	10.1250	.0130
Error	5.3333	8	.6667		
Total	56.25				

The regression approach yields identical values and provides greater insight into the calculations that produced them.

12.1.5 Probing an Interaction

Recall from Chap. 9 that an interaction tests whether the simple slope of a regression line is parallel across all values of another variable. If the regression lines are parallel, the interaction is not significant; if the regression lines are not parallel, the interaction is significant.

12.1.5.1 Simple Effects

With a 2 × 2 factorial design, we have only two values of each categorical variable to compare. We calculate the simple slopes using the same procedures we first learned in Chap. 9—create an **S** matrix of contrast weights, and then find the regression coefficients by computing **S′b** and their associated standard errors by taking the square root of the diagonal entries formed from **S′CS**.

Table 12.9 shows the **S** matrix we will use to calculate our simple effects.[4] There is only one difference between the procedure we used in Chap. 9 and the one we use here. Instead of defining high and low groups in terms of their standard deviation from the mean, we define our groups by using their contrast codes. To illustrate, to

Table 12.9 S Matrix to derive simple effects for a 2 × 2 factorial design

	Simple effect of A @ B_1	Simple effect of A @ B_2	Simple effect of B @ A_1	Simple effect of B @ A_2
b_0	0	0	0	0
b_1	1	1	0	0
b_2	0	0	1	1
b_3	−0.5	0.5	−0.5	0.5

---

[4] With categorical predictors, it is customary to refer to the simple slopes as simple effects. The terms are equivalent because, as noted in footnote 2, a simple slope represents the difference between two group means.

find the simple effect of cycling when no weights were lifted (see first column of contrast codes in Table 12.9), we assign a 1 to the b_1 coefficient (which represents the effect of cycling) and $-.5$ to the interaction coefficient, b_3. We use $-.5$ simply because that's the code we assigned to the "no-weights" condition. Had we used a different value (e. g.,$+.5$ or -1), we would have used that value instead. A similar algorithm was used to calculate the values in the remaining columns.

After calculating our coefficients and standard errors, we get the results shown in

		b	se_b	SS	t	p
Table 12.10 Simple effects for a 2×2 factorial design	A @ B_1	1.6667	.6667	4.1667	2.50	.0369
	A @ B_2	4.6667	.6667	32.6667	7.00	.0001
	B @ A_1	.6667	.6667	.6667	1.00	.3466
	B @ A_2	3.6667	.6667	20.1667	5.50	.0006

Table 12.10. Several points are noteworthy.

- First, let's examine the simple slope of A @ B_1 (1.6667) and the simple slope of A @ B_2 (4.6667). The difference between the two slopes equals the slope of the interaction coefficient in Table 12.7 ($b_3 = 3.00$). This is no coincidence, as the interaction term represents the difference between two simple slopes. And because interactions are symmetrical, the difference between the two simple slopes of B also equals 3.00.
- Also note that the simple effects of A are significant at both levels of B. To be sure, the effects of cycling are stronger when weights are lifted, but they are also significant when weights are not lifted. In this case, we have an "especially for" interaction, and it is appropriate to interpret the main effect.
- The situation is different when we consider the simple effects of B. Here we find that weightlifting does not affect muscle tone when subjects do not cycle (B @ A_1 is not significant), but does affect muscle tone when subjects do cycle (B @ A_2 is significant). In this case, we have an "only for" interaction and we should refrain from interpreting the main effect of B even though it was significant.
- Finally, let's add up the sum of squares for the two simple effects of A (4.1667 $+ 32.6667 = 36.8333$). If we then subtract the interaction sum of squares from this total, we get the sum of squares for the main effect of A ($SS_A = 36.0833 - 6.0 = 30.0833$). The same holds true for B ($.6667 + 20.1667 - 6.75 = 14.0833$). In short, with a balanced design, a main effect sum of squares equals the sum of its simple effects sum of squares minus the interaction sum of squares.

$$SS_{me} = \Sigma\left(SS_{simple}\right) - SS_{ab} \tag{12.2}$$

12.1.5.2 Understanding an Interaction

Before concluding this section, let's solidify our understanding of an interaction. Returning to Fig. 12.2, we see that the cross-training mean is higher than the other means. This fact alone, however, is not evidence of an interaction. The critical issue is not whether the combination of cycling and weightlifting produces the most benefits; the issue is whether cycling and weightlifting combine additively to affect muscle tone. If they combine additively, there is no interaction; if they don't, there is. In our example, their effects are not additive.

To appreciate this point, consider the hypothetical data in Table 12.11. The combination of weightlifting and cycling produces the greatest muscle tone in both examples (7.0), but they differ in an important way. In the first example, the effects are additive: Weightlifting leads to a 1.5 unit increase in muscle tone when subjects do not cycle $(5.0 - 3.5)$, and a 1.5 unit increase in muscle tone when subjects do cycle $(7.0 - 5.5)$. We could also say that cycling produces a 2-unit increase in muscle tone regardless of whether subjects lift weights.

Table 12.11 Additive and not additive (interactive) effects in a factorial design

		Example #1 Additive/no interaction				Example #2 Not additive/interaction	
		Weightlifting				Weightlifting	
		No	Yes			No	Yes
	No	3.5	5.0		No	3.0	4.0
Cycling	Yes	5.5	7.0	Cycling	Yes	5.0	7.0

The situation is different in the second example. Here the effects of each variable are not additive. Weightlifting produces a 1 unit increase in muscle tone when subjects do not cycle $(4.0 - 3.0)$, but a 2-unit increase in muscle tone when subjects do cycle $(7.0 - 5.0)$. We could also say that cycling produces a 2-unit increase in muscle tone when subjects don't lift weights $(5.0 - 3.0)$ but a 3-unit increase in muscle tone when subjects do lift weights $(7.0 - 4.0)$.

In sum, only the data on the right tell us that the combination of cycling and weightlifting produces benefits that go beyond the benefits each exercise provides alone. In this sense, we can say that the whole equals the sum of the parts when there is no interaction, but not when there is an interaction.

12.1.6 R Code: Factorial Design

```
grp <-c(1,1,1,2,2,2,3,3,3,4,4,4)
y <-c(3,4,3,5,3,4,5,6,4,9,9,8)
dum1 <-c(rep(0,3),rep(1,3),rep(0,6))
dum2 <-c(rep(0,6),rep(1,3),rep(0,3))
dum3 <-c(rep(0,9),rep(1,3))
summary(dum.reg <-lm(y~dum1+dum2+dum3))
tapply(y, factor(grp), mean)          #Calculate Group Means
```

(continued)

12.1.6   R Code: Factorial Design (continued)

```
#Effect coding
eff1 <-c(rep(-.5,6),rep(.5,6));
eff2 <-c(rep(c(rep(-.5,3),rep(.5,3)),2))

#ANOVA
anova.mod <-aov(y~factor(eff1)*factor(eff2))
summary(anova.mod)
barplot(tapply(y,list(eff2,eff1),mean),beside=T,
main = "Exercise and Muscle Tone", col = c("white", "gray"),
xlab = "Weightlifting", names = c("No", "Yes"),
ylab = "Muscle Tone", legend = c("No Bike", "Bike"),
args.legend = list(title = "Bike", x = "top", cex =1),ylim = c(0, 10))

#Regression model
reg.mod <-lm(y~eff1*eff2)
summary(reg.mod)
anova(reg.mod)

#Construct S Matrix for Simple Slopes
s0 <-rep(0,4)
s1 <-c(rep(1,2),rep(0,2))
s2 <-c(rep(0,2),rep(1,2))
s3 <-c(-.5,.5,-.5,.5)
S <- rbind(s0,s1,s2,s3)

#Simple Effecs
simp.slope <-t(S)%*%coef(reg.mod)
simp.cov <-t(S)%*%vcov(reg.mod)%*%S
simp.err <-sqrt(diag(simp.cov))
simples <-simp.slope/sqrt(diag(simp.cov))
df <-length(y)-nrow(S)
tvalues <-2*pt(-abs(simples),df=df)
crit <-abs(qt(0.025,df))
CI.low <-simp.slope-(crit*simp.err)
CI.high <-simp.slope+(crit*simp.err)
simp.table<-round(matrix(c(simp.slope,simp.err,simples,tvalues,
CI.low,CI.high),nrow=nrow(S),ncol=6),digits=5)
dimnames(simp.table)=list(c("a@b1","a@b2","b@a1","b@a2"),
c("slope", "stderr", "t","p","CI.low","CI.high"))
simp.table
```

12.2   Unbalanced Factorial Designs

The preceding discussion applies only to balanced designs (i.e., ones with equal or proportional cell sizes). Adjustments need to be made with unbalanced designs. As we learned in Chap. 11, these adjustments can take two forms: use unweighted means or weighted ones. This decision is even more important with a factorial design, as the same set of data can produce very different interpretations depending on which approach is taken.

   To help us appreciate the issues involved, let's pretend a public university is interested in knowing how many minutes/week math students spend studying matrix algebra as a function of their class standing and living arrangement. Having only limited resources to answer this question (this is a public university after all!), the administration passes out a survey to sophomores and seniors taking a course in matrix algebra asking (1) do you live in a dorm or in an off-campus apartment and (2) how many minutes/week do you study the material. The (mock) data appear in Table 12.12, and the means appear in Table 12.13, with the cell sizes in parentheses. Clearly, the cell sizes are neither equal nor proportional, signaling an unbalanced design.

Table 12.12   Small sample example for an unbalanced 2×2 factorial design

Subject	Class	Living arrangement	y	Contrast 1 (main effect of A)	Contrast 2 (main effect of B)	Contrast 3 (A $\times$ B interaction)
1	Sophomore	Dorm	62	−.5	−.5	0.25
2	Sophomore	Dorm	58	−.5	−.5	0.25
3	Sophomore	Dorm	62	−.5	−.5	0.25
4	Sophomore	Apartment	42	−.5	.5	−0.25
5	Sophomore	Apartment	40	−.5	.5	−0.25
6	Senior	Dorm	66	.5	−.5	−0.25
7	Senior	Dorm	64	.5	−.5	−0.25
8	Senior	Apartment	46	.5	.5	0.25
9	Senior	Apartment	44	.5	.5	0.25
10	Senior	Apartment	42	.5	.5	0.25
11	Senior	Apartment	46	.5	.5	0.25
12	Senior	Apartment	42	.5	.5	0.25

Table 12.13   Means for an unbalanced factorial design with cell sizes in parentheses

		Living arrangement		
		Dorm	Apartment	Marginal mean
Class standing	Sophomore	60.6667 (3)	41.00 (2)	50.8333
	Senior	65.00 (2)	44.00 (5)	54.50
	Marginal mean	62.83	42.50	52.6667

12.2.1   *Unweighted Means*

One way to analyze unbalanced data is to assign contrast codes to each group without taking cells size differences into account. The coding vectors in Table 12.12 follow this approach, producing the output shown in Table 12.14.[5]

Table 12.14 Regression output for an unbalanced factorial design using unweighted means

	SS	df	MS	R^2	F	p	
			Significance test of regression model				
Regression	1,117.00	3	372.3333	.9733	97.1304	.0000	
Residual	30.6667	8	3.8333				
Total	1,147.6667						

	b	se_b	SS	t	p	Explanation
			Regression coefficients			
b_0	52.6667	.6061		86.8939	.0000	Unweighted grand mean
b_1	3.6667	1.2122	35.0725	3.0248	.0164	Sophomore vs. senior
b_2	−20.3333	1.2122	1078.5507	16.7738	.0000	Dorm vs. apartment
b_3	−1.3333	2.4244	1.1594	.5500	.5974	Interaction

		Covariance matrix	
.3674	−.0639	−.0639	−.4472
−.0639	1.4694	−.4472	−.2556
−.0639	−.4472	1.4694	−.2556
−.4472	−.2556	−.2556	5.8778

12.2.1.1   Interpretation of Unweighted-Means Regression

- The overall regression equation is statistically significant, indicating that the variance among our groups is greater than would be expected if chance were the only operating factor.
- The intercept equals the unweighted grand mean (average of the group averages).
- Both main effects are significant, but the interaction is not.
- The main effect of A (b_1) indicates that sophomores study fewer hours per week (M = 50.8333) than do seniors (M = 54.50). Notice that the regression coefficient for this term equals the difference between the two marginal means ($b_1 = 3.6667$) (which do not take sample size into account).
- The main effect of B (b_2) indicates that students who live in the dorms study more hours per week ($M = 62.8333$) than do students who live off campus

---

[5] The values reported in Table 12.14 are the ones you get by default using the ANOVA program in most statistical packages. **R** is an exception, and the code you need to reproduce the table in **R** is provided at the end of this section.

(M $= 42.50$). Unsurprisingly, the difference between these marginal means equals the regression coefficient for this term ($b_2 = 20.3333$).

- The covariance matrix is not diagonal. Consequently, if we add up the sum of squares of each of our component effects, we do not reproduce SS_{reg}.

$$35.0725 + 1078.5507 + 1.1594 \neq 1117.00$$

12.2.1.2 Calculating the Sum of Squares from the Group Means

The values shown in Table 12.14 can be calculated from the group means displayed in Table 12.13. First, we use Eq. (11.9) to calculate the harmonic mean:

$$m_h = \frac{4}{\dfrac{1}{3} + \dfrac{1}{2} + \dfrac{1}{2} + \dfrac{1}{5}} = \frac{4}{1.5333} = 2.6087$$

and then we calculate the contrasts using Eq. (11.8). I will show the calculations for SS_A, and you can calculate the other terms if you desire.

$$SS_A = \frac{[(60.6667 * -.5) + (41.00 * -.5) + (65.00 * .5) + (44.00 * .5)]^2 * 2.6087}{(-.5^2 + -.5^2 + .5^2 + .5^2)} = 35.0725$$

12.2.2 *Weighted Means*

Performing an unweighted-means ANOVA with unbalanced data requires no adjustment to the contrast coefficients, but this is not true when weighted means are desired. There are some tricks we can use with a 2×2 design, but more complicated designs require one of the orthogonalization methods we discussed in Chap. 11. We will use the method of successive residuals, as shown in Table 12.15. Notice that the contrast codes comprise the seed vectors.

If we then perform the usual regression analysis, we get the output shown in Table 12.16.

12.2.2.1 Interpretation of Weighted-Means Regression

- The overall regression equation is identical to the one produced using the unweighted-means approach.
- The regression coefficients have changed from the earlier analysis.

  - Our intercept now equals the weighted grand mean (simple average) rather than the unweighted grand mean (average of the group averages).
  - As before, both of our main effects are significant, but our interaction is not. Notice, however, that the sign of the coefficient for the main effect of A is

Table 12.15 Orthogonal contrast codes for an unbalanced factorial design using weighted means

Seed 1	Seed 2	Seed 3	Subtract mean from all scores of seed 1 to find first contrast codes	Regress seed 2 on a vector of 1's and the preceding vector. The residuals are the second orthogonal contrast codes	Regress seed 3 on a vector of 1's and the two preceding vectors. The residuals are the third orthogonal contrast codes
−0.5	−0.5	0.25	−.583333	−.4	.217391
−0.5	−0.5	0.25	−.583333	−.4	.217391
−0.5	−0.5	0.25	−.583333	−.4	.217391
−0.5	0.5	−0.25	−.583333	.6	−.326087
−0.5	0.5	−0.25	−.583333	.6	−.326087
0.5	−0.5	−0.25	.416667	−.714286	−.326087
0.5	−0.5	−0.25	.416667	−.714286	−.326087
0.5	0.5	0.25	.416667	.285714	.130435
0.5	0.5	0.25	.416667	.285714	.130435
0.5	0.5	0.25	.416667	.285714	.130435
0.5	0.5	0.25	.416667	.285714	.130435
0.5	0.5	0.25	.416667	.285714	.130435

Table 12.16 Regression output for an unbalanced factorial design using weighted means

Significance test of regression model						
	SS	df	MS	R^2	F	p
Regression	1,117.00	3	372.3333	.9733	97.1304	.0000
Residual	30.6667	8	3.8333			
Total	1,147.6667					

Regression coefficients						
	b	se_b	SS	t	p	Explanation
b_0	51.1667	.5652		90.5294	.0000	Weighted grand mean
b_1	−2.80	1.1464	22.8667	2.4424	.0404	Sophomore vs. senior
b_2	−20.3913	1.2076	1092.9739	16.8856	.0000	Dorm vs. apartment
b_3	−1.3333	2.4244	1.1594	.5500	.5974	Interaction

Covariance matrix			
.3194	0	0	0
0	1.3143	0	0
0	0	1.4583	0
0	0	0	5.8778

opposite to the one we found using the unweighted approach. We will discuss this issue in greater detail momentarily.

- Notice also that the interaction term is identical to the one we found using an unweighted-means regression. Because this term includes all four cells, it is the same regardless of whether we use unweighted or weighted means.

- Finally, notice that our covariance matrix is diagonal, confirming that we have created orthogonal contrast weights. When we sum the various sum of squares, we find that they equal SS_{reg}.

$$22.8667 + 1092.9739 + 1.1594 = 1117.00$$

12.2.2.2   Comparing the Unweighted and Weighted-Means Analysis

The most interesting result from this analysis is that the main effect of Factor A is significant in both analyses, but opposite in sign. When we use the unweighted analysis, we find that sophomores study fewer hours per week ($M = 50.8333$) than seniors ($M = 54.50$), but when we use the weighted analysis, we find that sophomores study more hours per week ($M = 52.80$) than seniors ($M = 50.00$). Which conclusion is correct? The answer is it depends on the question you are asking.

With an unweighted analysis, we are taking living arrangement into account. Among dorm students and among those who live off campus, seniors study harder than sophomores. We can think of this conclusion as similar to a semi-partial correlation in which we are controlling for living arrangement before assessing the effects of class standing.

With a weighted analysis, we ignore living arrangement when calculating the effects of class standing. In this case, we conclude that sophomores study harder than seniors. The reversal comes about because we have very few seniors living in a dorm and very few sophomores living off campus, and the means of these two groups are very discrepant. The unweighted analysis treats these smaller cells as if they were the same size as the others, amplifying the amount of time seniors spend studying and reducing the amount of time sophomores spend studying. In contrast, the weighted analysis assigns these scores less weight because there are so few of them.

Finally, you should bear in mind that this contrived example is, well, contrived, and that reversals this dramatic are rare. But the issue we confront when deciding whether to use an unweighted or weighted approach still requires careful thought. Given that the whole point of using a factorial design is to take more than one factor into account, a weighted-means approach will usually be misleading. The only exception is when the two categorical variables are classificatory (rather than manipulated), and the imbalance in their cell weights reflects a genuine difference in the population at large. This is true in our example, so the weighted-means approach is a viable alternative to the usual, unweighted approach.

12.2.3   R Code: Unbalanced Factorial Design

```
g <-c(1,1,1,2,2,3,3,4,4,4,4,4)
y <-c(62,58,62,42,40,66,64,46,44,42,46,42)
#Effect coding
eff1 <-c(rep(-.5,5),rep(.5,7))
```

(continued)

12.2.3 R Code: Unbalanced Factorial Design (continued)

```
eff2 <-c(rep(-.5,3),rep(.5,2),rep(-.5,2),rep(.5,5))
tapply(y, factor(g), mean)          #Calculate Group Means

#Plot Means
barplot(tapply(y,list(eff2,eff1),mean),beside=T,
col = c("white", "gray"),
xlab = "Class", names = c("Sophomore", "Senior"),
ylab = "Study Time", legend = c("Dorm", "Apartment"),
args.legend = list(title = "Domicile", x = "topright", cex =.9),ylim =
c(0, 80))

#Unweighted Means
unweighted.mod <-lm(y~eff1*eff2)
summary(unweighted.mod)
vcov(unweighted.mod)
library(car)               #Attach car package for unweighted ANOVA
Anova(unweighted.mod,type=3)

#Weighted Means
eff3 <-eff1*eff2
b1.A <-eff1-mean(eff1)
b1.reg <-lm(eff2~eff1+b1.A)
b2.B <-resid(b1.reg)
b2.reg <-lm(eff3~b1.A+b2.B)
b3.AB <-resid(b2.reg)
wgts <-cbind(b1.A,b2.B,b3.AB)
wgts
weighted.mod <-lm(y~b1.A+b2.B+b3.AB)
summary(weighted.mod)
vcov(weighted.mod)
Anova(weighted.mod)         #No need to specify type with weighted means
```

12.3 Multilevel Designs

I have used a 2×2 factorial to introduce you to factorial designs because the logic and computations are best learned when the number of factors and their levels is small. But computationally, a 2×2 design is a special case, and it's important to be familiar with more complicated designs as well. In this section, we will learn how to analyze a 3×2 design. We still have only 2 factors, but the first factor now has 3 levels instead of 2. Consequently, we have 6 possible combinations and we need 2 coding vectors to represent our 3 level factor.

To make things less abstract, let's imagine that the university received some additional funds to conduct their research again. This time they are able to randomly assign three groups of math students (sophomores, juniors, and seniors) to one of two living conditions (dorm or off-campus apartment) for one semester. To keep things manageable, we will imagine we have a balanced design, with two subjects in each of the six (3 groups of students × 2 types of living arrangements) cells. The (invented) data, along with a coding scheme we will use to analyze them are presented in Table 12.17; Table 12.18 presents the means.

Table 12.17 Small sample example for a balanced 3 × 2 factorial design

Group 1	Group 2	y	V1	V2	V3 (Housing)	V4 = (V1 × V3)	V5 = (V2 × V3)
1	1	48	1	0	.5	.5	0
1	1	46	1	0	.5	.5	0
1	2	30	1	0	−.5	−.5	0
1	2	35	1	0	−.5	−.5	0
2	1	56	0	1	.5	0	.5
2	1	66	0	1	.5	0	.5
2	2	76	0	1	−.5	0	−.5
2	2	72	0	1	−.5	0	−.5
3	1	76	−1	−1	.5	−.5	−.5
3	1	68	−1	−1	.5	−.5	−.5
3	2	53	−1	−1	−.5	.5	.5
3	2	49	−1	−1	−5	.5	.5

Table 12.18 Means for a balanced 3 × 2 factorial design

		Living arrangement		
		Dorm	Apartment	Marginal mean
Class standing	Sophomore	47.00	32.50	39.75
	Junior	61.00	74.00	67.50
	Senior	72.00	51.00	61.50
	Marginal mean	60.00	52.50	56.25

12.3.1 Coding Scheme

Notice that we are using effect coding for this analysis. Vectors 1 and 2 represent our 2 *df* grouping variable of class standing, with the third group of seniors designated as the base group. The importance of this coding scheme will be explained momentarily. Notice also that we have used fractional contrast coding for living arrangement ($v3$). With 1 *df*, there is no difference between effect coding and contrast coding, but using fractional codes here will help us distinguish the 1 *df* main effect of living arrangement from the 2 *df* main effect of class standing.[6]

---

[6] You might notice that V_3 starts with a positive value (.5) not a negative one, as was true in our earlier examples. When a factor has only two levels, this decision is largely arbitrary.

Finally, notice that there are two cross-product terms. Each involves multiplying one of the Factor A vectors with the Factor B vector. We do not multiply the two vectors that represent Factor A.

12.3.2 Regression Analysis

Table 12.19 presents the regression output, along with an ANOVA table for reference. The regression equation shows that the six groups show significant variation, and each regression coefficient provides unique information:

Table 12.19 Regression output from a balanced 3×2 factorial design

			Significance test of regression model			
	SS	df	MS	R^2	F	p
Regression	2,525.75	5	505.1500	.9574	26.9413	.0005
Residual	112.50	6	18.7500			
Total	2,638.25					

			Regression coefficients			
	b	se_b	SS	t	p	Explanation
b_0	56.25	1.25		45.00	.0000	Unweighted grand mean
b_1	−16.50	1.7678	1,633.50	−9.3338	.0001	Sophomores— grand mean
b_2	11.25	1.7678	759.3750	6.3640	.0007	Juniors—grand mean
b_3	7.50	2.50	168.75	3.00	.0240	Dorm—apartment
b_4	7.00	3.5355	73.50	1.9799	.0950	Contrast 1 × contrast 3
b_5	−20.50	3.5355	630.3750	−5.7983	.0012	Contrast 2 × contrast 3

			Covariance matrix			
1.5625	0	0	0	0	0	
0	3.1250	−1.5625	0	0	0	
0	−1.5625	3.1250	0	0	0	
0	0	0	6.25	0	0	
0	0	0	0	12.50	−6.25	
0	0	0	0	−6.25	12.50	

			ANOVA table			
	SS	df	MS	F	p	
A	1,705.50	2	852.75	45.48	.0002	
B	168.75	1	168.75	9.00	.0240	
AB	651.50	2	325.75	17.3733	.0032	
Error	112.50	6	18.75			

- The intercept represents the unweighted grand mean (average of the group averages). With a balanced design, this value also equals the weighted grand mean, but it's best to think of this as an unweighted average rather than a weighted one.
- b_1 is the difference between the marginal mean for sophomores ($M = 39.75$) and the unweighted grand mean ($M = 56.25$).
- b_2 is the difference between the marginal mean for juniors ($M = 67.05$) and the unweighted grand mean ($M = 56.25$).
- b_3 is the difference between the marginal mean of dorm dwellers ($M = 60.00$) vs. those who live off campus ($M = 52.50$).
- b_4 is a difference score between two other difference scores: dorm sophomores—dorm mean ($47.00 - 60.00 = -13.00$) and apartment sophomores— apartment mean ($32.50 - 52.50 = -20.00$).

$$-13.00 - -20.00 = 7.00$$

- b_5 is also a difference score between two difference scores: dorm juniors—dorm mean ($61.00 - 60.00 = 1.00$) and apartment juniors—apartment mean ($74.00 - 52.50 = 21.50$).

$$1.00 - 21.50 = -20.50$$

- The covariance matrix is not diagonal, confirming that we have not used orthogonal contrast weights. When we sum the various sum of squares, they do not equal SS_{reg}.[7]

$$1633.50 + 759.3750 + 168.75 + 73.50 + 630.3750 \neq 2525.75$$

12.3.3   ANOVA Table

The final entry in Table 12.19 shows an ANOVA summary table. With a balanced design, there are some shortcuts we could use to recover our sum of squares when using effect coding, but these shortcuts won't work with an unbalanced design, so we will learn a more general method that will come in handy later in this text.

In the present case, we need to conduct three additional regression analyses. For each one, we omit one of the three terms (i.e., vectors associated with Factor A, vector associated with Factor B, and the cross-product vectors associated with the A × B interaction). Next, we subtract SS_{reg} from the reduced analysis from SS_{reg} from our full analysis to derive the sum of squares for each effect. Finally, we construct an F statistic.

---

[7] The highlighted portion of the covariance matrix will be used in a subsequent section to construct an augmented covariance matrix.

$$F = \frac{\left(SS_{reg_{full}} - SS_{reg_{reduced}}\right) / \left(df_{reg_{full}} - df_{reg_{reduced}}\right)}{SS_{res_{full}} / df_{res_{full}}} \qquad (12.3)$$

Before performing these operations, let's look more closely at the procedures themselves. To start, you might have noticed that the form of this equation is similar to the one we used to test the significance of a semi-partial correlation in Chap. 4 (see Eq. 4.13). This is no coincidence. Both equations use a subtraction method to test the increment in explained variance due to a predictor. Here we are using changes in SS_{reg} instead of changes in R^2, but we could just as easily use R^2.

Returning now to our data set, Table 12.20 repeats the pertinent information from the full analysis, followed by the SS_{reg}, R^2, and degrees of freedom from each of the three reduced analyses. To calculate the main effect of Factor A, we subtract the SS_{reg} value obtained when we omitted the two vectors that represent this factor from the overall analysis.

$$\text{Main Effect of A} = 2525.75 - 820.25 = 1705.50$$

Table 12.20 Reduced regression analyses for a balanced 3×2 factorial design

	Complete regression analysis					
	SS	df	MS	R^2	F	p
Regression	2,525.75	5	505.1500	.9574	26.9413	.0005
Residual	112.50	6	18.7500			
Total	2,638.25					
	Reduced regression analyses					
	SS_{reg}	df_{reg}	ΔSS_{reg}	Δdf_{reg}	F	p
Omit vectors for Factor A	820.25	3	1,705.50	2	45.48	.0002
Omit vector for Factor B	2357.00	4	168.75	1	9.00	.0240
Omit cross-product vectors	1,874.25	3	651.50	2	17.3733	.0032

We then construct an F test using Eq. (12.3).

$$F = \frac{1705.50/2}{112.50/6} = 45.48$$

The other two terms are calculated in a similar fashion.

$$\text{Main Effect of B} = 2525.75 - 2357.00 = 168.75$$

$$F = \frac{168.75/1}{112.5000/6} = 9.00$$

and

$$A \times B = 2525.75 - 1874.25 = 651.50$$

$$F = \frac{651.50/2}{112.5000/6} = 17.3733$$

When we refer these values to an F distribution with appropriate degrees of freedom, we find that all three terms are statistically significant, matching the results of the ANOVA table shown at the bottom of Table 12.19.

12.3.4   R Code: Multilevel Design

```
g <-c(1,1,2,2,3,3,4,4,5,5,6,6)
y=c(48,46,30,35,56,66,76,72,76,68,53,49)
g1 <-c(1,1,1,1,2,2,2,2,3,3,3,3)
g2 <-c(1,1,2,2,1,1,2,2,1,1,2,2)
tapply(y, factor(g), mean)         #Calculate Group Means

#ANOVA model
anova.mod <-aov(y~factor(g1)*factor(g2))
summary(anova.mod)

barplot(tapply(y,list(g2,g1),mean),beside=T,
main = "Unbalanced ANOVA", col = c("white", "gray"),
xlab = "Class Standing", names = c("Sophomores", "Juniors", "Seniors"),
ylab = "Study Time", legend = c("Dorm", "Apartment"),
args.legend = list(title = "Housing", x = "top", cex =1),ylim=c(0, 100))

#2df coding for Regression Model
v1 <-c(rep(1,4),rep(0,4),rep(-1,4))
v2 <-c(rep(0,4),rep(1,4),rep(-1,4))
v3 <-c(rep(c(rep(.5,2),rep(-.5,2)),3))
v4 <-v1*v3
v5 <-v2*v3
reg.mod <-lm(y~v1+v2+v3+v4+v5)
summary(reg.mod)

#Get Weighted Sum of Squares Regression for Each Model
ssreg <- sum(anova(reg.mod)[1:5,2] );
dfreg <- sum(anova(reg.mod)[1:5,1] )
ssres <- anova(lm(y~v1+v2+v3+v4+v5))[6,2]
dfres <- anova(lm(y~v1+v2+v3+v4+v5))[6,1]
msres = ssres/dfres

ss.omit.a <- sum(anova(lm(y~v3+v4+v5))[1:3,2])
df.omit.a <- sum(anova(lm(y~v3+v4+v5))[1:3,1])
ss.a <- ssreg-ss.omit.a;ss.a
f.a <-(ss.a/(dfreg-df.omit.a))/msres;f.a
```

(continued)

12.3.4   R Code: Multilevel Design (continued)

```
ss.omit.b <- sum(anova(lm(y~v1+v2+v4+v5))[1:4,2])
df.omit.b <- sum(anova(lm(y~v1+v2+v4+v5))[1:4,1])
ss.b <- ssreg-ss.omit.b;ss.b
f.b <-(ss.b/(dfreg-df.omit.b))/msres;f.b

ss.omit.ab <- sum(anova(lm(y~v1+v2+v3))[1:3,2])
df.omit.ab <- sum(anova(lm(y~v1+v2+v3))[1:3,1])
ss.ab <- ssreg-ss.omit.ab;ss.ab
f.ab <-(ss.ab/(dfreg-df.omit.ab))/msres;f.ab
```

12.3.5   Probing the Interaction

The ANOVA in Table 12.19 shows a significant interaction. Figure 12.3 plots the means, and it appears that sophomores and seniors study more when they live in a dorm than when they live in an apartment, but juniors do just the opposite.

Fig. 12.3 Mean studying time for a 3 × 2 factorial design

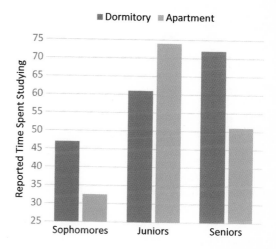

Visual impressions don't always translate into statistical significance. For this reason, we need to test the simple effects of our interaction. With a 3 × 2 design, there are 29 possible tests to conduct. That's a lot of tests! [8] In this section, you will learn a method that will allow you to perform these tests in a reasonably efficient manner. You won't need every test with any one data set, but once you learn how to

---

[8] To avoid making Type I errors, it is customary to adjust your alpha levels when you report multiple comparisons. A discussion of this issue can be found in most introductory statistics texts.

conduct all of them you can pick the ones that are relevant to the data you have collected. After working through the procedures, we will apply a few of these contrasts to our data set. A word of warning: There are quite a few steps to follow, so it's best to go slow and work through them carefully.

12.3.5.1   Augmented b Vector and Covariance Matrix

To utilize the new procedure we must start with effect codes (as we have done in our example). We then create a new column vector of regression coefficients we will designate $\mathbf{b}^+$ and a modified covariance matrix we will designate $\mathbf{C}^+$. Before we do, let's review the meaning of the regression coefficients when effect coding is used. As first discussed in Chap. 11, with effect coding, each coefficient represents the deviation of a group mean (designated with $1's$) vs. the grand mean. We can think of these scores as treatment effects, as they represent each group's unique deviation from the overall mean.

We already know our treatment effects for our first two groups (sophomores and juniors) because they equal the regression coefficients, b_1 and b_2, respectively (see Table 12.19). But we do not know the treatment effect for our third group of seniors because this was our base group (i.e., the one that received $-1's$ on both vectors). Fortunately, we can easily calculate the effect by taking advantage of a simple fact: The sum of the deviation scores from the grand mean must equal 0. Consequently, we can use Eq. (12.4) to compute our missing treatment effect.

$$b^+ = -\Sigma b_j \tag{12.4}$$

Gathering the values from Table 12.19 and applying the formula yields the following value for our group 3 coefficient (which we will designate b_{2+}):

$$b_{2+} = -(-16.50 + 11.25) = 5.25$$

Looking back to the means displayed in Table 12.18 confirms that this value represents the treatment effect for group 3.

Now we need to calculate an interaction coefficient for group 3. Here again, we take advantage of the fact that the three cross-product coefficients must sum to 0 to create another coefficient we will designate b_{5+}.

$$b_{5+} = -(7.00 + -20.50) = 13.50$$

As shown in Table 12.21, we then place these values, along with our original coefficients into a column vector to create an augmented vector of regression coefficients. (Notice that the augmented vector does not include the intercept.)

Our next step is to create an augmented covariance matrix. We begin by gathering the variance/covariance matrix from our overall regression analysis, omitting the intercept term (see Table 12.19). We then augment the matrix

Table 12.21 Augmented vector of regression coefficients

	Coefficient order	Computation	Value
$b^+ =$	b_1	b_1	-16.50
	b_2	b_2	11.25
	b_{2+}	$-(b_1+b_2)$	5.25
	b_3	b_3	7.50
	b_4	b_4	7.00
	b_5	b_5	-20.50
	b_{5+}	$-(b_4+b_5)$	13.50

following the same rules we used above, only this time we use variances and covariances rather than regression coefficients.

$$c^+ = -\Sigma c_j \tag{12.5}$$

The highlighted portion of the covariance matrix shown in Table 12.19 serves as our original covariance matrix, and Table 12.22 presents the augmented matrix. The shaded values have been calculated using Eq. (12.5). To illustrate, the shaded entries in the first row were calculated as follows:

Table 12.22 Augmented covariance matrix

	c_1	c_2	c_{2+}	c_3	c_4	c_5	c_{5+}
c_1	3.1250	-1.5625	-1.5625	.0000	.0000	.0000	.0000
c_2	-1.5625	3.1250	-1.5625	.0000	.0000	.0000	.0000
c_{2+}	-1.5625	-1.5625	3.1250	.0000	.0000	.0000	.0000
c_3	.0000	.0000	.0000	6.2500	.0000	.0000	.0000
c_4	.0000	.0000	.0000	.0000	12.5000	-6.2500	-6.2500
c_5	.0000	.0000	.0000	.0000	-6.2500	12.5000	-6.2500
c_{5+}	.0000	.0000	.0000	.0000	-6.2500	-6.2500	12.5000

$$c_{12+} = -(c_{12} + c_{13})$$
$$= -(3.1250 + -1.5625) = -1.5625$$

and

$$c_{15+} = -(c_{14} + c_{15})$$
$$= -(.0000 + -.0000) = .0000$$

After repeating these calculations for the second row, we find our third row as the negative sum of the first two augmented rows [i. e., $-(c_1+c_2)$]. The other shaded values in Table 12.22 were calculated in a similar fashion, with the final row found as $-(c_4+c_5)$.

12.3.5.2   S Matrix for Simple Effects

Having created these terms, we are ready to calculate 27 of the 29 possible simple effects.[9] As always, we begin by constructing an **S** matrix of contrast weights. The entries shown in Table 12.23 present the contrast weights we use, along with the rules used to generate each one. There is a lot of information here, but if you take the time to study things, I think it will be clear. To help you, I will provide an overview.

Table 12.23 **S** matrix rules for computing 27 simple effects for a 3×2 factorial design

Description	Simple effects of B @ each level of A	Factor A contrasts collapsed across B	Simple $A \times B$ interactions	Factor A contrasts @ B_1	Factor A contrasts @ B_2
#	3	6	6	6	6
Rules	Enter "1" for the B coefficient, and "1" for the $A \times B$ level of interest	Enter contrast coefficients in first three rows	Enter contrast coefficients in last three rows	Enter contrast coefficients in first three rows, and multiply them by the B_1 code in last 3 rows	Enter contrast coefficients in first three rows, and multiply them by the B_2 code in last 3 rows

		Example				
		$B @ A_1$	A_1 vs. A_3	A_1 vs. A_3 $\times$ B	A_1 vs. A_3 @ B_1	A_1 vs. A_3 @ B_2
b_1	A_1	0	1	0	1	1
b_2	A_2	0	0	0	0	0
b_{2+}	A_3^*	0	-1	0	-1	-1
b_3	B	1	0	0	0	0
b_4	$A_1 \times B$	1	0	1	.5	$-.5$
b_5	$A_2 \times B$	0	0	0	0	0
b_{5+}	$A_3^* \times B$	0	0	-1	$-.5$	.5

- First, we might wish to examine the effects of B at each level of A. Here we are asking whether living condition makes a difference among sophomores, juniors, and seniors. The coefficients needed to make this determination appear in the first column of coefficients in Table 12.23.

---

[9] We will discuss the other two comparisons after we complete our discussion here.

- We might also wish to assess the simple effects of Factor A collapsed across Factor B. Here we are comparing the marginal means of Factor A. There are 6 comparisons we could make:

  - Sophomore vs. juniors
  - Sophomore vs. seniors
  - Juniors vs. seniors
  - Sophomores vs. (juniors + seniors)
  - Juniors vs. (sophomores + seniors)
  - Seniors vs. (sophomores + juniors)

The second column of coefficients in Table 12.23 provides the rules for these test.

- We can also compute interaction contrasts. For example, do living conditions have a similar effect on sophomores and seniors? Here again, there are 6 possible comparisons, and the rules appear in the third column of coefficients in Table 12.23.
- Finally, we can make Factor A comparisons within each level of B. For example, we can compare sophomores vs. juniors in the dorm or juniors vs. seniors in an apartment. The final two columns in Table 12.23 provide the rules needed to make these comparisons.

Table 12.24 presents a complete S matrix with all 27 contrast weights. To conduct our tests, we find their associated regression coefficients by calculating $S'b^{+}$ and their associated standard errors by taking the square root of the diagonal entries of $S'C^{+}S$. We then create t statistics and refer each value to a t distribution with $N - k - 1$ degrees of freedom.

Table 12.25 presents the results of all 27 tests. As indicated earlier, you will certainly not report them all with any given data set, but all of them can help you better understand the data you have collected. Ultimately, you need to tell a coherent and compelling story with your data, using your simple effects to back up your narrative. There are no hard-and-fast rules regarding this process. You have to look at your data and think carefully about the pattern of effects.

Returning to Fig. 12.3, I would start by examining the most visually salient pattern: Sophomores and seniors study more when living in a dorm than when living in an apartment, but juniors do the opposite. The simple effects of B at each level of A test this pattern. The first row of coefficients in Table 12.25 confirms that sophomores 1 and seniors 3 study more when living in a dorm than when living in an apartment, but juniors 2 study more when they live in an apartment than when they live in a dorm.[10]

Next, I would make comparisons within pairs of student groups, starting with a comparison between sophomores and seniors. I would begin here because our earlier study used only these groups, and it is important to know whether our earlier survey findings were replicated with an experimental design. In our original study using unweighted means, we found that seniors studied more than sophomores and

---

[10] The boxed numbers reference the relevant comparison in Table 12.25.

Table 12.24 S matrix for computing simple effects in a 3×2 factorial design

B @ A

	B @ A_1	B @ A_2	B @ A_3
b_2	0	0	0
b_3	0	0	0
b_{3+}	0	0	0
b_4	1	1	1
b_5	1	0	0
b_6	0	1	0
b_{6+}	0	0	1

Factor A collapsed across B

	1 vs. 2	1 vs. 3	2 vs. 3	1 vs. (2, 3)	2 vs. (1, 3)	3 vs. (1, 2)
b_2	1	1	0	2	−1	−1
b_3	−1	0	1	−1	2	−1
b_{3+}	0	−1	−1	−1	−1	2
b_4	0	0	0	0	0	0
b_5	0	0	0	0	0	0
b_6	0	0	0	0	0	0
b_{6+}	0	0	0	0	0	0

Simple A × B interactions

	1 vs. 2	1 vs. 3	2 vs. 3	1 vs. (2, 3)	2 vs. (1, 3)	3 vs. (1, 2)
b_2	0	0	0	0	0	0
b_3	0	0	0	0	0	0
b_{3+}	0	0	0	0	0	0
b_4	0	0	0	0	0	0
b_5	1	1	0	2	−1	−1
b_6	−1	0	1	−1	2	−1
b_{6+}	0	−1	−1	−1	−1	2

Simple A @ B_1

	1 vs. 2	1 vs. 3	2 vs. 3	1 vs. (2, 3)	2 vs. (1, 3)	3 vs. (1, 2)
b_2	1	1	0	2	−1	−1
b_3	−1	0	1	−1	2	−1
b_{3+}	0	−1	−1	−1	−1	2
b_4	0	0	0	0	0	0
b_5	.5	.5	0	1	−.5	−.5
b_6	−.5	0	.5	−.5	1	−.5
b_{6+}	0	−.5	−.5	−.5	−.5	1

Simple A @ B_2

	1 vs. 2	1 vs. 3	2 vs. 3	1 vs. (2, 3)	2 vs. (1, 3)	3 vs. (1, 2)
b_2	1	1	0	2	−1	−1
b_3	−1	0	1	−1	2	−1
b_{3+}	0	−1	−1	−1	−1	2
b_4	0	0	0	0	0	0
b_5	−.5	−.5	0	−1	.5	.5
b_6	.5	0	−.5	.5	−1	.5
b_{6+}	0	.5	.5	.5	.5	−1

Table 12.25 27 1 *df* simple effects for a 3 × 2 factorial design

#		*b*	*se_b*	*SS*	*t*	*p*
			Simple effects of B at A			
1	B @ A$_1$	14.50	4.3301	210.25	3.3486	.0154
2	B @ A$_2$	−13.00	4.3301	169.00	−3.0022	.0239
3	B @ A$_3$	21.00	4.3301	441.00	4.8497	.0029
			Simple effects of A collapsed across B			
4	1 vs. 2	−27.75	3.0619	1540.125	−9.0631	.0001
5	1 vs. 3	−21.75	3.0619	946.125	−7.1035	.0004
6	2 vs. 3	6.00	3.0619	72.00	1.9596	.0978
7	1 vs. (2, 3)	−49.5	5.3033	1633.5	−9.3338	.0001
8	2 vs. (1, 3)	33.75	5.3033	759.375	6.3640	.0007
9	3 vs. (1, 2)	15.75	5.3033	165.375	2.9698	.0250
			Simple A × B interaction contrasts			
10	1 vs. 2	27.50	6.1237	378.125	4.4907	.0041
11	1 vs. 3	−6.50	6.1237	21.125	−1.0614	.3293
12	2 vs. 3	−34.00	6.1237	578.00	−5.5522	.0014
13	1 vs. (2, 3)	21.00	10.6066	73.50	1.9799	.0950
14	2 vs. (1, 3)	−61.50	10.6066	630.3750	−5.7983	.0012
15	3 vs. (1, 2)	40.50	10.6066	273.3750	3.8184	.0088
			Simple effects of A at B$_1$			
16	1 vs. 2	−14.00	4.3301	196.00	−3.2332	.0178
17	1 vs. 3	−25.00	4.3301	625.00	−5.7735	.0012
18	2 vs. 3	−11.00	4.3301	121.00	−2.5403	.0441
19	1 vs. (2, 3)	−39.00	7.50	507.00	−5.2000	.0020
20	2 vs. (1, 3)	3.00	7.50	3.00	.4000	.7030
21	3 vs. (1, 2)	36.00	7.50	432.00	4.8000	.0030
			Simple effects of A at B$_2$			
22	1 vs. 2	−41.50	4.3301	1,722.25	−9.5840	.0001
23	1 vs. 3	−18.50	4.3301	342.25	−4.2724	.0052
24	2 vs. 3	23.00	4.3301	529.00	5.3116	.0018
25	1 vs. (2, 3)	−60.00	7.50	1,200.00	−8.00	.0002
26	2 vs. (1, 3)	64.50	7.50	1,386.75	8.60	.0001
27	3 vs. (1, 2)	−4.50	7.50	6.75	-.60	.5705

that this effect was just as true in the dorms as in an apartment. To determine whether this pattern occurred in our current data set, we first examine the simple Group 1 × Group 3 comparison collapsed across living conditions. Table 12.25 shows the comparison is significant [5]. To see whether the effect is qualified by living conditions, we examine the simple Group 1 × Group 3 interaction. Table 12.25 shows the simple interaction is not significant [11]. Taken together, these findings replicate our earlier findings: Seniors study more than sophomores, and this effect does not vary across living conditions.

I would then repeat these analyses, comparing sophomores and juniors. Here, the simple Group 1 × Group 2 comparison collapsed across living conditions $\boxed{4}$ is qualified by a Group 1 × Group 2 simple interaction $\boxed{10}$. The interaction reflects the fact that juniors study more than sophomores in the dorm $\boxed{16}$, but this effect is even more pronounced in an apartment $\boxed{22}$. We can characterize this pattern as "an especially for" interaction: Juniors study more than sophomores, especially when they live in an apartment.

Finally, I would repeat the analyses comparing juniors and seniors. Here, the simple Group 2 × Group 3 comparison collapsed across living conditions is not significant $\boxed{6}$, but the simple Group 2 × Group 3 interaction contrast is significant $\boxed{12}$. Follow-up tests show that seniors study more than juniors in a dorm $\boxed{18}$, but juniors study more than seniors in an apartment $\boxed{24}$. This pattern represents a cross-over interaction.

12.3.5.3   Summary

We have covered a lot of ground here, so let's pause for a moment to take stock of what we have accomplished. Using effect codes with a few modifications, we were able to test all possible 1 df comparisons and contrasts. Neither orthogonal contrasts nor dummy codes offer this feature. For this reason, effect codes should be used when analyzing data with a factorial design. You have to augment the regression coefficients and covariance matrix in order to take full advantage of the flexibility they offer, but if you're willing to do so, you will possess a highly efficient tool for making group comparisons and contrasts.

12.3.5.4   Simple Effect of A at Each Level of B

We still have 2 simple effects to compute. This is because using an **S** matrix to compute simple effects works only when each contrast has a single degree of freedom. But suppose we wish to determine the simple effect of class standing (Factor A) at each level of living condition (Factor B). Each of these simple effects has 2 df, so we can't use an **S** matrix. With a balanced design, there are some shortcuts we could use, but it's useful to learn a more general approach that will work in all situations.

The easiest thing to do is to recode our $1df$ vector using dummy coding rather than effect coding. For example, to find the simple effect of A at B_1, we set dorm $= 0$ and apartment $= 1$. We then compute cross-products with our grouping variable using this new coding scheme and conduct a new regression analysis using only the new dummy variable and the newly formed cross-products (i.e., we omit the group A vectors after using them to compute the cross-products). By subtracting the obtained SS_{reg} from the SS_{reg} obtained in the original, overall analysis, we find the simple effect of A at B_1. To find the simple effect of A at B_2, we recode the dummy variable so that apartment $= 0$ and dorm $= 1$, and then follow the steps just described. The **R** code that accompanies this chapter performs these analyses, yielding the results displayed in (Table 12.26).[11]

Table 12.26 2 *df* Tests of *A* at each level of *B*

	SS	df	MS	F	p
Simple effect of A at B_1	628.00	2	314.00	16.7467	.0035
Simple effect of A at B_2	1,729.00	2	864.50	46.1067	.0002

12.3.6 Higher-Order Designs

In this chapter we have only considered two-factor designs, but higher-order ones are not uncommon. We analyze higher-order designs using the same statistical techniques we have learned here and interpret their higher-order interactions using methods discussed in Chap. 9. Although interpreting higher-order interactions requires considerable care, it is worth remembering that most effects occur within a limited range of conditions. Because factorial designs are well suited to identifying these limiting conditions, they provide a powerful statistical tool for identifying causal relations.

12.3.7 R Code: Simple Effects in Multilevel Design

```
grp <-c(1,1,2,2,3,3,4,4,5,5,6,6)
y <-c(48,46,30,35,56,66,76,72,76,68,53,49)
tapply(y, factor(grp), mean)        #Calculate Group Means

#2df coding
v1 <-c(rep(1,4),rep(0,4),rep(-1,4));v2 <-c(rep(0,4),rep(1,4),rep(-1,4))
v3 <-c(rep(c(rep(.5,2),rep(-.5,2)),3));v4 <-v1*v3;v5 <-v2*v3
reg.mod <-lm(y~v1+v2+v3+v4+v5);summary(reg.mod)

#Generate Augmented Coefficients and Covariance Matrix
b <-reg.mod$coef[2: length(reg.mod$coef)]
BB<-c(b[1:2],-(b[1]+b[2]),b[3:5],-(b[4]+b[5]));BB;
covar <-vcov(reg.mod)
cov <- covar[2: length(reg.mod$coef),2: length(reg.mod$coef)]
```

(continued)

---

[11] We would follow the same procedure if we had an A $\times$ B design of any size (e.g., 3×4). Use dummy coding for Factor B, setting one group to receive 0's on all vectors, and rerun your regression analysis. The regression coefficient is the simple effect of A at the level of B that received 0's on all vectors.

12.3.7   R Code: Simple Effects in Multilevel Design (continued)

```
rows<-function(cov,i,j)cbind(cov[i,1],cov[i,2],-(cov[i,1]+
cov[i,2]),cov[i,3],cov[i,4],cov[i,5],-(cov[i,4]+cov[i,5]))
aug <-rows(cov)
CC<-rbind(aug[1,],aug[2,],-(aug[1,]+aug[2,]),aug[3,],aug[4,],
aug[5,],-(aug[4,]+aug[5,]));CC

#Simple Effects
simple <-function(S){
simp.slope <-t(S)%*%BB;simp.err <-sqrt(diag(t(S)%*%CC%*%S))
df=(length(y)-length(coef(reg.mod)))
ttests <-simp.slope/simp.err;pvalues <-2*pt(-abs(ttests),df=df)
crit <-abs(qt(0.025,df))
CI.low <-simp.slope-(crit*simp.err);CI.high <-simp.slope+
(crit*simp.err)
simp.table<-round(matrix(c(simp.slope,simp.err,ttests,pvalues,
CI.low,CI.high),nrow=ncol(S),ncol=6),digits=5)
dimnames(simp.table)=list(c(),c("slope","stderr","t","p",
"CI.low","CI.high"))
return(list(S,simp.table))
}

#B at Levels of A
c1 <-c(0,0,0);c2 <-c(0,0,0);c3 <-c(0,0,0);c4 <-c(1,1,1);c5 <
-c(1,0,0);c6 <-c(0,1,0);c7 <-c(0,0,1);S <-rbind(c1,c2,c3,c4,c5,
c6,c7);simple(S)

SEFF <-function(a,b){
s1 <-c(1*a,1*a,0,2*a,-1*a,-1*a);s2 <-c(-1*a,0,1*a,-1*a,2*a,-1*a)
s3 <-c(0,-1*a,-1*a,-1*a,-1*a,2*a);s4 <-c(rep(0,6))
s5 <-c(1*b,1*b,0,2*b,-1*b,-1*b);s6 <-c(-1*b,0,1*b,-1*b,2*b,-1*b)
s7 <-c(0,-1*b,-1*b,-1*b,-1*b,2*b);S <-rbind(s1,s2,s3,s4,s5,s6,s7)
simple(S)}

#Simple Effects Function  #Codes must be specified
SEFF (1,0)            #A collapsed across B
SEFF (0,1)            #AxB interactions
SEFF (1,.5)          #Simple effects of A @ b1
SEFF (1,-.5)         #Simple effects of A @ b2

## 2df Simple Effects
dum.a1 <-ifelse(v3 == .5,0,v3);dum.a2 <-dum.a1*v1;dum.a3 <-dum.a2*v2
simpa <-lm(y~ dum.a1+dum.a2+dum.a3);F.A <-anova(simpa,reg.mod);F.A
dum.b1 <-ifelse(v3 == -.5,0,v3);dum.b2 <-dum.b1*v1;dum.b3 <-dum.b2*v2
simpb <-lm(y~ dum.b1+dum.b2+dum.b3);F.B <-anova(simpb,reg.mod);F.B
```

12.4   Chapter Summary

1. A complete factorial design occurs when two or more categorical variables are crossed, such that all levels of each variable are paired with all levels of all other variables. The number of combinations (or cells) is found by multiplying the levels of each factor.

2. In a balanced factorial design with equal or proportional cell sizes, the regression sum of squares can be partitioned into two sources of variance: main effects and interactions. Main effects examine the overall impact of a factor without taking other factors into account; interactions examine non additive effects of each unique pairing.

3. Simple effects describe the effect of one variable at a single level of another variable. Simple effects are similar to simple slopes discussed in Chap. 9, and a significant interaction occurs when the slopes of two simple effects are not parallel.

4. An unbalanced design occurs when cells sizes are not equal or proportional. There are two approaches to analyzing an unbalanced factorial design. When cell size discrepancies arise from random or nonsystematic factors, we use unweighted means that neutralize disparities in cell sizes; when cell size discrepancies represent meaningful differences, we use weighted means that preserve disparities in cell sizes.

5. Multiple vectors are needed to represent a factor with more than 2 levels. As the number of factors increase, the number of main effects and the number of interactions increase as well. Using effect coding, we can create an augmented coefficient and covariance matrix that enables the efficient testing of all 1 df simple effects.

Chapter 13
Analysis of Covariance

In Chaps. 11 and 12, we learned that multiple regression can be used to analyze data with categorical predictors. In the next three chapters, we will extend our coverage to include designs that combine categorical and continuous predictors. The role of our categorical predictor is straightforward. Either as a result of random assignment to conditions or subject classification, subjects are divided into groups based on discrete categories. We then use orthogonal, dummy, or effect coding to create vectors that represent group membership.

Unlike the simplicity of the categorical variable, the continuous predictor can assume various roles, each demanding its own statistical analysis. Figure 13.1 depicts three relations between a categorical predictor x, a continuous criterion y, and a continuous third variable z. In the left-hand portion of the figure, x and z independently predict y. When x is a grouping variable, we use analysis of covariance (ANCOVA) to analyze data like these. In this chapter, we examine this analytic strategy.

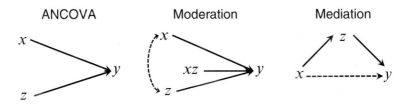

Fig. 13.1 Three possible relations between a categorical predictor (x), a continuous criterion (y), and a continuous third variable (z)

Electronic Supplementary Material: The online version of this chapter (doi: 10.1007/978-3-319-11734-8_13) contains supplementary material, which is available to authorized users

The middle portion of Fig. 13.1 portrays a second possibility, known as moderation. Here, x and z can be correlated, and we are interested not only in their independent ability to predict y but also in their interaction (as represented by a cross-product term). Chapter 14 examines this scenario.

The right-hand portion of Fig. 13.1 depicts a third possibility, termed mediation. Here, we assume that x predicts z and that z predicts y. In this case, the effect of x on y is said to be mediated through z. In some cases, x also has a direct effect on y, indicating a situation known as partial mediation (see dotted line). Chapter 15 explores mediation.[1]

Table 13.1 presents another way to think about these blended designs. Notice that each was developed for use in a particular situation for a particular purpose. These descriptions represent ideal circumstances that are not always realized in practice. As we will see, a design intended to answer one question is sometimes mistakenly used to answer another.

Table 13.1 Nature of a continuous predictor in a blended design

	ANCOVA	Moderation	Mediation
Was the continuous variable measured before subjects were assigned to experimental conditions?	Yes	Yes	No
Does the continuous variable interact with the grouping variable to predict the criterion?	No	Yes	No
Does the researcher assume that the continuous predictor explains the observed effects of the categorical variable on the criterion?	No	No	Yes
Primary purpose	Minimize error term by controlling extraneous sources of variance	Establish the boundary conditions of a categorical variable	Illuminate possible causal mechanisms

13.1 Introduction to ANCOVA

As originally developed, ANCOVA involves adding a continuous predictor (hereafter termed a covariate) to a typical analysis of variance (ANOVA). Ideally, the covariate is correlated with the criterion but uncorrelated with the categorical

---

[1] A fourth possibility, known as confounding, is not depicted in Fig. 13.1. With confounding, x and y are related because of their common association with z. This possibility exists only for correlational designs, not experimental ones. For example, shoe size (x) and vocabulary size (y) are correlated among elementary school children: the bigger your feet, the more words you know. But these variables are joined simply because they are both linked to a common third variable, age (z). Older kids have bigger feet and a larger vocabulary than younger kids, but there is no causal association in either direction between shoe size and vocabulary size. If we controlled for age by measuring the association between shoe size and vocabulary size *within* each age group, we would find no association between our predictor and our criterion.

variable. Given these conditions, including a covariate reduces the unexplained variance (SS_{res}), thereby increasing the power of a categorical predictor (SS_{reg}). Unfortunately, the technique is not always used in its intended form, leading to confusion. In this chapter, we will first discuss the correct approach and then consider problems that arise in the misuse of ANCOVA.

13.1.1   Mechanics

Following some preliminary analyses, ANCOVA compares two regression models: one in which the covariate predicts the criterion and one in which the categorical predictors and covariate predict the criterion. The difference in SS_{reg} between the two models indicates whether group differences in the covariance adjusted means are likely to be due to chance alone. The term "adjusted means" underscores that the means being analyzed are not the same as the observed means. The size of the difference between the observed and adjusted means depends on the strength of the association between the covariate and the criterion, and the magnitude of group differences for the covariate.

   To make our discussion less abstract, imagine that an educator has developed a method of teaching abstract reasoning to school children. The educator randomly assigns 12 students to one of three conditions: the new method the educator has developed, a traditional approach to teaching abstract reasoning, and a control condition involving no instruction. At the end of three months, she administers a standardized test of abstract reasoning. The (sham) data, along with a coding scheme we will use to analyze them, are shown in Table 13.2. As you can see, we are using effect coding, with the control condition as our base group.

Table 13.2 Small sample example for an analysis of covariance with age as the covariate

Method	Age	y	V1	V2	Age×V1	Age×V2
New	12	7	1	0	12	0
New	8	5	1	0	8	0
New	9	5	1	0	9	0
New	7	4	1	0	7	0
Traditional	13	6	0	1	0	13
Traditional	8	2	0	1	0	8
Traditional	11	5	0	1	0	11
Traditional	8	3	0	1	0	8
Control	12	4	−1	−1	−12	−12
Control	10	4	−1	−1	−10	−10
Control	7	1	−1	−1	−7	−7
Control	9	2	−1	−1	−9	−9

   Notice that Table 13.2 also includes each student's age. Because the ability to think abstractly increases with age during childhood, this variable is likely to influence scores on the reasoning test. With enough subjects, random assignment to conditions theoretically ensures that the experimental groups will not differ on this variable; nevertheless, failing to take its influence into account inflates the error term. However, if we statistically control for age by including it in our analysis, we can reduce the error term and increase the power of our statistical test. This is the intended use of an ANCOVA: Before assessing the impact of a categorical predictor, we include a covariate to decrease the magnitude of the error term and increase the power of our experimental treatment.

13.1.2   Preliminary Analyses

It is desirable to perform several preliminary analyses before conducting an ANCOVA. Not all of them are essential, but all provide useful information. Table 13.3 presents the results of various analyses we will examine.

13.1.2.1   Does the Covariate Predict the Criterion?

Because the goal of ANCOVA is to reduce our error term, we want to confirm that our covariate predicts our criterion. The first entry in Table 13.3 shows the results of a regression analysis in which the covariate predicts the criterion. A significant

Table 13.3 Regression analyses used in analysis of covariance

	SS	df	MS	R^2	F	p
a. Regress criterion on covariate						
Regression	15.5106	1	15.5106	.4562	8.3890	.0159
Residual	18.4894	10	1.8489			
Total	34.00					
b. Regress covariate on categorical vectors						
Regression	2.00	2	1.00	.0426	.2000	.8223
Residual	45.00	9	5.00			
Total	47.00					
c. Regress criterion on categorical vectors						
Regression	12.50	2	6.25	.3676	2.6163	.1272
Residual	21.50	9	2.3889			
Total	34.00					
d. Regress criterion on covariate, categorical vectors, and cross-product terms						
Regression	32.0180	5	6.4036	.9417	19.3854	.0012
Residual	1.9820	6	.3303			
Total	34.00					
e. Regress criterion on covariate and categorical vectors						
Regression	31.8389	3	10.6130	.9364	39.2871	.0000
Residual	2.1611	8	.2701			
Total	34.00					

effect is observed, indicating that age predicts abstract reasoning ability. Taking the square root of the R^2 value, we see that the correlation between the two continuous predictors $= .6754$, indicating a rather substantial association.

13.1.2.2   Do the Grouping Vectors Predict the Covariate?

Our next analysis examines whether our covariate is statistically independent of our categorical variables. To examine this issue, we regress the covariate on the categorical vectors. The second analysis in Table 13.3 shows the results. Here we see that the three groups do not differ significantly with regard to their average age. This is not surprising, as assignment to conditions was random. At the same time, the two predictors are not completely independent. This, too, is not surprising, as it is highly unlikely that random assignment would produce groups that are identical with respect to any variable with a limited number of subjects. An ANCOVA is appropriate as long as the association between the two variables is not greater than would be expected by chance alone.

13.1.2.3   Do the Grouping Vectors Predict the Criterion?

Our third analysis examines whether our grouping vectors predict our criterion before we have included the covariate. This test, which is simply an ANOVA on the criterion, is not necessary, but it is often informative. Analysis "c" in Table 13.3 shows that the grouping variable does not significantly predict the criterion before age is taken into account.

13.1.2.4   Testing the Homogeneity of Regression Lines

Our next analysis examines whether the covariate interacts with the categorical variable to predict the criterion. If it does, we cannot perform an ANCOVA. Why? An ANCOVA assumes that the slope of the regression line relating the covariate to the criterion is identical across groups. Formally, this assumption is known as the "homogeneity of regression lines," but it could just as easily be termed "parallelism of regression lines." Recall from Chap. 9 that parallel regression lines indicate that the simple slope relating x to y is the same at all levels of z. The same is true when x is a categorical predictor and z is a covariate. If this assumption is violated (i.e., the regression lines for each group are not parallel), we forgo conducting an ANCOVA and perform a moderated-regression analysis (to be discussed in Chap. 14).

Testing for the homogeneity of regression lines involves three steps:

1. First, we create cross-product terms between our categorical vectors and our covariate and regress our criterion on all of the predictors (covariate, categorical predictors, and cross-product terms).[2] Analysis "d" reports the results of this analysis.
2. Next, we perform another regression analysis omitting the cross-product terms (analysis "e" in Table 13.3).
3. Finally, we use the subtraction method discussed in Chap. 12 to test the homogeneity of regression lines. In the present case, we subtract SS_{reg} in analysis "e" from SS_{reg} in analysis "d" and then construct an F ratio with the mean difference in the numerator and the error term from analysis "d" in the denominator (see Eq. 12.3).

Plugging in our values tests the homogeneity of the regression lines:

$$SS_{dif} = 32.0180 - 31.8389 = .1791$$
$$F = (.1791/2)/.3303 = .2711$$

With 2 and 6 degrees of freedom, the F value is not significant ($p=.7714$), indicating that the cross-product terms do not add to the prediction of our criterion and that the regression lines are homogeneous.[3]

13.1.2.5   Inspecting the Regression Lines

It is useful to visually examine the regression lines to confirm their homogeneity. As these regression lines represent simple slopes, we follow the same procedures we learned in earlier chapters (i.e., we construct an **S** matrix and then calculate **S′ b**). The values we need for our **b** column vector are obtained from the regression equation that includes all of the predictors (analysis "d" in Table 13.3). Table 13.4 shows the regression coefficients from that analysis, and Table 13.5 shows the **S** matrix we use to find the simple intercepts and slopes. To find the simple intercepts, we assign a 1 to our intercept coefficient and the group contrast codes to b_2 and b_3; the remaining values are set to 0. To find the simple slopes, we assign a 1 to our covariate and the group contrast codes to the cross-product terms (b_4 and b_5); the remaining values are set to 0. When we multiply **S′ b**, we obtain the coefficients shown in Table 13.6.

---

[2] There is no need to center our continuous predictor as our concern is with the cross-product terms, which are unaffected by centering.

[3] We could also use changes in R^2 to assess whether the cross-product terms add significantly to the prediction of the criterion. Equation (4.13) presents the formula for performing this test.

Table 13.4 Unstandardized regression coefficients for a regression model with age, categorical vectors, and their cross-product terms predicting abstract reasoning scores

	b	se_b	SS	t	p
b_0	−2.1922	.8358		2.6228	.0394
b_1 (age)	.6492	.0865	18.6003	7.5039	.0003
b_2 (effect code 1)	2.2993	1.1676	1.2812	1.9694	.0964
b_3(effect code 2)	−1.0300	1.1567	.2620	.8905	.4075
b_4 (age* effect code 1)	−.0777	.1239	.1301	.6275	.5535
b_5 (age*effect code 2)	.0731	.1166	.1296	.6264	.5541

Table 13.5 S matrix for calculating simple regression lines (intercepts and slopes)

	S matrix for intercept			**S** matrix for slope		
	Group 1	Group 2	Group 3	Group 1	Group 2	Group 3
b_0	1	1	1	0	0	0
b_1	0	0	0	1	1	1
b_2	1	0	−1	0	0	0
b_3	0	1	−1	0	0	0
b_4	0	0	0	1	0	−1
b_5	0	0	0	0	1	−1

Table 13.6 Simple intercepts and slopes within each experimental condition

	b_0	b_1
New instructional method	.1071	.5714
Traditional instructional method	−3.2222	.7222
Control condition	−3.4615	.6538

Notice that the simple slopes are of similar magnitude, indicating that the regression slopes are homogeneous. We can visually confirm their homogeneity by using the coefficients to calculate predicted values for subjects scoring one standard deviation below and above the mean of age $(sd = 2.0671)$. Figure 13.2 plots the predicted values. As you can see, the lines relating age to reasoning ability are nearly parallel across the three experimental conditions, indicating that the differences in slope are negligible. This is precisely what it means to say that the regression lines are homogeneous. Notice, also, that the intercepts are not identical. As the variables are categorical, the intercepts represent mean differences among the three groups.

Fig. 13.2 Simple
regression lines relating age
to reasoning ability across
three instructional methods

13.1.3   Main Analysis

Having conducted a series of preliminary analyses, we are ready to perform an
ANCOVA by subtracting SS_{reg} in analysis "a" from SS_{reg} in analysis "e." We then
test the mean difference using the MS_{res} from analysis "e." The difference between
the two models represents the effect of the omitted predictors. In the present case,
analysis "e" includes the covariate and categorical vectors, but analysis "a"
includes only the covariate. So the categorical vectors are absent in analysis "a,"
and we gauge their importance by subtraction:

$$SS_{dif} = 31.8389 - 15.5106 = 16.3283$$

and

$$F = (16.3283/2)/.2701 = 30.2220$$

Referring this value to an F distribution with 2 and 8 degrees of freedom shows a
statistically significant effect (p=.0002), indicating that variations in our covariance adjusted means are unlikely to be due to chance alone. Table 13.7 summarizes
the analysis.[4]

Table 13.7 Summary
of the ANCOVA

	SS	df	MS	F	P
Age	15.5106	1	15.5106	8.3890	.0159
Error	18.4894	10	1.8489		
Groups	16.3283	2	8.1641	30.2220	.0002
Error	2.1611	8	.2701		

---

[4] Table 13.7 shows that the error term for the covariate differs from the error term for the treatment
effect. This difference reflects the hierarchical nature of an ANCOVA, with the significance of the
covariate tested before the categorical vectors are entered. Unfortunately, many textbooks and
statistical packages use the overall error term to evaluate the statistical significance of the
covariate. This practice is inconsistent with the principles underlying the analysis and should be
avoided (see Cohen and Cohen 1975, p. 349).

Perhaps you are wondering why the ANCOVA shows a significant effect when the original ANOVA (analysis "c" in Table 13.3) does not. If you examine the error terms from the two analyses you will understand why. Before we entered the covariate, the residual error term was substantial ($SS_{res} = 2.3889$ from analysis "c" in Table 13.3); after entering the covariate, the residual error term is greatly reduced ($SS_{res} = .2701$ from analysis "e" in Table 13.3). This reduction is largely responsible for the difference between the two analyses. To emphasize this point, consider the differences in the treatment sum of squares. Without the covariate, $SS_{bg} = 12.50$ (see analysis "c" in Table 13.3); after entering the covariate, SS_{bg} is only slightly larger ($SS_{bg} = 16.3283$). Thus, the reduction in the error term, not the increase of the treatment sum of squares, accounts for the greater power of the ANCOVA over an ANOVA in our example.

13.1.4   R Code: ANCOVA

```
grp <-c(1,1,1,1,2,2,2,2,3,3,3,3)
age <-c(12,8,9,7,13,8,11,8,12,10,7,9)
y <-c(7,5,5,4,6,2,5,3,4,4,1,2)
eff1 <-c(rep(1,4),rep(0,4),rep(-1,4))
eff2 <-c(rep(0,4),rep(1,4),rep(-1,4))

#Compute all 5 models
summary(mod.a <-lm(y~age))
summary(mod.b <-lm(age~eff1+eff2))
summary(mod.c <-lm(y~eff1+eff2))
summary(mod.d <-lm(y~age+eff1+eff2+age*eff1+age*eff2))
summary(mod.e <-lm(y~age+eff1+eff2))

#Homogeneity of regression lines
anova(mod.e,mod.d)

#Plot Regression Lines Using Separate Coefficients for Each Group
p0 <-c(rep(1,3))
p1 <-c(rep(0,3))
p2 <-c(1,0,-1)
p3 <-c(0,1,-1)
p4 <-c(rep(0,3))
p5 <-c(rep(0,3))
P <-round(rbind(p0,p1,p2,p3,p4,p5),digits=5)
intercept <-t(P)%*%coef(mod.d)
```

(continued)

13.1.4   R Code: ANCOVA (continued)

```
s0 <-c(rep(0,3))
s1 <-c(rep(1,3))
s2 <-c(rep(0,3))
s3 <-c(rep(0,3))
s4 <-c(1,0,-1)
s5 <-c(0,1,-1)
SS <-round(rbind(s0,s1,s2,s3,s4,s5),digits=5)
slope <-t(SS)%*%coef(mod.d)

#Predicted values one sd above and below age for each group
new.young <-(intercept[1]+(slope[1]*-(sd(age))))
new.old <-(intercept[1]+(slope[1]*(sd(age))))
trad.young <-(intercept[2]+(slope[2]*-(sd(age))))
trad.old <-(intercept[2]+(slope[2]*(sd(age))))
cont.young <-(intercept[3]+(slope[3]*-(sd(age))))
cont.old <-(intercept[3]+(slope[3]*(sd(age))))
preds<-cbind(c(new.young,new.old),c(trad.young,trad.old),
c(cont.young,cont.old))
matplot((preds), main = "Homogeneity of Regression Lines", type="l",
lwd=3,ylab = "Reasoning Ability", xlab = "Age")
legend("topleft",legend=c("New","Traditional","Control"),
lty=c(1,3,5),lwd=2,pch=21,col=c("black","red","darkgreen"),
ncol=1,bty="n",cex=1,
text.col=c("black","red","darkgreen"),
inset=0.01)

#ANCOVA by comparing two models
ancova <-anova(mod.a,mod.e)
```

13.1.5   Adjusted Means and Simple Effects

Earlier we noted that an ANCOVA examines group differences among adjusted means. The following formula is used to compute the adjusted means:

$$\bar{y}_{adj} = \bar{y}_g - b_c(\bar{z}_g - \bar{z}) \tag{13.1}$$

where $\bar{y}_g$ = a raw group mean, b_c = the common regression coefficient, $\bar{z}_g$ = the group mean on the covariate, and $\bar{z}_g$ = the overall mean of the covariate. The common regression coefficient comes from analysis "e" in Table 13.3, in which we predict the criterion using the covariate and two coded vectors. Table 13.8 presents the pertinent coefficients, and the displayed value for b_1 represents the common regression coefficient (b_c=.6556).

Table 13.8 Unstandardized regression coefficients for model "e" in Table 13.3

	b	se_b	SS	t	p
b_0	-2.2278	.7512		2.9657	.0180
b_1	.6556	.0775	19.3389	8.4610	.0000
b_2	1.5778	.2157	14.4545	7.3149	.0001
b_3	$-.3278$	.2157	.6238	1.5196	.1671

13.1.5.1 Understanding the Common Regression Coefficient

The common regression coefficient is a weighted average of the group slopes:

$$b_c = \frac{\Sigma(b_g * SS_g)}{\Sigma(SS_g)} \tag{13.2}$$

To compute this term, we need the regression coefficients and deviation sum of squares for the covariate within each group. We already calculated the regression coefficients (see Table 13.6), so we only need to find the deviation sum of squares for our covariate. These values, along with some others we will need to compute adjusted means, are shown in Table 13.9.

Table 13.9 Group means and deviation sum of squares for ANCOVA

	Age		Reasoning ability	
	Mean	SS	Mean	Adjusted mean
New	9.00	14.00	5.25	5.5778
Traditional	10.00	18.00	4.00	3.6722
Control	9.50	13.00	2.75	2.75
Average	9.50		4.0	

Substituting our values into Eq. (13.2) confirms the value shown in Table 13.8:

$$b_c = \frac{(.5714 * 14) + (.7222 * 18) + (.6538 * 13)}{(14 + 18 + 13)} = .6556$$

13.1.5.2 Adjusting the Means

To adjust our means, we apply Eq. (13.1):

$$\text{New Method of Instruction} = 5.25 - .6556(9 - 9.5) = 5.5778$$
$$\text{Traditional Instruction} = 4.00 - .6556(10 - 9.5) = 3.6722$$
$$\text{Control Group} = 2.75 - .6556(9.5 - 9.5) = 2.75$$

Notice what has happened here. When the covariate group mean falls below the covariate grand mean, the means are adjusted upward, but when the covariate group mean lies above the covariate grand mean, the means are adjusted downward.

No adjustment occurs when the covariate group mean equals the covariate grand mean. Notice also that in our example, the adjustment is not all that great. Yes, the adjusted means are farther apart than are the raw means, but the difference is slight. This point further underscores that the power of an ANCOVA lies in the reduction of the residual (error) term, rather than an increase in the spread of the adjusted means.

13.1.5.3   Contrasting Adjusted Means

The following formula can be used to test the different between any two adjusted means:

$$F = \frac{\left[\overline{Y}_{(1.adj)} - \overline{Y}_{(2.adj)}\right]^2}{MS_{res}} * \left[\frac{1}{n_1} + \frac{1}{n_2} + \frac{\left[\overline{X}_{(1)} - \overline{X}_{(2)}\right]^2}{SS_{res.X}}\right] \qquad (13.3)$$

The MS_{res} comes from analysis "e" in Table 13.3, and the final denominator term ($SS_{res.x}$) refers to the residual sum of squares observed when the covariate was regressed on the categorical predictors (see analysis "b" in Table 13.3). We will illustrate the equation by using the adjusted means to compare the new instructional method vs. the traditional method:

$$F = \frac{[5.5778 - 3.6722]^2}{.2701} * \left[\frac{1}{4} + \frac{1}{4} + \frac{[9 - 10]^2}{45.00}\right] = \frac{3.6311}{.1411} = 25.7342$$

Referring this value to an F distribution with 1 and 8 degrees of freedom reveals a significant effect ($p = .0010$). Consequently, we judge the new approach to be superior to the traditional method once age is statistically controlled.

13.1.5.4   Multiple Contrasts Using Regression Coefficients

Equation (13.3) works, but it is a bit unwieldy. Moreover, if we have many covariates or are using a factorial design, the algebra becomes tedious and prone to error. Fortunately, we have another solution. If we use effect coding (as we have done in our example), we can construct an augmented coefficient vector and covariance matrix using the procedures discussed in Chap. 12. Table 13.10 shows the original and augmented values. Looking first at the regression coefficients, we already have the values we need for Group 1 ($b_2 = 1.5778$) and Group 2 ($b_3 = -.3278$), so we apply Eq. (12.4) to calculate the coefficient for Group 3:

$$b_{3+} = -(1.5778 + -.3278) = -1.2500$$

The bottom portion of Table 13.10 shows the covariance matrix. We concern ourselves only with the shaded portion, as these values represent our categorical vectors. Applying Eq. (12.5), we calculate the missing values. To illustrate, the final value in column 1 was obtained as follows: $-(.0465 + -.0240) = -.0225$. The other values were calculated in a similar manner.

Table 13.10 Original and augmented **b** vector and covariance matrix

	Original regression coefficients					Augmented regression coefficients		
b_0	−2.2278				(**b+**)			
b_1	.6556				b_2	1.5778		
b_2	**1.5778**				b_3	−.3278		
b_3	**−.3278**				b_{3+}	−1.2500		
	Original covariance matrix					Augmented covariance matrix		
	c_0	c_1	c_2	c_3		(**C+**)		
c_0	.5643	−.0570	−.0285	.0285				
c_1	−.0570	.0060	.0030	−.0030		c_2	c_3	c_{3+}
c_2	−.0285	.0030	**.0465**	−.0240	c_2	.0465	−.0240	−.0225
c_3	.0285	−.0030	**−.0240**	**.0465**	c_3	−.0240	.0465	−.0225
					c_{3+}	−.0225	−.0225	.0450

We are now ready to test any contrast of interest by creating an **S** matrix and performing the calculations described in earlier chapters (i.e., we find our contrast coefficients by multiplying $\mathbf{S'b^+}$ and their associated standard errors by taking the square root of the diagonal entries of $\mathbf{S'C^+S}$). Table 13.11 shows all possible two-group comparisons and three-group contrasts. Notice how easy it is to construct these contrasts. For example, to compare Group 1 and Group 2, we assign a −1 to Group 1 and a 1 to Group 2. The other contrasts are formed in an equally intuitive way.

Table 13.11 **S** matrix with all possible 3-group comparisons and contrasts

	Group 1 vs. Group 2	Group 1 vs. Group 3	Group 2 vs. Group 3	Group 1 vs. (2, 3)	Group 2 vs. (1, 3)	Group 3 vs. (1, 2)
b_2	−1	−1	0	−2	1	1
b_3	1	0	−1	1	−2	1
b_{3+}	0	1	1	1	1	−2

Table 13.12 presents the results after performing the necessary matrix multiplication. Let's begin by looking at the first row, which compares the adjusted mean of Group 1 vs. the adjusted mean of Group 2. This is the same comparison we calculated algebraically, and if you square the observed t value, you will find that, within rounding error, it matches the F value we obtained using the algebraic formula ($5.0734^2 = 25.7394$). Our second comparison compares the new method

against the control condition. This comparison is also significant. Our third comparison pits the traditional approach vs. the control group, and the third row in Table 13.12 shows a significant difference. Thus, after controlling for age in our (fabricated) example, we find that both teaching methods are superior to the control condition, and the new approach is better than the traditional one. The remaining tests report the results of various "one group vs. the other two" contrasts.

Table 13.12 Comparisons and contrasts

	b	se_b	SS	t	p
Group 1 vs. Group 2	−1.9056	.3756	6.9533	5.0734	.0010
Group 1 vs. Group 3	−2.8278	.3696	15.8169	7.6519	.0001
Group 2 vs. Group 3	−.9222	.3696	1.6823	2.4955	.0372
Group 1 vs. (2, 3)	−4.7333	.6471	14.4545	7.3149	.0001
Group 2 vs. (1, 3)	.9833	.6471	.6238	1.5196	.1671
Group 3 vs. (1, 2)	3.7500	.6366	9.3750	5.8910	.0004

13.1.6   R Code: Adjusted Means and Simple Effects

```
grp <-c(1,1,1,1,2,2,2,2,3,3,3,3)
age <-c(12,8,9,7,13,8,11,8,12,10,7,9)
y <-c(7,5,5,4,6,2,5,3,4,4,1,2)
eff1  <-c(rep(1,4),rep(0,4),rep(-1,4));eff2  <-c(rep(0,4),rep(1,4),
rep(-1,4))
summary(mod.e <-lm(y~age+eff1+eff2))

#Compute Adjusted Means
b.common <-mod.e$coef[2]
grp.y <-tapply(y, factor(grp), mean)
grp.age <-tapply(age, factor(grp), mean)
mean.age <-mean(age)
adj.new <-(grp.y[1]-(b.common*(grp.age[1]-mean.age)))
adj.trad <-(grp.y[2]-(b.common*(grp.age[2]-mean.age)))
adj.cont <-(grp.y[3]-(b.common*(grp.age[3]-mean.age)))

#Graph Adjusted Means
barplot(c(adj.new,adj.trad,adj.cont),
main="Adjusted Means",col=c("black","gray","black"),
density= c(20,15,10), angle=c(30,20,0),xlab= "Conditions",
names = c("New","Traditional","Control"), ylim = c(0, 6),
ylab = "Reasoning Ability")
```

(continued)

13.1.6 R Code: Adjusted Means and Simple Effects (continued)

```
#Intercepts Using Common Regression Coefficient
int.1 <-(1*mod.e$coef[3])+(0*mod.e$coef[4])+mod.e$coef[1]
int.2 <-(0*mod.e$coef[3])+(1*mod.e$coef[4])+mod.e$coef[1]
int.3 <-(-1*mod.e$coef[3])+(-1*mod.e$coef[4])+mod.e$coef[1]
intercepts <-cbind(int.1,int.2,int.3)

#Augmented Coefficients and Covariance Matrix
BB <-c(mod.e$coef[3],mod.e$coef[4],-(mod.e$coef[3]+mod.e$coef[4]));BB
covar <-vcov(mod.e)
cov <- covar[3: length(mod.e$coef),3: length(mod.e$coef)]
rows<-function(cov,i,j)cbind(cov[i,1],cov[i,2],-(cov[i,1]+cov[i,2]))
aug <-rows(cov)
CC<-rbind(aug[1,],aug[2,],-(aug[1,]+aug[2,]));CC

#Simple Slopes
s1 <-c(-1,-1,0,-2,1,1);s2 <-c(1,0,-1,1,-2,1);s3 <-c(0,1,1,1,1,-2)
S <-rbind(s1,s2,s3);S
simp.slope <-t(S)%*%BB;simp.err <-sqrt(diag(t(S)%*%CC%*%S))
ttests <-simp.slope/simp.err;pvalues <-2*pt(-abs(ttests),
(df=(length(y)-length(coef(mod.e))))))
crit <-abs(qt(0.025,(df=(length(y)-length(coef(mod.e))))))
CI.low <-simp.slope-(crit*simp.err);CI.high <-simp.slope+(crit*simp.err)
simp.table<-round(matrix(c(simp.slope,simp.err,ttests,pvalues,CI.low,
CI.high),
nrow=ncol(S),ncol=6),digits=5)
dimnames(simp.table)=list(c(),c("slope","stderr","t","p","CI.low",
"CI.high"))
simp.table
```

13.2 Extensions to More Complex Designs

The procedures described in the preceding section can be used with multiple covariates and/or applied to factorial designs. Because more complex designs necessitate very few changes, I will quickly describe the procedures we use to perform a one-way ANCOVA with two covariates.

Imagine that our researcher repeats her study with a new sample of middle-school children, but this time she gathers information on age and IQ. The (fake) data are shown in Table 13.13, using the same effect coding scheme we used earlier.

Table 13.13 Small sample data with two covariates

Condition	Age	IQ	V1	V2	V3 (V1*age)	V4 (V2*age)	V5 (V1*IQ)	V6 (V2*IQ)	y
New	9	130	1	0	9	0	130	0	9
New	8	85	1	0	8	0	85	0	3
New	11	105	1	0	11	0	105	0	7
New	7	100	1	0	7	0	100	0	4
Traditional	12	110	0	1	0	12	0	110	6
Traditional	8	80	0	1	0	8	0	80	1
Traditional	11	90	0	1	0	11	0	90	5
Traditional	8	100	0	1	0	8	0	100	3
Control	13	105	−1	−1	−13	−13	−105	−105	5
Control	10	110	−1	−1	−10	−10	−110	−110	4
Control	7	85	−1	−1	−7	−7	−85	−85	1
Control	7	100	−1	−1	−7	−7	−100	−100	2

13.2.1 Preliminary Analyses

As before, we analyze these data by calculating a number of preliminary regression analyses, shown in Table 13.14.

13.2.1.1 Do the Covariates Predict the Criterion?

Analysis "a" in Table 13.14 shows that, collectively, age and IQ predict reasoning ability ($p=.0013$). Notice, also, however, that the regression coefficient for age falls short of statistical significance. Some textbook authors recommend dropping covariates that are not significant, but I disagree. Recall that regression coefficients are characterized by interdependence. If we drop age from the analysis, it is conceivable that IQ will no longer be significant; alternatively, if we drop IQ from the analysis, age might be significant. Suffice it to say that if the set of covariates is significant, you should retain all variables it contains.

13.2.1.2 Do the Grouping Vectors Predict the Covariate?

It's a good idea to confirm that the covariates are unrelated to the experimental conditions. Analysis "b" in Table 13.14 shows this to be the case for both covariates.

13.2.1.3 Do the Grouping Vectors Predict the Criterion?

As noted earlier, although it is not necessary to examine whether the experimental manipulation predicts the criterion before entering the covariates, it is often

Table 13.14 Preliminary regression analyses used in analysis of covariance (multiple predictors)

	SS	df	MS	R^2	F	p
a. Regress criterion on covariates						
Regression	49.0139	2	24.5070	.7699	15.0526	.0013
Residual	14.6528	9	1.6281			
Total	63.6667					
			b	se_b	F	p
		Age	.3831	.1970	1.9446	.0837
		IQ	.1219	.0299	4.0816	.0028

b. Regress covariates on categorical vectors

	SS	df	MS	R^2	F	p
Age						
Regression	2.00	2	1.00	.0415	.1946	.8265
Residual	46.25	9	5.1389			
Total	48.25					
IQ						
Regression	200.00	2	100.00	.0952	.4737	.6374
Residual	1,900.00	9	211.1111			
Total	2,100.00					

c. Regress criterion on categorical vectors

	SS	df	MS	R^2	F	p
Regression	16.1667	2	8.0833	.2539	1.5316	.2676
Residual	47.50	9	5.2778			
Total	63.6667					

d. Regress criterion on covariates, categorical vectors, and cross products

	SS	df	MS	R^2	F	p
Regression	62.8021	8	7.8503	.9864	27.2409	.0101
Residual	.8645	3	.2882			
Total	63.6667					

e. Regress criterion on covariates and categorical vectors

	SS	df	MS	R^2	F	p
Regression	60.6100	4	15.1525	.9520	34.6998	.0001
Residual	3.0567	7	.4367			
Total	63.6667					

informative. In our case, analysis "c" shows that the groups do not differ on reasoning ability before taking age and IQ into account ($p=.2676$).

13.2.1.4 Testing the Homogeneity of Regression Lines

Before entering our covariates, we must establish that the slopes relating the covariates to performance are parallel across experimental conditions. We test this effect by subtracting SS_{reg} from analysis "e" from SS_{reg} from analysis "d" and dividing the difference by the residual error term from analysis "d":

$$F = \frac{[(62.8021 - 60.6100)/4]}{.2882} = 1.9015$$

Referring this value to an F distribution with 4 and 3 degrees of freedom reveals a nonsignificant effect ($p=.3122$). Accordingly, we conclude that our regression lines are homogeneous and proceed to our main analysis.

13.2.2   Main Analysis

To perform the ANCOVA, we subtract analysis "a" from analysis "e" and divide the difference by the residual term from analysis "e":

$$F = \frac{[(60.6100 - 49.0139)/2]}{.4367} = 13.2770$$

Table 13.15 shows that, with 2 and 7 degrees of freedom, the effect is significant ($p=.0041$).

Table 13.15 Summary of the ANCOVA with multiple covariates

	SS	df	MS	F	p
Age	6.1568	1	6.1568	3.7816	.0837
IQ	27.1227	1	27.1227	16.6593	.0028
Error	14.6528	9	1.6281		
Groups	11.5960	2	5.7980	13.2777	.0041
Error	3.0567	7	.4367		

13.2.3   Adjusted Means

Our next task is to calculate adjusted means using the common regression coefficients from analysis "e" in Table 13.14. The coefficients, along with other relevant values, are displayed in Table 13.16.

Expanding the procedures described in Eq. (13.1) yields the adjusted means:

New : $5.75 - [.4917 * (8.75 - 9.25)] - [.1019 * (105 - 100)] = 5.4864$

Traditional : $3.75 - [.4917 * (9.75 - 9.25)] - [.1019 * (95 - 100)] = 4.0137$

Control : $3.00 - [.4917 * (9.25 - 9.25)] - [.1019 * (100 - 100)] = 3.00$

Table 13.16 Regression Coefficients

	b	se_b	SS	t	p
b_0	-10.5722	1.5458	20.4262	6.8394	.0002
b_1 (age)	.4917	.1091	8.8643	4.5055	.0028
b_2 (IQ)	.1019	.0170	15.6379	5.9843	.0006
b_3 (effect 1)	1.3197	.2954	8.7176	4.4681	.0029
b_4 (effect 2)	$-.1530$	.2954	.1172	.5181	.6204

Covariance Matrix

2.389464	$-.025548$	$-.021168$	.093063	$-.093063$
$-.025548$	.011912	$-.000846$	.010188	$-.010188$
$-.021168$	$-.000846$	.000290	$-.001873$	.001873
.093063	.010188	$-.001873$	.087238	$-.050849$
$-.093063$	$-.010188$	.001873	$-.050849$	.087238

Adjusted Means

	AGE	IQ	Raw Means	Adjusted
New	8.75	105	5.75	5.4864
Traditional	9.75	95	3.75	4.0136
Control	9.25	100	3.00	3.00
Average	9.25	100	4.1667	

Fig. 13.3 Raw and adjusted means for ANCOVA with two covariates

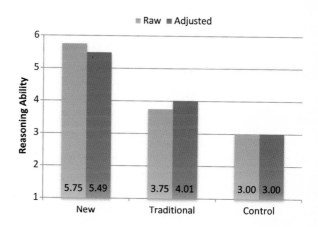

It is interesting to visually compare the raw means with the adjusted means. Figure 13.3 shows them side by side, and it's evident that there is very little difference between them. Yet an ANOVA on the raw means failed to show a significant effect ($p=.2676$), whereas the ANCOVA did ($p=.0041$). The disparity between these significance levels seems all the more remarkable when we consider how little the means differ, further underscoring that the power of an ANCOVA derives from the reduction in the error term, rather than an adjustment to the means. In this case, the error term from the ANOVA ($MS_e = 5.2778$) is nearly 12 times larger than the error term from the ANCOVA (MS_e=.4367). As always, you should bear in mind that the data set has been artificially created to make particular points (of which this is one). But the point remains valid: When properly used with random assignment to conditions, ANCOVA is designed to remove extraneous sources of variance from our criterion, not modify our categorical predictor.

13.2.4   Augmented Matrix and Multiple Comparisons

Finally, we can conduct multiple comparisons and contrasts using augmented matrices. The procedures are identical to the ones described earlier, so I will simply show you the output and refer you to the **R** code that accompanies this section for the calculations themselves (Table 13.17).

As you can see, the new method is superior to the traditional method (p=.0264) and the control condition (p=.0013), but the traditional method and control condition do not differ (p=.0738).

Table 13.17   Augmented matrices from Table 13.15, **S** matrix, and simple effects

Augmented **b** vector (b+)		Augmented covariance matrix (**C**+)		
Group 1	1.3197	.0872	−.0508	−.0364
Group 2	−.1530	−.0508	.0872	−.0364
Group 3	−1.1667	−.0364	−.0364	.0728

			S		
Group 1 vs. Group 2	Group 1 vs. Group 3	Group 2 vs. Group 3	Group 1 vs. (2, 3)	Group 2 vs. (1, 3)	Group 3 vs. (1, 2)
−1	−1	0	−2	1	1
1	0	−1	1	−2	1
0	1	1	1	1	−2

Simple effects					
	b	se_b	SS	t	p
Group 1 vs. Group 2	−1.4727	.5255	3.4294	2.8024	.0264
Group 1 vs. Group 3	−2.4864	.4825	11.5960	5.1532	.0013
Group 2 vs. Group 3	−1.0136	.4825	1.9273	2.1009	.0738
Group 1 vs. (2, 3)	−3.9591	.8861	8.7176	4.4681	.0029
Group 2 vs. (1, 3)	.4591	.8861	.1172	.5181	.6204
Group 3 vs. (1, 2)	3.5000	.8093	8.1667	4.3246	.0035

13.2.5   R Code: ANCOVA with Multiple Covariates

```
grp <-c(rep(1,4),rep(2,4),rep(3,4))
age <-c(9,8,11,7,12,8,11,8,13,10,7,7)
IQ <-c(130,85,105,100,110,80,90,100,105,110,85,100)
y <-c(9,3,7,4,6,1,5,3,5,4,1,2)
eff1 <-c(rep(1,4),rep(0,4),rep(-1,4))
eff2 <-c(rep(0,4),rep(1,4),rep(-1,4))


#Compute all 5 models
summary(mod.a <-lm(y~age+IQ))
anova(mod.b1 <-lm(age~eff1+eff2))
anova(mod.b2 <-lm(IQ~eff1+eff2))
summary(mod.c <-lm(y~eff1+eff2))
summary(mod.d <-lm(y~age+IQ+eff1+eff2+age*eff1+age*eff2+IQ*eff1+
IQ*eff2))
summary(mod.e <-lm(y~age+IQ+eff1+eff2))
ancova <-anova(mod.e,mod.a)
ancova


#Compute Adjusted Means
b.com1 <-mod.e$coef[2]
b.com2 <-mod.e$coef[3]
grp.y <-tapply(y, factor(grp), mean)
grp.age <-tapply(age, factor(grp), mean)
mean.age <-mean(age)
grp.IQ <-tapply(IQ, factor(grp), mean)
mean.IQ <-mean(IQ)
adj.new<-(grp.y[1]-(b.com1*(grp.age[1]-mean.age))-(b.com2*(grp.IQ
[1]-mean.IQ)))
adj.trad<-(grp.y[2]-(b.com1*(grp.age[2]-mean.age))-(b.com2*(grp.IQ
[2]-mean.IQ)))
adj.cont<-(grp.y[3]-(b.com1*(grp.age[3]-mean.age))-(b.com2*(grp.IQ
[3]-mean.IQ)))
adj.means <-cbind(adj.new,adj.trad,adj.cont);adj.means


#Graph Adjusted Means
barplot(c(adj.new,adj.trad,adj.cont),
main="Adjusted Means",col=c("black","gray","black"),density=
c(20,15,10), angle=c(30,20,0),xlab= "Condition", names =
c("New","Traditional","Control"), ylim = c(0, 6), ylab =
"Reasoning Ability")

#Generate Augmented Coefficients and Covariance Matrix
BB <-c(mod.e$coef[4],mod.e$coef[5],-(mod.e$coef[4]+mod.e$coef[5]))
BB
```

(continued)

13.2.5   R Code: ANCOVA with Multiple Covariates (continued)

```
covar <-vcov(mod.e)
cov <- covar[4: length(mod.e$coef),4: length(mod.e$coef)]
rows<-function(cov,i,j)cbind(cov[i,1],cov[i,2],-(cov[i,1]+cov[i,2]))
aug <-rows(cov)
CC<-rbind(aug[1,],aug[2,],-(aug[1,]+aug[2,]))
CC


#Simple Effects
simple <-function(S){
simp.slope <-t(S)%*%BB;simp.err <-sqrt(diag(t(S)%*%CC%*%S))
ttests <-simp.slope/simp.err;pvalues <-2*pt(-abs(ttests),
(df=(length(y)-length(coef(mod.e))))))
crit <-abs(qt(0.025,(df=(length(y)-length(coef(mod.e)))))))
CI.low <-simp.slope-(crit*simp.err);CI.high <-simp.slope+
(crit*simp.err)
simp.table<-round(matrix(c(simp.slope,simp.err,ttests,pvalues,
CI.low,CI.high),nrow=ncol(S),ncol=6),digits=5)
dimnames(simp.table)=list(c(),c("slope","stderr","t","p","CI.
low","CI.high"))
return(list(S,simp.table))
}
s1 <-c(-1,-1,0,-2,1,1);s2 <-c(1,0,-1,1,-2,1);s3 <-c(0,1,1,1,1,-2)
S <-rbind(s1,s2,s3)

simple(S)
```

13.3   Uses (and Misuses) of ANCOVA

Throughout this chapter, you have seen that ANCOVA creates a more powerful test of a grouping variable by reducing the error variance. Ideally, the covariate is assessed prior to the start of the experiment and is minimally associated with the experimental groups that have been created through random assignment. When these conditions are met, we can reasonably ask, "how different would the groups be on y if everyone in the sample had been equivalent on z?"

Considering its ability to transform nonsignificant findings into significant ones, ANCOVA seems to resemble a medieval alchemist. Alas, like other forms of magic, ANCOVA can sometimes produce less than meets the eye. In the final section of this chapter, we will identify two problems that limit ANCOVA's usefulness.

13.3.1   A Residualized Criterion

The first issue concerns the nature of the criterion. With an ANOVA, we ask a very straightforward question: "What percentage of y can be explained by our categorical predictor?" With ANCOVA, we ask a different question, namely, "What percentage of *y that is not explained by our covariate* can be explained by our categorical predictor?" Thus, the analyses predict to different criterions, and whether the residualized criterion represents something meaningful is an open question.

To illustrate the potential problem, consider that the ability to reason abstractly predicts important outcomes in life, such as GPA, level of educational attainment, and success at matrix algebra. But whether reasoning ability predicts anything of interest once age and IQ have been statistically removed is uncertain. Since we probably don't care whether a new method improves reasoning ability unless the criterion still predicts significant outcomes, it would seem important to establish that the residualized criterion is meaningfully related to some outcome of interest when performing an ANCOVA. Unfortunately, this is rarely done.

13.3.2 Association with the Predictor

A second problem with ANCOVA is less tractable and therefore more serious. ANCOVA was originally developed in the context of experimental research using random assignment to conditions. Doing so ensured that the predictor and covariate were largely independent. Over the years, the use of ANCOVA has broadened and it is now widely used in nonexperimental contexts with subject variables rather than randomly assigned ones (e.g., cultural groups, gender). Here, the attempt is to *equate* naturally occurring groups that are inherently nonequivalent. This use of ANCOVA is highly problematic, as the question now becomes: "What percentage of *y that is not explained by our covariate* can be explained by *that portion of our categorical predictor that is not explained by our covariate*?" It is evident that we have now moved far beyond the simple question with which we began.

The overlap between the predictor and the covariate creates interpretive problems as well. It is far from obvious that we can statistically equate groups that are intrinsically different. To paraphrase an anonymous observer, what does it mean to say that "this is what the groups would look like if they didn't look like what they do look like?"!

Finally, any overlap between the predictor and the covariate can lead to misleading conclusions. Imagine that a researcher believes that learning matrix algebra promotes abstract reasoning skills. To test his hypothesis, he administers a test of reasoning ability to two groups of students—ones who are enrolled in a matrix algebra course and ones who are not enrolled in a matrix algebra course. Recognizing that IQ also predicts reasoning ability, the researcher uses IQ scores as a covariate. Table 13.18 presents some (made-up) data, and Fig. 13.4 depicts the raw and covariance adjusted means.

If you work through the analysis (which you should do to solidify your understanding of an ANCOVA), you will discover something interesting: Before the covariate is added, students who study matrix algebra outperform those who do not study matrix algebra on the test of abstract reasoning $[F(1, 10) = 5.3312, p=.0436]$;

Table 13.18 Small sample
example for an ANCOVA
with a naturally occurring
categorical predictor

Enrolled in matrix algebra course	IQ	Effect Codes	y
No	90	1	2.0
No	93	1	3.1
No	93	1	2.9
No	93	1	2.5
No	89	1	1.4
No	94	1	2.9
Yes	108	−1	3.7
Yes	108	−1	3.4
Yes	110	−1	4.1
Yes	107	−1	3.1
Yes	108	−1	2.9
Yes	107	−1	4.0

Fig. 13.4 Raw and
adjusted means for
ANCOVA with a naturally
occurring categorical
predictor

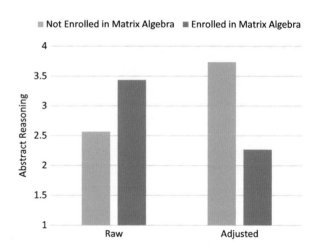

after the covariate is added, the reverse is true $[F(1, 9) = 7.4086, p=.0235]$. Were these real data (which, of course, they are not), we would conclude that taking a matrix algebra course impairs reasoning ability once IQ is taken into account.

The reversal comes about because, in this fictitious example, students who study matrix algebra have higher IQ scores ($M = 109.6667$) than those who don't ($M = 90.3333$), $[F(1, 10) = 54.2581, p=.0000]$, and IQ scores are strongly correlated with abstract reasoning scores ($r=.8106$). Consequently, the adjustment raises the scores of those who do not study matrix algebra and lowers the scores of those who do. More generally, whenever a covariate is highly associated with the categorical predictor and the criterion, an ANCOVA will produce a diminished estimate of the categorical predictor.

For these, and other reasons (see Miller and Chapman 2001), the use of ANCOVA with intact groups should be undertaken cautiously, and all results should be interpreted carefully. A better solution is to incorporate the covariate into a factorial design, treating it as a partner variable rather than a causally prior one. This approach, known as moderation, is the topic of our next chapter.

13.4   Chapter Summary

1. Categorical and continuous predictors can be combined in a single analysis. The form of the analysis depends on when the categorical predictor is measured and its assumed relation with the criterion.
2. An analysis of covariance (ANCOVA) is appropriate when a continuous variable is assessed before subjects are randomly assigned to experimental conditions. In this case, it can be used to reduce the error variance, thereby creating a more powerful statistical test of the categorical predictor.
3. An ANOCVA assumes that the slope of the regression line relating the covariate to the criterion is equal across experimental conditions. The assumption is tested by assessing the homogeneity of the regression lines.
4. A subtraction method is used to test the statistical significance of an ANCOVA. The regression sum of squares obtained from a model using only the covariate is subtracted from the regression sum of squares obtained from a model using the covariate and categorical vectors.
5. Covariance adjusted means are computed by subtracting a weighted difference score from each group mean. The difference score is found by calculating the discrepancy between the group mean on the covariate and the overall covariate mean and then weighting this difference by the common regression coefficient.
6. ANCOVAs are appropriate only when the covariate and the categorical variable are uncorrelated. When the covariate and categorical variable are correlated, a straightforward interpretation of the ANCOVA is problematic and the method should be abandoned in favor of moderated regression.

Chapter 14
Moderation

In Chap. 13, we discussed analysis of covariance (ANCOVA). With ANCOVA, we enter a continuous predictor into a regression model before adding a categorical one. The presence of the continuous predictor is designed to reduce the error variance, thereby increasing the power of an experimental treatment or grouping variable.

ANCOVA is not the only way to combine categorical and continuous predictors in a regression analysis, however. Continuous predictors can also be used to examine the generality of an experimental manipulation by studying their interaction with a categorical variable. Such an analysis is termed moderated regression analysis (or, simply, moderation), because our interest is in whether a continuous predictor moderates the impact of a categorical one.

14.1 Moderated Regression

The procedures we follow when conducting a moderated regression analysis are similar to the ones we used in Chap. 9 when analyzing continuous predictors. In both cases, we create cross-product terms from our predictors to model an interaction and then conduct simple slope tests or locate regions of significance if the interaction is significant. The use of a categorical predictor necessitates some changes to the implementation of the analysis, but the underlying logic is the same.

Testing whether a continuous variable and a categorical variable interact to predict a criterion is also identical to the procedure we used in Chap. 13 to test the homogeneity of regression lines with an ANCOVA. The only difference between this test and a test of moderation is our perspective. When testing the homogeneity of

Electronic Supplementary Material: The online version of this chapter (doi: 10.1007/978-3-319-11734-8_14) contains supplementary material, which is available to authorized users

regression lines, we examine whether the association between the covariate and the criterion is consistent across experimental conditions; when performing a test of moderation, we examine whether the association between the experimental manipulation and the criterion is consistent across all levels of the covariate. Because interactions are symmetrical, the same analysis answers both questions.

Despite these apparent similarities, ANCOVA and moderation address very different issues. When the homogeneity assumption is upheld, ANCOVA asks, "What effect would our manipulation have if everyone in our sample had the same score on the covariate?" With moderation, we ask, "What effect does our manipulation have on subjects with different scores on the covariate?" Thus, whereas ANCOVA attempts to statistically remove the influence of a continuous variable, moderation makes its impact focal.

14.1.1   Illustration

An example will further clarify the difference between ANCOVA and moderation. Let's pretend our researcher from Chap. 13 now wishes to test the generality of her new method of teaching abstract reasoning by broadening her sample to include adults of different ages. Having heard the adage "you can't teach an old dog new tricks," she suspects her new teaching method might benefit younger adults, but not older ones. Formally, the researcher is interested in testing the interaction between a categorical variable (teaching method) and a continuous one (age).

To test her hypothesis, she randomly assigns 12 individuals ranging in age from 21 to 57 to one of three experimental conditions: her new method of teaching abstract reasoning, a traditional method of teaching abstract reasoning, or a control condition with no instruction. Three weeks later, she tests their abstract reasoning ability. The (concocted) data, along with the same effect coding scheme we used in Chap. 13, are shown in Table 14.1. Notice that we are creating cross-product terms using mean-centered scores for our continuous variable (Dev_{age}). As first noted in Chap. 9, you don't need to center the continuous variable when using it to form a cross-product term, but doing so makes it easier to interpret lower-order effects. For this reason, I have centered the continuous variable (age) around its mean ($M = 38.00$). I have also included its standard deviation in Table 14.1, as we will use this information later.

14.1.2   Implementation

Moderated regression involves testing various models using the subtraction method introduced in Chap. 12 (see Eq. 12.3). We begin with a model that includes all of the predictors. We then subtract predictors one at a time and, with each subtraction, we calculate the difference between the full model and the reduced one to

Table 14.1 Small sample example for moderation

Instructional Method	Age	y	Dev_{age}	V_1	V_2	$Dev_{age} * V_1$	$Dev_{age} * V_2$
New Instructional Method	29	7	−9	1	0	−9	0
New Instructional Method	27	9	−11	1	0	−11	0
New Instructional Method	39	5	1	1	0	1	0
New Instructional Method	53	2	15	1	0	15	0
Traditional Instructional Method	57	5	19	0	1	0	19
Traditional Instructional Method	32	7	−6	0	1	0	−6
Traditional Instructional Method	45	6	7	0	1	0	7
Traditional Instructional Method	21	5	−17	0	1	0	−17
Control Condition	47	4	9	−1	−1	−9	−9
Control Condition	55	3	17	−1	−1	−17	−17
Control Condition	29	3	−9	−1	−1	9	9
Control Condition	22	4	−16	−1	−1	16	16
Mean	**38.00**	**5.00**					
Standard Deviation	**13.0384**						

determine whether the subtracted predictor makes a significant contribution to the prediction of our criterion. Notice that these procedures differ a bit from the ones we used with ANCOVA. With ANCOVA, we are interested in whether the categorical predictor explains variance that the continuous variable does not; with moderation, we are interested in whether the cross-product terms explain variance that the continuous and categorical vectors do not.

Analysis "a" in Table 14.2 shows that our overall regression equation is significant. Notice, also, that the R^2 value is extremely large. As always, I created this data set to make some specific points, so please remember that it is purposefully unrealistic.

Analysis "b" shows the results of a regression analysis with the cross-product terms omitted. To test the significance of the Groups × Age interaction, we subtract the SS_{reg} value for analysis "b" from the SS_{reg} from our complete analysis, divide the term by the differences in the degrees of freedom between the two analyses, and test the difference using the residual term from the complete analysis:

$$F = \frac{[(39.1608 - 21.0672)/2]}{.8065} = 11.2167$$

Referring this value to an F distribution with 2 and 6 degrees of freedom reveals a significant effect ($p=.0094$). Consequently, we conclude that age moderates the relation between teaching method and reasoning ability.

At this point, we would usually examine the form of the interaction. However, in the interest of being thorough, we'll test the significance of the continuous predictor

Table 14.2 Regression analyses used to perform moderated regression

a. Regress criterion on categorical vectors, continuous variable, and cross-product vectors

	SS	df	MS	R^2	F	p
Regression	39.1608	5	7.8322	.8900	9.7108	.0077
Residual	4.8392	6	.8065			
Total	44.00					

b. Omit cross-product vectors

	SS	df	MS	R^2	F	p
Regression	21.0672	3	7.0224	.4788	2.4497	.1383
Residual	22.9328	8	2.8666			
Total	44.00					

c. Omit continuous variable

	SS	df	MS	R^2	F	p
Regression	25.4118	4	6.353	.5775	2.39242	.1481
Residual	18.5882	7	2.6555			
Total	44.00					

d. Omit categorical vectors

	SS	df	MS	R^2	F	p
Regression	26.9768	3	8.9923	.6131	4.2259	.0458
Residual	17.0232	8	2.1279			
Total	44.00					

Analysis summary

	SS	df	MS	ΔR^2	F	p
Groups (a–d)	12.1839	2	6.0920	.2769	7.5532	.0230
Age (a–c)	13.7489	1	13.7489	.3125	17.0468	.0062
Interaction (a–b)	18.0935	2	9.0468	.4112	11.2167	.0094
Residual	4.8392	6	.8065			

by subtracting SS_{reg} from analysis "c" from the complete analysis and testing the difference as described earlier:

$$F = \frac{[(39.1608 - 25.4118)/1]}{.8065} = 17.0468$$

This value is also significant ($p=.0062$), indicating that age uniquely predicts reasoning ability. Finally, we can test the significance of the categorical vectors by subtracting analysis "d" from analysis "a":[1]

---

[1] When performing these analyses, some textbooks recommend a hierarchical approach, using a different error term to test the significance of the lower-order terms. I disagree. When an interaction is the focus of an investigation, I believe all terms should be tested using the overall MS_{res} with the cross-product term included. Consequently, the order you use to perform these analyses is arbitrary.

$$F = \frac{[(39.1608 - 26.9768)/2]}{.8065} = 7.5532$$

This effect is also significant (p=.0230), indicating that instructional method predicts reasoning ability.

Considering that both lower-order effects are statistically significant, it is tempting to think of them as analogous to main effects in an analysis of variance. This interpretation is not warranted, however. As first noted in Chap. 9, with a cross-product term in the analysis, lower-order effects are conditional simple slopes, not general ones. Moreover, the value for the continuous variable can change depending on the coding scheme we choose. For these, and other reasons, it's best to refrain from equating lower-order effects with main effects.

14.1.3 Regression Coefficients

Having established that age and instructional method interact to predict reasoning ability, we will now examine the regression coefficients from analysis "a" in Table 14.2. Table 14.3 presents the relevant values, using the effect coding scheme shown in Table 14.1.

Table 14.3 Regression coefficients from analysis "a" in Table 14.2

	b	se_b	t	SS	p
b_0	4.9216	.2598	18.9433		.0000
b_1	−.0886	.0215	4.1288	13.7489	.0062
b_2	.5831	.3679	1.5850	2.0263	.1641
b_3	.8358	.3673	2.2755	4.1762	.0632
b_4	−.1567	.0331	4.7361	18.0910	.0032
b_5	.0787	.0288	2.7361	6.0382	.0339

The easiest way to understand where these numbers come from is to compute separate regression equations within each experimental condition. The easiest way to find these values is to create two **S** matrices, one for the intercept and one for the slope. Table 14.4 shows the values we use. For the intercepts, we enter a "1" for b_0 and the group codes for b_2 and b_3; all other values receive 0's. For the slopes, we enter a "1" for b_1 and the group codes for b_4 and b_5; all other values receive 0's. When we multiply **S'b** using the coefficients shown in Table 14.3, we get the values shown in the bottom of Table 14.4. I have included the averages as well, as these values provide important information.

Table 14.4 S matrix to derive intercept and slopes within each experimental condition

	Intercepts			Slopes		
b_0	1	1	1	0	0	0
b_1	0	0	0	1	1	1
b_2	1	0	−1	0	0	0
b_3	0	1	−1	0	0	0
b_4	0	0	0	1	0	−1
b_5	0	0	0	0	1	−1

	b_0	b_1
Group 1	5.5047	−.2453
Group 2	5.7574	−.0099
Group 3	3.5027	−.0106
Average	**4.9216**	**−.0886**

We are now in a position to understand the meaning of the coefficients shown in Table 14.3. First, b_0 is the average of the group intercepts ($b_0 = 4.9216$) and b_1 is the average of the group slopes ($b_1 = -.0886$). These values would be different had we used dummy coding instead of effect coding, which is one reason we shouldn't interpret regression coefficients as main effects. The value for b_2 is the deviation of Group 1's intercept from the average intercept ($b_2 = 5.5047 - 4.9216 = .5831$) and b_3 is the deviation of Group 2's intercept from the average intercept ($b_3 = 5.7574 - 4.9216 = .8358$). Similarly, the value for b_4 is the deviation of Group 1's slope from the average slope ($b_4 = -.2453 - -.0886 = -.1567$) and b_5 is the deviation of Group 2's slope from the average slope ($b_5 = -.0099 - -.0886 = .0787$). These values would be different had we used a different coding scheme or designated a different group to be the base group.

14.1.4 Plotting Predicted Values

With a significant interaction, it is a good idea to plot the regression lines by generating predicted values. In keeping with earlier chapters, we will plot values for subjects scoring 1 standard deviation below the mean in age, at the mean, and 1 standard deviation above the mean in age using the **P** matrix shown in Table 14.5. The values were derived as follows:

- Enter a 1 for the intercept (b_0).
- Enter one standard deviation below the mean, zero, or one standard deviation above the mean for the continuous predictor (b_1).
- Enter the group codes for b_2 and b_3.
- Multiply the b_1 values by the group codes to form the cross-product vectors ($b_4 = b_1 * b_2$; $b_5 = b_1 * b_3$).

Table 14.5 **P** matrix to derive predicted values

	Low age			Average age			High age		
b_0	1	1	1	1	1	1	1	1	1
b_1	−13.0384	−13.0384	−13.0384	0	0	0	13.0384	13.0384	13.0384
b_2	1	0	−1	1	0	−1	1	0	−1
b_3	0	1	−1	0	1	−1	0	1	−1
b_4	−13.0384	0	13.0384	0	0	−0	13.0384	0	−13.0384
b_5	0	−13.0384	13.0384	0	0	−0	0	13.0384	−13.0384

When we multiply **P'b**, we get the values plotted in Fig. 14.1. Remembering that nonparallel regression lines signal the presence of an interaction, Fig. 14.1 shows that age is negatively related to reasoning ability with the new method, but not predictive of reasoning ability with the traditional method or in the control condition. Notice also that the slope for the traditional method is (roughly) parallel to the slope for the control condition but that the intercept is higher in the traditional group. We will have more to say about these effects momentarily.

Fig. 14.1 Simple slopes for moderated regression relating a categorical variable (instructional method) and a continuous variable (age) to reasoning ability

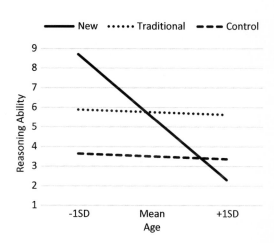

14.1.5 Crossing Point

We can determine the point at which the lines shown in Fig. 14.1 intersect using the following formula:

$$CP = \frac{a_1 - a_2}{b_2 - b_1} \tag{14.1}$$

where *a* and *b* refer to intercepts and slopes, respectively, and the subscripts denote the group number.[2] To illustrate, we will calculate the crossing point of the new method (Group 1) and the traditional method (Group 2):

$$CP_{12} = \frac{5.5047 - 5.7574}{-.0099 - (-.2453)} = -1.0735$$

When we convert the obtained deviate score to a raw one by adding 38, we find that the simple slopes cross at ~37 years of age. This value is well within the range of observed values, so we conclude that our interaction is disordinal.[3] Our crossing point also tells us that the new method is superior to the traditional method for subjects who are younger than ~37, inferior to the traditional method for subjects who are older than ~37, and equivalent in effectiveness for subjects who are ~37. Bear in mind, however, that these characterizations only describe the data; they do not tell us at what point these differences become statistically significant. The final section of this chapter will consider this issue.

Continuing on, we find the crossing point of Group 1 and Group 3:

$$CP_{13} = \frac{5.5047 - 3.5027}{-.0106 - (-.2453)} = 8.53$$

After converting to raw scores, we find that this value (~46.5) is also within the range of observed scores, so this interaction is also disordinal.

Glancing back to Fig. 14.1, it is apparent that the slopes for Groups 2 and 3 are nearly parallel; consequently, there is no reason to calculate their crossing point.

14.1.6   Testing the Simple Slopes

Testing the significance of the regression lines displayed in Fig. 14.1 is our next step. We already calculated their slopes (see bottom portion of Table 14.4), so we need only find their standard errors to test their significance. We do so by taking the square root of the diagonal elements of $\mathbf{S'CS}$, using the $\mathbf{S}$ matrix of slopes shown in the upper right-hand portion of Table 14.4, and the complete covariance matrix from analysis "a" in Table 14.2, which is shown below in Table 14.6.[4]

---

[2] Equation 9.11 presents an alternative formula for calculating the crossing point with two continuous predictors.

[3] See Chap. 9 for a discussion of disordinal vs. ordinal interactions.

[4] Later, we will use the highlighted portion of the covariance matrix to create an augmented covariance matrix.

Table 14.6 Covariance matrix from analysis "a" in Table 14.2

.067500	.000088	.000346	−.000082	.000546	−.000363
.000088	.000460	.000546	−.000363	.000174	−.000094
.000346	.000546	.135346	−.067764	.000722	−.000271
−.000082	−.000363	−.067764	.134918	−.000271	−.000187
.000546	.000174	.000722	−.000271	.001095	−.000541
−.000363	−.000094	−.000271	−.000187	−.000541	.000827

When we perform the required calculations, we obtain the values shown in Table 14.7. As suggested by Fig. 14.1, the simple slope of age is significant with the new teaching method, but not with the traditional method or in the control condition.

Table 14.7 Simple slopes within each experimental group

	b	se_b	SS	t	p
New	−.2453	.0436	25.5094	5.6239	.0014
Traditional	−.0099	.0332	.0717	.2982	.7756
Control	−.0106	.0338	.0796	.3141	.7641

14.1.7 R Code: Moderation—Simple Slopes

```
grp <-c(rep(1,4),rep(2,4),rep(3,4))
age <-c(29,27,39,53,57,32,45,21,47,55,29,22)
y <-c(7,9,5,2,5,7,6,5,4,3,3,4)
eff1 <-c(rep(1,4),rep(0,4),rep(-1,4));eff2 <-c(rep(0,4),rep(1,4),
rep(-1,4))
devage <-age-mean(age)

summary(mod.a <-lm(y~devage+eff1+eff2+eff1*devage+eff2*devage))
summary(mod.b <-lm(y~devage+eff1+eff2))
summary(mod.c <-update(mod.a, .~. -devage))  #Use update function to
#delete continuous variable but retain cross-products
summary(mod.d <-update(mod.a, .~. -eff1-eff2))   #Use update function
#to delete categorical terms but retain cross-products
anova(mod.b,mod.a)      #test interaction
anova(mod.c,mod.a)      #test continuous variable
anova(mod.d,mod.a)      #test categorical variable

##Plot Regression Lines Using Separate Coefficients for Each Group
#Predicted Values and Simple Slopes
i0 <-c(rep(1,3));i1 <-c(rep(0,3));i2 <-c(1,0,-1);i3 <-c(0,1,-1)
```

(continued)

14.1.7   R Code: Moderation—Simple Slopes (continued)

```
i4 <-c(rep(0,3));i5 <-c(rep(0,3))
II <-round(rbind(i0,i1,i2,i3,i4,i5),digits=5);intercept <-t(II)%*%
coef(mod.a)

s0 <-c(rep(0,3));s1 <-c(rep(1,3));s2 <-c(rep(0,3));s3 <-c(rep(0,3))
s4 <-c(1,0,-1);s5 <-c(0,1,-1);SS <-round(rbind(s0,s1,s2,s3,s4,s5),
digits=5)
slope <-t(SS)%*%coef(mod.a);both <-cbind(intercept,slope);both
mean(both[,1])        #average intercept = intercept in mod.a
mean(both[,2])        #average slope = slope in mod.a


#Predicted Values for Ss one standard deviation above and below the mean
p0 <-c(rep(1,9))
p1 <-c(rep(-sd(age),3),rep(0,3),rep(sd(age),3))
p2 <-c(rep(c(1,0,-1),3))
p3 <-c(rep(c(0,1,-1),3))
p4 <-c(-sd(age),0,sd(age),rep(0,3),sd(age),0,-sd(age))
p5 <-c(0,-sd(age),sd(age),rep(0,3),0,sd(age),-sd(age))
P <-rbind(p0,p1,p2,p3,p4,p5)
pred.val <-t(P)%*%coef(mod.a); pred.val


#Plotting Predicted Values 1
byrow <-rbind(c(pred.val[1:3]),c(pred.val[4:6]),c(pred.val[7:9]))
matplot((byrow), main = "Simple Slopes Relating Age to Reasoning
Across Three Conditions", type="l",ylab = "Reasoning Ability", xlab =
"DEV_age",lwd=2)
legend("topright",legend=c("New","Traditional","Control"),
lty=1,lwd=2,pch=21,col=c("black","red","darkgreen"),
ncol=1,bty="n",cex=0.8,
text.col=c("black","red","darkgreen"),
inset=0.01)


#Crossing Points
cross.12 <-(intercept[1]-intercept[2])/(slope[2]-slope[1]);cross.12
cross.13 <-(intercept[1]-intercept[3])/(slope[3]-slope[1]);cross.13
cross.23 <-(intercept[2]-intercept[3])/(slope[3]-slope[2]);cross.23


#Test Simple Slopes
simple.slope <-function(S,B,C){
simp.slope <-t(S)%*%B
simp.cov <-t(S)%*%C%*%S
simp.err <-sqrt(diag(simp.cov))
simples <-simp.slope/sqrt(diag(simp.cov))
df <-length(y)-nrow(S)
tvalues <-2*pt(-abs(simples),df=df)
```

(continued)

14.1.7 R Code: Moderation—Simple Slopes (continued)

```
crit <-abs(qt(0.025,df))
CI.low <-simp.slope-(crit*simp.err)
CI.high <-simp.slope+(crit*simp.err)
simp.table<-round(matrix(c(simp.slope,simp.err,simples,tvalues,
CI.low,CI.high),nrow=ncol(S),ncol=6),digits=5)
dimnames(simp.table)=list(c(),c("slope","stderr","t","p",
"CI.low","CI.high"))
return(list(S,simp.table))
}
simple <-simple.slope(SS,coef(mod.a),vcov(mod.a))
simple
```

14.2 Simple Effects

Testing the simple slope of a continuous predictor at each level of a categorical variable is one way to probe an interaction, but the primary purpose of a moderated regression analysis is to determine whether the effects of a categorical variable vary across levels of a continuous predictor. In terms of our example, we wish to know whether the effects of instructional method vary across age groups. To help us address this issue, Fig. 14.2 shows our predicted values, organized by experimental condition. Here, we have plotted the effects of the experimental manipulation at two levels of age: one standard deviation below the mean (younger subjects) and one standard deviation above the mean (older subjects).[5] Looking the figure over, the following conclusions seem warranted:

- The new method improves performance among younger subjects but impairs performance among older subjects.
- Age makes little difference in the control condition or with the traditional method.
- Across age groups, the traditional method is uniformly better than the control condition.

Specific analyses are required to support each of these interpretations. In the following section, we will examine the matrices and operations used to test these comparisons, as well as a few others.

---

[5] I have omitted plotting the predicted value for subjects of average age because the means fall halfway between those that are displayed.

Fig. 14.2 Predicted values
for moderated regression

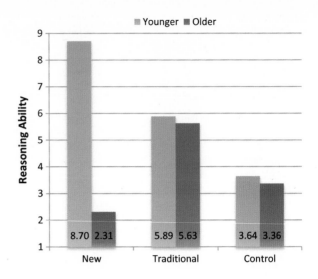

14.2.1   *Augmented b Vector and C Matrix*

Using procedures first described in Chap. 12, we begin by creating an augmented
b vector to find our Group 3 coefficients ($\mathbf{b}^+$). The original regression coefficients
come from Table 14.3 (carried out to 6 decimals to minimize rounding error), and
the calculated values are shown in Table 14.8. Because the intercept is not involved
in any calculations, we omit b_0 and find our coefficients for Group 3 using
Eq. (12.4), reproduced below:

$$b^+ = -\Sigma b_j \tag{14.2}$$

Looking only at the first column in Table 14.8, the two unshaded values were found
as follows:

$$b_{3+} = -(.583120 + .835824) = -1.418944$$

Table 14.8 Augmented **b** vector and covariance matrix for simple effects (unshaded entries are
augmented values)

	$\mathbf{b}^+$	$\mathbf{C}^+$					
b_2	.583120	.135346	−.067764	−.067581	.000722	−.000271	−.000451
b_3	.835824	−.067764	.134918	−.067154	−.000271	−.000187	.000458
b_{3+}	−1.418944	−.067581	−.067154	.134735	−.000451	.000458	−.000007
b_4	−.156687	.000722	−.000271	−.000451	.001095	−.000541	−.000554
b_5	.078702	−.000271	−.000187	.000458	−.000541	.000827	−.000287
b_{5+}	.077984	−.000451	.000458	−.000007	−.000554	−.000287	.000841

and

$$b_{5+} = -(-.156687 + .078702) = .077984$$

The value for b_{3+} represents Group 3's deviation from the average intercept, and the value of b_{5+} represents Group 3's deviation from the average slope. You can verify the correctness of these values by referring to the entries in Table 14.4.

Our next step is to create an augmented covariance matrix ($\mathbf{C^+}$). The shaded entries in Table 14.8 come from the covariance matrix shown in Table 14.6. We then augment the entries (unshaded entries) to find our variances and covariances for Group 3 using Eq. (12.5), reproduced below:

$$\mathbf{c^+} = -\Sigma c_j \tag{14.3}$$

To illustrate, the first two values in the final column of Table 14.8 were found as follows:

$$-(.000722 + -.000271) = -.000451$$

and

$$-(-.000271 + -.000187) = .000458$$

14.2.2 S Matrix

Having created these augmented matrices, we now create an $\mathbf{S}$ matrix that will allow us to compute a large number of simple effects. We won't use them all, but it's useful to calculate them for didactic purposes. Table 14.9 presents the coefficients used to compute 24 simple effects. The rules used to form the coefficients follow the ones we used in Chap. 12 for a 3×2 ANOVA (see Table 12.23). The only difference is that we substitute the standard deviation of the continuous variable ($sd_{age} = 13.0384$) for the group codes when forming our coefficients.[6] For each matrix shown in Table 14.9, we find our test coefficients by computing $\mathbf{S'b^+}$ and our standard errors by taking the square root of the diagonal elements of $\mathbf{S'C^+S}$.

---

[6] When comparing Table 14.9 to Table 12.23, you might also notice that only Table 12.23 reports the coefficients for computing the simple effects of B at each level of A. I omitted these simple slopes here because we computed them earlier (see Table 14.7).

Table 14.9 S matrices for a three-group moderated regression

	a. Simple group contrasts collapsed across B					
	1 vs. 2	1 vs. 3	2 vs. 3	1 vs. (2,3)	2 vs. (1,3)	3 vs. (1,2)
b_2	1	1	0	2	−1	−1
b_3	−1	0	1	−1	2	−1
b_{3*}	0	−1	−1	1	−1	2
b_4	0	0	0	0	0	0
b_5	0	0	0	0	0	0
b_{5*}	0	0	0	0	0	0
	b. Simple A × B interaction contrasts					
	1 vs. 2	1 vs. 3	2 vs. 3	1 vs. (2,3)	2 vs. (1,3)	3 vs. (1,2)
b_2	0	0	0	0	0	0
b_3	0	0	0	0	0	0
b_{3*}	0	0	0	0	0	0
b_4	1	1	0	2	−1	−1
b_5	−1	0	1	−1	2	−1
b_{5*}	0	−1	−1	−1	−1	2
	c. Simple group contrasts at one standard deviation below the mean of B					
	1 vs. 2	1 vs. 3	2 vs. 3	1 vs. (2,3)	2 vs. (1,3)	3 vs. (1,2)
b_2	1	1	0	2	−1	−1
b_3	−1	0	1	−1	2	−1
b_{3*}	0	−1	−1	−1	−1	2
b_4	−13.0384	−13.0384	0	−26.0768	13.0384	13.0384
b_5	13.0384	0	−13.0384	13.0384	−26.0768	13.0384
b_{5*}	0	13.0384	13.0384	13.0384	13.0384	−26.0768
	d. Simple group contrasts at one standard deviation above the mean of B					
	1 vs. 2	1 vs. 3	2 vs. 3	1 vs. (2,3)	2 vs. (1,3)	3 vs. (1,2)
b_2	1	1	0	2	−1	−1
b_3	−1	0	1	−1	2	−1
b_{3*}	0	−1	−1	−1	−1	2
b_4	13.0384	13.0384	0	26.0768	−13.0384	−13.0384
b_5	−13.0384	0	13.0384	−13.0384	26.0768	−13.0384
b_{5*}	0	−13.0384	−13.0384	−13.0384	−13.0384	26.0768

14.2.3   Specific Tests of Interest

We are now ready to test some specific effects of interest. In presenting these tests, I will pretend our data were real in order to illustrate how researchers build an argument to make sense of the data they have collected.

14.2.3.1   Simple Interaction Contrasts

We begin by returning to the hypothesis that motivated our (imaginary) research. Recall that our researcher believed "you can't teach an old dog new tricks," leading

Table 14.10 Simple comparisons and contrasts for a three-group moderated regression

#		b	se_b	SS	t	p
	a. Simple group contrasts collapsed across age					
1	Group 1 vs. Group 2	−.2527	.6370	.1269	.3967	.7053
2	Group 1 vs. Group 3	2.0021	.6366	7.9775	3.1450	.0199
3	Group 2 vs. Group 3	2.2548	.6356	10.1506	3.5476	.0121
4	Group 1 vs. Groups (2,3)	1.7494	1.1037	2.0263	1.5850	.1641
5	Group 2 vs. Groups (1,3)	2.5075	1.1019	4.1762	2.2755	.0632
6	Group 3 vs. Groups (1,2)	−4.2568	1.1012	12.0524	3.8657	.0083
	b. Simple interaction contrasts					
7	Group 1 vs. Group 2	−.2354	.0548	14.8817	4.2955	.0051
8	Group 1 vs. Group 3	−.2347	.0552	14.5943	4.2538	.0054
9	Group 2 vs. Group 3	.0007	.0473	.0002	.0152	.9884
10	Group 1 vs. Groups (2,3)	−.4701	.0993	18.0910	4.7361	.0032
11	Group 2 vs. Groups (1,3)	.2361	.0863	6.0382	2.7361	.0339
12	Group 3 vs. Groups (1,2)	.2340	.0870	5.8334	2.6893	.0361
	c. Simple effects of group at low age					
	2 df F test			23.4774	14.5544	.0050
13	Group 1 vs. Group 2	2.8164	.9424	7.2027	2.9884	.0244
14	Group 1 vs. Group 3	5.0618	.9383	23.4707	5.3945	.0017
15	Group 2 vs. Group 3	2.2454	.9022	4.9954	2.4887	.0472
16	Group 1 vs. Groups (2,3)	7.8782	1.6502	18.3817	4.7740	.0031
17	Group 2 vs. Groups (1,3)	−.5710	1.5887	.1042	.3594	.7316
18	Group 3 vs. Groups (1,2)	−7.3072	1.5814	17.2206	4.6207	.0036
	d. Simple effects of group at high age					
	2 df F test			10.6628	6.6102	.0304
19	Group 1 vs. Group 2	−3.3218	.9718	9.4239	3.4182	.0142
20	Group 1 vs. Group 3	−1.0577	.9822	.9352	1.0768	.3229
21	Group 2 vs. Group 3	2.2641	.8695	5.4681	2.6038	.0405
22	Group 1 vs. Groups (2,3)	−4.3795	1.7499	5.0517	2.5027	.0464
23	Group 2 vs. Groups (1,3)	5.5859	1.5608	10.3302	3.5788	.0117
24	Group 3 vs. Groups (1,2)	−1.2065	1.5803	.4701	.7634	.4742

Note: 2 df tests are underlined

her to hypothesize that the new method of teaching abstract reasoning would benefit younger subjects but not older ones. We can test the researcher's hypothesis by conducting interaction contrasts, followed up by simple effects tests within each age group. First, we will perform an interaction contrast comparing the new method (Group 1) with the control condition (Group 3). Locating the results of this test in Table 14.10, we find that the predicted interaction contrast is significant $\boxed{8}$, indicating that when the new method is compared with the control condition, its effectiveness varies by age.[7]

We then follow up this result by conducting simple comparisons within each age group. As the researcher hypothesized, the new method is superior to the control condition among younger subjects $\boxed{14}$, but not among older subjects $\boxed{20}$. In this case, we have an "only for" interaction: the new method only benefits younger subjects.

We can repeat these analyses comparing the new method (Group 1) with the traditional method (Group 2). Table 14.10 shows that this interaction contrast is also significant $\boxed{7}$, confirming that when the new method is compared with the traditional method, its effectiveness varies by age. When we follow up this result by conducting simple comparisons within each age group, we find that the new method is superior to the traditional method among younger subjects $\boxed{13}$, but inferior to the traditional method among older subjects $\boxed{19}$. In this case, we have a crossover interaction.

Finally, we can compare Groups 2 and 3. This interaction contrast is not significant $\boxed{9}$, indicating that the difference between the two groups does not vary with age.

14.2.3.2 Simple Group Contrasts Collapsed Across Age

With a significant interaction, we ordinarily would not be concerned with simple contrasts collapsed across a continuous variable. However, we have found there is no interaction when we compare the traditional method (Group 2) with the control condition (Group 3). It is of interest, then, to ask whether Group 2 is generally different from Group 3. Analysis $\boxed{3}$ in Table 14.10 shows that the traditional method is, in fact, superior to the control condition.

Don't forget that this is a conditional simple slope. As first discussed in Chap. 9, with a cross-product term in the equation, lower-order effects represent a test of slope differences when the continuous predictor $= 0$. Because we centered our continuous predictor, the simple main effect tells us that the traditional method is superior to the control group among subjects of average age.

14.2.3.3 Simple Main Effects of Group at Low and High Age

The preceding analyses have involved 1 *df* contrasts, but we might also wish to test whether the simple main effect of group is significant within each "one standard deviation away from the mean" age group. This is a 2 *df* test, as now we are asking

---

[7] The boxed values reference the relevant analysis in Table 14.10.

whether the three instructional conditions produce different results in reasoning ability among younger and older subjects.

As first noted in Chap. 9, an easy way to answer this question is to rerun our analyses after recentering our data so that 0 represents one standard deviation below or above the mean on age (see Sect. 9.3.4). The following steps are involved to find the simple main effect of group among younger subjects:

- Add the standard deviation for age (13.0384) to the mean-centered scores used in the original analysis, and form cross products between the modified scores and the group codes.
- Omit the group codes, and run a regression using the modified scores and the cross products formed from them.
- Subtract the obtained SS_{reg} from the SS_{reg} from the original complete analysis (analysis "a" in Table 14.2), and test the difference:

$$F = \frac{[(39.1608 - 15.6833)/2]}{.8065} = 14.5544$$

Referring this value to an F distribution with 2 and 6 degrees of freedom shows a significant effect ($p=.0050$), confirming that the variance in performance across the three groups for younger subjects is unlikely to be due to chance alone.

We repeat the steps for older subjects, except now we subtract the standard deviation from the deviate scores so that 0 corresponds to one standard deviation above the mean:

$$F = \frac{[(39.1608 - 28.4979)/2]}{.8065} = 6.6106$$

Referring this value to an F distribution with 2 and 6 degrees of freedom also reveals a significant effect ($p=.0304$).

14.2.4 R Code: Moderation—Simple Effects

```
grp <-c(rep(1,4),rep(2,4),rep(3,4))
age <-c(29,27,39,53,57,32,45,21,47,55,29,22)
y <-c(7,9,5,2,5,7,6,5,4,3,3,4)
eff1 <-c(rep(1,4),rep(0,4),rep(-1,4))
eff2 <-c(rep(0,4),rep(1,4),rep(-1,4))
devage <-age-mean(age)
summary(mod.a <-lm(y~devage+eff1+eff2+eff1*devage+eff2*devage))

#Predicted Values for Ss one standard deviation above and below the mean
p0 <-c(rep(1,9))
p1 <-c(rep(-sd(age),3),rep(0,3),rep(sd(age),3))
p2 <-c(rep(c(1,0,-1),3))
```

(continued)

14.2.4 R Code: Moderation—Simple Effects (continued)

```
p3 <-c(rep(c(0,1,-1),3))
p4 <-c(-sd(age),0,sd(age),rep(0,3),sd(age),0,-sd(age))
p5 <-c(0,-sd(age),sd(age),rep(0,3),0,sd(age),-sd(age))
P <-rbind(p0,p1,p2,p3,p4,p5)
pred.val <-t(P)%*%coef(mod.a); pred.val

#Create Bar Graph
bar.graph<-c(pred.val[1],pred.val[7],pred.val[2],pred.val[8],pred.val[3],
pred.val[9])
mat1 <- matrix(bar.graph, 2)
barplot(mat1,beside=T,
main = "Moderated Regression", col = c("white", "gray"),
xlab = "Treatment", names = c("New", "Traditional", "Control"),
ylab = "Reasoning Ability", legend = c("Younger", "Older"),
args.legend = list(title = "Age", x = "top", cex =1),ylim = c(0, 10))

#Create Augmented Matrices
aug.b <-function(mod,start){
coef <-coef(mod)
BB<-c(coef[start],coef[start+1],-(coef[start]+coef[start+1]),coef
[start+2],coef[start+3],-(coef[start+2]+coef[start+3]))
}
BB <-aug.b(mod.a,3)

aug.c <-function(mod,start){
covar <-vcov(mod)
cov <-covar[start:(start+3),start:(start+3)]
rows<-function(cov,i,j)cbind(cov[i,1],cov[i,2],-(cov[i,1]+cov
[i,2]),cov[i,3],cov[i,4],-(cov[i,3]+cov[i,4]))
aug <-rows(cov)
CC<-rbind(aug[1,],aug[2,],-(aug[1,]+aug[2,]),aug[3,],aug[4,],
-(aug[3,]+aug[4,]))
}
CC <-aug.c(mod.a,3)

#Simple Slopes
simple <-function(S){
simp.slope <-t(S)%*%BB;simp.err <-sqrt(diag(t(S)%*%CC%*%S))
df=(length(y)-length(coef(mod.a)))
ttests <-simp.slope/simp.err;pvalues <-2*pt(-abs(ttests),df=df)
crit <-abs(qt(0.025,df))
CI.low <-simp.slope-(crit*simp.err);CI.high <-simp.slope+(crit*simp.err)
simp.table<-round(matrix(c(simp.slope,simp.err,ttests,pvalues,
CI.low,CI.high),nrow=ncol(S),ncol=6),digits=5)
```

(continued)

14.2.4   R Code: Moderation—Simple Effects (continued)

```
dimnames(simp.table)=list(c(),c("slope","stderr","t","p","CI.
low","CI.high"))
return(list(S,simp.table))
}


smat <-function(a,b){
s1 <-c(1*a,1*a,0,2*a,-1*a,-1*a);s2 <-c(-1*a,0,1*a,-1*a,2*a,-1*a)
s3 <-c(0,-1*a,-1*a,-1*a,-1*a,2*a);s4 <-c(1*b,1*b,0,2*b,-1*b,-1*b)
s5 <-c(-1*b,0,1*b,-1*b,2*b,-1*b);s6 <-c(0,-1*b,-1*b,-1*b,-1*b,2*b)
S <-rbind(s1,s2,s3,s4,s5,s6);simple(S)
}


simple.grp <-smat(1,0);simple.grp
simple.inter <-smat(0,1);simple.inter
simple.young <-smat(1,-sd(age));simple.young
simple.old <-smat(1,sd(age));simple.old


#2df Tests
dum.lo <-devage+sd(age)
eff1.lo <-eff1*dum.lo
eff2.lo <-eff2*dum.lo
summary(atlow <-lm(y~dum.lo+eff1.lo+eff2.lo))
anova(atlow,mod.a)


dum.hi <-devage-sd(age)
eff1.hi <-eff1*dum.hi
eff2.hi <-eff2*dum.hi
summary(athigh <-lm(y~dum.hi+eff1.hi+eff2.hi))
anova(athigh,mod.a)
```

14.3   Regions of Significance

In Chap. 9, we noted that plotting and testing the simple slopes at particular points is not the only way to probe an interaction. Another approach is to identify regions of significance using the Johnson-Neyman technique. Unfortunately, the Johnson-Neyman formula was only designed to compare two groups. But suppose we want to know the age at which the new method of instruction will be superior to the average of the other two methods. To answer questions of this nature, we need to modify the Johnson-Neyman technique (Bauer and Curran 2005, p. 380). As you will see, the modification will allow us to identify regions of significance for any 1 *df* contrast.

14.3.1   Reviewing the Johnson-Neyman Method

As you might remember, the Johnson-Neyman technique finds values of x that produce statistically significant simple slopes:

$$\frac{b_i + b_j x}{\sqrt{c_{ii} + 2x c_{ij} + x^2 c_{jj}}} \geq t_{critical} \qquad (14.4)$$

The following steps are involved:

- Determine the critical value. With $\alpha=.05$ and 6 degrees of freedom, the critical t value $= 2.4469$.
- Eliminate the square root in the denominator by squaring all terms so we are now working with an F value rather than a t value:

$$F_{critical} = t^2_{critical} = 5.9874$$

- Compute three preliminary terms:

$$A = \left[F_{critical} * c_{jj} \right] - b_j^2$$
$$B = 2 * \left\{ \left[F_{critical} * c_{ij} \right] - b_i b_j \right\} \qquad (14.5)$$
$$C = \left[F_{critical} * c_{ii} \right] - b_i^2$$

and enter them into the quadratic formula to find two values of x that produce a statistically significant simple slope:

$$x = \frac{-B \pm \sqrt{B^2 - 4AC}}{2A}$$

14.3.2   Extending the Johnson-Neyman Method

Instead of individually testing all possible comparisons, we can do things more efficiently using our augmented coefficient and covariance matrices and a modified S matrix. The augmented matrices were computed earlier (see Table 14.8), and Table 14.11 shows the modified S matrix. There are two vectors for each contrast: the first describes the group contrast (s_g) and the second models the cross-product term (s_p).

After calculating these terms, we use them to find A, B, and C and then solve for x using the quadratic formula.

Table 14.11   **S** matrix and matrix operations for regions of significance for 1*df* comparisons using the augmented **b** vector (**b**⁺) and augmented covariance matrix (**C**⁺) from Table 14.8

	s vectors for six possible 1*df* contrasts										
	1 vs. 2		1 vs. 3		2 vs. 3		1 vs. (2,3)		2 vs. (1,3)		3 vs. (1,2)
	s_g	s_p	s_g	s_p	s_g	s_p	s_g	s_p	s_g	s_p	s_g	s_p
b_2	1	0	1	0	0	0	2	0	−1	0	−1	0
b_3	-1	0	0	0	1	0	−1	0	2	0	−1	0
b_{3+}	0	0	−1	0	−1	0	−1	0	−1	0	2	0
b_4	0	−1	0	−1	0	0	0	−2	0	1	0	1
b_5	0	1	0	0	0	−1	0	1	0	−2	0	1
b_{5+}	0	0	0	1	0	1	0	1	0	1	0	−2

Term	Matrix operations	Illustration using Group 1 vs. Group 2
b_i	$-(s_g' b^+)$	.252704
b_j	$s_p' b^+$	.235389
c_{ii}	$s_g' C^+ s_g$	.405792
c_{ij}	$-(s_g' C^+ s_p)$	.001077
c_{jj}	$s_p' C^+ s_p$	.003003

14.3.3   *Illustration*

To illustrate, we will work through the calculations for our first comparison (i.e., Group 1 vs. Group 2). Just as a reminder, we are looking for the ages for which this comparison will produce a significant result. The final column in Table 14.11 shows the result of the matrix multiplication, and inserting these values into our simple slope formula leaves us with the following equation:

$$2.4469 = \frac{.252704 + .235389x}{\sqrt{.405792 + (2x * .001077) + (x^2 * .003003)}}$$

After squaring all values, we solve for A, B, and C:

$$A = (5.9874 * .003003) - .235389^2 = -.037428$$
$$B = 2 * \{(5.9874 * .001077) - (.252704 * .235389)\} = -.106074$$

and

$$C = (5.9874 * .405792) - .252704^2 = 2.36578$$

Entering these values into the quadratic formula produces the region of significance:

Table 14.12 Regions of significance for six contrasts

Comparison	Lower	Upper
Group 1 vs. Group 2	−9.4927	6.6586
Group 1 vs. Group 3	1.7699	24.2561
Group 2 vs. Group 3	14.7210	−13.4888
Group 1 vs. (Group 2, Group 3)	−2.0589	12.6927
Group 2 vs. (Group 1, Group 3)	−108.6889	.8101
Group 3 vs. (Group 1, Group 2)	5.6032	205.6813

$$x = \frac{-(-.106074) + \sqrt{-.106074^2 - [4 * (-.037428) * (2.36578)]}}{2 * (-.037428)} = -9.4927$$

and

$$x = \frac{-(-.106074) - \sqrt{-.106074^2 - [4 * (-.037428) * (2.36578)]}}{2 * (-.037428)} = 6.6586$$

Converting these deviate scores to raw scores, we predict that subjects who are younger than ~28.50 years of age will perform significantly better with the new method than with the traditional method and subjects who are older than ~44.66 years of age will perform significantly better with the traditional method than with the new method. Subjects falling between these age differences should perform equally well in both conditions.

Table 14.12 provides the remaining values, expressed in deviate form. To convert to raw scores, add 38.00 to each value.

14.3.4   R Code: Regions of Significance

```
grp <-c(rep(1,4),rep(2,4),rep(3,4))
age <-c(29,27,39,53,57,32,45,21,47,55,29,22)
y <-c(7,9,5,2,5,7,6,5,4,3,3,4)
eff1 <-c(rep(1,4),rep(0,4),rep(-1,4))
eff2 <-c(rep(0,4),rep(1,4),rep(-1,4))
devage <-age-mean(age)
summary(mod.a <-lm(y~devage+eff1+eff2+eff1*devage+eff2*devage))


#Create function for augmented vector
aug.b <-function(mod,start){
coef <-coef(mod)
BB<-c(coef[start],coef[start+1],-(coef[start]+coef[start+1]),coef
[start+2],coef[start+3],-(coef[start+2]+coef[start+3]))
}
BB <-aug.b(mod.a,3) #Enter model and first categorical coefficient
```

(continued)

14.3.4 R Code: Regions of Significance (continued)

```
#Create function for augmented covariance matrix
aug.c <-function(mod,start){
covar <-vcov(mod)
cov <-covar[start:(start+3),start:(start+3)]
rows<-function(cov,i,j)cbind(cov[i,1],cov[i,2],-(cov[i,1]+cov
[i,2]),cov[i,3],cov[i,4],-(cov[i,3]+cov[i,4]))
aug <-rows(cov)
CC<-rbind(aug[1,],aug[2,],-(aug[1,]+aug[2,]),aug[3,],aug[4,],-(aug
[3,]+aug[4,]))
}
CC <-aug.c(mod.a,3) #Enter model and first categorical coefficient

#Create Johnson-Neyman Function -- specify group comparisons
jnfunc <-function(a,b){
c = -(a+b)
j1 <-c(a,b,c,0,0,0)
j2 <-c(0,0,0,-a,-b,-c)
bg <-(-(t(j1)%*%BB))
bp <-t(j2)%*%BB
cgg <-t(j1)%*%CC%*%j1
cgp <-(-(t(j1)%*%CC%*%j2))
cpp <-t(j2)%*%CC%*%j2
df <-length(y)- length(coef(mod.a))
t.crit <-abs(qt(.025,df))
A <-(t.crit^2*cpp)-bp^2
B <-2*((t.crit^2*cgp-(bg*bp)))
C <-(t.crit^2*cgg)-bg^2
lower <-(-B+sqrt(B^2-4*A*C))/(2*A)
upper <-(-B-sqrt(B^2-4*A*C))/(2*A)
return <-cbind(lower,upper)
}

comp1 <-jnfunc(1,-1);comp2 <-jnfunc(1,0);comp3 <-jnfunc(0,1)
comp4 <-jnfunc(2,-1);comp5 <-jnfunc(-1,2);comp6 <-jnfunc(-1,-1)

regions <-rbind(comp1,comp2,comp3,comp4,comp5,comp6)
regions
raw.regions <-regions+mean(age);raw.regions
```

14.4 Chapter Summary

1. Moderation occurs when a continuous variable qualifies the effects of a categorical one.

2. Moderation is tested by creating cross-product terms between a continuous variable and the vectors that represent a categorical predictor. A complete model with all terms is then compared with one that excludes the cross-product terms. Moderation is established if the complete model fits the data significantly better than the reduced one.
3. Simple effects tests can be conducted to illuminate the nature of moderation. These tests are comparable to ones that probe the form of an interaction in an analysis of variance.
4. Regions of significance can be identified by modifying the Johnson-Neyman technique. Using effect coding with an augmented matrix constitutes the easiest way to perform these tests.

Chapter 15
Mediation

When scientists find an association between two or more variables, they are rarely content knowing only that the variables are related. Instead, they want to know why they are related; by what causal mechanism does one variable affect another? Unfortunately, the search for causes is fraught with difficulty. As the Scottish philosopher David Hume noted more than 250 years ago, causes are always inferred, never observed. For this reason, causal assumptions are subject to error, leading scientists to function as detectives, examining clues to construct a plausible (though not unassailable) causal argument.

In this chapter, we will examine a research strategy designed to illuminate the causal pathway from a categorical predictor to a continuous criterion. Let's begin by reviewing some material we first discussed in Chap. 13. Figure 15.1 (reproduced from Fig. 13.1) depicts three possible relations between a categorical predictor (x), a continuous criterion (y), and a continuous third variable (z):

- In the first example, x and z independently predict y. This is the intended use of an analysis of covariance (ANCOVA), covered in Chap. 13.
- In the second case, z moderates the relation between x and y. We discussed this research strategy in Chap. 14.
- The final example shows the case where x causes z, and z causes y. Here, z mediates (goes between) the predictor and the criterion in a presumed causal chain. This possibility, known as mediation, is the subject of the present chapter.

Electronic Supplementary Material: The online version of this chapter (doi: 10.1007/978-3-319-11734-8_15) contains supplementary material, which is available to authorized users

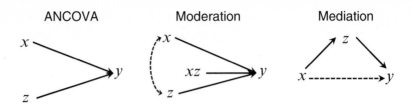

Fig. 15.1 Three possible relations between a categorical predictor (x), a continuous criterion (y), and a continuous third variable (z) (Reproduced from Fig. 13.1)

Before considering the steps involved in a test of mediation, let's discuss a more general issue, namely, research design and causal inference. Broadly speaking, there are two types of research designs: experimental research and correlational research. The most important difference between them is that subjects are randomly assigned to conditions in the former design but not in the latter. Because of this variation, only experimental research offers evidence of causality. Correlational research can suggest causal hypotheses to be tested, but, as is often said, "correlation is not causality."

It follows from the preceding discussion that mediational analyses must use the experimental method if they are to illuminate causal relations. Regrettably, this is not always so. Many investigations combine mediation with correlational methods, creating interpretive problems. In this chapter, we will first examine situations in which mediational analyses are appropriately paired with random assignment to conditions; afterward, we will consider problems that arise when mediational analyses are conducted with correlational data.

15.1   Simple Mediation

We will begin with a small sample example using a single mediator. This type of analysis is often called simple mediation. Imagine that our researcher from Chaps. 13 and 14 is now interested in finding out why her new method of teaching abstract reasoning works so well with young adults. She suspects that enhanced creativity is the key: Her new method leads young adults to think "outside the box," and creativity improves their performance on tests of abstract reasoning. Notice the causal argument here: Instructional method (X) affects creativity (henceforth designated M for mediator), and creativity affects performance (Y). Figure 15.2 illustrates the causal model, with lowercase letters (a, b, and c') identifying various pathways. The meaning of these designations will be clarified momentarily.

Fig. 15.2 Simple mediational model with a categorical predictor (X), a mediating variable (M), and a criterion (Y)

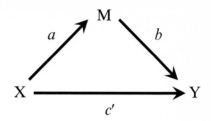

To test her mediation model, the researcher randomly assigns 12 young adults (age 30 and younger) to one of two conditions: the new instructional method or a traditional instructional method. She then measures creativity and, afterward, performance at a test of abstract reasoning. The (fictitious) data, along with a coding scheme, appear in Table 15.1.[1]

Table 15.1 Small sample example for a mediational analysis using two experimental groups and one mediator

Subject	Instructional method	Dummy coding X	Creativity M	Abstract reasoning Y	Random seed	X	M	Y
			Raw data		One bootstrap sample			
1	New	0	6	7	2	0	5	7
2	New	0	5	7	12	1	1	3
3	New	0	9	9	2	0	5	7
4	New	0	7	8	11	1	2	2
5	New	0	8	8	6	0	4	6
6	New	0	4	6	10	1	8	7
7	Traditional	1	5	6	5	0	8	8
8	Traditional	1	2	4	4	0	7	8
9	Traditional	1	3	4	10	1	8	7
10	Traditional	1	8	7	1	0	6	7
11	Traditional	1	2	2	7	1	5	6
12	Traditional	1	1	3	6	0	4	6

15.1.1 Analytic Strategy

A test of mediation involves partitioning the total effect of X on Y into two components: a direct (unmediated) effect and an indirect (mediated) effect. One way to produce this partitioning is to conduct a series of regression analyses that address five questions. Table 15.2 shows the regression analyses and questions, along with notation that is commonly used with tests of mediation.

---

[1] Later in this chapter, we will add the control condition and older adults to round out the experimental design used in Chap. 14.

15.1.1.1   Preliminary Analyses

Our first question is whether the categorical variable predicts the criterion. This question is answered by testing the statistical significance of the regression coefficient (c) in the first regression analysis.[2] Replicating our (phony) research from Chaps. 13 and 14, the top portion of Table 15.3 shows that the instructional method is superior to the traditional method ($c = -3.1667, p=.0046$). The left-hand portion of Fig. 15.3 depicts the effect.

Table 15.2 Regression analyses and tests of mediation

Question	Regression analysis	Ordinary notation	Mediation notation
Does the categorical variable predict the criterion?	Regress Y on X and test significance of X	$\hat{Y} = b_0 + b_1 X$	$\hat{Y} = c_0 + cX$
Does the categorical variable predict the mediator?	Regress M on X and test significance of X	$\hat{M} = b_0 + b_1 X$	$\hat{M} = a_0 + aX$
Does the mediator predict the criterion after controlling for the categorical variable?	Regress Y on X and M, and test significance of M	$\hat{Y} = b_0 + b_1 M + b_2 X$	$\hat{Y} = b_0 + bM + c'X$
Does the categorical variable predict the criterion after controlling for the mediator?	Regress Y on X and M, and test significance of X	$\hat{Y} = b_0 + b_1 M + b_2 X$	$\hat{Y} = b_0 + bM + c'X$
Does the categorical variable predict significantly less variance in the criterion after controlling for the mediator?			$c - c' = ab$

Mediational analyses are undertaken to illuminate why groups differ. In our case, the researcher suspects that creativity explains why the new method is better than the traditional one. To effectively make this argument, she first needs to establish that instructional method affects creativity. The second analysis in Table 15.3 confirms that creativity scores are significantly higher following the new instructional method than the traditional method ($a = -3.00, p=.0442$). The middle portion of Fig. 15.3 displays the effect.

In our third regression analysis, we regress the criterion on the categorical variable and the presumed mediator. Doing so allows us to determine whether the mediator predicts the criterion after the categorical variable has been statistically controlled. The final analysis in Table 15.3 shows that creativity does, in fact,

---

[2] Using statistical significance to answer this question assumes that the power of the experiment is sufficient to detect an effect. Clearly, this is unlikely to be the case with a sample size as small as ours. But since our data are fictitious anyway, we can also pretend that the sample size is large.

predict performance after statistically controlling for the categorical variable ($b=.6176$, $p=.0001$).

Taken together, the analyses we have performed provide preliminary evidence of mediation: Instructional method predicts performance and creativity, and creativity predicts variance in performance that instructional method does not.

Having established a plausible mediational argument, we can ask two additional questions of our data. First, does the categorical variable predict the criterion after the mediator is statistically controlled? If it does not, we characterize the form of mediation as "total mediation," as here the mediator seems to completely explain all of the benefits of the new instructional method; if it does, we say that we have "partial mediation," because creativity does not fully explain the benefits of the new instructional method. The final analysis in Table 15.3 shows evidence of partial mediation in our data set, as the categorical variable continues to predict

Table 15.3 Regression analyses in tests of mediation using mediation notation

1. Predict criterion from categorical vector						
	SS	df	MS	R^2	F	p
Regression	30.0833	1	30.0833	.5685	13.1752	.0046
Residual	22.8333	10	2.2833			
Total	52.9167					
	b	se_b	t	p		
c_0	7.50	.6169	12.1577	.0000		
c	-3.1667	.8724	3.6298	.0046		
2. Predict mediator from categorical vector						
	SS	df	MS	R^2	F	p
Regression	27.00	1	27.00	.3462	5.2941	.0442
Residual	51.00	10	5.10			
Total	78.00					
	b	se_b	t	p		
a_0	6.50	.9220	7.0502	.0000		
a	-3.00	1.3038	2.3009	.0442		
3. Predict criterion from mediator and categorical vector						
	SS	df	MS	R^2	F	p
Regression	49.5392	2	24.7696	.9362	66.0044	.0000
Residual	3.3775	9	.3753			
Total	52.9167					
	b	se_b	t	p		
b_0	3.4853	.6111	5.7034	.0003		
b	.6176	.0858	7.2003	.0001		
c'	-1.3137	.4374	3.0035	.0149		

performance even after controlling for the mediator ($c' = -1.3137$, $p=.0149$). The final columns in Fig. 15.3 display the nature of the effect.[3]

---

[3] The adjusted means shown in Fig. 15.3 were calculated using Eq. (13.1).

Fig. 15.3 Abstract
reasoning, creativity, and
adjusted reasoning as a
function of instructional
method

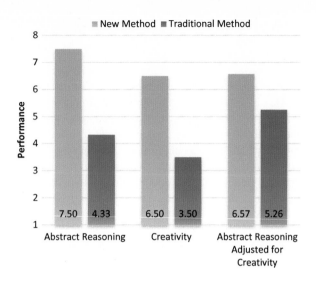

15.1.1.2   Partitioning the Variance

Figure 15.4 depicts our mediational model, but we still have one more question to
ask: Is the impact of the categorical variable significantly reduced after the mediator
is statistically controlled? It might seem that the most straightforward way to answer
this question is to test the statistical significance of $c - c'$, but there is an equivalent
but numerically superior solution: compute ab and test its statistical significance.

Fig. 15.4 Simple
mediational model with a
categorical predictor (X),
a mediating variable (M),
and a criterion (Y)

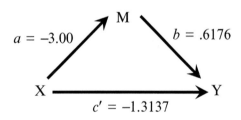

Partitioning the variance will convince you that $c - c' = ab$. Earlier, we noted that
a mediational analysis assumes that the total effect of a categorical variable on a
criterion is the sum of two components: an unmediated, direct effect and a mediated,
indirect effect. The total effect is represented by c, the direct effect is represented by
c', and the mediated effect is found by multiplying the coefficient linking the
categorical variable to the mediator (a) by the coefficient representing the unique
effect of the mediator on the criterion with the categorical variable controlled (b):

$$c = c' + ab \tag{15.1}$$

Rearranging these terms reveals that

$$ab = c - c'$$ (15.2)

and simple arithmetic confirms their equivalence:

$$c - c' = (-3.1667 - -1.3137) = -1.8529$$

and

$$ab = -3.00 * .6176 = -1.8529$$

15.1.2 Assessing the Importance of the Mediated Effect

There are at least four ways to judge the importance of a mediated effect: (1) test its statistical significance, (2) calculate confidence intervals, (3) compute its effect size, or (4) compare it with the size of other effects. All four methods require that we first compute the standard error of the mediated effect. The standard error, you will recall, represents the standard deviation of a distribution formed from repeatedly sampling a parameter. For example, if we conducted our study 10,000 times, we could form a distribution of all 10,000 mediated effects and compute the distribution's standard deviation. We call that standard deviation the standard error. Of course, we don't actually conduct a study 10,000 times; instead, we use our knowledge of probability to calculate a standard error that matches the values we expect to find had we performed all of the replications.

15.1.2.1 Calculating the Standard Error of the Mediated Effect

Because the mediated effect is the product of two terms computed from separate regression equations, we cannot calculate its standard error in the usual fashion. Fortunately, there is a work-around. Using a procedure known as the multivariate delta method, Sobel (1982) derived an approximation of the standard error of a mediated effect. The procedure starts by forming a column vector of partial derivatives. As we have learned, a partial derivative represents the instantaneous rate of change of a function as one of its input values changes while the others are held constant.

With respect to mediation, we assume that y is a function of a (the path from x to m), b (the path from m to y with x in the regression equation), and c' (the path from x to y with m in the regression equation):

$$\hat{y} = ab + c'$$ (15.3)

Because we are interested only in the mediated effect, we disregard the path from the categorical predictor to the criterion (c'), leaving $\hat{y} = ab$.

To calculate the partial derivative of ab with respect to a, we replace a with "1," leaving b:

$$\frac{\partial ab}{\partial a} = b \tag{15.4}$$

Table 15.4 Covariance matrices for Analysis 2 and Analysis 3 in Table 15.3

	a_0	a		b_0	b	c'
a_0	.8500	−.8500	b_0	.3734328	−.0478288	−.2060319
a	−.8500	1.7000	b	−.0478288	.0073583	.0220748
			c'	−.2060319	.0220748	.1913153

To calculate the partial derivative of ab with respect to b, we replace b with "1," leaving a:

$$\frac{\partial ab}{\partial b} = a \tag{15.5}$$

We then place these values into a column vector of partial derivatives we designate **d**:

$$\mathbf{d} = \begin{bmatrix} b \\ a \end{bmatrix} \tag{15.6}$$

Inserting the values from Table 15.3 produces the vector of partial derivatives:

$$\mathbf{d} = \begin{bmatrix} .6176 \\ -3.00 \end{bmatrix}$$

Next, we form a covariance matrix that includes values from both analyses. The covariance matrices for Analysis 2 (predict M from X) and Analysis 3 (predict Y from M and X) appear in Table 15.4, with the relevant variances shaded for emphasis.[4]

Because a and b come from different analyses, their covariance equals zero; consequently, we form a diagonal covariance matrix we designate **C**:

$$\mathbf{C} = \begin{bmatrix} \sigma_a^2 & 0 \\ 0 & \sigma_b^2 \end{bmatrix} \tag{15.7}$$

Inserting values from Table 15.4 produces our covariance matrix. Notice that although b appears first in the column vector **d**, the variance of a appears first in **C**:

---

[4] The shaded variances can also be calculated by squaring the relevant standard errors shown in Table 15.3.

$$C = \begin{bmatrix} 1.70 & 0 \\ 0 & .0073583 \end{bmatrix}$$

Now, we compute the standard error in the usual manner by finding the square root of $\mathbf{d'Cd}$:[5]

$$se_{ab} = \sqrt{\mathbf{d'Cd}} \tag{15.8}$$

Plugging in our values yields our standard error:

$$se_{ab} = .845431$$

15.1.2.2 Testing the Statistical Significance of the Mediated Effect

The standard error is used to test the statistical significance of the mediated effect. The test, known as Sobel's test, is normally distributed, and absolute values greater than 1.96 are statistically significant:

$$Z_{ab} = \frac{ab}{se_{ab}} \tag{15.9}$$

Plugging in our values yields Sobel's test:

$$|Z_{ab}| = \frac{-1.8529}{.845431} = 2.1917, \ p = .0284$$

On the basis of this test, we would conclude that the mediated path from instructional method to performance via creativity is unlikely to be a chance finding. And since testing ab is the same as testing $c - c'$, we can also say that the effect of instructional method on performance is significantly reduced when creativity is taken into account. These equivalent conclusions lend credence to the researcher's hypothesis that her new instructional method works, at least in part, because it promotes creativity.

---

[5] The formula in Eq. (15.8) computes a first-order standard error. It can also be found using simple algebra, dropping the last term when the variables are uncorrelated:

$$se_{ab} = \sqrt{b^2\sigma_a^2 + a^2\sigma_b^2 + 2ab\sigma_{ab}}$$

Additionally, some researchers compute a more conservative, second-order standard error by adding the variances to the first-order term. We will only use the formula for the first-order standard error in this book because it is the one used by most statistical packages:

$$se_{ab} = \sqrt{b^2\sigma_a^2 + a^2\sigma_b^2 + 2ab\sigma_{ab} + \sigma_a^2\sigma_b^2}$$

We can also gauge the magnitude of the mediated effect by constructing 95 % confidence limits around it:

$$-1.8529 - (1.96 * .845431) = -3.51$$

and

$$-1.8529 + (1.96 * .845431) = -.1959$$

If we were to conduct our study $10,000$ times, we would expect to find a mediated effect that falls between these extremes in $9,500$ of them; the remaining 500 samples would (probably) produce mediated effects that fall outside this range.

15.1.2.3   Bootstrapping

Sobel's test assumes that the mediated effect is normally distributed, but this is not always true when sample sizes are small ($< \sim 200$). Consequently, the test is not always accurate and the confidence intervals are not always symmetric (MacKinnon 2008). There are several ways to remedy this situation, but the most common is to use a technique known as bootstrapping. With bootstrapping, we use the data we collected to simulate conducting hundreds of studies. The following steps are involved:

- Randomly selecting *with replacement* from the data you collected, create hundreds of samples, each the same size as your original sample.
- Calculate the mediated effect from each sample, and construct 95 % confidence limits by computing the average of the bottom and top 2.5 % of the calculated effects.

The last four columns of Table 15.1 show one bootstrap sample. As you can see, observations 2, 6, and 10 are represented twice, and observations 3, 8, and 9 are missing. The mediated effect from this sample $= -.4914$. If you then repeat this process hundreds of times, you create bootstrapped standard errors. Due to its repetitive nature, bootstrapping is tedious to perform with a spreadsheet, and I would encourage you to use a statistical package when using this technique. The **R** code that accompanies this chapter provides a function you can use, and the values shown below represent the average of the top and bottom 2.5 % of the calculated values for our mediated effect when I performed these analyses:[6]

$$\text{Lower Bootstrapping Confidence Limit} = -3.767714$$
$$\text{Upper Bootstrapping Confidence Limit} = .1489967$$

---

[6] Because the resamples are drawn randomly, the results you get using **R** will not precisely match the ones reported here.

Because our sample size is so small, these confidence intervals are quite different from the ones we obtained using the standard error. With a larger sample, the two estimates are ordinarily quite similar.

15.1.3 Effect Sizes

Effect sizes can also be used to characterize the magnitude of a mediated effect. Currently, there is no accepted standard for computing the effect size of a mediated effect, but several options are available (Preacher and Kelly 2011). Table 15.5 displays some of the effect size measures that have been suggested, along with their values using our data set. In general, effect sizes with absolute values $>.25$ are considered large, and all of the values listed in Table 15.5 far exceed this standard.

Table 15.5 Measures of effect size for the mediated effect

Proportion of total effect that is mediated	$\dfrac{ab}{c}$	.5851
Ratio of direct effect to total effect	$\dfrac{c'}{c}$	.4149
Ratio of mediated effect to direct effect	$\dfrac{ab}{c'}$	1.4104
Standardized effect	$\dfrac{ab}{\sigma_y}$	$-.8448$

15.1.4 Contrasts

A final way to gauge the importance of the mediated effect is to test it against other effects. For example, we could compare the size of the mediated effect against the direct effect $(ab - c')$. To find the standard error of the contrast, we create a vector of partial derivatives $\mathbf{d}$:[7]

$$\mathbf{d} = \begin{bmatrix} \dfrac{\partial ab}{\partial a} = b \\[2ex] \dfrac{\partial ab}{\partial b} = a \\[2ex] \dfrac{\partial c'}{\partial c'} = -1 \end{bmatrix} \tag{15.10}$$

---

[7] The partial derivative of c' carries a negative sign in Eq. (15.10) because we are contrasting the effect against the mediated effect.

and a covariance matrix $\mathbf{C}$ (shown below):

$$\mathbf{C} = \begin{bmatrix} \sigma_a^2 & 0 & 0 \\ 0 & \sigma_b^2 & \sigma_{bc'} \\ 0 & \sigma_{c'b} & \sigma_{c'}^2 \end{bmatrix} \tag{15.11}$$

Inserting the relevant values from Tables 15.3 and 15.4 produces the vector and matrix:

$$\mathbf{d} = \begin{bmatrix} .6176 \\ -3.00 \\ -1 \end{bmatrix} \quad \mathbf{C} = \begin{bmatrix} 1.70 & 0 & 0 \\ 0 & .007358 & .022075 \\ 0 & .022075 & .191315 \end{bmatrix}$$

We then find the square root of the contrast in the usual fashion:

$$se_{ab-c'} = \sqrt{\mathbf{d}'\mathbf{C}\mathbf{d}} = 1.0191$$

and construct a Z test to assess its significance:

$$|Z_{ab-c'}| = \frac{ab - c'}{se_{ab-c'}} = \frac{-1.8529 - (-1.3137)}{1.0191} = .5291, \, p = .5967$$

As you can see, the mediated effect is not significantly greater than the direct effect. Note, however, that the standard error of the contrast is not normally distributed, so bootstrapping should be used when sample sizes are small. In this case, you compute the contrast $(ab - c')$ across hundreds of bootstrap samples and construct confidence intervals using the bottom and top 2.5 % of your observed values. If zero falls within the confidence limits, you fail to reject the null hypothesis.

15.1.5 Summary

Mediational analyses are designed to illuminate causal pathways. Because we measured creativity after randomly assigning subjects to experimental conditions, it is reasonable to attribute any effect of creativity on performance to the causal chain initiated by the instructional method, and to take this indirect effect into account when considering the overall impact of instructional method on performance.

Whether this conclusion would be warranted if the total effect of instructional method on performance had fallen short of statistical significance is debatable. Some researchers believe that a mediational analysis is appropriate only when the total effect is statistically significant (Baron and Kenny 1986), whereas others believe this requirement is too restrictive (Shrout and Bolger 2002). The latter position rests on two facts: (1) a mediator can sometimes act as a suppressor variable, obscuring the effects of X on Y, and (2) the power to detect a total effect can be less than the power to detect a mediated one (Kenny and Judd 2014). When these circumstances can be shown to exist, testing for mediation in the absence of a total effect of X on Y makes sense.

Power considerations are relevant to a related issue. Because they are based solely on statistical significance, characterizations of "full mediation" vs. "partial mediation" are problematic. By simply increasing the sample size, an effect formerly described as representing "full mediation" could become one of "partial mediation." For this reason, it is best to forgo using this designation or offer the description as appropriate only for the sample at hand.

15.1.6 R Code: Simple Mediation

```
X <-c(rep(0,6),rep(1,6))
M <-c(6,5,9,7,8,4,5,2,3,8,2,1)
Y <-c(7,7,9,8,8,6,6,4,4,7,2,3)
summary(mod.1 <-lm(Y~X))
summary(mod.2 <-lm(M~X))
summary(mod.3 <-lm(Y~M+X))
a <-mod.2$coef[2]
b <-mod.3$coef[2]
c <-mod.1$coef[2]
cdir <-mod.3$coef[3]

#Sobel's Test
ab <-a*b
d <-rbind(mod.3$coef[2],mod.2$coef[2])
C    <-    matrix(c(vcov(mod.2)[2:2,2:2],0,0,vcov(mod.3)[2:2,2:2]),
nrow=2)
SE <-sqrt(t(d)%*%C%*%d)
Z.ab <-ab/SE
pvalue <-2*(1-pnorm(abs(Z.ab)))
CI.low <-ab-(1.96*SE)
CI.high <-ab+(1.96*SE)
Sobel<-round(matrix(c(ab,SE,Z.ab,pvalue,CI.low,CI.high),nrow=1,
ncol=6),digits=5)
dimnames(Sobel)=list(c(""),c("ab",    "SE",    "Z","p","CI.low","CI.
high"))
Sobel

#Bootstrap Function
med.boot <- function(X, M, Y, reps, ci){
 ab_vector = NULL
 for (i in 1:reps){
 s = sample(1:length(X), replace=TRUE)
 Xs = X[s]
 Ys = Y[s]
 Ms = M[s]
 M_Xs = lm(Ms ~ Xs)
```

(continued)

15.1.6   R Code: Simple Mediation (continued)

```
Y_XMs = lm(Ys ~ Xs + Ms)
a = M_Xs$coefficients[2]
b = Y_XMs$coefficients[3]
ab = a*b
ab_vector = c(ab_vector, ab)
}
sorted <-sort(ab_vector)
num=reps*(ci/2)
CI.low <-mean(sorted[1:num])
CI.high <-mean(sorted[(length(sorted)-(num-1)):length(sorted)])
CI <-cbind(CI.low,CI.high)
return=CI
}
bootstrap <-med.boot(X,M,Y,1000,.05);bootstrap #Specify variables,
#sample size, and confidence interval

#Effect Size Measures
prop <-ab/c
ratio.1 <-cdir/c
ratio.2 <-ab/cdir
standard <-ab/sd(Y)
effect <-cbind(prop,ratio.1,ratio.2,standard)
effect

#Contrasts
d <-rbind(b,a,-1)
c1 <-c(vcov(mod.2)[2:2,2:2],0,0)
c2 <-c(0,vcov(mod.3)[5],vcov(mod.3)[6])
c3 <-c(0,vcov(mod.3)[8],vcov(mod.3)[9])
CC <-cbind(c1,c2,c3)
se.cont <-sqrt(t(d)%*%CC%*%d)
Z.contrast <-(ab-cdir)/se.cont
Z.contrast
2*(1-pnorm(abs(Z.contrast)))
```

15.2   Higher-Order Designs

The procedures we have examined can be extended for use with more complicated designs. We cannot cover all possibilities here, but I will discuss three extensions in this section: (1) mediation with three groups and one mediator, (2) mediation with two groups and two mediators, and (3) mediation with moderation.

15.2.1 *Mediation with Three Groups*

The first extension we will consider uses a categorical predictor with three groups, although the procedures apply to multiple groups of all sizes. Imagine our researcher adds a control condition to her original design. After finding another sample of 12 young adults, she randomly assigns them to one of three conditions: her new instructional method, the traditional method, or a control (no instruction) condition. She then measures creativity, followed by performance on a test of abstract reasoning ability. The (concocted) data, along with a coding scheme we will use to analyze them, appear in Table 15.6. Notice that we are using two dummy-coded vectors to represent our three groups. The first compares the new method with the traditional method, and the second compares the new method with the control condition.

Table 15.6 Small sample example for a mediational analysis using three experimental groups and one mediator

Instructional Method	Creativity M	X_1	X_2	Performance Y
New	7	0	0	7
New	9	0	0	8
New	7	0	0	7
New	6	0	0	7
Traditional	5	1	0	4
Traditional	5	1	0	4
Traditional	3	1	0	3
Traditional	7	1	0	5
Control	3	0	1	4
Control	2	0	1	3
Control	1	0	1	3
Control	2	0	1	2

15.2.1.1 Regression Analyses

As with a two-group design, we begin a test of mediation by conducting three regression analyses. The regression equations appear below:

$$\hat{Y} = c_0 + c_1 X_1 + c_2 X_2$$
$$\hat{M} = c_0 + c_1 X_1 + c_2 X_2 \qquad (15.12)$$
$$\hat{Y} = b_0 + bM + c_1' X_1 + c_2' X_2$$

and the corresponding analyses shown in Table 15.7 support the following conclusions:

- Both categorical vectors are significant in the first analysis, indicating that abstract reasoning skills are greater with the new instructional method than with the traditional instructional method or no instruction.

- Similarly, both categorical vectors are significant in the second analysis, indicating that creativity scores are greater with the new instructional method than with the traditional method or no instruction.
- All terms are significant in the final analysis, indicating that the mediated effect is significant even after controlling for the categorical vector and that mediation in this sample is partial, not total.

Table 15.7 Regression analyses for a test of mediation with a 2 df predictor and one mediator

1. Predict criterion from categorical vectors						
	SS	df	MS	R^2	F	p
Regression	39.50	2	19.75	.8927	37.4211	.0000
Residual	4.75	9	.5278			
Total	44.25					
	b	se_b	t	p		
c_0	7.25	.3632	19.9592	.0000		
c_1	−3.25	.5137	6.3266	.0001		
c_2	−4.25	.5137	8.2733	.0000		
2. Predict mediator from categorical vectors						
	SS	df	MS	R^2	F	p
Regression	55.50	2	27.75	.7900	16.9322	.0009
Residual	14.75	9	1.6389			
Total	70.25					
	b	se_b	t	p		
a_0	7.25	.6401	11.3264	.0000		
a_1	−2.25	.9052	2.4856	.0347		
a_2	−5.25	.9052	5.7996	.0003		
3. Predict criterion from mediator and categorical vectors						
	SS	df	MS	R^2	F	p
Regression	42.5890	3	14.1963	.9625	68.3741	.0000
Residual	1.6610	8	.2076			
Total	44.25					
	b	se_b	t	p		
b_0	3.9322	.8898	4.4190	.0022		
b (creativity)	.4576	.1186	3.8571	.0048		
c_1'	−2.2203	.4184	5.3065	.0007		
c_2'	−1.8475	.7013	2.6344	.0300		

15.2.1.2 Partitioning of Effects

Figure 15.5 shows the mediational model. Computing an omnibus mediated effect is not recommended, so we form two equations, one for each categorical vector. Because we have used dummy coding, each vector represents a comparison between the new instructional method and one of the other groups. The left-hand side of Fig. 15.5 examines whether creativity explains differences in abstract reasoning between

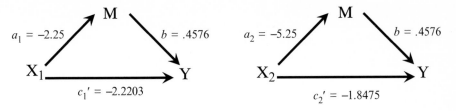

Fig. 15.5 Mediational model with a 2 *df* predictor

the new method and the traditional method, and the right-hand side of Fig. 15.5 examines whether creativity explains differences in abstract reasoning between the new method and the control condition. Note, however, that the value for b (path from mediator to criterion) is identical in both analyses. Even though we are separating the categorical vectors to form our equations, both vectors are included when performing the regression analyses, so the path from the mediator to the criterion does not change.

After performing the regression analyses, we compute two mediated effects, one for each vector:

$$ab_1 = -2.25 * .4576 = -1.0297$$
$$ab_2 = -5.25 * .4576 = -2.4025$$

We can then compute standard errors of each effect and calculate confidence intervals, tests of statistical significance, and effect sizes using the procedures described earlier. Table 15.8 presents the vector of partial derivatives and covariance matrix for each categorical vector, first for a Sobel's test and then for a contrast in which the mediated path is compared with the direct effect.

Table 15.8 Vectors and matrices for standard errors used in Sobel's test and a comparison of the mediated effect vs. the direct effect

Test	d	C	Standard error = $\sqrt{d'\,Cd}$
Vector 1—Sobel's test	$\begin{bmatrix} b \\ a_1 \end{bmatrix} = \begin{bmatrix} .4576 \\ -2.25 \end{bmatrix}$	$\begin{bmatrix} \sigma_{a_1}^2 & 0 \\ 0 & \sigma_b^2 \end{bmatrix} = \begin{bmatrix} .819444 & 0 \\ 0 & .014076 \end{bmatrix}$	.4928
Vector 1—ab vs. c'	$\begin{bmatrix} b \\ a_1 \\ -1 \end{bmatrix} = \begin{bmatrix} .4576 \\ -2.25 \\ -1 \end{bmatrix}$	$\begin{bmatrix} \sigma_{a_1}^2 & 0 & 0 \\ 0 & \sigma_b^2 & \sigma_{bc_1'} \\ 0 & \sigma_{bc_1'} & \sigma_{c_1'}^2 \end{bmatrix} = \begin{bmatrix} .819444 & 0 & 0 \\ 0 & .014076 & .031672 \\ 0 & .031672 & .175075 \end{bmatrix}$	.7486
Vector 2—Sobel's test	$\begin{bmatrix} b \\ a_2 \end{bmatrix} = \begin{bmatrix} .4576 \\ -5.25 \end{bmatrix}$	$\begin{bmatrix} \sigma_{a_2}^2 & 0 \\ 0 & \sigma_b^2 \end{bmatrix} = \begin{bmatrix} .819444 & 0 \\ 0 & .014076 \end{bmatrix}$	.7481
Vector 2—ab vs. c'	$\begin{bmatrix} b \\ a_2 \\ -1 \end{bmatrix} = \begin{bmatrix} .4576 \\ -5.25 \\ -1 \end{bmatrix}$	$\begin{bmatrix} \sigma_{a_2}^2 & 0 & 0 \\ 0 & \sigma_b^2 & \sigma_{bc_2'} \\ 0 & \sigma_{cb_2'} & \sigma_{c_2'}^2 \end{bmatrix} = \begin{bmatrix} .819444 & 0 & 0 \\ 0 & .014076 & .073901 \\ 0 & .073901 & .491795 \end{bmatrix}$	1.3518

Table 15.9 presents the tests of significance. As you can see, both mediated effects are significant, suggesting that the newer method is better than each of the other methods, in part, because it fosters creativity and creativity fosters performance. Note also, however, that neither mediated effect is significantly greater than its corresponding direct effect.

Table 15.9 Tests of mediated effects in a 2 *df* mediational model

	Point estimate	*se*	|Z|	*p*	CI_{low}	CI_{High}
a_1b	−1.0297	.4928	2.0893	.0367	−1.9956	−.0637
a_1b vs. c_1'	1.1907	.7486	1.5904	.1117	−.2767	2.65802
a_2b	−2.4025	.7481	3.2117	.0013	−3.8687	−.9363
a_2b vs. c_2'	−.5551	1.3518	.4106	.6813	−3.2046	2.09443

15.2.1.3   Summary

Using a 2 *df* predictor poses no challenge to a mediational analysis. Both categorical vectors are included in all regression analyses, but we treat them separately when performing tests of mediation. We used dummy coding because it makes the most sense in this context to compare the first group against each of the others, but we could have used orthogonal contrast codes or effect codes. And if we use effect codes, we can augment our vectors and matrices as we learned to do in Chap. 12, yielding all possible 1 *df* comparisons and contrasts.

15.2.2   R Code: Mediation with Three Groups

```
X1 <-c(rep(0,4),rep(1,4),rep(0,4))
X2 <-c(rep(0,8),rep(1,4))
M <-c(7,9,7,6,5,5,3,7,3,2,1,2)
Y <-c(7,8,7,7,4,4,3,5,4,3,3,2)
summary(mod.1 <-lm(Y~X1+X2))
summary(mod.2 <-lm(M~X1+X2))
summary(mod.3 <-lm(Y~M+X1+X2))
b <-mod.3$coef[2]


#Sobel
Sobel <-function(x){
a <-mod.2$coef[1+x];ab <-a*b
d <-rbind(b,a)
C <- matrix(c(vcov(mod.2)[5],0,0,vcov(mod.3)[6]), nrow=2)
SE <-sqrt(t(d)%*%C%*%d)
Z <-ab/SE
pvalue <-2*(1-pnorm(abs(Z)))
```

(continued)

15.2.2 R Code: Mediation with Three Groups (continued)

```
CI.low <-ab-(1.96*SE)
CI.high <-ab+(1.96*SE)
Sobel<-round(matrix(c(ab,SE,Z,pvalue,CI.low,CI.high),nrow=1,
ncol=6),digits=5)
dimnames(Sobel)=list(c(""),c("ab",    "SE",    "Z","p","CI.low","CI.
high"))
Sobel
}
sobel.1 <-Sobel(1)
sobel.2 <-Sobel(2)
Sobel.table <-rbind(sobel.1,sobel.2);Sobel.table

#Contrasts
contrast <-function(x){
a <-mod.2$coef[1+x];ab <-a*b;c <-mod.3$coef[2+x];minus <-ab-c
d <-rbind(b,a,-1)
c1 <-c(vcov(mod.2)[2:2,2:2],0,0);
c2 <-c(0,vcov(mod.3)[6],vcov(mod.3)[6+x])
c3 <-c(0,vcov(mod.3)[6+x],vcov(mod.3)[6+x+(4*x)])
CC <-cbind(c1,c2,c3)
SE <-sqrt(t(d)%*%CC%*%d)
Z <-minus/SE
pvalue <-2*(1-pnorm(abs(Z)))
CI.low <-minus-(1.96*SE)
CI.high <-minus+(1.96*SE)
cont<-round(matrix(c(minus,SE,Z,pvalue,CI.low,CI.high),nrow=1,
ncol=6),digits=5)
dimnames(cont)=list(c(""),c("ab",    "SE",    "Z","p","CI.low","CI.
high"))
cont
}
cont.1 <-contrast(1)
cont.2 <-contrast(2)
contrasts <-rbind(cont.1,cont.2)
contrasts
```

15.2.3 Multiple Mediators

Our next analysis uses two mediators. Let's imagine that our researcher now believes that her new instructional method increases motivation as well as creativity and that motivation independently improves performance. So now we have two mediators: creativity and motivation. To test her hypothesis, she repeats her first, two-group only experiment and assesses creativity and motivation before measuring performance. The (fictional) data, along with a coding scheme, appear in Table 15.10.

Table 15.10 Small sample example for a mediational analysis using two mediators

Experimental condition	Creativity M_1	Motivation M_2	Dummy coding X	Performance Y
New	9	6	0	9
New	7	9	0	8
New	8	7	0	6
New	1	6	0	3
New	6	5	0	7
New	9	6	0	9
Traditional	2	3	1	3
Traditional	2	6	1	6
Traditional	1	7	1	5
Traditional	8	1	1	7
Traditional	2	2	1	2
Traditional	1	1	1	1

15.2.3.1 Regression Analyses

With two mediators, we conduct four regression analyses instead of three:

$$\begin{aligned}
\hat{Y} &= c_0 + cX \\
\hat{M}_1 &= a_{01} + a_1X \\
\hat{M}_2 &= a_{02} + a_2X \\
\hat{Y} &= b_0 + b_1M_1 + b_2M_2 + c'X
\end{aligned} \qquad (15.13)$$

The pertinent results from these analyses are shown in Table 15.11, and Fig. 15.6 displays the mediational model. Because the procedures used to analyze the data are similar to the ones used with simple mediation, only the pertinent coefficients are reported and discussed.

Before turning to the mediated effects, we'll examine the probability values shown in Table 15.11. Here we see that the total effect of X on Y is significant ($p=.0493$), the path leading from X to each mediator is significant ($p=.0349$ and $p=.0243$), and each mediator is an independent predictor of performance ($p=.0005$ and $p=.0165$) in a regression model that includes the categorical predictor. Note, however, that the direct effect of X on Y is not significant ($p=.1832$).

Table 15.11 Regression coefficients and variance/covariance analyses for a test of mediation with two mediators

	Path	b	se_b	t	p	Variances/covariances			
Regress Y on categorical vector	c	−3.00	1.3416	2.2361	.0493				
Regress Mediator 1 on categorical vector	a_1	−4.00	1.6398	2.4393	.0349	a_1	2.68889		
Regress Mediator 2 on categorical vector	a_2	−3.1667	1.1949	2.6502	.0243	a_2	1.42778		
Regress Y on mediators and categorical vector							b_1	b_2	c'
	b_1	.7136	.1283	5.5628	.0005	b_1	.01646	.00474	.08083
	b_2	.5323	.1760	3.0237	.0165	b_2	.00474	.03099	.11710
	c'	1.5402	1.0570	1.4572	.1832	c'	.08083	.11710	1.11719

Fig. 15.6 Mediation model with two mediators

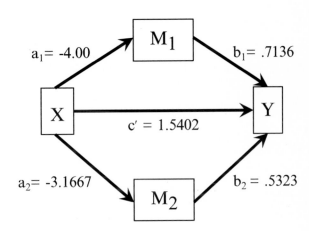

15.2.3.2   Mediated Effects and Their Standard Errors

We calculate our mediated effects in the usual manner. First, we multiply the relevant regression coefficients:

$$a_1 b_1 = (-4.00 * .7136) = -2.8544$$

and

$$a_2 b_2 = (-3.1667 * .5323) = -1.6857$$

Then we calculate the combined mediated effect by adding the separate paths:

$$\text{combined mediated effect} = -2.8544 + -1.6857 = -4.5401$$

To compute the standard errors of these terms, we need to find the partial derivatives. Because we are also going to perform some additional contrasts, we will create a large **D** matrix that will allow us to calculate many terms at once. Using values from Table 15.11, Table 15.12 presents the coefficients for calculating seven standard errors. The first two columns find the standard errors of the two mediated effects; the third column finds the standard error of the combined mediated effect; the fourth column contrasts the two mediated effects; the fifth and sixth columns contrast each mediated effect against the direct effect; and the final column contrasts the total mediated effect against the direct effect. To find our standard errors, we multiply **D′CD** and take the square root of the diagonal entries. The point estimates are derived using simple arithmetic.

Table 15.12 **D** matrix of partial derivatives and covariance matrix (**C**) for seven effects and contrasts with multiple mediators

	Sobel tests and contrasts						
	First mediator	Second mediator	Combined mediation	Compare mediators	First mediator vs. direct effect	Second mediator vs. direct effect	Total mediation vs. direct effect
	a_1b_1	a_2b_2	$a_1b_1 + a_2b_2$	$a_1b_1 - a_2b_2$	$a_1b_1 - c'$	$a_2b_2 - c'$	$(a_1b_1 + a_2b_2) - 2c'$
	Rules for forming **D** matrix of partial derivatives						
b_1	b_1	0	b_1	b_1	b_1	0	b_1
a_1	a_1	0	a_1	a_1	a_1	0	a_1
b_2	0	b_2	b_2	$-b_2$	0	b_2	b_2
a_2	0	a_2	a_2	$-a_2$	0	a_2	a_2
c'	0	0	0	0	-1	-1	-2
	Inserted values for **D** matrix of partial derivatives						
b_1	.7136	0	.7136	.7136	.7136	0	.7136
a_1	-4.00	0	-4.00	-4.00	-4.00	0	-4.00
b_2	0	.5323	.5323	-.5323	0	.5323	.5323
a_2	0	-3.1667	-3.1667	3.1667	0	-3.1667	-3.1667
c'	0	0	0	0	-1	-1	-2

	Covariance matrix (**C**)				
	a_1	b_1	a_2	b_2	c'
a_1	2.68889	0	0	0	0
b_1	0	.01646	0	.00474	.08083
a_2	0	0	1.42778	0	0
b_2	0	.00474	0	.03099	.11710
c'	0	.08083	0	.11710	1.11719

Table 15.13 presents the results of these contrasts. Since the data are fictitious, we will forget that we should be using bootstrapping with such a small sample and go ahead and interpret the results as if they were accurate. The findings show that

each mediator is significant when considered separately and that their combination is also significant. They are not, however, significantly different from each other. Finally, we can see that separately and in combination, they are significantly different from the direct effect.

Table 15.13 Sobel tests and contrasts with multiple mediators

Test	Notation	b	se_b	$|Z|$	p
Indirect effect of first mediator	a_1b_1	−2.8545	1.2777	2.2340	.0255
Indirect effect of second mediator	a_2b_2	−1.6857	.8458	1.9930	.0463
Indirect effect of both mediators	$a_1b_1 + a_2b_2$	−4.5402	1.5710	2.8900	.0039
Contrast Mediator 1 vs. Mediator 2	$a_1b_1 - a_2b_2$	−1.1688	1.4926	.7830	.4336
Contrast Mediator 1 vs. direct effect	$a_1b_1 - c'$	−4.3947	1.8430	2.3846	.0171
Contrast Mediator 2 vs. direct effect	$a_2b_2 - c'$	−3.2259	1.6044	2.0106	.0444
Contrast both mediators vs. direct effect	$(a_1b_1 + a_2b_2) - 2c'$	−7.6205	3.1166	2.4451	.0145

15.2.4 R Code: Multiple Mediators

```
M1 <-c(9,7,8,1,6,9,2,2,1,8,2,1)
M2 <-c(6,9,7,6,5,6,3,6,7,1,2,1)
X <-c(rep(0,6),rep(1,6))
Y <-c(9,8,6,3,7,9,3,6,5,7,2,1)
summary(mod.1 <-lm(Y~X))
summary(mod.2 <-lm(M1~X))
summary(mod.3 <-lm(M2~X))
summary(mod.4 <-lm(Y~M1+M2+X))
c <-mod.1$coef[2]
a1 <-mod.2$coef[2]
a2 <-mod.3$coef[2]
b1 <-mod.4$coef[2]
b2 <-mod.4$coef[3]
cdir <-mod.4$coef[4]
coef <-rbind(c,a1,a2,b1,b2,cdir)
coef
a1b1 <-a1*b1
a2b2 <-a2*b2

#D matrix
d1 <-c(b1,0,b1,b1,b1,0,b1)
d2 <-c(a1,0,a1,a1,a1,0,a1)
d3 <-c(0,b2,b2,-b2,0,b2,b2)
d4 <-c(0,a2,a2,-a2,0,a2,a2)
```

(continued)

15.2.4   R Code: Multiple Mediators (continued)

```
d5 <-c(0,0,0,0,-1,-1,-2)
D <-rbind(d1,d2,d3,d4,d5)

c1 <-c(vcov(mod.2)[4],rep(0,4))
c2 <-c(0,vcov(mod.4)[6],0,vcov(mod.4)[7],vcov(mod.4)[8])
c3 <-c(0,0,vcov(mod.3)[4],0,0)
c4 <-c(0,vcov(mod.4)[10],0,vcov(mod.4)[11],vcov(mod.4)[12])
c5 <-c(0,vcov(mod.4)[14],0,vcov(mod.4)[15],vcov(mod.4)[16])
C <-rbind(c1,c2,c3,c4,c5)

#Function for Simple Slopes
simple.slope <-function(D,C){
simp.slope       <-c(a1*b1,a2*b2,a1*b1+a2*b2,a1*b1-a2*b2,a1*b1-cdir,
a2*b2-cdir,(a1*b1+a2*b2)-2*cdir)
simp.cov <-t(D)%*%C%*%D
simp.err <-sqrt(diag(simp.cov))
simples <-simp.slope/sqrt(diag(simp.cov))
zvalues <-2*(1-pnorm(abs(simples)))
CI.low <-simp.slope-(1.96*simp.err)
CI.high <-simp.slope+(1.96*simp.err)
simp.table<-round(matrix(c(simp.slope,simp.err,simples,zvalues,
CI.low,CI.high),nrow=length(simp.slope),ncol=6),digits=5)
dimnames(simp.table)=list(c("a1b1","a2b2","a1b2+a2b2","a1b1-
a2b2","a1b1-cdir","a2b2-cdir","a1b1+a2b2-2*cdir"),c
("slope","stderr","Z","p","CI.low","CI.high"))
return(list(D,simp.table))
}
simple <-simple.slope(D,C)
simple
```

15.2.5   Mediation and Moderation

In Chap. 14, you learned that moderation occurs when a continuous variable qualifies the effect of a categorical one. In this section, you will learn how to combine moderation and mediation in a single analysis. Consider the information presented in Fig. 15.7. The model shows four variables: a categorical predictor X, a continuous moderator Z, a continuous mediator M, and a continuous criterion Y. The moderator can intrude at three points:

1. First, it can moderate the relation between the predictor and the mediator (path a_{xz}).
2. Second, it can moderate the relation between the mediator and the criterion (path b_{mz}).

3. Third, it can moderate the direct effect of the predictor on the criterion (path c'_{xz}).
   None, some, or all of these relations can occur.[8]

Fig. 15.7 Mediation
and moderation

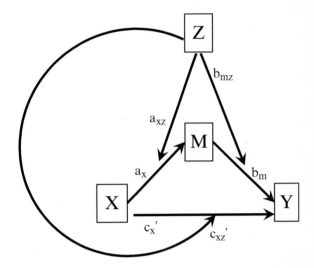

In this section, we will examine a model with all three moderated paths. Recall
from Chap. 14 that our researcher's new instructional method was less effective
with older adults than with younger ones. This is an example of moderation. We say
"age moderates the effectiveness of the experimental manipulation." Now imagine
our researcher wants to know why older adults don't benefit from the new instructional method. She suspects that anxiety is the key. Older adults get anxious when
they are asked to learn something new, and anxiety impairs their performance.
Notice the logic here: The new instructional is more likely to create anxiety in older
adults than in younger ones (path a_{xz}), and anxiety impairs performance (path b_m).
We characterize this situation as "mediated moderation," because a mediator
explains moderation.

Now suppose that the researcher further assumes that the new instructional
method not only produces greater anxiety in older adults than in younger ones but
that anxiety is also more likely to disrupt the performance of older adults than
younger ones.[9] Now we have an example of moderated mediation, because the
mediator's influence is moderated by age (path b_{mz}).

To test these hypotheses, the researcher recruits 12 adults of various ages and
randomly assigns them to the new instructional method or the traditional method.
She then measures anxiety (the presumed mediator) and, afterward, performance on

---

[8] A moderator can also influence the total effect of X on Y (c). This effect is assessed prior to
including the mediator, so it is not depicted in Fig. 15.7.

[9] This effect is not far-fetched. As many aging golfers will tell you, the older you get, the harder it
gets to control your nerves on the putting green.

Table 15.14 Small sample example illustrating mediation and moderation

Experimental condition	Age	Anxiety	Effect coding	Dev_{age}	Effect coding $\times Dev_{age}$	$Dev_{anxiety}$	$Dev_{anxiety} \times Dev_{age}$	Y
New	62	8	0.5	17.8333	8.9167	3.25	57.9583	3
New	40	4	0.5	−4.1667	−2.0833	−0.75	3.1250	8
New	22	1	0.5	−22.1667	−11.0833	−3.75	83.1250	9
New	65	9	0.5	20.8333	10.4167	4.25	88.5417	1
New	43	5	0.5	−1.1667	−.5833	0.25	−.2917	7
New	32	7	0.5	−12.1667	−6.0833	2.25	−27.3750	8
Traditional	56	2	−0.5	11.8333	−5.9167	−2.75	−32.5417	6
Traditional	63	1	−0.5	18.8333	−9.4167	−3.75	−70.6250	7
Traditional	33	1	−0.5	−11.1667	5.5833	−3.75	41.8750	6
Traditional	62	5	−0.5	17.8333	−8.9167	0.25	4.4583	4
Traditional	27	5	−0.5	−17.1667	8.5833	0.25	−4.2917	5
Traditional	25	9	−0.5	−19.1667	9.5833	4.25	−81.4583	2

a test of abstract reasoning. The (phony) data appear in Table 15.14. Three aspects of the data merit mention:

- First, we have centered the moderator (Age) and the mediator (Anxiety) around their respective means.
- Second, we are using effect coding rather than dummy coding for our $1df$ group vector.
- Third, we have added two cross-product terms (Effect Coding $\times Dev_{age}$) and $(Dev_{anxiety} \times Dev_{age})$.[10]

15.2.5.1 Preliminary Analyses

As before, we begin by conducting three regression analyses:

$$
\begin{aligned}
\hat{Y} &= c_0 + c_1 X + c_2 Z + c_3 XZ \\
\hat{M} &= a_0 + a_1 X + a_2 Z + a_3 XZ \\
\hat{Y} &= b_0 + c_1' X + c_2' Z + c_3' XZ + b_1 M + b_2 MZ
\end{aligned}
\tag{15.14}
$$

Table 15.15 presents the results of these analyses, as well as some simple slope vectors (**s**) we will use to perform some additional contrasts at three age levels (i.e., one standard deviation below the mean, at the mean, and one standard deviation above the mean). To help us keep the analyses straight, the numbering below matches the numbering in Tables 15.15 and 15.16:

---

[10] We are using effect coding rather than dummy coding because it is easier to interpret the lower-order effects of a mean-centered continuous variable when effect coding is used for the categorical variable.

Table 15.15 Regression coefficients and covariance matrices for tests of moderation and mediation

			1. Predict performance from group, age, and group C^+ age				
	Path	b	se_b	t	p	s_c	
Intercept		5.4807	.4207	13.0282	.0000		
Group	c_1	.9776	.8414	1.1620	.2787	1	
Age	c_2	−.0671	.0266	2.5212	.0357	0	
Group × Age	c_3	−.2317	.0532	4.3535	.0024	z	
		Covariance matrix					
Group		.707887	.000236	.000030			
Age		.000236	.000708	.000089			
Group × Age		.000030	.000089	.002834			

			2. Predict anxiety from group, age, and group × age				
	Path	b	se_b	t	p	s_a	
Intercept		.0196	.7183	.0272	.9789		
Group	a_1	1.8429	1.4365	1.2829	.2354	1	
Age	a_2	.0288	.0454	.6329	.5445	0	
Group × Age	a_3	.2347	.0909	2.5820	.0325	z	
		Covariance matrix					
Group		2.063629	.000688	.000087			
Age		.000688	.002065	.000260			
Group × Age		.000087	.000260	.008261			

		3. Predict performance from group, age, group × age, anxiety, and anxiety × age					
	Path	b	se_b	t	p	s_b	$s_{c'}$
Intercept		5.5780	.2262	24.6553	.0000		
Group	c_1'	3.0211	.6833	4.4215	.0045	0	1
Age	c_2'	−.0441	.0151	2.9304	.0263	0	0
Group × Age	c_3'	−.0702	.0446	1.5733	.1667	0	b_3
Anxiety	b_1	−.6015	.1276	4.7135	.0033	1	0
Anxiety × Age	b_2	−.0161	.0061	2.6260	.0393	z	0
		Covariance matrix					
Group		.466887	.002622	.016124	−.052987	−.002912	
Age		.002622	.000227	.000199	−.000608	−.000025	
Group × Age		.016124	.000199	.001993	−.004322	−.000140	
Anxiety		−.052987	−.000608	−.004322	.016286	.000396	
Anxiety × Age		−.002912	−.000025	−.000140	.000396	.000038	

1. In the first analysis, we predict performance from group, age, and the group × age cross-product term. The cross-product term is significant ($p=.0024$), replicating the moderator effect we found in Chap. 14. The rules for constructing the **s** vector needed to probe the interaction appear in the final column of Table 15.15, and Table 15.16 presents the simple slopes. As you can see, the simple slope of instructional condition is significantly positive for younger adults ($p=.0042$), not significant for adults of "average" age ($p=.2787$), and significantly negative for older adults ($p=.0471$).

Table 15.16 Simple slopes and standard errors for terms in an analysis of mediated moderation

1. Simple slopes of instruction predicting performance (c)				
Age	b	se_b	t	p
−1 SD	4.8110	1.217	3.9517	.0042
Mean	.9776	.8414	1.1620	.2787
+1 SD	−2.8557	1.218	2.3441	.0471
2. Simple slopes of instruction predicting anxiety (a)				
Age	b	se_b	t	p
−1 SD	−2.0389	2.079	.9809	.3554
Mean	1.8429	1.4365	1.2829	.2354
+1 SD	5.7247	2.08	2.7522	.0250
3. Simple slopes of anxiety predicting performance with instruction in the equation (b)				
Age	b	se_b	t	p
−1 SD	−.3352	.1161	2.8863	.0278
Mean	−.6015	.1276	4.7135	.0033
+1 SD	−.8679	.1992	4.3578	.0048
4. Simple slopes of instruction predicting performance with anxiety in the equation (c′)				
Age	b	se_b	t	p
−1 SD	4.1829	.6919	6.0455	.0009
Mean	3.0211	.6833	4.4215	.0045
+1 SD	1.8594	1.2432	1.4957	.1854

15.2.5.2   Testing for Mediated Moderation

Our next step is to determine whether anxiety explains why the new instructional method is less effective with older adults than with young adults.

2. To begin, we ask whether older adults experience more anxiety than younger adults when using the new instructional method (path *a* in mediation notation). The second analysis in Table 15.15 addresses this issue, and the cross-product term is significant, indicating that age moderates the effects of instructional method on anxiety ($p=.0325$). Moreover, the simple slopes shown in Table 15.16 confirm that older adults experienced more anxiety with the new method than with the traditional method ($p=.0250$), but instructional method did not affect the anxiety levels of young adults ($p=.3554$) or adults of average age ($p=.2354$).
3. Next, we examine the path from the mediator to the criterion controlling for instructional method (path *b* in mediation notation). The third analysis in Table 15.15 shows that anxiety ($p=.0033$) and the anxiety × age interaction ($p=.0393$) are significant predictors of performance, and the simple slope tests in Table 15.16 confirm that, although anxiety negatively impacts performance across age levels (all $p's < .005$), its effects are most damaging among older adults.

15.2.5.3 Calculating Mediated Simple Effects

Our next task is to calculate the mediated simple effects using the a and b values shown in Table 15.16:

$$ab_L = (-2.0389) * (-.3352) = .6834$$
$$ab_M = (1.8429) * (-.6015) = -1.1086$$
$$ab_H = (5.7247) * (-.8679) = -4.9683$$

These values represent the simple mediated effect at each of the three age levels. To test their significance, we construct a vector of partial derivatives and a covariance matrix using the estimates from each simple effect. Table 15.17 presents the relevant vectors and calculations. As you can see, the simple mediated effect is significant for older adults ($p=.0200$), but not for young adults ($p=.3530$) or those of average adult age ($p=.2158$). This pattern fits the researcher's intuitions: Older adults fail to benefit from the new instructional method because learning something new is especially likely to make them anxious, and anxiety is especially likely to impair their performance.

Table 15.17 Simple mediated effects at three age levels

	d		**C**			**Sobel's test**		
			One standard deviation below the mean					
b_L	−.3352	σ^2_{aL}	4.320940	0	ab	se_{ab}	$|Z|$	p
a_L	−2.0389	σ^2_{bL}	0	.013485	.6834	.7358	.9287	.3530
	d		**C**			**Sobel's test**		
			Mean					
b_M	−.6015	σ^2_{aM}	2.063629	0	ab	se_{ab}	$|Z|$	p
a_M	1.8429	σ^2_{bM}	0	.016286	−1.1086	.8955	1.2379	.2158
	d		**C**			**Sobel's test**		
			One standard deviation above the mean					
b_H	−.8679	σ^2_{aH}	4.326671	0	ab	se_{ab}	$|Z|$	p
a_H	5.7247	σ^2_{bH}	0	.039663	−4.9683	2.1351	2.3270	.0200

15.2.5.4 Direct Effects

The only effects left to examine are the direct effects (c'). Table 15.15 shows that the group × age interaction is no longer significant once anxiety and the anxiety × age interaction are controlled ($p=.1667$). If we had sufficient power to detect this effect (which we clearly don't with such a small sample), we would conclude that anxiety can fully explain why the new instructional method is less effective with older adults than with younger ones.

Note that this is true even though the simple direct effects are significant for young adults and those of average age, but not for older adults (see Analysis 4 in Table 15.16). The critical issue is not whether some effects are significant but others are not; the critical issue is whether the differences among the effects are likely to be due to chance. If you look at the coefficients for the simple direct effects, you will see that they are all positive, indicating that subjects of all ages benefit from the new method once anxiety and its interaction with age are statistically controlled. These results, which are graphically presented in Fig. 15.8, provide firm support for the researcher's hypotheses. Older adults not only experience more anxiety than younger adults when learning the new method, but their performance also suffers more from the anxiety they do experience. In combination, these effects explain why they do not benefit from the new instructional method.[11]

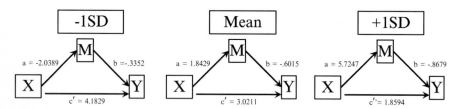

Fig. 15.8 Simple slopes at three age levels

15.2.6   R Code: Mediation and Moderation

```
Z <-c(62,40,22,65,43,32,56,63,33,62,27,25)
M <-c(8,4,1,9,5,7,2,1,1,5,5,9);
X <-c(rep(.5,6),rep(-.5,6))
Y <-c(3,8,9,1,7,8,6,7,6,4,5,2)
devZ <-Z-mean(Z);devM <-M-mean(M);XZ <-X*devZ;MZ <-devM*devZ
summary(mod.1 <-lm(Y~X*devZ))
summary(mod.2 <-lm(M~X*devZ))
summary(mod.3 <-lm(Y~X+devZ+XZ+devM+MZ))

#Simple slopes for predicting Y and M, respectively, from GRP, AGE,
#GRP*AGE
```

(continued)

---

[11] We can also use the Johnson-Neyman technique to identify ages at which the mediated effect is significant. With two moderated paths, the calculations for performing the test are complicated. As a work-around, I have included a root-finding function with the **R** code that accompanies this chapter. Implementing the function, we find that the mediated effect in our example will be significant for subjects who are older than 50.60 years of age and younger than 285.25 years of age.

15.2.6 R Code: Mediation and Moderation (continued)

```
simp.mod <-function(model){
s0 <-c(0,0,0);s1 <-c(1,1,1);s2 <-c(0,0,0);s3 <-c(-sd(Z),0,sd(Z))
S <-rbind(s0,s1,s2,s3)
simp.slope <-t(S)%*%model$coef
simp.err <-sqrt(diag(t(S)%*%vcov(model)%*%S))
simple <-simp.slope/simp.err
tvalues <-2*pt(-abs(simple),df=(length(X)-nrow(S)))
crit <-abs(qt(0.025, 8))
CI.low <-simp.slope-(crit*simp.err)
CI.high <-simp.slope+(crit*simp.err)
simp.table<-round(matrix(c(simp.slope,simp.err,simple,tvalues,
CI.low,CI.high),nrow=length(simp.slope),ncol=6),digits=5)
dimnames(simp.table)=list(c("group", "age", "group*age"),c("slope",
"stderr", "t","p","CI.low","CI.high"))
simp.table
}
simp.c <-simp.mod(mod.1);simp.c
simp.a <-simp.mod(mod.2);simp.a

#Simple slopes for predicting Y from GRP, AGE, GRP*AGE, ANXIETY,
#GRP*ANXIETY
simp.med <-function(a,b){ # enter 0,1 for b values; 1,0 for cdirect
s0 <-c(0,0,0);s1 <-c(1*a,1*a,1*a);s2 <-c(0,0,0);s3 <-c(-sd(Z)*a,0,
sd(Z)*a)
s4 <-c(1*b,1*b,1*b);s5 <-c(-sd(Z)*b,0,sd(Z)*b)
S <-rbind(s0,s1,s2,s3,s4,s5)
simp.slope <-t(S)%*%mod.3$coef
simp.err <-sqrt(diag(t(S)%*%vcov(mod.3)%*%S))
simple <-simp.slope/simp.err
tvalues <-2*pt(-abs(simple),df=(length(X)-nrow(S)))
crit <-abs(qt(0.025, 6))
CI.low <-simp.slope-(crit*simp.err)
CI.high <-simp.slope+(crit*simp.err)
simp.table<-round(matrix(c(simp.slope,simp.err,simple,tvalues,
CI.low,CI.high),nrow=length(simp.slope),ncol=6),digits=5)
dimnames(simp.table)=list(c("group", "age", "group*age"),c("slope",
"stderr", "t","p","CI.low","CI.high"))
simp.table
}
simp.b <-simp.med(0,1)
simp.b
simp.cdir <-simp.med(1,0)
simp.cdir

d.lo <-rbind(simp.b[1],simp.a[1])
d.md <-rbind(simp.b[2],simp.a[2])
```

(continued)

15.2.6   R Code: Mediation and Moderation (continued)

```
d.hi <-rbind(simp.b[3],simp.a[3])
ab.lo <-simp.b[1]*simp.a[1]
ab.md <-simp.b[2]*simp.a[2]
ab.hi <-simp.b[3]*simp.a[3]

s0 <-c(0,0,0);s1 <-c(1,1,1);s2 <-c(0,0,0);s3 <-c(-sd(Z),0,sd(Z))
S <-rbind(s0,s1,s2,s3)

cov.a <-diag(t(S)%*%vcov(mod.2)%*%S)
s.lo <-c(1,-sd(Z));s.md <-c(1,0);s.hi <-c(1,sd(Z))
SS <-cbind(s.lo,s.md,s.hi)
cov.b <-t(SS)%*%vcov(mod.3)[5:6,5:6]%*%SS
cov.lo <-matrix(c(cov.a[1],0,0,cov.b[1]),nrow=2)
cov.md <-matrix(c(cov.a[2],0,0,cov.b[5]),nrow=2)
cov.hi <-matrix(c(cov.a[3],0,0,cov.b[9]),nrow=2)
std.lo <- sqrt(t(d.lo)%*%cov.lo%*%d.lo)
Z.lo <-ab.lo/std.lo
Z.lo.p <-2*(1-pnorm(abs(Z.lo)))
std.md <- sqrt(t(d.md)%*%cov.md%*%d.md)
Z.md <-ab.md/std.md
Z.md.p <-2*(1-pnorm(abs(Z.md)))
std.hi <- sqrt(t(d.hi)%*%cov.hi%*%d.hi)
Z.hi <-ab.hi/std.hi
Z.hi.p <-2*(1-pnorm(abs(Z.hi)))

med.mod <-matrix(c(ab.lo, std.lo, Z.lo,Z.lo.p,ab.md, std.md, Z.md,
Z.md.p,ab.hi, std.hi, Z.hi,Z.hi.p ),nrow=3,byrow=TRUE)
dimnames(med.mod)=list(c("young",  "medium",  "old"),c("ab",  "std.
err", "Z","p"))
med.mod

#Johnson-Neyman Regions of Significance Using Uniroot Function
JN <-function(q){
S.2<-c(0,1,0,q)
slope.2 <-t(S.2)%*%mod.2$coef
std.err.2 <-sqrt(t(S.2)%*%vcov(mod.2)%*%S.2)
S.3 <-c(0,0,0,0,1,q)
slope.3 <-t(S.3)%*%mod.3$coef
std.err.3 <-sqrt(t(S.3)%*%vcov(mod.3)%*%S.3)
ab <-slope.2*slope.3
D <-rbind(slope.3,slope.2)
S.4 <-c(0,1,0,q)
cov.a <-diag(t(S.4)%*%vcov(mod.2)%*%S.4)
S.5 <-c(1,q)
cov.b <-t(S.5)%*%vcov(mod.3)[5:6,5:6]%*%S.5
```

(continued)

15.2.6   R Code: Mediation and Moderation (continued)

```
cov.m <-matrix(c(cov.a,0,0,cov.b),nrow=2)
std.m <- sqrt(t(D)%*%cov.m%*%D)
Z <-ab/std.m
P <-2*(1-pnorm(abs(Z)))
JN <-1.96-abs(Z)
}
jn.lo <-uniroot(JN,c(-100,100));jn.lo$root+mean(Z) #Use low starting
#values
jn.hi <-uniroot(JN,c(100,500));jn.hi$root+mean(Z)  #Use high starting
#values
```

15.3   Mediation and Causal Inference

In this chapter, we have seen that mediational analysis offers a powerful method for testing causal hypotheses. If we randomly assign subjects to conditions and measure our mediator before measuring our criterion, we can reasonably establish that the effect of X on Y depends, in part, on its association with M.

These benefits cannot be realized when correlational methods are used. In the absence of randomly assigning subjects to conditions, we cannot assume that an observed association between X and Y is transmitted through M. To understand why this is so, we need to appreciate the difference between a mediator and a confound variable. As first discussed in Chap. 13 (see Footnote 1), a confound creates an association between two variables that are not causally related at all. Consider, for example, the oft-cited claim that family dinners provide numerous benefits.

> Having meals together helps strengthen family relationships because children have the opportunity to learn more about their family's history and are encouraged to remain connected to their extended family, ethnic heritage, and community of faith. ... Being with family also encourages children to think critically and helps them feel as though they have some control over their environment. Family meals predict a child's behavior even more than church attendance or school grades. While family meals are correlated to a child's reading readiness and linked to positive outcomes such as emotional stability, academic success, psychological adjustment, higher self-esteem and higher family functioning, they are also strongly related to lower incidence of negative outcomes such as low GPAs, depression, suicide, and teenage alcohol and drug use.
>   http://www.parentingnow.net/documents/ThePowerofFamilyMeals.revised.pdf

Notice that this article assumes that there is a causal pathway from family dinners to positive life outcomes and that this pathway has a variety of mediators (e.g., opportunity to learn about family heritage, cultivate critical thinking skills, and develop responsibility). These causal pathways are certainly plausible, but there is another possibility—namely, stable, loving families eat dinner together and enjoy a high level of psychological well-being, but the dinners themselves produce no psychological benefits.

Fig. 15.9 Contrasting
mediation and confounding

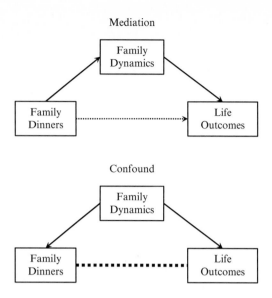

Figure 15.9 illustrates both possibilities. With mediation, we assume that family dinners foster a more favorable family environment and that a positive family environment confers benefits. In contrast, confounding assumes that family dynamics determine dining habits and life outcomes but that no causal association exists between dining habits and life outcomes. Note, then, that while both models assume that family dynamics influence outcomes, only the mediational model assumes that family dynamics are a consequence of eating habits.

Without random assignment to eating conditions, there is no way to dismiss the possibility that any observed association between family dinners, family dynamics, and life outcomes is confounded. And this is true whenever mediational analyses are undertaken with correlational data. Consequently, claims about mediation made using correlational data should always be viewed with skepticism. The data can be consistent with a causal claim, but the claim itself is always more than the data can substantiate.

Why is this important? Presumably, we'd all like to promote positive life outcomes in people. But encouraging families to eat together will fulfill this goal only if dining habits determine family dynamics. Unfortunately, unless we randomly assign some families to eat dinner together and some to eat alone, we can never conclude that dining habits initiate the causal chain.

15.4   Chapter Summary

1. A mediational analysis is performed to test whether a continuous predictor M explains the association between a categorical variable X and a criterion Y. Ideally, the categorical variable was formed through random assignment to conditions and the mediator was assessed before the criterion was measured.

2. A test of mediation involves partitioning the total effect of X on Y (c) into two components: a direct effect (c') and an indirect (mediated) effect (ab). The effects are commonly calculated from three regression analyses:

   2a. Predict the criterion from the categorical variable (c).
   2b. Predict the mediator from the categorical variable (a).
   2c. Predict the criterion from the mediator (b) and categorical variable (c').

   The mediated effect is found as ab.

3. Confidence intervals, tests of statistical significance, and effect sizes can be computed for the mediated effect. Because the standard errors needed to produce these effects come from two regression equations, the calculations are specialized.
4. Mediational analyses can involve multiple groups and/or multiple mediators.
5. Mediation and moderation can be combined. Mediated moderation occurs when a mediator explains a moderator effect; moderated mediation occurs when a mediated effect is modified by a third variable.
6. Mediational analyses illuminate causal relations only when random assignment to conditions is used to initiate the presumed causal chain. When all variables are measured, it is not possible to distinguish mediational effects from confounds.

References

Anscombe, F. J. (1973). Graphs in statistical analysis. *American Statistician, 27*, 17–21.

Baron, R. M., & Kenny, D. A. (1986). The moderator-mediator variable distinction in social psychological research: Conceptual, strategic, and statistical considerations. *Journal of Personality and Social Psychology, 51*, 1173–1182.

Bauer, D. J., & Curran, P. J. (2005). Probing interactions in fixed and multilevel regression: Inferential and graphical techniques. *Multivariate Behavioral Research, 40*, 373–400.

Belsley, D. A., Kuh, E., & Welsch, R. E. (1980). *Regression diagnostics: Identifying influential data and sources of collinearity*. New York: Wiley.

Box, G. E. P. (1979). Robustness in the strategy of scientific model building. In R. L. Launer & G. N. Wilkinson (Eds.), *Robustness in statistics* (pp. 201–236). New York: Academic Press.

Box, G. E. P., & Cox, D. R. (1964). An analysis of transformations. *Journal of the Royal Statistical Society, Series B, 26*, 211–252.

Box, G. E. P., & Tidwell, P. W. (1962). Transformation of the independent variables. *Technometrics, 4*, 531–550.

Breusch, T. S., & Pagan, A. R. (1979). Simple test for heteroscedasticity and random coefficient variation. *Econometrica, 47*, 1287–1294.

Buskirk, T. D., Willoughby, L. M., & Tomazic, T. J. (2013). Nonparametric statistical techniques. In T. D. Little (Ed.), *The Oxford handbook of quantitative methods in psychology* (Statistical analysis, Vol. 2, pp. 106–141). New York: Oxford University Press.

Cleveland, W. S. (1979). Robust locally weighted regression and smoothing scatterplots. *Journal of the American Statistical Association, 74*, 829–836.

Cohen, J. (1988). *Statistical power analysis for the behavioral sciences* (2nd ed.). Mahwah: Erlbaum.

Cohen, J., & Cohen, P. (1975). *Applied multiple regression/correlation analysis for the behavioral sciences*. Mahwah, NJ: Erlbaum.

Cohen, J., Cohen, P., West, S. G., & Aiken, L. S. (2003). *Applied multiple regression/correlation analysis for the behavioral sciences*. Mahwah: Erlbaum.

Cook, R. D. (1977). Detection of influential observations in linear regression. *Technometrics, 19*, 15–18.

De Boor, C. (1972). On calculating with B splines. *Journal of Approximation Theory, 6*, 50–62.

Draper, N. R., & Smith, H. (1998). *Applied regression analysis* (3rd ed.). New York: Wiley.

Eilers, P. H. C., & Marx, B. D. (1996). Flexible smoothing with B-splines and penalties. *Statistical Science, 11*, 89–121.

Fox, J. (2000). *Nonparametric simple regression: Smoothing scatterplots*. Thousand Oaks: Sage.

© Springer International Publishing Switzerland 2014
J.D. Brown, *Linear Models in Matrix Form*, DOI 10.1007/978-3-319-11734-8

Fox, J. (2008). *Applied regression analysis and generalized linear models* (2nd ed.). Newbury Park: Sage.

Golub, C. H., & Van Loan, C. F. (2013). Matrix computations (4th ed.). Baltimore: Johns-Hopkins University Press.

Gu, C. (2002). *Smoothing spline ANOVA models.* New York: Springer.

Hayes, A. F., & Li, C. (2007). Using heteroscedasticity-consistent standard error estimators in OLS regression: An introduction and software implementation. *Behavior Research Methods, 37*, 709–722.

Kenny, D. A., & Judd, C. M. (2014). Power anomalies in testing mediation. *Psychological Science, 25*, 334–339.

Long, J. S., & Ervin, L. H. (2000). Using heteroscedasticity consistent standard errors in the linear regression model. *The American Statistician, 54*, 217–224.

MacKinnon, D. P. (2008). *Introduction to statistical mediation analysis.* New York: Taylor & Francis.

Manning, W. G., & Mullahy, J. (2001). Estimating log models: To transform or not to transform? *Journal of Health Economics, 20*, 461–494.

McClelland, G. H., & Judd, C. M. (1993). Statistical difficulties of detecting interactions and moderator effects. *Psychological Bulletin, 114*, 376–390.

Miller, G. A., & Chapman, J. P. (2001). Misunderstanding analysis of covariance. *Journal of Abnormal Psychology, 110*, 40–48.

Mosteller, F., & Tukey, J. W. (1977). *Data analysis and regression: A second course in statistics.* New York: Pearson.

Newey, W. K., & West, K. (1987). A simple, positive semi-definite, heteroskedasticity and autocorrelation consistent covariance matrix. *Econometrica, 55*, 703–708.

Preacher, K. J., & Kelly, K. (2011). Effect size measures for mediation models: quantitative strategies for communicating indirect effects. *Psychological Methods, 16*, 93–115.

Ruppert, D., Wand, M. P., & Carroll, R. J. (2003). *Semiparametric regression.* London: Cambridge Press.

Seber, G. A. F., & Lee, A. J. (2003). Linear regression analysis (2nd ed). New York: Wiley.

Shrout, P. E., & Bolger, N. (2002). Mediation in experimental and nonexperimental studies: New procedures and recommendations. *Psychological Methods, 7*, 422–445.

Sobel, M. E. (1982). Asymptotic confidence intervals for indirect effects in structural equation models. *Sociological Methodology, 13*, 290–312.

Stewart, G. W. (2001). *Matrix algorithms* (Eigensystems, Vol. 2). Philadelphia: SIAM.

Strang, G. (2009). *Linear algebra and its applications* (4th ed.). Cambridge, MA: Wellesley-Cambridge Press.

Trefethen, L. N., & Schreiber, R. S. (1990). Average-case stability of Gaussian elimination. *Society for Industrial and Applied Mathematics, 11*, 335–360.

Watkins, D. S. (2010). *Fundamentals of matrix computations* (3rd ed.). New York: Wiley.

White, H. (1980). A heteroskedasticity-consistent covariance matrix estimator and a direct test for heteroskedasticity. *Econometrica, 48*, 817–838.

Wood, S. (2006). *Generalized additive models: An introduction with R.* London: Taylor & Francis.

Xiao, X., White, E. P., Hooten, M. B., & Durham, S. L. (2011). On the use of log-transformation vs. nonlinear regression for analyzing power laws. *Ecology, 92*, 1887–1894.

Index

A

Additivity assumption, 186, 261
Adjugate matrix, 31, 36, 138, 139, 142, 145
Alpha (power determination), 338
Analysis of covariance (ANCOVA), 443–467, 469–471, 493
    adjusted means, 445, 450, 452–457, 460–462, 465–467, 497
Analysis of variance (ANOVA), 266, 310, 377–382, 384, 386, 400, 409, 415, 416, 421, 422, 427–431, 444, 447, 451, 461, 462, 464, 473, 481, 492
Arrays (and spreadsheet functions), 7
Asymptotic equivalence, 86, 94, 98
Augmented b vector, 432–433, 455, 462, 480–481, 489
Augmented covariance matrix, 428, 432, 433, 455, 462, 476, 481, 489
Augmented design matrix, 374
Augmented hat matrix, 367
Augmented normal equations, 367
Autocorrelation consistent covariance matrix (ACCM), 250, 254–258
Autocorrelations, 244–260
Autoregressive parameter, 245
Average expected value, 45, 52, 60, 63–67

B

Balanced design, 412, 415, 420, 426, 428, 438
Ballentine Venn diagram, 127, 128
Bandwidth, 288, 291–294, 298, 299
Base (effect coding), 388, 390
Best linear unbiased estimator (BLUE), 48
Beta (power determination), 338

Between groups sum of squares, 379, 380
Bootstrapping, 502, 504, 514
Box-Cox transformation, 210–212, 227
Box-Tidwell transformation, 282–284
Breusch-Godfrey test, 249–250, 259
Breusch-Pagan test, 235–237, 259, 276
B-spline basis, 360
Bulging rule, 270

C

Calculus, 73, 76–93, 96, 100
Categorical predictors, 266, 377–409, 414, 416, 443, 445, 447, 454, 462, 465–467, 469, 491, 493–495, 498, 499, 507, 512, 516
Change in R^2, 129, 131, 134, 145
Characteristic equation, 151–153, 183
Cholesky decomposition, 177–182, 184
Coding schemes, 377–392, 401, 404, 405, 408, 409, 426–427, 438, 457, 470, 473, 474, 507, 512
Coefficient of determination, 56–57, 67, 110, 140–141, 203, 235, 337, 372
Cofactor matrix, 22–28, 35, 36, 142–143
Collinearity, 186, 213–226, 307, 360
Common regression coefficient, 452, 453, 460, 467
Comparing two regression coefficients, 200, 445
Comparing two simple slopes, 306, 309, 317, 328–329, 417
Conditional invariance, 303
Condition index, 219–223
Condition number, 221–223

Confidence intervals, 59–60, 64, 65, 67, 91,
    122–124, 145, 244, 259, 316, 346, 499,
    502, 504, 527
Conformability (and matrix multiplication), 6
Confound, 525–527
Contrast codes, 387, 389, 395, 399–407, 416,
    417, 422, 448
Cook's D, 198, 201, 202, 205,
    209, 212, 226
Correlation matrix, 13–14, 35, 36, 53, 107,
    157–158, 183, 217, 219, 220
Correlation model, 53
Covariance matrix, 11–14, 23, 35, 36
Covariate, 39, 444–448, 450–454, 457–461,
    463–467, 470
COVRATIO, 198, 202
Cramer's rule, 18–21, 30, 31, 35, 119, 142–143
Crossing point (cp), 318, 475–476
Cross-product terms and interactions,
    303–340, 349
Cubic polynomial, 348–349, 357–358, 360,
    361, 375
Curvilinear relations, 341–343, 348, 349

D

DE. *See* Deviance explained (DE)
Degrees of freedom, 12, 59–61
Deleted studentized residuals, 195, 196
Delta method (standard errors of mediated
    effect), 499
Density functions, 70, 294, 295
Departures from normality, 204–213
Derivative function, 77, 79–80, 84
Derivatives, 76–84, 89, 91–92, 94, 100–102,
    275, 314, 345, 346
Design matrix, 147, 226, 366, 367, 370, 374
Determinant, 15–25, 27–29, 31, 32, 35–37, 39,
    118–119, 138–143, 145, 150–152, 156,
    157, 176, 179, 183, 202, 214, 215
Deterministic processes, 40
Deviance explained (DE), 372, 374
Deviate matrix, 35, 36
Deviate scores, 11, 12, 14, 47, 61, 118, 208
Deviation sum of squares, 37, 95, 107, 110,
    120, 138, 143, 145, 188, 219, 453
DEVSQ, 37, 138
DFBETA, 198–201
DFFIT, 197–199, 370
Diagonal matrix, 14, 25–27
Differentiation, 76, 81, 84–87, 96
Direct (unmediated) effect, 495, 498
Discrepant observations, 187

Disordinal interaction, 318, 476
Disturbance, 40, 186, 245, 246, 259, 261, 262,
    286, 301, 341
Dummy coding, 387, 388, 390, 414, 438, 439,
    474, 495, 508, 510, 512, 518
Durbin-Watson test, 248–249, 259

E

Effect coding, 388–390, 426, 428, 432, 438,
    441, 443, 445, 454, 457, 470, 473, 474,
    492, 518
Effect size, 336–340, 503, 509, 527
Eigen decomposition, 147–160, 173, 182, 184,
    367, 368, 373
Eigen pairs, 147, 156–158, 167–172, 175, 176,
    183, 221, 222, 367
Eigenvalues, 147, 150–159, 168–171, 173,
    175–177, 180–184, 191, 219, 221,
    222, 367
Eigenvectors, 147, 149–150, 152–156, 158,
    159, 171, 173–176, 181–183, 219,
    221, 367
Elasticity, 282
Epanechnikov kernel, 295
Errors, 42–51, 185–186, 227–259
Euclidean norm, 154, 161
Expected values, 63–65, 89, 90, 95, 100, 227,
    228, 249
Exponential function, 272, 274, 276–279, 301

F

Factorial designs, 409–441, 454, 457, 467
First order autoregression process, 259
First order derivatives, 80
Fisher's information matrix, 89–90, 96, 97
Fisher's method of scoring, 96–99
Fitted values, 44–50, 52, 56, 58, 64, 67,
    109–110
Fixed variable, 53
Forecasting, 63–67, 122–124
Functional relations, 40, 301, 348

G

Gaussian kernel, 295, 296
Generality (without loss of), 362
Generalized cross validation, 370, 374, 375
Generalized least squares estimation, 231–232,
    250–254, 259
General linear model, 377
Geometric mean, 210, 211

Global fitting (polynomial regression), 357–358
Gram Schmidt orthogonalization, 393–396
Grid-search method, 73–74, 76, 96, 100, 251, 374
Group comparisons, 438

H
Harmonic mean, 385, 386, 402, 422
Hat matrix, 190–193, 197, 205, 226, 230, 241, 292, 299, 367, 370
HCCM. *See* Heteroscedasticity consistent covariance matrix (HCCM)
Helmert contrasts, 392
Hessian matrix (and standard errors), 89, 95
Heteroscedasticity, 232–244, 250, 258, 259, 275–277, 279, 280
Heteroscedasticity consistent covariance matrix (HCCM), 241–242, 254, 259
Hierarchical regression, 344, 349
Homogeneity of regression lines, 447–448, 459–460, 469
Homoscedasticity, 242
Householder transformation, 161–167

I
Idempotent matrix, 191, 226
Identity matrix, 25–26, 33, 35, 36, 151, 158, 161, 162, 165, 220, 229
Independent and identically distributed errors, 228, 230
Indirect (mediated) effect, 495, 498, 515, 527
Influential observations, 186–204, 226
Instantaneous rate of change (derivative), 76–80, 84, 117, 263
Integrated squared second derivative, 365, 366
Interactions, 303–340, 349, 352, 375, 412–418, 420–423, 428, 431–439, 441, 444, 469–472, 474–476, 479, 482–484, 487, 492, 519–522
Intercept, 43–48, 60–61
Interpolating spline, 364, 375

J
Jarque-Bera test, 207–208, 212, 226, 227
Johnson-Neyman technique, 320–322, 340, 487, 488, 492, 522

K
Kernel function, 294, 296
Kernel regression, 288, 289, 294–297, 301
Kurtosis, 208

L
Lack of fit test (for assessing linearity), 301
Lagged weights (for Newey West procedure), 255
Laplace (aka cofactor) expansion, 22
Learning curve, 271, 275, 276
Least squares estimation, 48, 67, 110, 147, 173, 182–184, 225, 238
Leverage, 187, 190–193, 195, 197, 201, 202, 225
Likelihood function, 71–73, 84, 85, 87, 93, 100, 251, 252
Limit method of differentiation, 78
Linear equations, 1–9, 18–21, 25, 27, 30–36, 43, 67, 105, 106, 144, 147, 357, 361
Linear function (aka linear model), 262
Linear in the parameters, 262, 272, 274, 301, 305
Linear in the variables, 262, 301
Linearity assumption, 50, 186, 228
Linearizing transformations, 261–301
Linear model (aka linear function), 261
Linear regression, 39–67
Linear relation, 158, 261, 262, 264–267, 279
Linear transformations, 148–150, 269, 285–286
Line of best fit, 47–48, 67, 187, 188, 194, 369
Local fitting (polynomial regression), 298, 356, 358, 375
Locally weighted regression, 298–299, 301
Locally weighted scatterplot smoother (LOWESS), 298, 301
Logarithmic model, 270–274
Log likelihood function, 72–73, 84, 85, 87, 88, 92, 93, 97, 100–104, 251–253
Lower order effects, 307, 310–311, 470, 473, 484
Lower triangular matrix, 177, 179

M
Main effects, 310, 311, 412–415, 417, 420–422, 424, 427, 429, 441, 473, 474, 484–485
Mathematical models, 40–42, 270
Matrix
    inverse, 28–34, 50, 90, 94, 106, 118, 167, 182, 222
    multiplication, 4–9, 15, 25–29, 35, 36, 64, 148–150, 290, 489
    operations, 3–5, 7–9, 19, 37, 137, 489
    properties, 1–37
    recomposition, 158–159
    transpose and sum of squares, 9–15
Maximum likelihood estimation, 69–104

Maximum point of a curvilinear relation, 347
MDETERM, 21, 37, 138, 139
Mean centering, 307, 310
Mean square residual, 58, 194
Mediated effect, 495, 498–504, 508–510,
    512–515, 521, 522, 527
Mediation, 444, 493–527
    and causal inference, 525–526
    and moderation, 444, 516–525
Minors, 22, 23, 35, 36, 118, 139
MINVERSE, 27, 37, 139
Mixed derivatives, 80
Moderated regression, 447, 467, 469–480,
    482, 483
Monotonic transformations, 72
Multilevel designs, 425–440
Multiple group contrasts, 381, 408
Multiple mediators, 511–516, 527
Multiple regression, 101–143

N

Nadaraya-Watson estimator, 295, 297, 301
Natural cubic splines, 360–374
Nearest neighbors, 288–289, 292, 294, 298,
    299, 301
Neighborhood size, 288–290
Newey-West, 255–258, 260, 294
Newton-Raphson, 96
Nonlinear transformations, 269
Nonparallel simple slopes, 318
Nonparametric functions, 300
Nonparametric smoothers. *See* Scatterplot
    smoothers
Normal distribution, 69–76, 92, 95, 101, 102,
    206, 226
Normal equations, 32, 35, 94, 144, 167
Normality assumption, 204–205, 208–210, 227
Normalized residuals, 229
Normalizing, 219–221
Normal probability plot, 206–207
Null hypothesis testing, 55
Numerical methods, 96–100

O

Ordinal interaction, 318, 476
Orthogonal contrast codes, 378, 381–388,
    391–399, 408, 423, 510
Orthogonalization, 393–396, 422
Orthogonal polynomial coefficients, 403
Orthonormal matrix, 160, 161, 167, 183
Outlier, 187, 193–196, 201, 225

P

Parameter covariance matrix, 61–62, 67, 95,
    118, 121–123, 145, 167, 179, 196, 202,
    227–231, 237, 241, 242, 244–246, 255,
    257, 259, 316, 325, 384, 385, 388–390,
    395, 396, 398, 401, 404, 405, 410, 411,
    414, 415, 421–424, 427, 428, 432–433,
    438, 441, 454, 455, 461, 476, 477, 480,
    481, 500, 503, 514,519, 521
Partial correlation, 128, 130, 132–134,
    143, 145
Partial derivatives, 80–86, 93, 94, 96, 97,
    100–104, 117, 263–264, 305, 314, 340,
    346, 347, 399, 500, 503, 509, 514, 521
Partial mediation, 444, 497
Partitioning the sum of squares in an analysis of
    variance, 55–56, 127–136, 498–499
Penalized natural cubic spline, 364–374
Piecewise polynomials, 356–374
Plotting predicted values, 313–317, 474–475
P matrix of predicted values, 20, 312
Point estimates, 59, 67, 80, 87, 510
Polynomial coefficients, 396, 406
Polynomial interactions, 349–356
Polynomial regression, 341–375
Population estimation, 55–63
Population parameters, 42, 48, 50, 69–71,
    87–89, 91, 94, 96, 100, 144, 237
Power
    function, 79, 210, 279–282, 301
    rule, 79, 81, 101–104, 346
    series, 159, 356, 375
Prais-Winsten, 251–253
Predicted value, 52, 60, 63–67, 78, 123, 124,
    238, 312–313, 319–320, 326, 344–345,
    351, 352, 449, 475, 479, 480
Prediction interval, 65, 124
Probability density function, 70

Q

QQ plot, 206–207, 227
QR algorithm for finding the eigenpairs,
    168–172
QR decomposition, 160–173, 177, 182–184, 190
Quadratic equation, 152, 321
Quadratic polynomial, 152
Quantiles, 206, 208, 365–367, 372, 373

R

Random variable, 41, 50, 53, 67, 70
Rank deficiency, 157

Rank of a matrix, 157
Recentering variables to calculate simple
    slopes, 332–336
Reciprocal transformation, 211, 212, 283, 284
Recoding categorical variables, 518
Reference group (dummy coding), 387,
    390, 409
Regions of significance, 320–322, 340, 469,
    487–492
Regression coefficients, 46, 92, 105, 167, 186,
    228, 265, 303, 341, 382, 411, 448,
    473–474, 496
Regression diagnostics, 185, 188, 198, 203,
    204, 218
Replicates and lack of fit test, 266
Residuals, 44–45, 93, 109–110, 175, 194–196,
    227–260, 264, 309, 342, 379, 411, 446,
    471, 497
Residual sum of squares, 45, 47, 56, 67,
    140–141, 201, 210, 211, 236, 252, 266,
    301, 379, 380, 386, 454
Root finding function, 522
Rules of differentiation, 84, 340
Running average, 288–292, 301
Running line smoother, 292–294

S
Sample estimates of population parameters,
    80, 86
Sample quantiles, 206, 372, 373
Sample size and power, 42, 57, 86, 188, 189,
    192, 193, 196
Scatterplot, 47, 48
Scatterplot smoothers, 287, 292, 298, 300, 301
Second derivatives, 80, 83, 87–88, 91, 365,
    366, 375
Seed vector. See Gram Schmidt
    orthogonalization
Self-efficacy beliefs, 42, 55, 57, 63, 92, 106,
    116, 117, 121, 122, 124, 125
Semipartial correlation, 128–134, 143, 145,
    309, 337
Serial correlations. See Autocorrelations
Silverman's rule, 294
Simple effects, 375, 414, 416–417, 431,
    434–441, 452–457, 462, 479–487, 492,
    521
Simple slopes, 304, 306, 307, 309–323,
    326–330, 332–336, 340, 344–347,
    353–354, 366, 375, 416, 417, 441,
    447–449, 469, 473, 475–479, 481, 484,
    487–489, 518–520, 522

Singularity and invertibility, 27, 34, 146, 168,
    173, 216–217
Singular value decomposition, 173–177, 183,
    219, 226
Skew, 207, 208
S matrix for simple slopes, 314, 328, 344
Smoother matrix, 290, 366
Smoothing coefficient in a penalized cubic
    spline, 365
Sobel's test, 501, 502, 509, 514, 515, 521
Span, 288
Specification errors, 185, 186
Spectral decomposition. See Eigen
    decomposition
Splines, 292, 300, 356–375
Spreadsheet functions, 27, 59, 70, 122, 138,
    206, 210, 249
Squared semipartial correlation, 128–134, 145,
    309, 337
Square matrix, 3, 10, 11, 14–17
Standard error
    approximation in mediation, 499
    of the mean, 87, 90, 95
    of the mediated effect, 499–501, 513–515
    and second partial derivatives, 87, 88, 96
    of simple slopes, 315–316, 327, 328,
        346, 520
    with multiple regression, 120–121
Standardized regression coefficients, 51–54,
    57, 112, 134, 143
Standardized residual, 194, 195, 210
Standard scores, 13, 14, 36, 51
Statistical significance, 55–65, 67
Stochastic processes, 40
Studentized residual, 195, 196, 234, 265, 357
Sum of squares, 9–15
SUMSQ, 37, 249
Symmetric matrix, 11, 32, 35, 36, 89, 137, 171,
    177, 180
Symmetric nearest neighbors, 289, 290, 294,
    298, 301

T
Tangent line and derivatives, 346
Testing simple slopes, 313–317, 326–328,
    345–347
Testing the significance of interactions, 471
Testing the statistical significance of simple
    slopes, 316–317
Tolerance, 217, 218, 338
Total sum of squares, 55, 56, 380
Trace, 3, 15, 35, 36, 156–158, 183, 221

Transformation matrix (gamma for WLS), 240
TRANSPOSE, 37
Transpose, 9–15, 27, 32, 35–37, 161, 255
Trend analysis, 396–398, 408
Tricube function, 298, 299
Truncated power basis, 359, 360

U
Unbalanced designs, 399–407, 412, 420,
        428, 441
Unbalanced factorial designs, 412,
        420–425, 441
Unbiased coefficients, 48
Unequal spaced polynomial codes, 397, 398
Unit length, 154, 220, 362
Unstandardized regression coefficient, 51, 52,
        108–109, 125, 134, 142, 143, 200, 304,
        449, 453
Unweighted grand mean, 383, 389, 395,
        397, 398, 401, 402, 414, 421, 422,
        427, 428

Unweighted means, 383, 400–403, 405, 408,
        421–423, 437, 441
Upper triangular matrix, 160, 161, 164, 166,
        167, 177, 183

V
Variance inflation factor (VIF), 217–219, 226
Variance proportion decomposition, 219–223

W
Weighted grand mean, 383, 404, 405, 422, 423
Weighted least squares (WLS) estimation,
        237–242, 250, 254, 259, 298, 301
Weighted means, 383, 400, 403–406, 408,
        422–425, 441
White's test of heteroscedasticity, 235, 236,
        240, 259
Window size, 292
Within groups sum of squares, 11, 182,
        221, 379